Practical Approaches to Biological Inorganic Chemistry

Practical Approaches to Biological Inorganic Chemistry

Edited by

Robert R. Crichton
Batiment Lavoisier
Université Catholique de Louvain
Louvain-la-Neuve, Belgium

Ricardo O. Louro
ITQB, Universidade Nova de Lisboa
Oeiras, Portugal

ELSEVIER

AMSTERDAM • WALTHAM • HEIDELBERG • LONDON • NEW YORK • OXFORD
PARIS • SAN DIEGO • SAN FRANCISCO • SYDNEY • TOKYO

Elsevier
Radarweg 29, PO Box 211, 1000 AE Amsterdam, The Netherlands
The Boulevard, Langford Lane, Kidlington, Oxford OX5 1GB, UK

British Library Cataloguing in Publication Data
A catalogue record for this book is available from the British Library

Library of Congress Cataloging-in-Publication Data
A catalog record for this book is available from the Library of Congress

ISBN: 978-0-444-56351-4

For information on all Elsevier publications visit
our website at www.store.elsevier.com

Printed and bound in China

12 13 11 10 9 8 7 6 5 4 3 2 1

Contents

Preface

Shrouded in the mists of scientific antiquity (things move so quickly that even a decade or two seems a long time), in reality a little less than 30 years ago — the Federation of European Biochemical Societies, better known by its acronym FEBS, invited the Belgian Biochemical society to organise their annual Congress in Belgium. For the first time in the history of these meetings (since the inaugural Congress, in London in 1964), two half day symposia were organised on the subject of metalloproteins. At the end of the second of these, a group of what in those days were called inorganic biochemists met to enjoy a drink together in the bar of the Sheraton Hotel. The outcome was that two of those present, one of whom is co-editor of the present volume, together with Cees Veeger were entrusted with the task of organising a FEBS Workshop Course on Inorganic Biochemistry. The first of these was held at the Hotel Etap in Louvain-la Neuve at the end of April, 1985. The origins of this book can be traced back to the long series of Advanced Courses which have followed that pioneering start.

At that very first Course, the pattern was established of organising lectures to introduce the subject and to present a theoretical background to the methods which could be used to study metals in biological systems, together with practical sessions in smaller groups. The final lectures were then devoted to specific examples. It is interesting, and perhaps not too surprising, that after an introduction to ligand field theory by Bob Williams, and metal coordination in biology by Jan Reedijk, X-ray, EPR, NMR, Mössbauer and EXAFS spectroscopy of metalloproteins were on the programme. The practicals included NMR, EPR and Mössbauer as well as Cees Veeger's favourite, biochemical analysis of Fe and S in Fe—S proteins. There was an evening lecture by Helmut Beinert (then on sabbatical in Konstanz) entitled 'Limitations of Spectroscopic Studies on Metalloproteins and Chemical Analysis of Metals in Proteins'. While the lecturers were shuffled around from year to year, Fred Hagen, Antonio Xavier, Alfred Trautwein, and Dave Garner represented the cornerstone of the spectroscopic part of the course over the early years.

Since then, over the period from 1985 until now we have organised some 20 courses, and trained over 800 students, most of whom were doctoral or post-doctoral students when they came on the course. It is a source of great pride and satisfaction that many of the former students still enjoy active and distinguished careers in the area of Biological Inorganic Chemistry, as we now call the subject. Even more rewarding are the number of former participants who now form the staff of the course, notably the other co-editor, who has also taken on the mantle of co-organiser of the most recent courses. Indeed, with the exception of Rob Robson, who taught the Molecular Biology lectures and practical for many years, the other authors contributing to this book, Frank Neese, Fred Hagen, Eckhard Bill, Martin Feiters, Christophe Leger and Margarida Archer are all alumni of the 'Louvain-la-Neuve' course.

Our intention in editing this volume is that it can serve as a starting point for any student who wants to study metals in biological systems. The presentations by the authors represent a distillation of what they have taught over a number of years in the advanced course. We begin with an overview of the roles of metal ions in biological systems, which we hope will serve as taster for the reader, who will find a much more detailed account in the companion work to this volume (Crichton, 2012). Thereafter, after an introduction to that most erudite of discipline (at least for non-inorganic chemists) ligand field theory, augmented by a good dose of how molecular orbital theory can predict the properties of catalytic metal sites. This leads naturally into a sequence which describes the physicochemical methods which can be used to study metals in biology, concluding with an overview of the application of the powerful methods of modern genetics to metalloproteins.

The considerations expressed by that pioneer of analytical precision Helmut Beinert in his 1985 evening lecture in Louvain-la-Neuve are as relevant today as they were then. Use as many techniques as possible to analyse your sample – the more information from different approaches you have, the better we will understand your protein. Do not waste expensive and sensitive methods on shoddy impure samples, and conversely do not employ primitive technical means to analyse highly purified samples, which have required enormous investment to obtain them. And above all recognise that the key to metalloprotein characterisation is collaboration. Do not think you can simply phagocytise a technique from the laboratory of a colleague who knows the method inside out – it is much richer to collaborate, incorporating his or her know-how into your research. And you will be the richer for it.

Bonne chance, good luck, boa sorte – and we look forward to greet you on one of the courses which will, we hope, continue into the future. Hopefully, this little introductory text will not only whet your appetite, but help you to find your way about the myriad practical methods which can be used to study metals in biological systems.

Robert R. Crichton and Ricardo O. Louro
Louvain-la-Neuve, July, 2012

Chapter 1

An Overview of the Roles of Metals in Biological Systems

Robert R. Crichton

Batiment Lavoisier, Université Catholique de Louvain, Louvain-la-Neuve, Belgium

Chapter Outline

INTRODUCTION: WHICH METALS IONS AND WHY?

In the companion book to this one, 'Biological Inorganic Chemistry 2nd edition' (Crichton, 2011), we explain in greater detail why life as we know it would not be possible with just the elements found in organic chemistry – namely carbon, oxygen, hydrogen, nitrogen, phosphorus and sulfur. We also need components of inorganic chemistry as well, and in the course of evolution nature has selected a number of metal ions to construct living organisms. Some of them, like sodium and potassium, calcium and magnesium, are present at quite large concentrations, constituting the so-called 'bulk elements', whereas others, like cobalt, copper, iron and zinc, are known as 'trace elements', with dietary requirements that are much lower than the bulk elements.

Just six elements – oxygen, carbon, hydrogen, nitrogen, calcium and phosphorus – make up almost 98.5% of the elemental composition of the human body by weight. And just 11 elements account for 99.9% of the human body (the five others are potassium, sulfur, sodium, magnesium and chlorine). However, between 22 and 30 elements are required by some, if not all, living organisms, and of these are quite a number are metals. In addition to the four metal ions mentioned above, we know that cobalt, copper, iron, manganese, molybdenum, nickel, vanadium and zinc are essential for humans, while tungsten replaces molybdenum in some bacteria. The essential nature of chromium for humans remains enigmatic.

Just why these elements out of the entire periodic table (Figure 1.1) have been selected will be discussed here. However, their selection was presumably based not only on suitability for the functions that they are called upon to

Practical Approaches to Biological Inorganic Chemistry, 1st Edition. http://dx.doi.org/10.1016/B978-0-444-56351-4.00002-6.

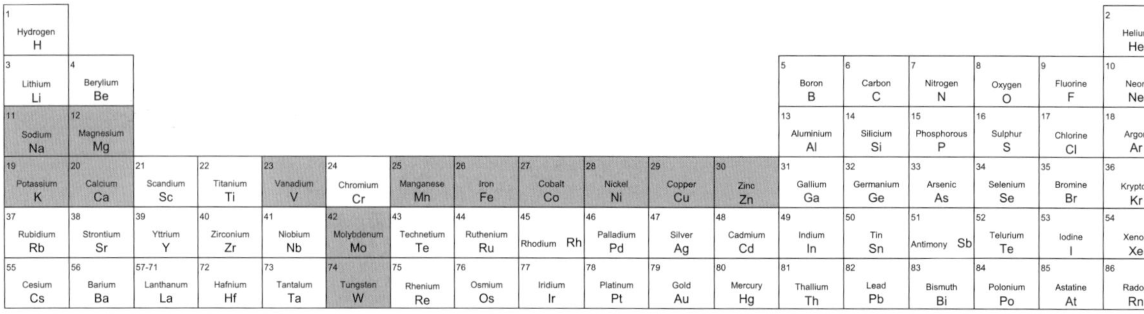

FIGURE 1.1 An abbreviated periodic table of the elements showing the metal ions discussed in this chapter.

play in what is predominantly an aqueous environment, but also on their abundance and their availability in the earth's crust and its oceans (which constitute the major proportion of the earth's surface).

The 13 metal ions that we will discuss here fall naturally into four groups based on their chemical properties. In the first, we have the alkali metal ions Na^{+} and K^{+}. Together with H^{+} and Cl^{-}, they bind weakly to organic ligands, have high mobility, and are therefore ideally suited for generating ionic gradients across membranes and for maintaining osmotic balance. In most mammalian cells, most K^{+} is intracellular, and Na^{+} extracellular, with this concentration differential ensuring cellular osmotic balance, signal transduction and neurotransmission. Na^{+} and K^{+} fluxes play a crucial role in the transmission of nervous impulses both within the brain and from the brain to other parts of the body.

The second group is made up by the alkaline earths, Mg^{2+} and Ca^{2+}. With intermediate binding strengths to organic ligands, they are, at best semi-mobile, and play important structural roles. The role of Mg^{2+} is intimately associated with phosphate, and it is involved in many phosphoryl transfer reactions. Mg-ATP is important in muscle contraction, and also functions in the stabilisation of nucleic acid structures, as well as in the catalytic activity of ribozymes (catalytic RNA molecules). Mg^{2+} is also found in photosynthetic organisms as the metal centre in the light-absorbing chlorophylls. Ca^{+} is a crucial second messenger, signalling key changes in cellular metabolism, but is also important in muscle activation, in the activation of many proteases, both intra- and extracellular, and as a major component of a range of bio-minerals, including bone.

Zn^{2+}, which is arguably not a transition element,[1] constitutes the third group on its own. It is moderate to strong binding, is of intermediate mobility and is often found playing a structural role, although it can also fulfil a very important function as a Lewis acid. Structural elements, called zinc fingers, play an important role in the regulation of gene expression.

The other eight transition metal ions, Co, Cu, Fe, Mn, Mo, Ni, V and W form the final group. They bind tightly to organic ligands and therefore have very low mobility. Since they can exist in various oxidation states, they participate in innumerable redox reactions, and many of them are involved in oxygen chemistry. Fe and Cu are constituents of a large number of proteins involved in electron transfer chains. They also play an important role in oxygen-binding proteins involved in oxygen activation as well as in oxygen transport and storage. Co, together with another essential transition metal, Ni, is particularly important in the metabolism of small molecules like carbon monoxide, hydrogen and methane. Co is also involved in isomerisation and methyl transfer reactions. A major role of Mn is in the catalytic cluster involved in the photosynthetic oxidation of water to dioxygen in plants, and, from a much earlier period in geological time, in cyanobacteria. Mo and W enzymes contain a pyranopterindithiolate cofactor, while nitrogenase, the key enzyme of N_2 fixation contains a molybdenum–iron–sulfur cofactor, in which V can replace Mo when Mo is deficient. Other V enzymes include

1. IUPAC defines a transition metal as "an element whose atom has an incomplete d sub-shell, or which can give rise to cations with an incomplete d sub-shell."

haloperoxidases. To date no Cr-binding proteins have been found, adding to the lack of biochemical evidence for a biological role of the enigmatic Cr.

SOME PHYSICOCHEMICAL CONSIDERATIONS ON ALKALI METALS

Before considering, in more detail, the roles of the alkali metals, Na^+ and K^+, and the alkaline earth metals, Mg^{2+} and Ca^{2+}, it may be useful to examine some of their physicochemical properties (Table 1.1). We can observe, for example that Na^+ and K^+ have quite significantly different unhydrated ionic radii, whereas, the hydrated radii are much more similar. It therefore comes as no surprise that the pumps and channels which carry them across membranes, and which can easily distinguish between them, as we will see shortly, transport the unhydrated ions. Although not indicated in the table, it is clear that Na^+ is invariably hexa-coordinate, whereas K^+ and Ca^{2+} can adjust to accommodate 6, 7 or 8 ligands. As we indicated above, both Na^+ and K^+ are characterised by very high solvent exchange rates (around 10^9/s), consistent with their high mobility and their role in generating ionic gradients across membranes. In contrast, the mobility of Mg^{2+} is some four orders of magnitude slower, consistent with its essentially structural and catalytic. Perhaps surprisingly, Ca^{2+} has a much higher mobility (3×10^8/s), which explains why it is involved in cell signalling via rapid changes on Ca^{2+} fluxes.

The selective binding of Ca^{2+} by biological ligands compared to Mg^{2+} can be explained by the difference in their ionic radius, as we pointed out above. Also, for the smaller Mg^{2+} ion, the central field of the cation dominates its coordination sphere, whereas for Ca^{2+}, the second and possibly even the third, coordination spheres have an important influence resulting in irregular coordination geometry. This allows Ca^{2+}, unlike Mg^{2+} to bind to a large number of centres at once.

The high charge density on Mg^{2+} as a consequence of its small ionic radius ensures that it is an excellent Lewis acid in reactions notably involving phosphoryl transfers and hydrolysis of phosphoesters. Typically, Mg^{2+} functions as a Lewis acid, either by activating a bound nucleophile to a more reactive anionic form (e.g. water to hydroxide anion), or by stabilising an intermediate. The invariably hexacoordinate Mg^{2+} often participates in structures where the metal is bound to four or five ligands from the protein and a phosphorylated substrate. This leaves one or two coordination positions vacant for occupation by water molecules, which can be positioned in a particular geometry by the Mg^{2+} to participate in the catalytic mechanism of the enzyme.

NA^+ AND K^+ – FUNCTIONAL IONIC GRADIENTS

How, we might ask, do the pumps and channels responsible for transport across membranes distinguish between Na^+ and K^+ ions? Studies over the last 50 years or so of synthetic and naturally occurring small molecules which bind ions have established the basic rules of ion selectivity. Two major factors appear to be of capital importance,

TABLE 1.1 Properties of Common Biological Cations

Cation	Ionic radius (Å)	Hydrated radius (Å)	Ionic volume (Å^3)	Hydrated volume (Å^3)	Exchange rate (sec^{-1})	Transport number
Na^+	0.95	2.75	3.6	88.3	8×10^8	7–13
K^+	1.38	2.32	11.0	52.5	10^9	4–6
Mg^{2+}	0.65	4.76	1.2	453	10^5	12–14
Ca^{2+}	0.99	2.95	4.1	108	3×10^8	8–12

(From Maguire and Cowan, 2002).

namely the molecular composition and the stereochemistry (essentially the size) of the binding site. Synthetic molecules have been created which selectivity bind Li^+ (radius 0.60 Å), Na^+ (0.95 Å), K^+ (1.35 Å) and Rb^+ (radius 1.48 Å) by simply adjusting the cavity size to match the ion (Dietrich, 1985). Now that we have the crystal structures of membrane transport proteins, we can begin to understand how ion selectivity is accomplished (MacKinnon, 2004; Gouax and MacKinnon, 2005). The Na^+-selective binding sites in the Na^+-dependant leucine transporter LeuT and the K^+-selective binding sites in the K^+ channel have been determined, providing a direct comparison of selectivity for Na^+ and K^+. The Na^+ and K^+ ions are completely dehydrated, both the Na^+ and the K^+ sites contain oxygen ligands, but by far the most important factor distinguishing Na^+ and K^+ sites is the size of the cavity formed by the binding site, which agrees well with the rules already learned from host/guest chemistry. What determines alkali metal cation selectivity, similar to that observed in ion binding by small molecules, is that the protein selects for a particular ion, Na^+ or K^+, by providing an oxygen-lined binding site of the appropriate cavity size.

Mammalian cells maintain a high intracellular K^+ (around 140 mM) and low intracellular Na^+ (around 12 mM) through the action of the Na^+, K^+-ATPase present in the plasma membrane. The overall reaction catalysed is:

$$3Na^+(in) + 2K^+\ (out) + ATP + H_2O \Leftrightarrow 3Na^+\ (out) + 2K^+\ (in) + ADP + P_i$$

The extrusion of three positive charges for every two which enter the cell, results in a transmembrane potential of 50–70 mV, which has enormous physiological significance, controlling cell volume, allowing neurons and muscle cells to be electrically excitable, and driving the active transport of important metabolites such as sugars and amino acids. More than one-third of ATP consumption by resting mammalian cells is used to maintain this intracellular $Na^+ - K^+$ gradient (in nerve cells this can rise to up to 70%).

This thermodynamically unfavourable exchange is achieved by ATP-mediated phosphorylation of the Na^+,K^+-ATPase followed by dephosphorylation of the resulting aspartyl phosphate residue, which drives conformational changes that allow ion access to the binding sites of the pump from only one side of the membrane at a time. The ATPase exists in two distinct conformations, E_1 and E_2, which differ in their catalytic activity and their ligand specificity (Figure 1.2). The E_1 form, which has a high affinity for Na^+, binds Na^+, and the $E_1.3Na^+$ form then reacts with ATP to form the "high-energy" aspartyl phosphate ternary complex $E_1 \sim P.3Na^+$. In relaxing to its "low-energy" conformation E_2-P, the bound Na^+ is released outside the cell. The E_2-P, which has a high affinity for K^+, binds $2K^+$, and the aspartyl phosphate group is hydrolysed to give $E_2.2K^+$, which then changes conformation to the E_1 form, releasing its $2K^+$ inside the cell. The structures of a number of P-type ATPases, including the Na^+ - K^+-ATPase and the Ca^{2+}–ATPase of the Sarcoplasmic reticulum have been determined and are shown in Figure 1.3.

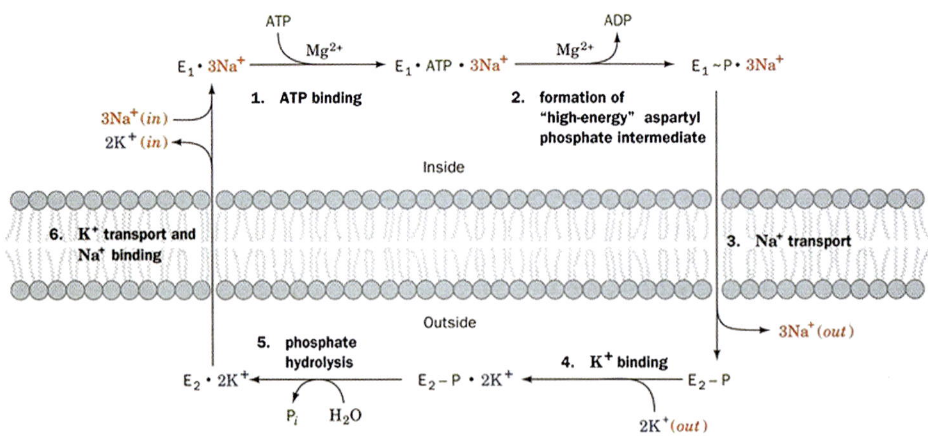

FIGURE 1.2 A model for the active transport of Na^+ and K^+ by the Na^+-K^+-ATPase.

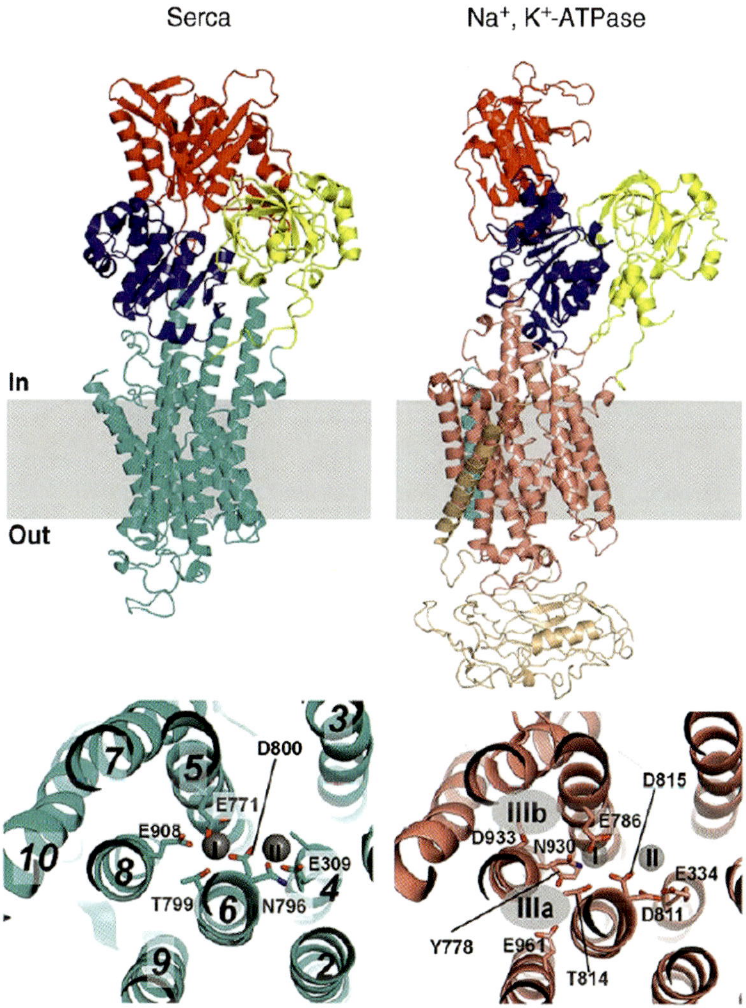

FIGURE 1.3 Overall structures and ion-binding site architectures of two P-type ATPases, rabbit sarcoplasmic reticulum Ca^{2+}-ATPase (SERCA) and pig Na^+,K^+-ATPase. The upper panel depicts rabbit SERCA (E1 Protein Data Base [PDB] entry 1T5S) and pig Na^+-K^+-ATPase (E2:Pi, PDB entry 3KDP). N-, P-, and A-domains are coloured red, blue and yellow, respectively; the β-subunit and γ-subunit of Na^+,K^+-ATPase wheat and cyan. The lower panel depicts the ion-binding sites, viewed approximately perpendicular to the membrane plane from the extracytoplasmic side, in the E1 state. Ion liganding residues are shown as sticks, transmembrane helices and calcium ions in SERCA are indicated by numbers and grey spheres, respectively, and the sites superposed as transparent spheres onto the Na^+,K^+-ATPase model. Putative binding sites for the third sodium ion in the Na^+,K^+-ATPase are indicated as grey ellipses. (From Bublitz et al., 2010. *Reproduced Copyright 2010 with permission from Elsevier).*

MG^{2+} – PHOSPHATE METABOLISM

The intracellular concentration of free Mg^{2+} is about 5×10^{-3} M, so that although Mg^{2+}-binding to enzymes is relatively weak (K_a not more than $10^5 M^{-1}$) and most Mg^{2+}-dependent enzymes have adequate local concentrations of Mg^{2+} for their activity. Mg^{2+} is the most abundant divalent cation in the cytosol of mammalian cells, binds strongly to ATP and ADP, and is therefore extensively involved in intermediary metabolism and in nucleic acid metabolism. However, like Zn^{2+}, it is a difficult metal ion to study, since it is spectroscopically silent, with the consequence that many spectroscopic studies on Mg^{2+} enzymes utilise Mn^{2+} as a replacement metal ion.

Of the five enzymes selected in the Enzyme Function Initiative, recently established to address the challenge of assigning reliable functions to enzymes discovered in bacterial genome projects, but for which functions have not yet been attributed (Gerlt et al., 2011), three of them are Mg^{2+}-dependent. We discuss two of them briefly here.

The haloalkanoic acid dehalogenase superfamily (HADSF) (>32,000 nonredundant members) catalyse a diverse range of reactions that involve the Mg^{2+}-dependent formation of a covalent intermediate with an active site Asp. Despite being named after a dehalogenase, the vast majority are involved in phosphoryl transfer reactions (Allen and Dunaway-Mariano, 2004, 2009). While ATPases and phosphatases are the most prevalent, the haloacid dehalogenase (HAD) family can carry out many different metabolic functions, including membrane transport, signal transduction and nucleic-acid repair. Their physiological substrates cover an extensive range of both size and shape, ranging from phosphoglycolate, the smallest organophosphate substrate, to phosphoproteins, nucleic acids, phospholipids, phosphorylated disaccharides, sialic acids and terpenes.

In HAD enzymes, Asp mediates carbon-group transfer to water (in the dehalogenases) and phosphoryl-group transfer to a variety of acceptors. Thus, the HAD superfamily is unique in catalysing both phosphoryl-group transfer (top) and carbon-group transfer (bottom) (Figure 1.4a). The roles of the four loops that comprise the catalytic scaffold are shown in Figure 1.4b. The activity 'switch' is located on loop 4 of the catalytic scaffold (yellow) which positions one carboxylate residue to function as a general base for the dehalogenases and either two or three carboxylates to bind the Mg^{2+} cofactor essential for the phosphotransferases. CO represents the backbone carbonyl oxygen of the moiety that is two residues downstream from the loop 1 nucleophile (red). The side-chain at this position is also used as an acid-base catalyst by phosphatase and phosphomutase HAD members. Loop 2 (green) and loop 3 (cyan) serve to position the nucleophile and substrate phosphoryl moiety. Figure 1.4c presents a ribbon diagram of the fold supporting the catalytic scaffold of phosphonatase.

The members of another large superfamily of Mg^{2+} enzymes, the enolase superfamily (with more than 6000 nonredundant members) catalyse diverse reactions, including β-eliminations (cycloisomerisation, dehydration and deamination) and 1,1-proton transfers (epimerisation and racemisation). The three founder members of the family are illustrated by mandelate racemase, muconate lactonising enzyme and enolase (Figure 1.5). They all catalyse reactions in which the α-proton of the carboxylate substrate is abstracted by the enzyme, generating an enolate anion intermediate. This intermediate, which is stabilised by coordination to the essential Mg^{2+} ion of the enzyme, is then directed to different products in the enzyme active sites.

CA^{2+} AND CELL SIGNALLING

Calcium ions play a major role as structural components of bone and teeth, but are also crucially important in cell signalling. To prevent the precipitation of phosphorylated or carboxylated calcium complexes, many of which are insoluble, the cytosolic levels of Ca^{2+} in unexcited cells must be kept extremely low, much lower than that in the extracellular fluid and in intracellular Ca^{2+} stores. This concentration gradient gives cells the opportunity to use Ca^{2+} as a metabolic trigger – the cytosolic Ca^{2+} concentration can be abruptly increased for signalling purposes by transiently opening Ca^{2+} channels in the plasma membrane or in an intracellular membrane. These increases in intracellular free Ca^{2+} concentration can regulate a wide range of cellular processes, including fertilisation, muscle contraction, secretion, learning and memory and ultimately cell death, both apoptotic and necrotic.

Extracellular signals often act by causing a transient rise in cytosolic Ca^{2+} levels, which, in turn, activates a great variety of enzymes through the action of Ca^{2+}-binding proteins like calmodulin, as we will discuss in detail below: this triggers such diverse processes as glycogen breakdown, glycolysis and muscle contraction. In the phosphoinositide cascade (Figure 1.6), binding of the external signal (often referred to as the agonist[2] when it provokes a positive response) to the surface receptor R (step 1) activates phospholipase C, either through a G

2. Many drugs have been developed either as agonist or antagonists to receptor-mediated signalling pathways, e.g. β-blockers block the action of the endogenous catecholamines adrenaline (epinephrine) and noradrenaline (norepinephrine) on β-adrenergic receptors.

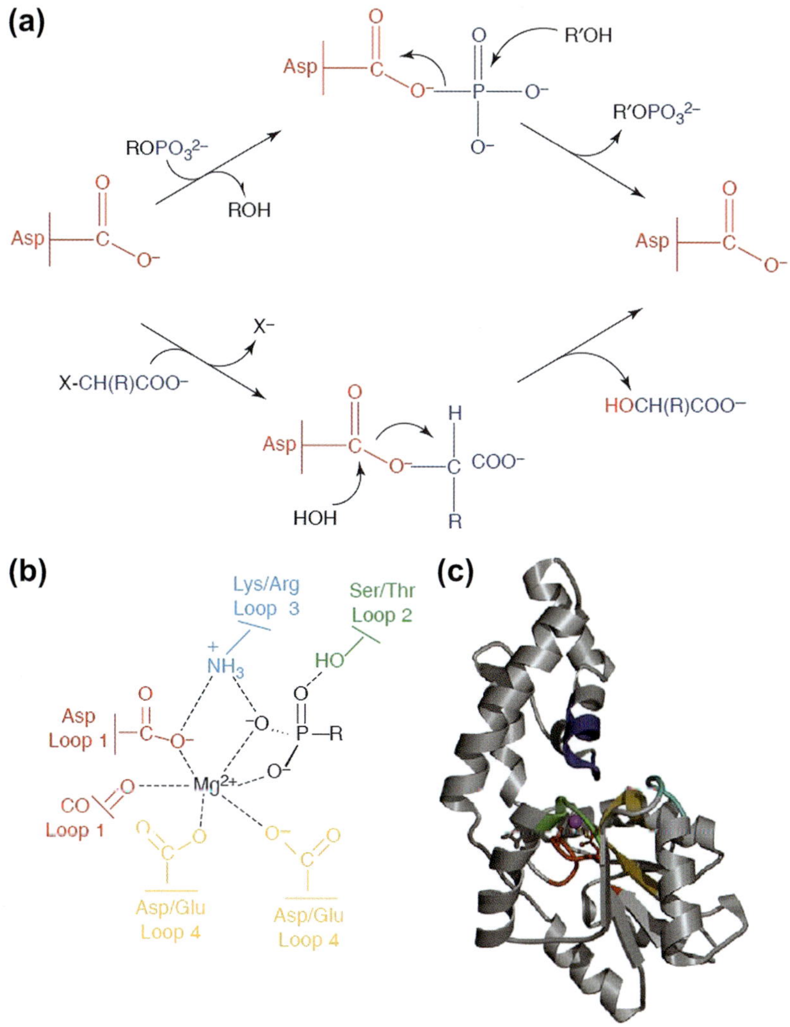

FIGURE 1.4 The catalytic scaffold in the haloacid dehalogenase (HAD) family of phosphotransferases. **(a)** In HAD enzymes, Asp mediates carbon-group transfer to water (in the dehalogenases) and phosphoryl-group transfer to a variety of acceptors. Thus, the HAD superfamily is unique in catalyzing both phosphoryl-group transfer (top) and carbon-group transfer (bottom). **(b)** Schematic of the roles of the four loops that comprise the catalytic scaffold. The activity 'switch' is located on loop 4 of the catalytic scaffold (yellow) which positions one carboxylate residue to function as a general base for the dehalogenases and either two or three carboxylates to bind the Mg^{2+} cofactor essential for the phosphotransferases. CO represents the backbone carbonyl oxygen of the moiety that is two residues downstream from the loop 1 nucleophile (red). The side-chain at this position is also used as an acid-base catalyst by phosphatase and phosphomutase HAD members. Loop 2 (green) and loop 3 (cyan) serve to position the nucleophile and substrate phosphoryl moiety. **(c)** Ribbon diagram (core domain: loop 1, red; loop 2, cyan; loop 3, green; loop 4, yellow; cap domain: specificity loop, blue) of the fold supporting the catalytic scaffold of phosphonatase (1FES). *(From Allen and Dunaway-Mariano, 2004. Copyright 2004, with permission from Elsevier).*

protein which uses the energy of guanosine triphosphate hydrolysis to liberate a subunit capable of activating the next partner in the cascade (2) or alternatively (not shown) by activating a tyrosine kinase. The activated phospholipase C, then hydrolyses phosphatidylinositol-4,5-bisphosphate (PIP_2) in the plasma membrane to $InsP_3$ (IP_3 in the figure) and diacylglycerol (DG) (3). $InsP_3$ stimulates the release of Ca^{2+}, sequestered in the endoplasmic reticulum (4), and this in turn activates numerous cellular processes through Ca^{2+}-binding proteins, such as

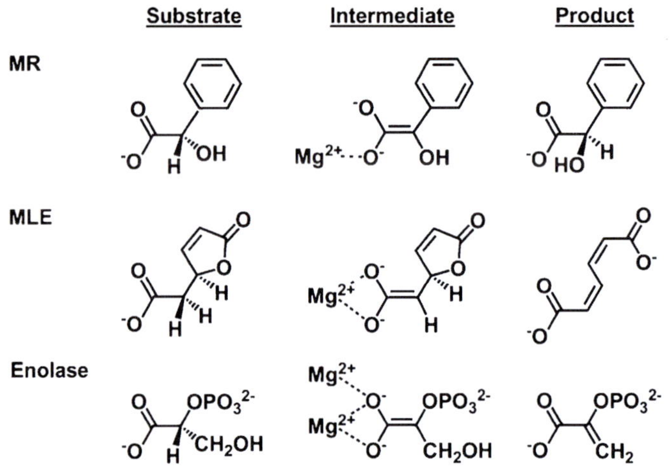

FIGURE 1.5 The substrates, enolate anion intermediates, and products of the MR; MLE and enolase reactions *(adapted from Gerlt et al., 2005).*

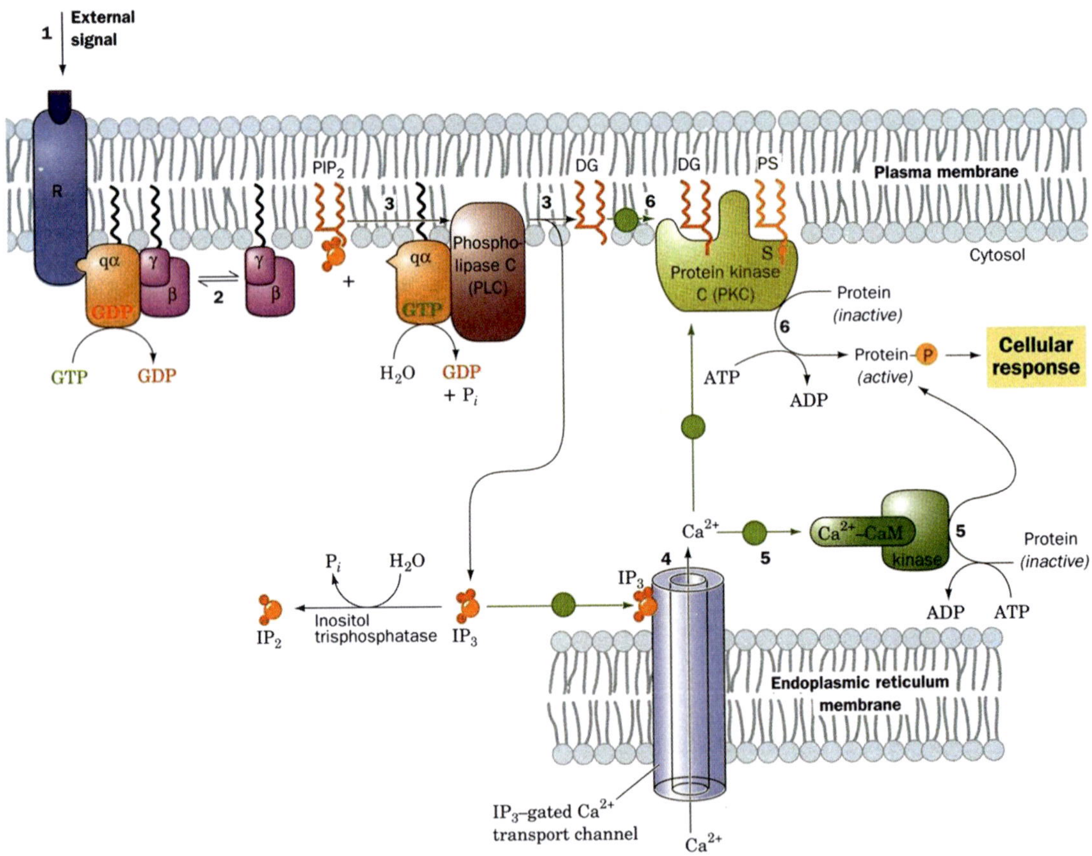

FIGURE 1.6 The phosphoinositide cascade.

calmodulin (CaM) (5). The membrane-associated DG activates protein kinase C (6) to phosphorylate and activate other enzymes, like glycogen phosphorylase. This step also requires Ca^{2+}.

Calmodulin is a small, dumbbell-shaped protein, abundant in the cytoplasm of most cells of higher organisms, which has been highly conserved throughout evolution It is made up of two globular domains each of which can bind two Ca^{2+} ions, connected by a flexible linker. When all four Ca^{2+} sites are filled, the linker forms a flexible seven turn long α-helix, and the protein undergoes a change in conformation, which does not alter its overall dimensions, but opens up its two Ca^{2+}-binding lobes, exposing previously hidden hydrophobic residues, particularly Met. In Figure 1.7a, the structure on the left shows calmodulin without calcium, while that on the right

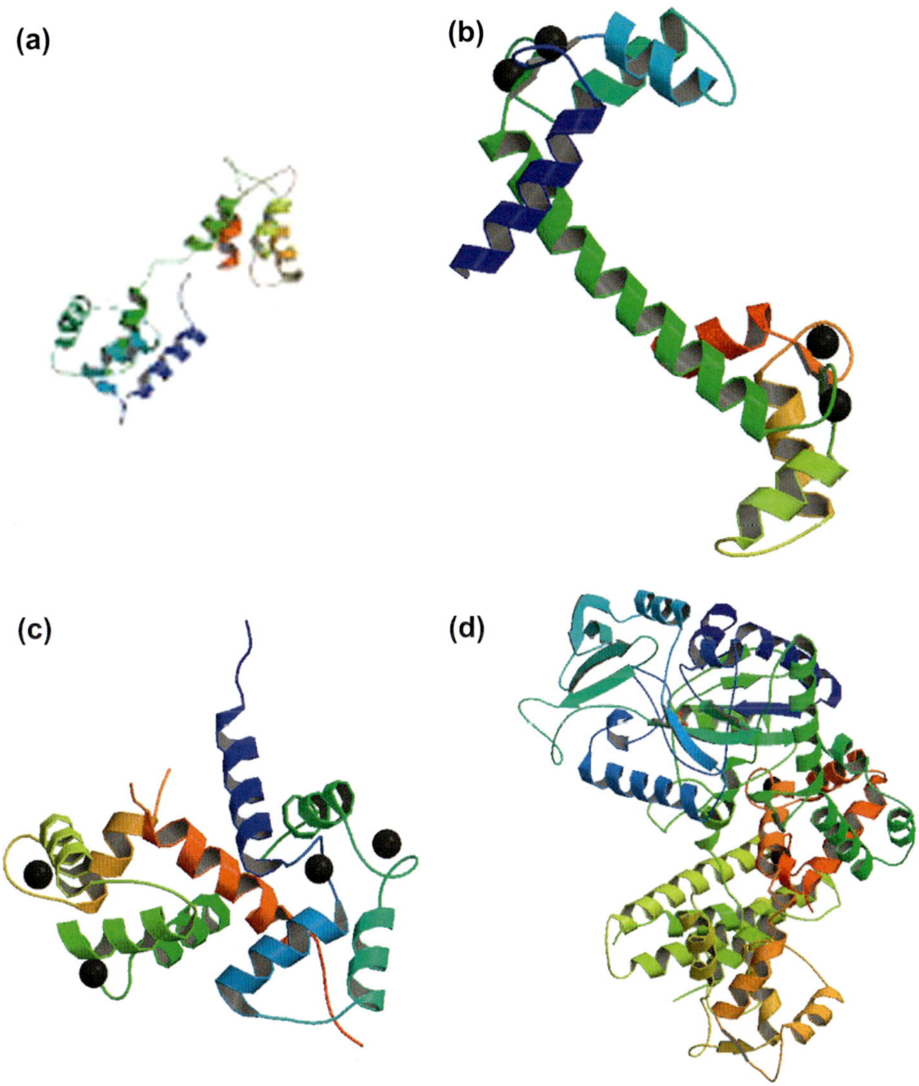

FIGURE 1.7 (a) Calmodulin without calcium (left, PDB 1cfd) and after calcium binds (right, PDB 1cll). The linker region between the two Ca^{2+}-binding domains is in pink and the Ca^{2+} ions are shown in light blue. (b) Left calmodulin bound to two different target enzymes: calmodulin-dependent protein kinase II-alpha (top, PDB 1cm1) and myosin light chain kinase (bottom, PDB 2bbm). Only a small piece of the target protein chain (in red) is included. Right calmodulin bound to anthrax bacteria oedema factor toxin (PDB 1k93). The entire toxin protein is shown in red.

shows calmodulin after calcium binds. The two hydrophobic regions are represented in green and yellow (C and Met S atoms, respectively), and we can see that with calcium bound, the hydrophobic residues form two grooves (red stars), waiting to grip around the target protein, while the linker (pink) has formed a long alpha-helix separating the two calcium-binding domains. A second, and much more dramatic conformation change then occurs, collapsing the elongated structure of calmodulin to a hairpin conformation, which enables it to wrap around the binding domain of the target enzyme, gripping the target protein between its two globular domains. This is illustrated in Figure 1.7b (left panel) for calmodulin bound to calmodulin-dependent protein kinase II-alpha (top) and myosin light chain kinase (bottom). Only a small piece of the target protein chain (red) is included, with the flexible linker of calmodulin, (purple), allowing it to adopt to the slightly different shapes of the two targets. In the case of the oedema factor toxin from the anthrax bacteria shown in Figure 1.7b (left panel), we see a quite different binding geometry. This time, the whole toxin protein is shown; once calmodulin binds, it induces a conformational change in the toxin which activates its adenylyl cyclase activity, thereby depleting the host cell's energy stores.

An unusual feature of calmodulin is that, unlike other Ca^{2+}-binding proteins which usually only interact with a specific target protein, calmodulin interacts with a wide range of targets. A comparison of amino acid sequences of calmodulin-binding domains of target proteins suggests that calmodulin principally recognises positively charged amphipathic helices. Upon binding to the target peptide (compare Figure 1.7a and b), the long central helix of uncomplexed calmodulin unwinds and bends to form a globular structure that encloses the target polypeptide within a hydrophobic tunnel.

ZINC – LEWIS ACID AND GENE REGULATOR

After iron, zinc is the second most abundant trace element in the human body: an average adult has about 3 g of Zn. Some 95% of Zn is intracellular. It is essential for growth and development in all forms of life, has been proposed to have beneficial therapeutic and preventative effects on infectious diseases, including a shortening of the length of the common cold in man. Zn is found in more than 300 enzymes, where it plays both a catalytic and a structural role. It is the only metal to have representatives in each of the six fundamental classes of enzymes recognised by the International Union of Biochemistry—**oxidoreductases** (e.g. alcohol dehydrogenase): **transferases** (RNA polymerase): **hydrolases** (carboxypeptidase A): **lyases** (carbonic anhydrase): **isomerases** (phosphomannose isomerase): and **ligases** (pyruvate carboxylase, aminoacyl-tRNA synthases). Zinc is involved in enzymes in both a catalytic and a structural role. Many nucleic acid-binding proteins have essential Zn atoms in characteristic structures called 'zinc fingers' which are widely involved in the regulation of the transcription and translation of the genetic message. Figure 1.8 is a representation of the Cys_2His_2 zinc finger motif, consisting of an α helix and an antiparallel β sheet. The zinc ion (green) is coordinated by two histidine residues and two cysteine residues.

The first zinc enzyme to be discovered was carbonic anhydrase in 1940; it represents the archetype of mono-zinc enzymes, with a central catalytically active Zn^{2+} atom bound to three protein ligands, and the fourth distorted tetrahedral site occupied by a water molecule. The mechanism of action of mononuclear zinc enzymes depends on the Zn^{2+}-OH_2 centre, which can participate in the catalytic cycle in three distinct ways (Figure 1.9) – either by ionisation, to give zinc-bound hydroxyl ion (in carbonic anhydrase), polarisation by a general base (as in carboxypeptidases) or displacement of the $-OH_2$ ligand by the substrate (in alkaline phosphatase). In the case of carbonic anhydrase, the zinc ion functions as a powerful electrophilic catalyst by providing some or all of the following properties: (i) an activated water molecule for nucleophilic attack, (ii) polarisation of the carbonyl of the bond to be cleaved, (iii) stabilisation of the negative charge which develops in the transition state.

The coordination chemistry and the main features of the mechanism of carbonic anhydrase are illustrated in Figure 1.10, and involve the following steps: (i) deprotonation of the coordinated water molecule with a $pK_a \sim 7$, in a process facilitated by general base catalysis involving His 64. This residue is too far away from the Zn^{2+}-bound water to directly remove its proton, but it is linked to it by two intervening water molecules, forming

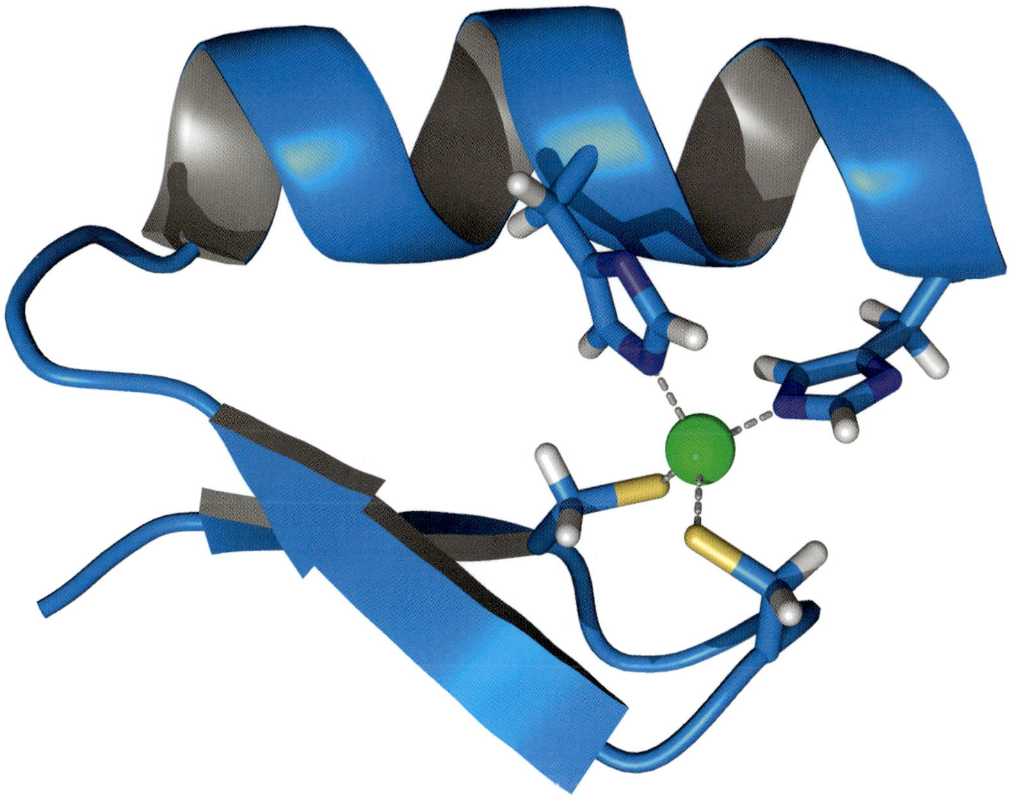

FIGURE 1.8 Cartoon representation of the Cys2His2 zinc finger motif, consisting of an α helix and an antiparallel β sheet. The zinc ion (green) is coordinated by two histidine residues and two cysteine residues.

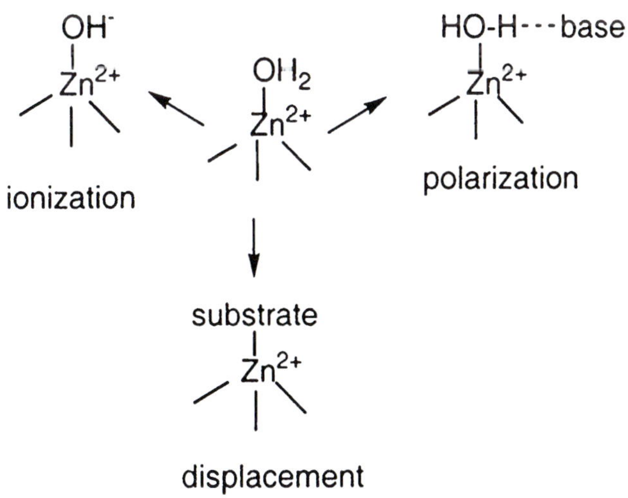

FIGURE 1.9 The zinc-bound water can either be ionized to zinc-bound hydroxide, polarised by a general base to generate a nucleophile for catalysis, or displaced by the substrate.

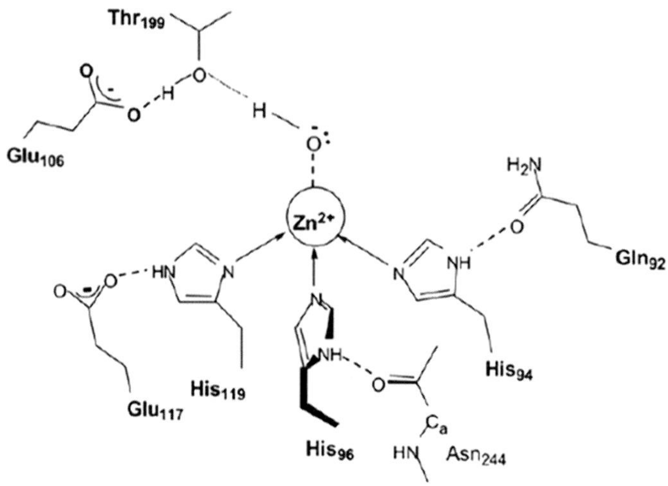

FIGURE 1.10 (**a**) The active site of human carbonic anhydrase and (**b**) the main features of the mechanism of action of carbonic anhydrase.

a hydrogen-bonded network which acts as a proton shuttle, (ii) the zinc-bound hydroxide then carries out a nucleophilic attack on the carbon dioxide substrate to generate a hydrogen carbonate intermediate $[(His)_3Zn\text{-}OCO_2H]^+$ which (iii) is displaced by H_2O to release bicarbonate and complete the catalytic cycle. The key to understanding the role of the Zn^{2+} ion is that its charge makes the bound water molecule more acidic than free H_2O, and enables it to act as a source of OH^- even at neutral pH values.

IRON AND COPPER – DEALING WITH OXYGEN

Both iron and copper play a very important role in the living world, and both seem to be essential for life, although iron may not be essential for lactic acid bacteria. On the basis of their chemistry and biochemistry, it seems probable that the early chemistry of life in an oxygen-free environment used water soluble Fe(II), whereas copper was present essentially as highly insoluble sulfides of Cu(I). The advent of oxygen was a catastrophic event for most living organisms, and can be considered to be the first general irreversible pollution of the earth. The oxidation of iron resulted in the loss of its bioavailability as Fe(II) was replaced by insoluble Fe(III), whereas the oxidation of insoluble Cu(I) led to soluble Cu(II). Further, the advent of an oxidising atmosphere exposed the potential toxicity of both elements through their capacity to generate oxygen-free radicals. A new iron biochemistry became possible after the advent of oxygen, with the development of chelators of Fe(III), which rendered iron once again accessible, and with the control of the potential toxicity of iron by its storage in a water soluble, non-toxic, bio-available storage protein (ferritin). Biology also discovered that whereas enzymes involved in anaerobic metabolism were designed to operate in the lower portion of the redox spectrum (attaining values of close to +0.6 V for iron itself), the arrival of dioxygen created the need for a new redox active metal which could attain higher redox potentials. Copper, now bioavailable, was ideally suited to exploit the oxidizing power of dioxygen. The arrival of copper also coincided with the development of multicellular organisms which had extracellular cross-linked matrices capable of resisting attack by oxygen-free radicals. After the initial 'iron age', subsequent evolution moved, not towards a 'copper age', but rather to an 'iron-copper' age.

The extensive range of biological functions carried out by both of these metal ions, range through oxygen transport by haemoglobins, haemerythrins and haemocyanins, through electron transfer by cytochromes, Fe-S proteins and plastocyanins, to multiple reactions involving oxygen activation and detoxification – the list is seemingly endless. However, here we focus briefly on one enzyme which combines the powerful redox chemistry of iron and copper acting in concert, cytochrome oxidase, CcO, the terminal component of the respiratory chain in aerobic organisms, CcO catalyses the one electron reduction of four reduced cytochrome c (c^{2+}) molecules and the four electron reduction of dioxygen to water:

$$O_2 + 4H^+ + 4c^{2+} \rightarrow 2H_2O + 4c^{3+}$$

In mammals, CcO spans the mitochondrial inner membrane and catalyses the reduction of molecular dioxygen to water at the rate of up to 250 molecules of O_2 per second. The energy released in this process is coupled to the translocation of protons, which in turn contributes to the chemiosmotic gradient required for ATP synthesis. Since the electrons and protons are taken up from the opposite sides of the membrane, the reaction results in a net charge separation across the membrane which, together with the coupled proton pumping, corresponds to the overall translocation of two positive charges across the membrane per electron transferred to O_2 from the negative (*N*) side to the positive (*P*) side of the membrane.

$$O_2 + 8H_N{}^+ + 4c_p{}^{2+} \rightarrow 2H_2O + 4H_N{}^+ + 4c^{3+}$$

While the structure of the mammalian enzyme, with between 8 and 13 different subunits has been determined, it presents a very much more complex problem than the enzyme from *Rhodobacter sphaeroides*. This contains only the two catalytic subunits – subunit I with 3 redox-active centres, containing haem *a*, and the catalytic site made up of haem a_3 and Cu_B, where dioxygen is reduced, and subunit II with the Cu_A redox centre made up of two copper ions, together with two other non-catalytic subunits (Qin et al., 2006). Its overall structure and the location of the different electron transfer components are shown in Figure 1.11a.

Electrons coming from cytochrome c enter the CcO complex via the dinuclear copper centre (Cu_A), and are then transferred consecutively one at a time to haem *a*, and then to the catalytic site of CcO, the dinuclear haem-copper centre (haem a_3-Cu_A). This is the primary oxygen-binding site, involving a haem iron, haem a_3, together with a copper ion, Cu_B, and it is at this dinuclear metal site that dioxygen is reduced. A tyrosine residue, Y(I-288), which is covalently cross-linked to one of the Cu_B ligands (His 240), is also a part of the active site. A more detailed view of the redox-active cofactors and amino acid residues involved in the proton transfer pathways is shown in Figure 1.11b (Brzezinski and Johansson, 2010).

The individual steps of oxygen binding and its subsequent reduction by CcO are presented in Figure 1.12 (Brzezinski and Johansson, 2010). In oxidised CcO (designated $\mathbf{O}^0$), both haem a_3 and Cu_B are oxidised. Transfer of the first and second electrons to the catalytic site results in the formation of states $\mathbf{E}^1$ and $\mathbf{R}^2$ (Fe_{a3}^{2+} and Cu_B^+), each step associated with proton uptake to the catalytic site and proton pumping, and O_2 then binds to haem a_3 in the $\mathbf{R}^2$ state. In the next step, the O-O bond is broken, with four electrons being donated, two from haem a_3, forming an oxo-ferryl state $Fe_{a3}^{4+}{=}O_2^{2-}$, one from Cu_B with a hydroxide ion bound, and one electron (and a proton) from residue Tyr288, located within the catalytic site, which forms a tyrosyl radical, $Tyr288^{\bullet}$. Formation of this $\mathbf{P}^2$ state is not linked to any proton uptake from solution and, both the protons and the electrons are only relocated locally within the catalytic site resulting in oxidation of haem a_3 and Cu_B. In the next step, an electron is transferred to the Tyr radical, accompanied by proton uptake to form state $\mathbf{F}^3$ and again, protons are pumped. In the final step, the last electron is transferred to the catalytic site, forming state $\mathbf{O}^4$ also accompanied by proton uptake and proton pumping. The $\mathbf{O}^4$ state is equivalent to $\mathbf{O}^0$ as the enzyme becomes fully oxidised when four electrons have been transferred to O_2.

Ni AND Co – EVOLUTIONARY RELICS

Both nickel and cobalt, together with iron, have the characteristic that they are electron-rich. They are further distinguished by the fact that in lower oxidation states some of their 3d electrons are forced into exposed σ-(or π–) orbitals: the outcome is that tetragonal Co(II) or Ni(III) are reactive-free radicals, able to give or take an odd electron, like σ-organic-free radicals. So, like iron, cobalt functions in free-radical reactions, such as the transformation of ribonucleotides into their corresponding deoxy derivatives. When one examines the kinds of reactions catalysed by nickel and cobalt enzymes and their evolutionary distribution, one arrives at the conclusion that these

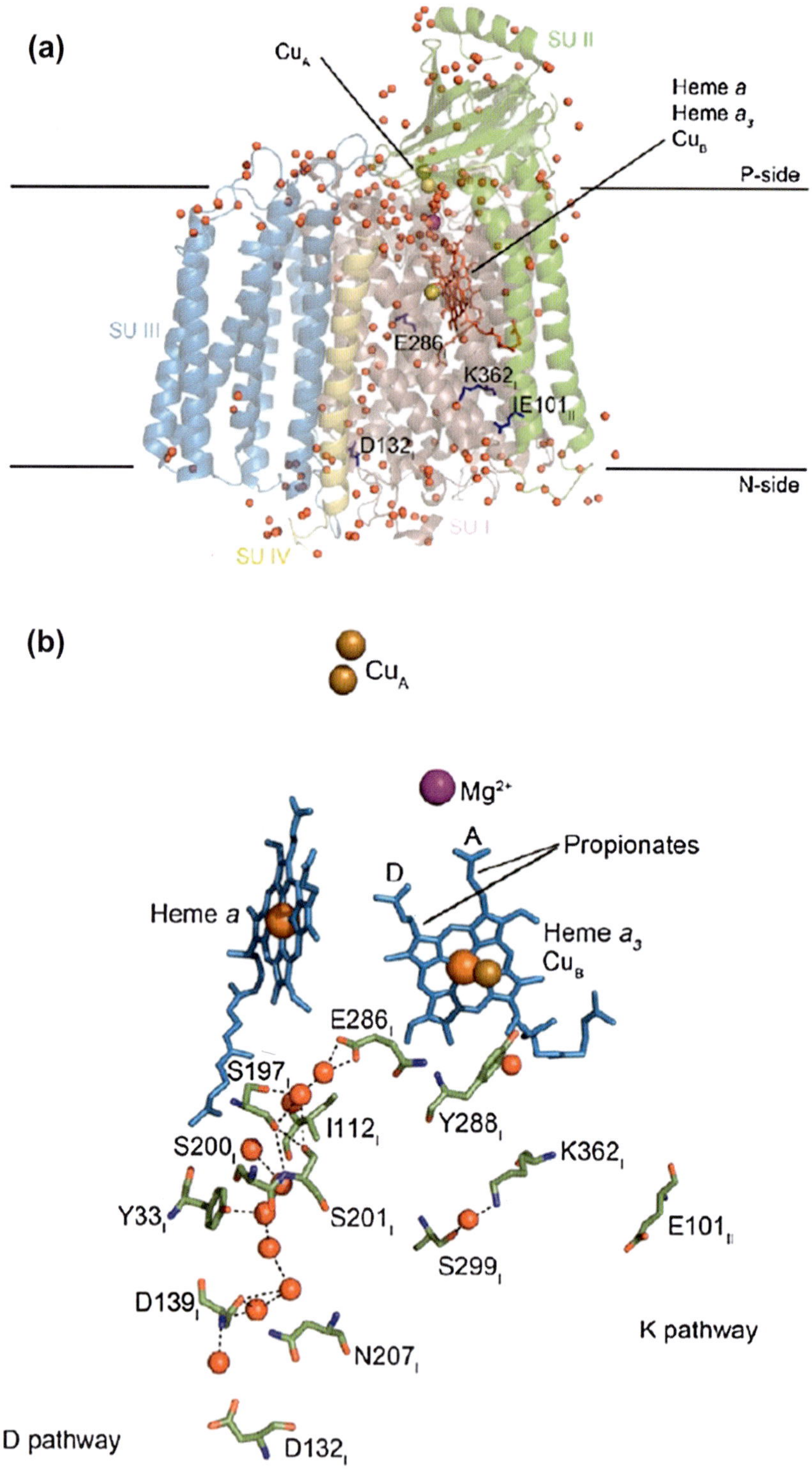
(a)
CuA
SU II
Heme a
Heme a3
CuB
P-side
SU III
E286I
K362I
E101II
D132I
N-side
SU I
SU IV
(b)
CuA
Mg2+
A
Propionates
D
Heme a
Heme a3
CuB
E286I
S197I
I112I
Y288I
S200I
K362I
Y33I
S201I
E101II
S299I
D139I
K pathway
N207I
D pathway
D132I

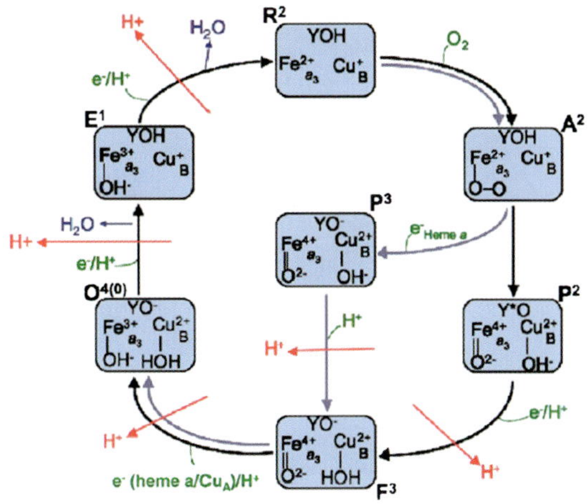

FIGURE 1.12 The catalytic cycle of cytochrome c oxidase. The electrons (e^-) and protons at the arrows (in green) are those transferred to the catalytic site, while the protons indicated by arrows (in red) perpendicular to the reaction arrows indicate pumped protons. Y is Tyr288 (see text), whereas Y × O indicates the tyrosyl radical. The reaction pathway along the blue arrows is that observed during reaction of the fully reduced CytcO (with four electrons) with O_2. *(From Brzezinski and Johansson, 2010. Copyright 2010, with permission from Elsevier).*

two elements were particularly important in the metabolism of chemicals like methane, carbon monoxide and hydrogen, all particularly abundant in the pre-oxygen evolutionary era. This is reflected in the high levels of both elements in a number of anaerobic bacteria. In contrast, the level of both metals in mammalian serum is less than 100-fold that of zinc, iron or copper. Nonetheless, cobalt, through its involvement in a number of important vitamin B_{12}-dependent enzymes continued to be used in higher organisms, including mammals. In contrast, with the exception of the plant enzyme urease, nickel proteins are virtually unknown in higher eukaryotes.

The Ni-Fe hydrogenases which play an important role in microbial energy metabolism catalyse the reversible oxidation of hydrogen:

$$H_2 \rightleftharpoons 2H^+ + 2e^-$$

Whereas in some anaerobic microorganisms, production of hydrogen serves as a mechanism to get rid of excess reducing potential, in many others hydrogen consumption is coupled to the reduction of carbon dioxide, oxygen, sulfate, or other electron acceptors while simultaneously generating a proton gradient for use in ATP production. [NiFe] hydrogenases have an unusual Ni-Fe active site (Figure 1.13) which required a combination of both spec troscopic and crystallographic studies to identify the three non-protein diatomic ligands – a good example of why one must use as many techniques as possible when studying metal ions in proteins. The unusual coordination of cyanide and carbon monoxide ligands to the 2Fe subcluster could only be established by spectroscopic methods, since the electron density of carbon, nitrogen and oxygen does not permit their differentiation by X-ray crystallography. It is thought that the single CO and two CN^- ligands maintain iron in its low spin ferrous state.

Vitamin B_{12} is a tetrapyrrole cofactor in which the central hexacoordinate cobalt atom is coordinated by four equatorial nitrogen ligands donated by the pyrroles of the corrin ring (Figure 1.14). The fifth Co ligand is

FIGURE 1.11 **(a)** The structure of cytochrome c oxidase from *R. sphaeroides* (PDB code 1M56). The four subunits of the enzyme are coloured as indicated in the figure. Haems *a* and a_3 are shown in red and the copper centres Cu_A and Cu_B in yellow. The red spheres are water molecules resolved in the structure. Residues Glu286, Asp132, Lys362, all in SU I, and Glu101 in SU II, are shown in the figure (the subscript indicates the subunit number). The approximate position of the membrane is indicated by the solid lines, where the *p*- and *n*-sides are the more positively and negatively charged sides of the membrane, respectively. The purple sphere is a non-redox-active Mg^{2+} ion found in the structure. **(b)** The D and K proton pathways shown in more detail. Also the haem a_3 propionates are indicated. *(From Brzezinski and Johansson, 2010. Copyright 2010, with permission from Elsevier).*

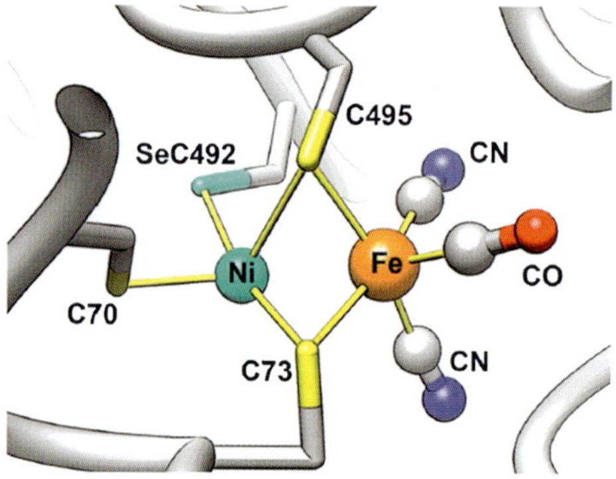

FIGURE 1.13 Structure of the NiFe active site of the hydrogenase from *Desulfomicrobium baculatum. (From Garcin et al., 1999, PDB 1CC1).*

a nitrogen atom from a 5,6-dimethylbenzimidazole nucleotide (Dmb) covalently linked to the corrin D ring. The sixth ligand in vitamin B_{12} is $-$CN. In the coenzyme B_{12} (AdoCbl), this ligand is 5′-deoxyadenosine, while in the other biologically active alkylcobalamine (MeCbl), it is a methyl group. This sixth ligand is unusual in that it forms a C-Co bond – carbon-metal bonds are rare in biology.

The principal role of essentially all AdoCbl-dependent enzymes is to facilitate the interchange of a group **X** and a hydrogen atom (**H**) on adjacent carbon atoms of the substrate. The identity of the migrating species **X** can be a small carbon-skeleton fragment or a small heteroatom-containing group like OH or NH_2, depending on the enzyme. Figure 1.15 outlines the generally accepted mechanism for these reactions and shows how radical intermediates play a crucial mechanistic role. In the first step, the substrate (**1**) binding induces homolytic cleavage of the Co–C bond of AdoCbl generating the 5′-deoxyadenosyl radical (Ado$^{\bullet}$) plus cob(II)alamine. Hydrogen abstraction by Ado$^{\bullet}$ from **1** then occurs to form 5′-deoxyadenosine (Ado-H) plus a substrate-derived radical **2** (step **A**). The rearrangement of **2** gives the product-related radical **3** (step **B**), which is followed by H-atom transfer from Ado-H to **3** to form the product **4** and to regenerate Ado$^{\bullet}$ (step **C**), which is able to recombine with cob(II)alamine, thereby completing the catalytic cycle. In some cases, elimination of H_2O or NH_4^+ from **4** results in production of an aldehyde (**5**, step **D**).

Mn – WATER SPLITTING AND OXYGEN GENERATION

The particular biological importance of manganese might be considered to reside in the tetranuclear Mn cluster which is involved in oxygen production in photosynthetic plants, algae and cyanobacteria. However, it also plays a key role in a number of mammalian enzymes like the key enzyme of the urea cycle, arginase and the mitochondrial superoxide dismutase. Most of manganese biochemistry can be explained on the one hand by its redox activity, and on the other by its analogy to Mg^{2+}. Mn has an extraordinarily important role in the photosystem II (PSII), which uses solar energy to power the oxidation of water to oxygen in photosynthetic plants, algae and cyanobacteria. The overall reaction catalysed by PSII is:

$$2Q + 2\,H_2O \xrightarrow{\text{light}} O_2 + 2\,QH_2$$

where Q represents plastoquinone and QH_2 represents plastoquinol. The electrons required to convert the oxidised quinone to the reduced quinol are extracted from the oxidation of two molecules of water, generating molecular dioxygen. This latter reaction takes place at a special centre, often called the oxygen evolving complex, which contains a tetranuclear Mn complex.

Organisms which produce oxygen, use chlorophyll a in their PSII reaction centre. This can generate a redox potential as oxidizing as +1 V, allowing the evolution of machinery that can oxidise water (redox potential

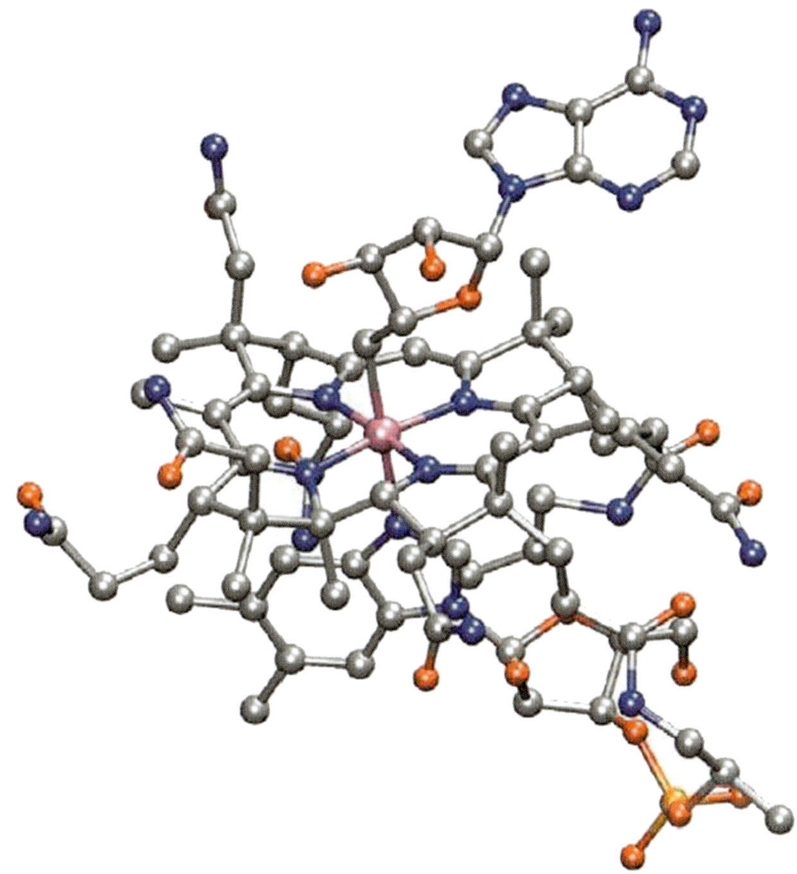

FIGURE 1.14 Ball and stick representation of adenosylcobalamin. *(From Reed, 2004. Copyright 2004 with permission from Elsevier).*

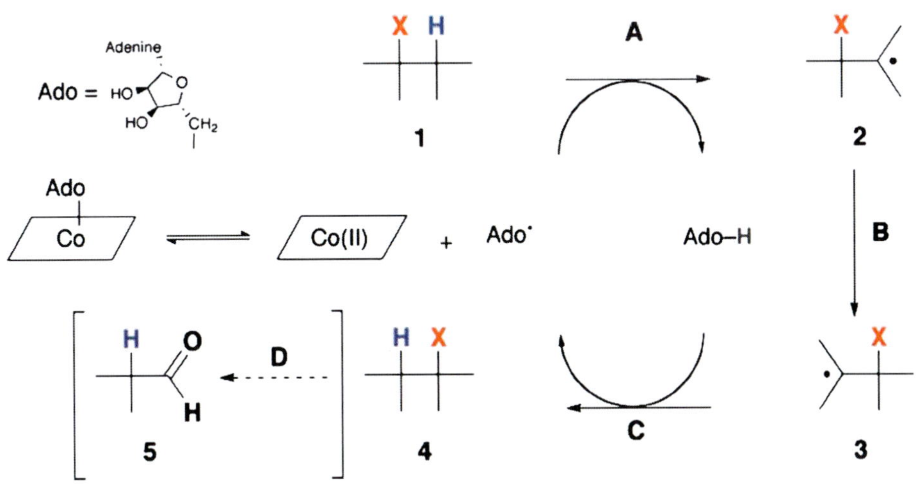

FIGURE 1.15 Mechanism for the rearrangements catalysed by AdoCbl-dependent enzymes *(adapted from Sandala et al., 2010).*

+0.9 V), by extracting four electrons from two water molecules to yield a molecule of dioxygen. The pathway of electron transfer in PSII is generally agreed to be:

$$H_2O \rightarrow [Mn_4CaC1] \rightarrow Y_z/Y_z^* \rightarrow P680/P680^+ \rightarrow Pheo_a/Pheo_a^- \rightarrow Q_A/Q_A^- \rightarrow Q_B/Q_B^-$$

where $[Mn_4CaCl]$ is the manganese cluster, Y_Z is the tyrosine residue that mediates electron transfer between the manganese cluster and the chlorophyll pair P680, $Pheo_a$ is a pheophytin, and Q_A and Q_B are plastoquinones (Figure 1.16).

The photochemistry of PSII begins when the special pair of chlorophyll molecules in PS II, often called *P680*, absorbs light at 680 nm and transfers an electron to a nearby pheophytin molecule, from where it is transferred through other electron carriers to an exchangeable plastoquinone pool (Figure 1.16). A long-lived charge separation between the positively charged species which is formed on the special pair, $P680^+$ (a powerful oxidant) and the plastoquinone Q_B, some 26A away, means that each time a photon of light kicks an electron out of *P680*, $P680^+$ extracts an electron from water molecules bound at the Mn centre, which is transferred through the redox-active TyrZ to reduce $P680^+$ back to *P680* for yet another photosynthetic cycle. In classic experiments using an oxygen electrode and short flashes of light, it was established that four flashes were required for every molecule of oxygen that was released, and the features of this were rationalised into a kinetic model, known as the S-state cycle. In this model, five states, designated S_n, of the enzyme are proposed to exist, with *n* 0–4, where each state corresponds to a different level of oxidation of the tetra-Mn centre. When the most oxidised state, S_4 is generated, it reacts in less than a microsecond to release dioxygen and return to the most reduced form of the enzyme, S_0.

While the structure of PSII from the cyanobacterium *T. elongatus* has been elucidated by X-ray crystallography at 3.5 and 2.9 Å resolution (Ferreira et al., 2004; Guskov et al., 2009), the precise positions of the Mn ions and water molecules in the photosynthetic water-splitting Mn_4Ca^{2+} cluster remain uncertain. This is due to the low resolutions of the crystal structures, and the possibility of radiation damage at the catalytic centre. Very recently, the structure of PSII from another cyanobacterium *T. vulcanus* has been determined at 1.9 Å resolution (Kawakami et al., 2011), which has yielded a detailed picture of the Mn_4CaO_5-cluster for the first time. In the high-resolution structure (Figure 1.17), the Mn_4CaO_5-cluster is arranged in a distorted chair form, with a cubane-like structure formed by 3 Mn and 1 Ca, four oxygen atoms as the distorted base of the chair, and 1 Mn and 1 oxygen atom outside of the cubane as the back of the chair. In addition, four water molecules were associated with the cluster, among which, two are associated with the terminal Mn atom and two are associated with the Ca atom. Some of these water molecules may therefore serve as the substrates for water-splitting.

Mo AND V – NITROGEN FIXATION

With the exception of bacterial nitrogenase, whose Fe-Mo-cofactor will be discussed in detail below, all other Mo enzymes contain the molybdenum pyranopterindithiolate cofactor (MoCo), which is the active component of their catalytic site (and of tungsten enzymes, in organisms which do not use molybdenum). They can be divided into three families, the xanthine oxidase, sulfite oxidase and the DMSO reductase families.

A relatively limited number of anaerobic microorganisms are capable of converting atmospheric dinitrogen into ammonia which can then be incorporated into amino acids glutamate and glutamine, and from there into other nitrogen-containing molecules. This represents about 10^8 tons/year, about the same as is produced by the industrial Haber-Bosch process – which functions at both high pressures (150–350 atm) and high temperatures (350–550 °C).

All nitrogenases consist of two types of subunit, one of which contains a special Fe-S cluster, known as the P-cluster, and a second, which contains an iron- and sulfur-containing cofactor which includes a heterometal. The heterometal is usually molybdenum, hence the cofactor is known as FeMoCo. In some species, under conditions of particular metal bio-availability, Mo can be replaced by vanadium or iron. These "alternative" nitrogenases contain vanadium instead of molybdenum (when Mo levels are low and V is available) and another which contains

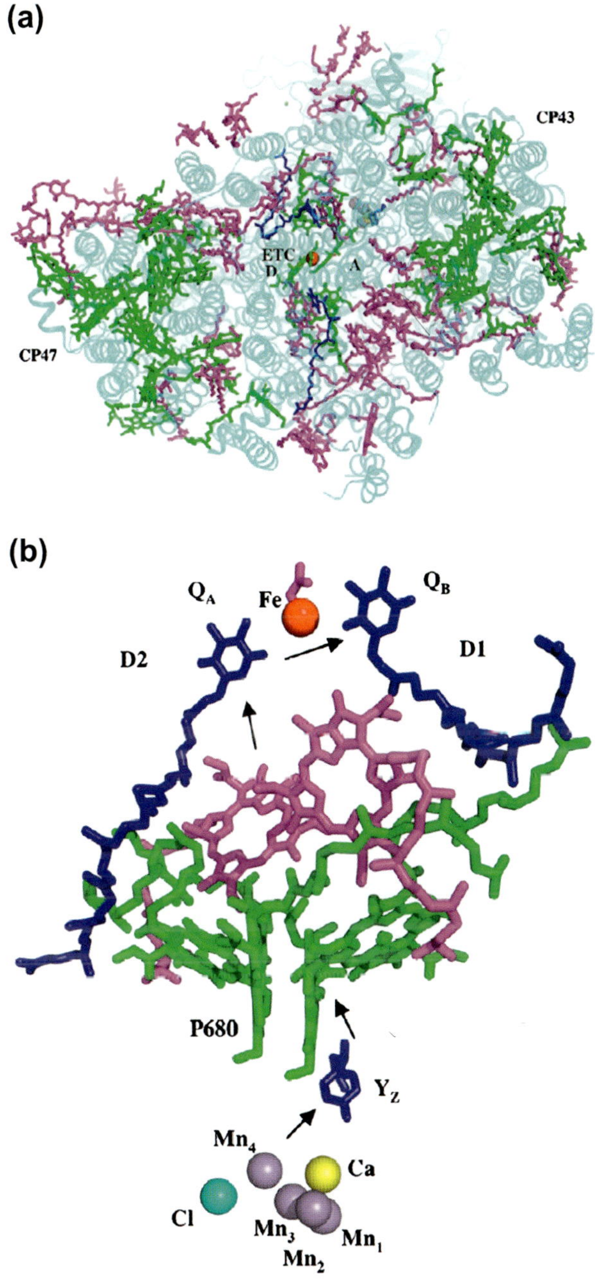

FIGURE 1.16 A structural model of the cyanobacterial PSII. (**a**) View from the cytoplasm of the PSII monomer. The model was constructed according to Guskov et al., PDB 3BZ1. Green, chlorophylls; cyan, all the other pigments and prosthetic groups. ETC is coloured as in (**b**). (**b**) Side view of the electron transfer components and the oxygen-evolving complex. Green, chlorophylls; magenta, pheophytin; blue, plastoquinone; red, iron; blue, Y_Z — tyrosine 161/D1; cyan, chloride; yellow, calcium; purple, manganese. *(From Nelson, 2011. Copyright 2011 with permission from Elsevier).*

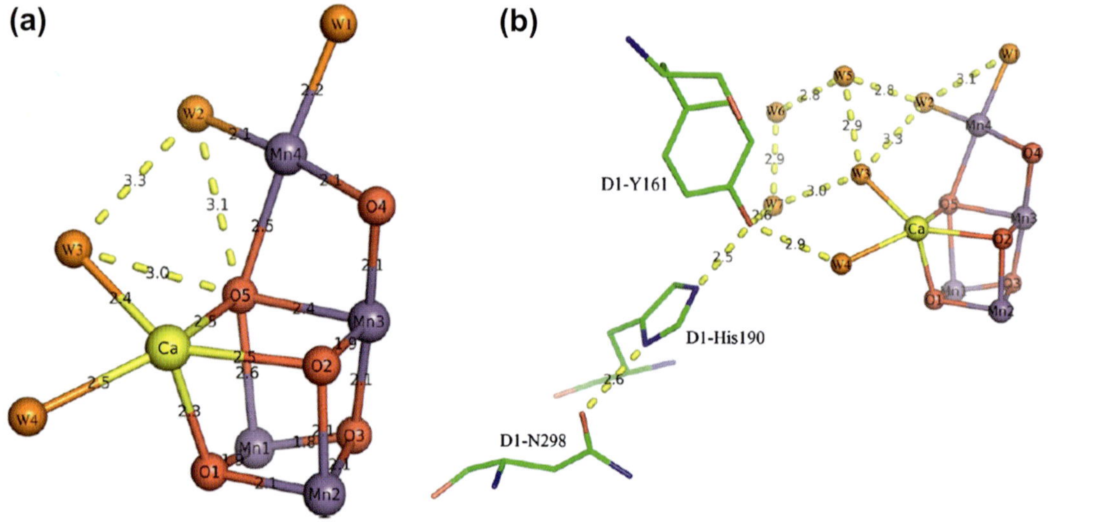

FIGURE 1.17 Structure of the Mn_4CaO_5 cluster determined at 1.9 Å resolution. **(a)** Structure of the metal cluster with oxo-bridges and water ligands. The bond distances are shown in Å. Hydrogen bonds were depicted as dashed lines. **(b)** Hydrogen-bond network linking the Mn_4CaO_5 cluster and Y_Z, and further from Y_Z to the opposite side of PSII. *(From Kawakami et al., 2011. Copyright 2011 with permission from Elsevier).*

only iron (when both Mo and V levels are low). However, by far the greatest advances in our understanding of the structure and mechanism of nitrogenases have come from studies on the MoFe-nitrogenases from free-living nitrogen-fixing bacteria like *Azotobacter*, *Clostridium* and *Klebsiella*.

The overall reaction catalysed by nitrogenases is:

$$N_2 + 8H^+ + 8e^- + 16ATP + 16H_2O \rightarrow 2NH_3 + H_2 + 16ADP + 16P_i$$

Nitrogen fixation is extremely energy-intensive, requiring both large amounts of ATP and of reducing equivalents. The nitrogenase is made up of two proteins (Figure 1.18), termed the MoFe protein and the Fe protein. The $\alpha_2\beta_2$ heterotetrameric MoFe protein contains both the FeMo-cofactor and the so-called P-cluster, with the functional unit constituted by $\alpha\beta$ dimer, containing one FeMo-cofactor and one P-cluster. In contrast, the Fe protein is a homodimer, which binds a single [4Fe-4S] cluster at the interface between the two subunits. Unlike many other multiple—electron transfer reactions in biochemistry, each individual electron transfer between the Fe protein and the MoFe protein requires the binding and hydrolysis of at least two ATP molecules. The basic mechanism of nitrogenase involves the following steps:

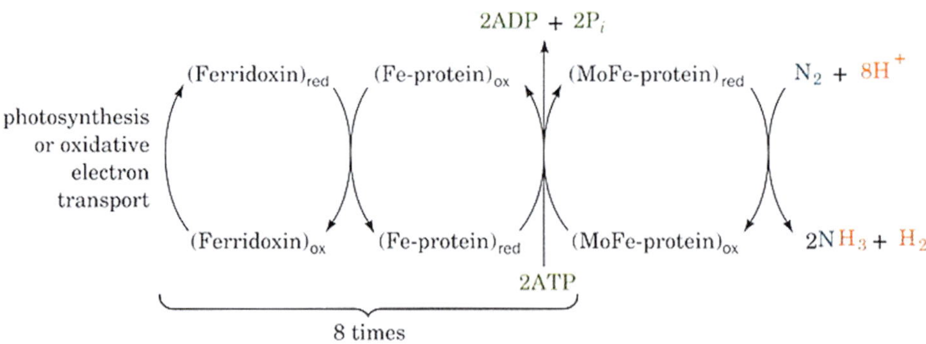

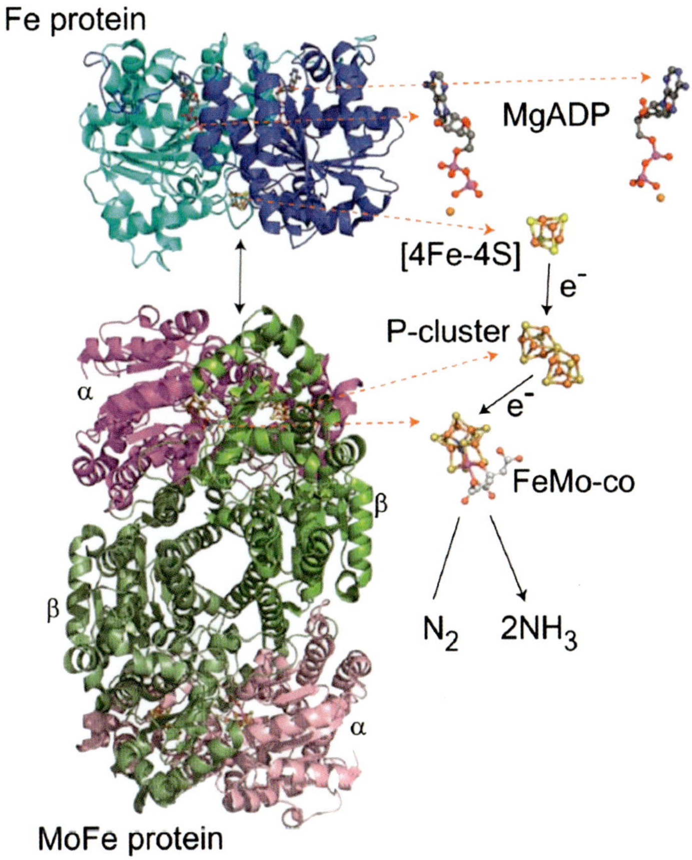

FIGURE 1.18 Structures of the nitrogenase MoFe and Fe proteins. The MoFe protein is an $\alpha_2\beta_2$ tetramer, with the alpha subunits shown in magenta and the beta subunits shown in green. The Fe protein is a γ_2 dimer, with each subunit shown in blue. A MoFe protein binds two Fe proteins, with each $\alpha\beta$ unit being a catalytic unit. One Fe protein is shown associating with one $\alpha\beta$ unit of the MoFe protein. The relative positions and structures of two bound MgADP molecules, the Fe protein [4Fe-4S] cluster, and MoFe protein P-cluster (8Fe-7S), and FeMo cofactor (7Fe-Mo-9S-homocitrate-X) are shown. Each is highlighted to the right. The flow of electrons is from the [4Fe-4S] cluster to the P-cluster to the FeMo cofactor. The element colour scheme is C grey, O red, N blue, Fe rust, S yellow, and Mo magenta. Graphics were generated with the program Pymol using the Protein Data Base (PDB) files 1M1N for the MoFe protein and 1FP6 for the Fe protein. *(From Seefeldt et al., 2009. Copyright 2009 with permission from Annual Reviews, Inc.).*

(i) complex formation between the MoFe protein and the reduced Fe protein with two molecules of ATP bound, (ii) electron transfer between the two proteins coupled with hydrolysis of ATP, (iii) dissociation of the oxidised Fe protein from the complex accompanied by its re-reduction and exchange of the 2ADPs for ATPs, (iv) repetition of this cycle of association, reduction, ATP hydrolysis and dissociation to transfer one electron at a time to the MoFe protein. Once a sufficient number of electrons and protons have been accumulated, available substrates can be reduced. Usually, when eight reducing equivalents have been accumulated, and 16 molecules of ATP hydrolysed, the enzyme can bind and reduce the very stable triple bond of a dinitrogen molecule to two molecules of ammonia. Concomitantly, two protons and two electrons are converted to gaseous hydrogen. Electrons derived from photosynthesis or from the mitochondrial electron transport chain are transferred to the Fe protein.

REFERENCES

Allen, K. N., & Dunaway-Mariano, D. (2004). Phosphoryl group transfer: evolution of a catalytic scaffold. *Trends Biochem. Sci., 29*, 495–503.

Allen, K. N., & Dunaway-Mariano, D. (2009). Markers of fitness in a successful enzyme superfamily. *Curr. Opin. Struct. Biol., 19*, 658–665.

Brzezinski, P., & Johansson, A. L. (2010). Variable proton-pumping stoichiometry in structural variants of cytochrome c oxidase. *Biochim. Biophys. Acta, 1797*, 710–713.

Bublitz, M., Poulsen, H., Morth, J. P., & Nissen, P. (2010). In and out of the cation pumps: P-type ATPase structure revisited. *Curr. Opin. Struct. Biol., 20*, 431–439.

Crichton, R. R. (2011). *Biological Inorganic Chemistry: A New Introduction to Molecular Structure and Function* (2nd ed.), p. 455. Oxford: Elsevier.

Dietrich, B. (1985). Coordination chemistry of alkali and alkaline-earth cations with macrocyclic ligands. *J. Chem. Edu., 62*, 954–964.

Ferreira, K. N., Iverson, T. M., Maghlaoui, K., Barber, J., & Iwata, S. (2004). Architecture of the photosynthetic oxygen-evolving center. *Science, 303*, 1831–1838.

Garcin, E., Vernede, X., Hatchikian, E. C., Volbede, A., Frey, M., & Fontecilla-Camps, J. C. (1999). Crystal structure of a reduced active form of an [NiFeSe] hydrogenase provides an image of the activated catalytic center. *Structure Fold. Des., 7*, 557–566.

Gerlt, J. A., Allen, K. N., Almo, S. C., Armstrong, R. N., Babbitt, P. C., Cronan, J. E., et al. (2011). The enzyme function initiative. *Biochemistry, 50*, 9950–9962.

Gouax, E., & MacKinnon, R. (2005). Principles of selective ion transport in channels and pumps. *Science, 310*, 1461–1465.

Guskov, A., Kern, J., Gabdulkhakov, A., Broser, M., Zouni, A., & Saenger, W. (2009). Cyanobacterial photosystem II at 2.9-A resolution and the role of quinones, lipids, channels and chloride. *Nat. Struct. Mol. Biol., 16*, 334–342.

Hodgkin, A. L., & Huxley, A. F. (1952). A quantitative description of membrane current and its application to conduction and excitation in nerve. *J. Physiol., 117*, 500–544.

Kawakami, K., Umena, Y., Kamiya, N., & Shen, J. R. (2011). Structure of the catalytic, inorganic core of oxygen-evolving photosystem II at 1.9 Å resolution. *J. Photochem. Photobiol. B, 104*, 9–18.

MacKinnon, R. (2004). Potassium channels and the atomic basis of selective ion conduction (Nobel Lecture). *Angew. Chem. Int. Edn., 43*, 4265–4277.

Maguire, M. E., & Cowan, J. A. (2002). Magnesium chemistry and biochemistry. *Biometals, 15*, 203–210.

Nelson, N. (2011). Photosystems and global effects of oxygenic photosynthesis. *Biochim. Biophys. Acta, 180*, 856–863.

Qin, L., Hiser, C., Mulichak, A., Garavito, R. M., & Ferguson-Miller, S. (2006). Identification of conserved lipid/detergent-binding sites in a high-resolution structure of the membrane protein cytochrome c oxidase. *Proc. Natl. Acad. Sci. U S A., 103*, 16117–16122.

Reed, G. H. (2004). Radical mechanisms in adenosylcobalamin-dependent enzymes. *Curr. Opin. Chem. Biol., 8*, 477–483.

Sandala, G. M., Smith, D. M., & Radom, L. (2010). Modeling the reactions catalysed by coenzyme B12-dependent enzymes. *Acc. Chem. Res., 43*, 642–651.

Seefeldt, L. C., Hoffman, B. M., & Dean, D. R. (2009). Mechanism of Mo-dependent nitrogenase. *Annu. Rev. Biochem., 78*, 701–722.

Chapter 2

Introduction to Ligand Field Theory

Frank Neese
Max-Planck Institut für Bioanorganische Chemie, Stifstr. 34-36, D-45470 Mülheim an der Ruhr, Germany

Chapter Outline

INTRODUCTION

Transition metal ions play a key role in all forms of life where they are used most prominently to drive a number of central catalytic reactions. Studying the roles that these transition metal centres play in biomolecules, foremost enzymes, is at the heart of bioinorganic chemistry. However, the chemistry of transition metal ions is fairly different from that of the main group compounds that are more familiar to biochemists. The reactivity of transition metal ions is very finely tuned by the protein environment. For example, heme iron can either serve as dioxygen transporter (as in hemoglobin and myoglobin) or react with dioxygen to form "hot" intermediates that hydroxylate unactivated C-H bonds (as in cytochrome P450). These dramatic differences in reactivity are intimately linked to the electronic structure of the transition metal centre. Thus, some understanding of the electronic structure of transition metal ions is necessary in order to appreciate their complex, yet fascinating, chemical behaviour.

In this chapter, an introduction to the important language of "crystal field theory" (CFT) and "ligand field theory" (LFT) will be provided. These closely related theories have been established over the course of several decades in the inorganic chemistry community and form the basis for most contemporary discussions of transition metal electronic and geometric structure. In fact, the structure and bonding of molecules are commonly taught in chemistry and biochemistry curricula in terms of the concept of Lewis structures and electron pair bonds. This is an extremely powerful formalism and it is at the heart of chemical thinking. However, when it comes to transition metals, this formalism is hardly, if at all, applicable. Rather, the structure, bonding, spectroscopy and reactivity of transition metal complexes are commonly discussed in the fairly distinct language of CFT/LFT.

The purpose of this chapter is to provide a non-mathematical overview of the language of LFT in a biochemical context. It is aimed at experimentally oriented graduate students. It is hoped that the chapter will provide enough introductory background to assist newcomers to CFT/LFT to read and understand research papers that make use of this language and perhaps spawn enough curiosity to study the subject in more depth.

It is appropriate to introduce the subject by first discussing what CFT and LFT is *not*: CFT and LFT are *not* first principles theories that lets one predict the properties of transition metal complexes from first physical principles without recourse to experimental data. On the contrary, CFT and LFT are *models*. The nature of a model is that it

Practical Approaches to Biological Inorganic Chemistry, 1st Edition. http://dx.doi.org/10.1016/B978-0-444-56351-4.00003-8.

creates an oversimplified picture that captures the essence of certain aspects of reality without ever attempting to provide a comprehensive description. Models in science are characterised by semi-empirical parameters that are determined from fitting experimental data. In this way, the models become semi-quantitative. However, much more important than their semi-quantitative nature is the fact that models create a *language*. This language is formulated in a number of elementary and intuitively appealing concepts. The existence of such a language lets scientists communicate about *classes* of substances on a common footing, rather than having to discuss specific details about each and every molecule individually. This is the key feature of a successful model.

It is important to understand and appreciate that the goal of LFT is very different from the goals of the nowadays widely available first principles electronic structure theory (e.g. true *ab initio* methods or density functional theory). In such electronic structure calculations, one treats individual molecules. Each molecule, irrespective of how minor the changes are compared to a previously investigated molecule, is a completely new case. One aims in these calculations at obtaining reliable numbers without any input from experiment (hence the term "*ab initio*"). Such calculations may provide accurate predictions for energies or molecular properties. However, they may not be easily interpretable in terms of simple, intuitive concepts. Hence, it is highly rewarding to view these calculations using "ligand field goggles" since then one might be able to obtain the best features of both worlds – the insight provided by LFT together with the predictive power of modern electronic structure theory.

This chapter is organised into two main parts. The first part provides an overview of traditional CFT and its extension to LFT. In part two, the limitations of ligand theory are discussed in the context of its logical extensions – molecular orbital (MO) theory and its relation to CFT and LFT. Quantitative MO methods will, however, not be covered. Instead, some pointers to the literature will be provided.

CRYSTAL FIELD AND LIGAND FIELD THEORY

States Versus Orbitals

From a physicist's perspective, any molecule can be thought of as consisting of a number of negatively charged electrons and positively charged point nuclei that interact according to the laws of quantum mechanics (Figure 2.1).

According to the laws of quantum mechanics, a system can assume a variety of discrete ***states*** with well-defined energy. This energy is the "inner energy" of a molecule and represents the energy that is needed to separate

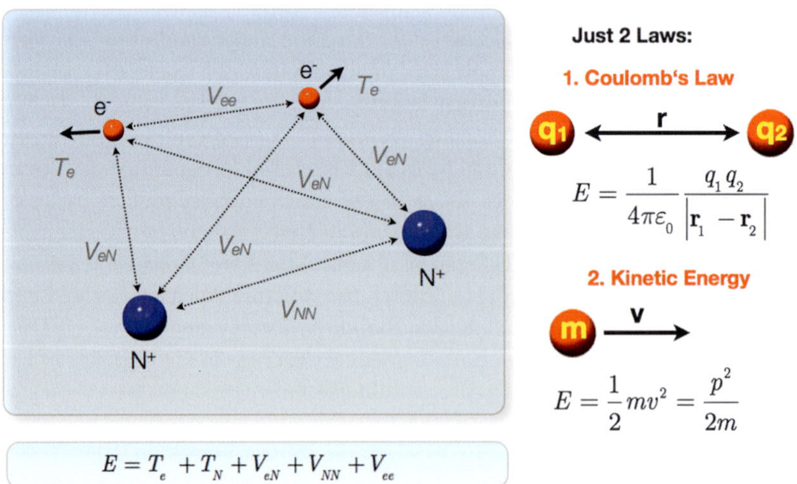

FIGURE 2.1 Fundamental interactions in molecules. The physics of molecular structure is based on the Coulombic interactions between electrons and nuclei and among each other together with the kinetic energy of the electrons.

all particles to infinite distance. The energy of any given state changes with the arrangement of the nuclei, this defining a "potential energy surface". Minima on this surface pertain to equilibrium structures (as can be determined through X-ray diffraction) while saddle points pertain to transition states of chemical reactions. Spectroscopic transitions can be induced by electromagnetic radiation between different states of a molecule.

It is important to appreciate that all observable properties are governed by the states of the system. These states always involve all electrons of the system and all nuclei. Calculating these states require the solution of the many particle Schrödinger equation. This, however, is a hopeless endeavour for all but the simplest one-electron systems, like the hydrogen atom. While fairly accurate approximation to the many particle wave functions can nowadays be calculated their complexity is extremely high. Hence, multi-particle wave functions, being multi-dimensional functions, cannot be pictured in an intuitive way.

Chemists, on the other hand, are used to thinking in terms of ***orbitals***. Orbitals are one-particle wave functions. In the case of one-electron systems, states and orbitals obviously coincide. For many particle systems, e.g. all "real-life" molecules, this connection is not obvious. Let us take here for granted that it is possible to *approximate* many particle wave functions in a suitable way with the help of orbitals. It is then important to appreciate that all that the orbitals are good for is to build up an approximate many particle wave function. Thus, the *orbitals* themselves are never observable — only the states that are built from them. Much confusion arises from not properly distinguishing between the observable properties of a molecule (derived from *states*) and the auxiliary quantities (*orbitals*) that are used to build up states. If the beginning student keeps this important difference in mind, further study of electronic structure will be greatly simplified.

Orbitals have the benefit that they can be readily visualised since they are only functions of three space coordinates, e.g. the three Cartesian coordinates x, y, z. An orbital is a function that is normalised to unity and can be positive or negative. The square of an orbital evaluated at a given point $\boldsymbol{r}$ represents the probability density for finding the electron in an infinitesimal volume element around this point.

In the framework of CFT and LFT, the all-important orbitals are the real-valued d-orbitals of a given metal ion (we exclude f elements in this chapter). Their shapes are shown in Figure 2.2. In this figure, red areas represent positive probability amplitude, yellow areas of negative probability amplitude. Since only the square of the probability amplitude is physically meaningful (it gives a probability density, e.g. a probability per unit volume) red and yellow areas can be interchanged which amounts to the multiplication of the orbital by a physically meaningless factor of -1. The figures are drawn such that the red or yellow areas represent a constant value for the function, e.g. ± 0.03. An electron that occupies such an orbital is essentially confined to the volume inside the probability amplitude "blobs".

Spin and Orbital Angular Momentum

In classical physics, a charge in motion creates a magnetic dipole moment. A useful model system is the current passing through a circle shaped wire. Subjecting this wire to a magnetic field a force will be exerted that will act to position the circle perpendicular to the direction of the magnetic field. In atoms, the motion of electrons can be regarded to be circular. Hence, there is a magnetic dipole moment associated with the ***orbital motion*** of an electron. In quantum mechanics, magnetic dipole moments are associated with angular momentum. Hence, it is ***orbital angular momentum*** that gives rise to observable magnetic dipole moments. However, quantum mechanics also demands that angular momentum is ***quantised***. Thus, unlike a classical rotor that can have any angular momentum, electrons can only have angular momenta that are multiples of Planck's constant h divided by 2π ($\hbar = h/2\pi = 6.58211928 \times 10^{-16}$ eV s). Thus, associated with the orbital motion of electrons there is a ***angular momentum quantum number***, l, that can take values $l = 0, 1, 2, \ldots$ in integer steps. In classical physics, of course, the angular momentum is a vector with three components. Another fundamental result of quantum mechanics is that for angular momenta one can only ever know the "length" of the vector ($= l(l+1)\hbar$) and the value of ith projection onto the z-axis (or any other chosen axis of quantisation). This projection is associated with a second quantum number, the "magnetic quantum number" m_l which can assume values from l to $-l$ in integer steps.

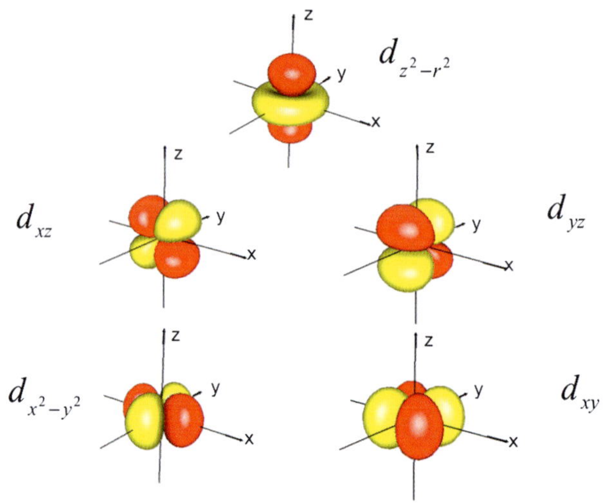

FIGURE 2.2 The shapes of the five d-orbitals and their labels. Red corresponds to positive, yellow to negative phase.

Hence, for each state with angular momentum l, there are $2l+1$ substates with $m_l = l, l-1, \ldots, -l$. Different atomic orbitals have different angular momenta. In fact the quantum number l is identical to the number of nodes that the orbitals have. For s-orbitals, $l = 0$, for p-orbitals, $l = 1$, for d-orbitals, $l = 2$, etc. This explains, why there is one s-orbital, three p-orbitals, five d-orbitals etc. For a system of many electrons, the individual angular momenta of electrons must be coupled to give a total angular momentum L, and a total projection M_L. In molecules, the angular momentum of electrons is largely quenched and the quantum numbers L and M_L are no longer "good" quantum numbers. However, it is important to appreciate the angular momentum rules in order to obtain some understanding of the electronic structures of atoms and ions.

However, in quantum mechanics of electrons, there is a second type of angular momentum that must be considered. Its origin lies deep in the foundation of quantum mechanics itself and there is no classical analogue for this type of angular momentum. This angular momentum is related to another fundamental property of electrons: ***the electron spin***. An electron can be thought of as a point charge with a certain probability distribution that follows the laws of quantum mechanics. However, at the same time, an electron is not only a point charge but also a bar magnet. The magnetic moment of the electron is caused by so-called electron spin. It can be thought of as an "internal" angular momentum. As any charge in motion creates a magnetic field, so does an electron from its "inner motion". However, the electron does behave like a quantum mechanical bar magnet. If a magnetic field is present a classical bar magnet can assume any orientation with respect to the magnetic field. For quantum mechanical bar magnets, this is not possible and only two orientations are quantum mechanically allowed. These two states of an electron are loosely referred to as "spin up" and "spin down". Thus, in addition to its position in space, the states of an electron are characterised as "spin up" or "spin down". Associated with the spin of an electron is the spin quantum number $s = 1/2$ and the projection quantum numbers $M_s = \pm 1/2$ corresponding to the "spin-up" and "spin-down" states of the electron. Again, in many electron systems, the individual spins of the electrons must be coupled to a total spin S and ith projection M_S. Much confusion arises from not properly distinguishing between the spin of individual electrons and the total spin of the entire system. Hence, it is of great importance to have an awareness of this important distinction.

Hence, an electron moves in a four dimensional space (disregarding time), where the fourth dimension is the rather strange "spin space". The spin of the electron is, of course, fundamentally related to all magnetic properties of molecules. The fact that there are the macroscopic phenomena of paramagnetism, ferromagnetism or diamagnetism are all related to the electron spin and its behaviour. In this chapter, magnetic properties will not be treated and the interested reader is referred to the chapter about electron paramagnetic resonance spectroscopy.

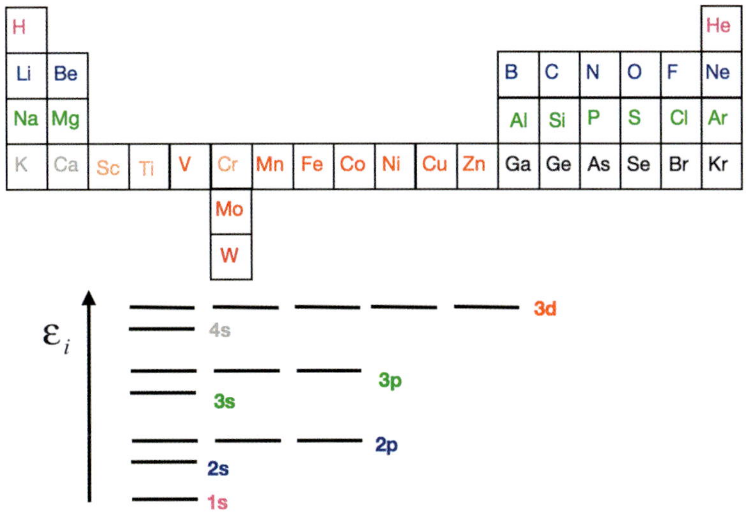

FIGURE 2.3 The periodic system of the elements. Only the transition metals that are highlighted are occurring in natural systems. These are the first row ions with the exception of Sc, Ti and Cr together with the second row element Mo and the third row element W.

The Crystal Field Model

The first key concept in LFT is the notion of a d^N configuration. Transition metal ions from all three transition rows (Figure 2.3) have accessible d-orbitals that are partially filled as well as an s-shell that is filled by one or two electrons in the electronic ground state of the neutral atom.

States or Atoms or Ions

As mentioned in the preceding paragraphs, the observable properties of any system (molecule, atom or ion) derive from the many particle states that the system can assume. Hence, we start by focussing on the states that a given atom or ion with N valence electrons can assume. The easiest case is obviously met for systems with a single electron. In this case, the electron can occupy any of the five equivalent d-orbitals or the valence s-orbital. Since the d-orbitals and the s-orbitals have different energies, the states have different energies. One electron in any of the five d-orbitals is at the same energy because in the absence of external fields or ligands all five d-orbitals are energetically degenerate. In addition, in the absence of a magnetic field the two spin states that a single electron can assume are equivalent. Hence, there are ten equivalent states arising from the (d^1s^0) configuration and two equivalent states arising from the (d^0s^1) configuration.

These states can be further classified according to their total spin- and angular momentum quantum numbers. Since there is one unpaired electron, the total spin must amount to $S = 1/2$ in any of the 12 states. The orbital angular momentum in a state with a singly occupied d-orbital is $L = 2$ and in a singly occupied s-orbital is $L = 0$. Hence, the ten (d^0s^1) states are all energetically equivalent and belong to a ($L = 2$, $S = 1/2$) combination while the two (d^0s^1) states belong to a ($L = 0$, $S = 1/2$) combination. For such states, it is customary to employ the so-called ***Russell-Saunders*** or LS notation in which an upper left index denotes the spin multiplicity $M = 2S + 1$ and the main letter refers to the spectroscopic designation of the L-quantum number (i.e. $L = 0,1,2,3,4$ correspond to letters S,P,D,F,G). Thus, according to these rules, the 2 ($=2S + 1$) $\times$ 5 ($=2L + 1$) $=$ 10 (d^0s^1) states belong to a Russell-Saunders ^{2}D *term*, while the two (d^0s^1) states belong to a ^{2}S term. In an actual atomic spectrum, these are the two states that are observable. For example, for the Ti^{3+}-ion (electronic configuration (Ar)($3d^1 4s^0$)), the ground state corresponds to the ^{2}D term and the excited ^{2}S state is located at 80388.92 cm^{-1} above it. A wealth of data can be obtained from the NIST data tables (http://physics.nist.gov/PhysRefData/ASD/levels_form.html).

The next complicated situation obviously arises from distributing two d-electrons over the available d- and s-orbitals, for example in Ti^{2+} (electronic configuration (Ar)($3d^2 4s^0$)). First one must couple the spins of the two electrons to a given total spin. For two electrons, one can form either singlet ($S = 0$) or triplet ($S = 1$) states. However, this is not possible for every spatial configuration of the electrons as a triplet state can only be realised if the two electrons occupy different orbitals. Likewise, the orbital angular momenta of the two electrons can be coupled to a total orbital angular momentum which according to the so-called ***Clebsch–Gordan rules*** can be $L = 4,3,2,1$ or 0. There are methods and rules that allow for the construction of all of the terms that arise from the various configurations, but this is beyond the scope of this chapter. The interested student should consult the literature for further study. It suffices to say that the (d^2s^0) configuration gives rise to ^3F, ^3P, ^1D, ^1G and ^1S states in accordance with the Clebsch–Gordon series and the electron spin coupling arguments eluded to above. These energetic ordering of these states is subject to ***Hund's rules***. The three rules can be stated as follow:

(1) *Completely occupied shells carry neither spin nor orbital angular momentum*
(2) *For a given configuration states with higher total spin S are lower in energy*
(3) *For a given configuration and maximum spin multiplicity states with higher total angular momentum L are lower in energy.*

These rules allow one to guess the energetic ordering of a set of states arising from a given configuration. In the example of the (d^2s^0) configuration according to rule (2), triplets must be lower than singlets and according to rule (3) ^3F must lower than ^3P. Hence, the ground state is expected to be ^3F and this is in accord with experiment for all ions with the indicated configuration. Among the singlet states, the energetic order is *not* determined by Hund's rules because the spin multiplicity is lower than maximum. Whether any of the singlet states is lower in energy than the excited ^3P state is also not evident from Hund's rules. For the chosen example of the Ti^{2+}-ion, the energetic ordering of the states relative to the ^3F ground term is 8473.5 cm^{-1} (^1D), 10583.4 cm^{-1} (^3P), 13397.6 cm^{-1} (^1G) and 32474.5 cm^{-1} (^1S). Thus, substantial energy differences arise from the various ways to couple spin- and angular momenta of electrons. For the (d^1s^1) configuration, the situation is relatively simple. The Pauli principle puts no constraint on the spin coupling since the electrons occupy different orbitals. Then the total angular momentum must be $L = 2$ from the single electron in the d-shell and consequently one expects (and observes) ^3D and ^1D states in this energetic order. In the Ti^{2+} ion, they are found at 38064.4 and 41704.3 cm^{-1}, respectively.

Obviously, the Russell-Saunders coupling becomes more involved the more electrons are present. If, however, a shell is more than half filled the available states can be deduced from the *holes* rather than the electrons in the system. Hence, in terms of (d^Ns^0) configurations, the highest complexity arises for $N = 5$ while the d^4/d^6, d^3/d^7, d^2/d^8, and d^1/d^9 cases are equivalent. Again, the NIST tables provide invaluable experimental information about the states of atoms and ions across the entire periodic table.

Atoms or Ions in a "Crystal Field"

Introduction

The valence shell s-electrons are the first to be ionized and hence the s-orbital remains empty in all transition metal complexes that have a charge of at least +2. In the transition metal complexes that contain a metal(I) ion, all valence electrons usually occupy the d-orbitals. The chemistry and spectroscopy of the transition metal elements can hence be understood by focussing attention on the metal d-orbitals as these are the orbitals that donate or receive electrons in chemical reactions or are being populated or depopulated in spectroscopic transitions.

An interesting situation arises when a transition metal ion is placed in the electrostatic field of a set of ligands. The model of CFT (as opposed to LFT; the distinction will be clarified in section The Nephelauxetic Effect, Covalency and LFT) rests upon the extremely bold assumption that it is adequate to represent the ligands by point charges. Obviously, this falls very far short of acknowledging the chemical nature of a given ligand. Nevertheless, already this grossly oversimplified model provides important and useful insights into the behaviour of transition metal ions in coordination complexes and hence also in bioinorganic chemistry.

Let us adopt the ligand field model and treat the most straightforward case where six identical ligands are placed along the $\pm$x, $\pm$y, and $\pm$z axes at a given distance R with the transition metal ion located at the origin of the coordinate system. The ligands are replaced by negative point charges with point charge q. The immediate consequence of this arrangement is that the spherical symmetry of the atom or ion is raised. The consequence of this situation is two-fold: (a) orbitals that are energetically degenerate in the free ion are no longer degenerate and (b) the orbital angular momentum is said to be quenched and hence the total angular momentum L is no longer a good quantum number. Hence, we must find a different classification scheme for states and orbitals. This classification scheme is realised using the powerful language of ***group theory***.

Group Theory

The application of group theory to chemistry in its most elementary form classifies a given molecule according to the symmetry operations (rotations, inversion, reflections, improper rotations) that turn the nuclear framework back into itself or an indistinguishable configuration where equivalent atoms occupy equivalent positions (for examples see Figure 2.4). All symmetry operations that exist for a given nuclear arrangement form a "point group". In the case of the octahedral arrangement eluded to above, the appropriate group is the so-called octahedral group that is given the symbol O_h. Other arrangements give rise to other point groups. For example, a tetrahedral arrangement conforms to the group T_d, a square planar arrangement to group D_{4h}, a trigonal bipyramidal arrangement to D_{3h}, a tetragonal pyramidal arrangement to C_{4v} etc. Explanations for assigning point groups to molecules can be found in the literature collected at the end of the chapter.

The behaviour of states and orbitals under the various symmetry operations of a given group is determined by the so-called irreducible representations (***irreps***) of the group. The subject is fairly involved and hence we can only scratch the surface in this chapter. A given state or orbital is said to "transform under the irreps of a given group". The irreps have names that indicate how a given objects behaves under various symmetry operations. For example, if a centre of inversion exists in the group (as it does in O_h) labels 'g' and 'u' indicate 'gerade' or 'ungerade' behaviour under inversion. This means that the transformed object changes sign ('ungerade') or not ('gerade'). The five d-orbitals, for example are all 'gerade' as they do not change sign upon inversion (refer to Figure 2.2). Likewise

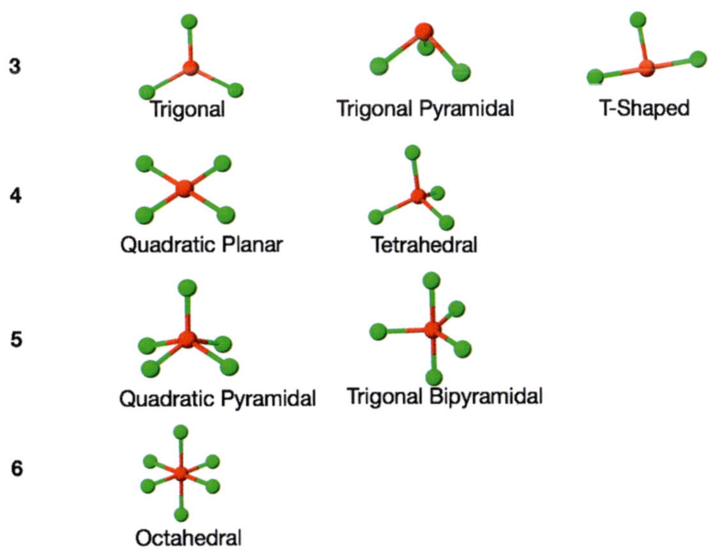

FIGURE 2.4 Typical coordination geometries.

labels 'a' or 'b' indicate 'gerade' or 'ungerade' behaviour with respect to the main rotation axis, a ' or '' indicates 'gerade' or 'ungerade' behaviour with respect to the main plane of reflection (if no centre of inversion exists). Importantly, in cases where there is at least an axis of rotation with a higher value than 2 (a C_2 axis corresponds to a rotation by 180°) the irreps might be more than one-dimensional. This means that there must always be groups of two or three objects that behave identically under all operations of the point group. Thus, symbols 'e' and 't' refer to two- and three-dimensional irreps, while 'a' and 'b' refer to mono-dimensional point groups.

While this nomenclature will invariably appear to be fairly obscure to the beginning student, it should be stressed that group theory is an extremely useful language that is indispensable for the serious interpretation of spectroscopic properties of molecules. There are excellent textbooks that should be consulted for a thorough and systematic exposition of the subject without dwelling too much on mathematical subtleties that are of lesser concern to chemistry.

Terms and Term Symbols

Importantly, to avoid confusion, a widely – but not universally adopted – convention is to label orbitals (or more generally one-electron quantities) with lowercase labels and states (or more generally many electron quantities) with uppercase labels. On an operational level, the states of a many electron system are labelled the irrep the states correspond to, rather than the total angular momentum L. Thus, the equivalent of a Russell-Saunders term for atoms consists of a ***molecular term symbol*** of the form $^{2S+1}\Gamma$, where Γ denotes the irrep of the state.

Let us apply these concepts to the easiest case, a single electron in a ligand field. According to the rules of group theory, the five d-orbitals are distributed over two irreps of the O_h point group. Specifically, the $d_{x2\text{-}y2}$ and d_{z2} orbitals transform under the irrep e_g while the d_{xy}, d_{xz} and d_{yz} orbitals transform under the irrep t_{2g}. If one is familiar with group theory this can be readily derived. However, it can also simply be looked up in group tables that accompany almost every textbook on the subject. If the single electron occupies an e_g orbital a ^2E_g *state* arises, while a $^2T_{2g}$ *state* arises from the single occupation of the t_{2g} shell. Importantly, the ^2E_g and $^2T_{2g}$ are said to be two- and three-fold orbitally degenerate because the occupation of any orbital within each set is energetically equivalent. Both states are two-fold spin-degenerate because in the absence of a magnetic field (or other magnetic perturbation), the two spin-states of the single electron are energetically equivalent. Hence, the total degeneracy of the ^2E_g state is four while that of the $^2T_{2g}$ state is six.

Crystal Field Splittings

One of the major achievements of CFT is to provide a means of estimating whether the $^2T_{2g}$ or the ^2E_g state is lower in energy. The main ideas can be visualised by a plausibility argument: the transition metal ions are positively charged while the ligands are invariably negatively charged or carry a partial negative charge at the coordinating atoms. Hence, if negatively charged electrons at the transition metal centre move close to a (partially) negatively charged ligand, they are repelled. The repulsion is stronger if the average distance of the electron to the ligand is smaller. If one investigates the two groups of d-orbitals in Figure 2.2 one notices that the e_g orbitals ($d_{x2\text{-}y2}$ and d_{z2}) have their probability density essentially located *on* the coordinates axes, while the three t_{2g} orbitals (d_{xy}, d_{xz} and d_{yz}) have their probability density located *between* the coordinate axes. Thus, given, that the point charges representing the ligands are situated *on* the coordinate axes, it follows that from the point of view of CFT, the e_g orbitals must be higher in energy than the t_{2g} orbitals (Figure 2.5).

The energy difference between the two sets of orbitals is called the ***crystal field splitting***. It is given the symbol Δ, or, for historic reasons, is called '10Dq'. CFT provides an expression for this quantity which is given by:

$$\Delta \equiv 10\text{Dq} = \frac{5}{3}\frac{q}{R^5}\langle r^4 \rangle \tag{2.1}$$

Here, q is the charge at the ligand, R is the distance between metal and ligand and $\langle r^4 \rangle$ is the mean value of the fourth power of the distance of an electron occupying a d-orbital from the metal nucleus. Obviously, the crystal

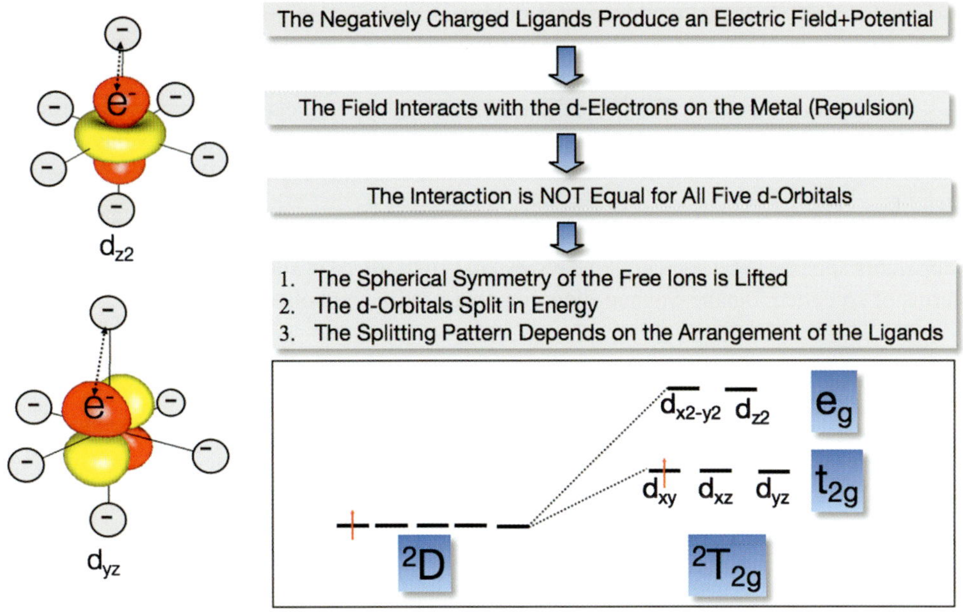

FIGURE 2.5 The origin of crystal field splittings according to the electrostatic interpretation.

field splitting is expected to be stronger, the more negatively charged a given ligand is. Whether this prediction of CFT holds in practice will be investigated later. It would be a misunderstanding of the nature of a qualitative model, like CFT, to try to obtain accurate predictions by calculating all quantities that occur in Eqn (2.1) from first principles. In fact, the results of such calculations are far from reality. Rather, one should acknowledge that CFT predicts correctly, that the five d-orbitals split into t_{2g} and e_g sets in an octahedral field and that the t_{2g} orbitals are lower in energy than the e_g orbitals.

Many Electrons in a Ligand Field

Obviously, the preceding example is rather straightforward because it involved only a single electron. Following crystal field logic, the energy differences between the t_{2g} and e_g orbitals are identical to the energy difference between the $^2T_{2g}$ and ^2E_g states. It is a peculiarity of CFT in the case of a single electron that an orbital energy difference matches a state energy differences. In general, orbital energy differences do not approximate state energy differences. The underlying reason for this is the fact that electrons repel each other and as soon as more than one electron is involved it is necessary to properly take into account the differences in electron−electron repulsion energy. Properly calculating electron−electron repulsion from first principles is a subject that has occupied theoretical chemistry for the past 80+ years. In CFT, a semi-empirical solution is found to the problem that is simple and elegant. It will be explained later.

As in the case of several electrons occupying equivalent orbitals in free atoms or ions, the distribution of electrons among the available d-orbitals gives rise to a series of states that will have different energies and will be observable in actual experiments. These are referred to as ***multiplet states***. The enumeration and classification of these multiplet states is, again, somewhat involved and can only be properly derived if one is familiar with the techniques of group theory. However, the general principles are readily appreciated. Using the so-called 'direct product' tables of group theory it follows which states can be formed from a given configuration. For example, for two electrons in the e_g shell, it is obvious that there are three different ways of distributing the electrons $(d^2_{x2\text{-}y2}, d^0_{z2})$, $(d^0_{x2\text{-}y2}, d^2_{z2})$ and $(d^1_{x2\text{-}y2}, d^1_{z2})$. In the first two cases, the spins of the electrons must be antiparallel but in the third case either singlet or triplet could result. Looking at the appropriate tables one finds that the states that arise from this distribution are $A_{1g} + A_{2g} + E$ (thus $2 \times 2 = 4$ states arise from putting two electrons into two degenerate orbitals). With a little more effort, one finds that the spin couplings are such that the permissible states are $^1A_{1g}$, $^3A_{2g}$ and ^1E_g. In the case of the t_{2g}^2 configuration a slightly more elaborate procedure results in terms of symmetry $T_{1g} + T_{2g} + A_{1g} + E_g$. After taking into account spin, the permissible states are classified as $^3T_{1g}$, $^1T_{2g}$, ^1E_g and $^1A_{1g}$. The ($t_{2g}^1e_g^1$) is more straightforward and gives rise to $^3T_{1g}$, $^3T_{2g}$ as well as the corresponding singlet-coupled states $^1T_{1g}$, $^1T_{2g}$.

It is important to emphasise that all of these states may be observable in actual experiments. While the theoretical apparatus appears to be somewhat heavy on first sight, there is a very useful compilation of the results of CFT in terms of so-called Tanabe-Sugano diagrams that will be explained below.

Ligand Field Stabilization Energy

If we disregard the subtleties that arise from interelectronic repulsion for a moment one can obtain some insight into the thermochemistry of transition metal complexes from CFT. Imagine taking a metal 2+ ion and moving it from the gas phase to aqueous solution where it will form an octahedral hexaquo transition metal complex $[M(H_2O)_6]^{2+}$. Associated with the process is a net gain in energy resulting from bond formation. This is the hydration enthalpy.

It is expected that the hydration enthalpies increase along the transition series due to the increasing effective nuclear charge of the transition metal ion as one moves from left to right across the periodic table. Increased effective positive charge will lead to tighter binding to the water molecules that will bind to the metal ion via the negative end of the dipole (the oxygen atom) that is associated with water. The increased effective nuclear charge arises from the fact that with each consecutive ion a proton and an electron is added. However, that additional electron in the d-shell of the transition metal ion will shield the additional positive charge of the nucleus only incompletely. Hence, the force that a given test charge in the vicinity of the transition metal nucleus experiences will increase towards the right of the transition series. This behaviour is indeed observed experimentally. However, superimposed over the trend to higher hydration enthalpies there is a peculiar 'double bowl' behaviour (Figure 2.6).

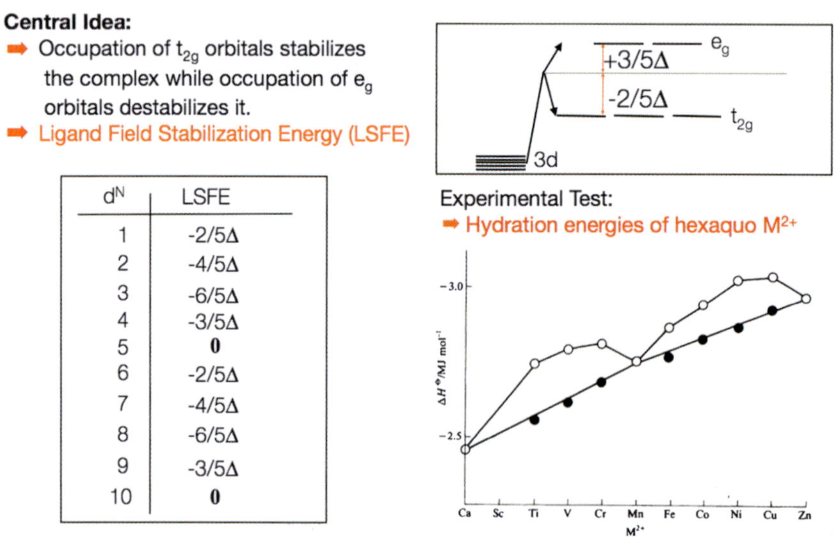

d^N	LSFE
1	-2/5Δ
2	-4/5Δ
3	-6/5Δ
4	-3/5Δ
5	0
6	-2/5Δ
7	-4/5Δ
8	-6/5Δ
9	-3/5Δ
10	0

FIGURE 2.6 Illustration of the ligand field stabilization energy (LSFE).

The explanation for the double bowl behaviour is readily provided by CFT. The first step consists of subtracting a straight line from the observed hydration enthalpy curve in order to account for the effective nuclear charge. What remains is the double bowl behaviour. It arises from unequal occupation of the d-orbitals of the transition metal. If all d-orbitals were at the same energy a straight-line behaviour would be expected. However, *relative to the centre of gravity* of the d-orbital energies the t_{2g} orbitals are stabilised while the e_g orbitals are destabilised. It is important to emphasise 'relative to the centre of gravity' because overall the d-orbitals are all strongly destabilised by the presence of a negatively or partially negatively charged ligand. Nevertheless, if the splitting between the d-orbitals is equated with 10Dq then the t_{2g} lie at −4Dq and the eg orbitals at +6Dq relative to the centre of gravity.

Obviously, the first electrons to enter the d-shell will go into the t_{2g}. Thus, Sc^{2+}, Ti^{2+} and V^{2+} have the electronic ground state configurations ($t_{2g}^1e_g^0$), ($t_{2g}^2e_g^0$) and ($t_{2g}^3e_g^0$), respectively. Following the logic outlined above, they are associated with ***ligand field stabilization energies (LFSE)*** (relative to the centre of gravity) of −4Dq, −8Dq and −12Dq, respectively. Assuming that Hund's rule holds, the next electron has to enter the e_g which consequently leads to a reduction of the LFSE by 6Dq to +6Dq for Cr^{2+}. The next electron leads to a half filled d-shell at Mn^{2+} in which case the LSFE vanishes. Indeed, the hydration enthalpy of Mn^{2+} nicely falls onto the straight-line behaviour. The second half of the transition series mirrors the behaviour of the first half as the remaining electrons also enter the t_{2g} and e_g shells in the same order. Hence, there is maximal LSFE for Ni^{2+} and no LSFE for Zn^{2+}.

The Spectrochemical Series

Given the empirical success of the crystal field model, it is tempting to apply it not only to thermochemistry but also to spectroscopy. Clearly, the most convenient way to 'measure' the ligand field splitting is optical spectroscopy. Providing a photon of correct energy an electron can be promoted from the t_{2g} to the e_g shell. Since the ligand field splittings are of the order of a few eV (or several tens of thousands of wave numbers), the associated transitions fall into the near-infrared, visible and near ultraviolet regions of the spectrum.

In the case of a single d-electron, the energy difference between the ($t_{2g}^1e_g^0$) and ($t_{2g}^0e_g^1$) configurations (corresponding to $^2T_{2g}$ and ^2E_g states, respectively) is simply given by 10Dq. The example of the $[Ti(H_2O)_6]^{3+}$ complex is shown in Figure 2.7.

Obviously, there is an optical transition centred around 20,000 cm^{-1} that corresponds to the transition in question. It is, however, asymmetric. The reason for this behaviour is well understood, but its explanation would

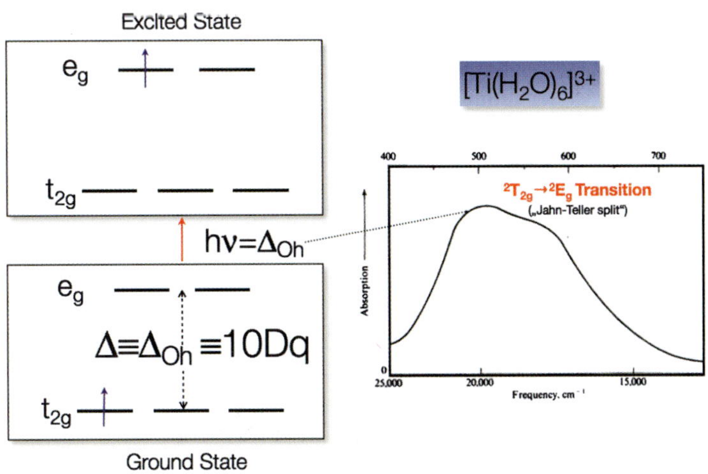

FIGURE 2.7 Optical measurement of the ligand field splitting by optical spectroscopy as exemplified for the case of the hexaquo Ti^{3+} ion.

A „Chemical“ Spectrochemical Series

$I^- < S^{2-} < F^- < OH^- < H_2O < NH_3 < NO_2^- < CN^- < CO \sim NO < NO^+$

Δ SMALL Δ LARGE

FIGURE 2.8 The spectrochemical series.

lead us too far astray. For the purposes of this section, the spectrum serves as an illustration that the ligand field splitting can be measured by optical spectroscopy.

Obviously, the ligand field splitting depends on the nature of the coordinating ligands. Hence, much work has gone into investigating the spectra of many transition metal complexes. It has been found that for any given metal, there is a regularity in the ligand splittings that depend only on the nature of the ligand. Thus, prototypical ligands can be ordered in a spectrochemical series according to increasing ligand field splitting. The spectrochemical series is shown in Figure 2.8.

Ligands to the left of the spectrochemical series are referred to as ***weak field ligands***, whereas ligands to the right are referred to as ***strong field ligands***. From an electrostatic point of view, the spectrochemical series is difficult to understand. The most striking irregularity is the fact that OH^- appears to be a weaker field ligand than H_2O. If the origin of the ligand field splitting really was only electrostatic the opposite behaviour would have been anticipated. Hence, a different explanation is required and will be given in conjunction with MO theory in the section Optical Spectra of Coordination Complexes Revisited.

High-Spin and Low-Spin Complexes

In this section, we will come back to section Many Electrons in a Ligand Field and acknowledge that in addition to the ligand field splitting the inter-electronic repulsion contributes to the total energy of a given d^N configuration. As explained in that section, this gives rise to a series of multiplets that are all spectroscopically observable. Perhaps the most striking consequence is the occurrence of high- and low-spin complexes.

The physical situation is readily appreciated even if the mathematical details of the underlying theory become more and more involved. In general, one should distinguish three types of interelectronic repulsion (Figure 2.9): (1) the repulsion of two electrons that occupy the same orbital (they necessarily will have opposite spin); (2) the

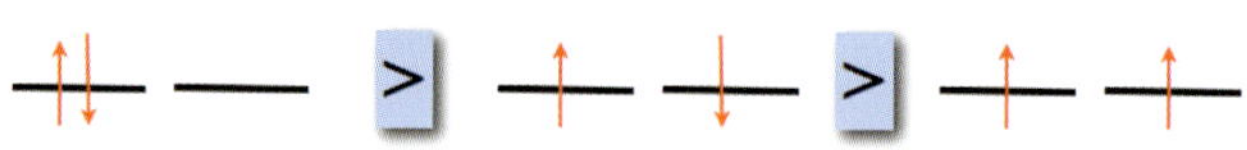

Rules:

- Electrons in the SAME orbital repel each other most strongly.
- Electrons of oppsite spin repel each other more strongly than electrons of the same spin.

Consequences:

- In degenerate orbitals electrons enter first with the same spin in different orbitals (→Hund's Rules in atoms!)
- A given configuration produces several states with different energies

Ligand Field Theory:

- Electron repulsion can be taken care of by ONE PARAMETER: B (~1000 cm^{-1})

FIGURE 2.9 Illustration of the three basic electronic repulsion cases to be considered in ligand field theory.

repulsion of two electrons in different orbitals but with opposite spin; (3) the repulsion of two electrons in different orbitals but with the same spin. In general, the interelectronic repulsion decreases in the same order. Thus, two electrons that occupy the same orbital repel each other more strongly than two electrons in different orbitals. This is simply related to the fact that the average interelectronic distance is smaller in the second case. The fact that electrons with the same spin repel each other less strongly than electrons with opposite spin is less evident and is known as *Fermi-correlation*.

Using these relationships, it is evident that electrons will preferentially enter different orbitals with the same spin as long as the orbitals are energetically degenerate. Whenever electrons are forced to be of opposite spin and in particular when they are forced to enter the same orbital there is an associated energetic penalty. This is called the ***spin pairing energy***. The ligand field stabilization and spin pairing energies can be in competition with each other as far as the determination of the electronic ground state is concerned.

For the d^1 to d^3 configurations, no ambiguity arises – the electrons will fill the t_{2g} shell with parallel spins thus leading to $^2T_{2g}$, $^3T_{1g}$ and $^4A_{2g}$ electronic ground states, respectively. However, for the d^4 configuration, the situation is less clear cut. The fourth electron could enter the t_{2g} shell. This would maximise the LSFE, but lead to a penalty in terms of the spin-pairing energy since the fourth electron is forced to pair with another electron in the t_{2g} subshell. This would lead to a $^3T_{1g}$ state. Alternatively, the fourth electron could enter the e_g-subshell in which case the spin could be kept parallel to the remaining electrons (^5E_g state). The second alternative minimises the interelectronic repulsion at the expense of a reduced LSFE.

The interelectronic repulsion only depends to a very small degree on the actual nature of the ligand but increases to the right of the transition series and also increases with increasing oxidation state of the metal. On the other hand, the ligand field splitting depends strongly on the nature of the ligand. Hence, by moving along the spectrochemical series, it must be possible to traverse a certain critical ligand field strength at which it is energetically preferably to 'pay' the spin pairing energy in order to gain LSFE. In this case, Hund's so-called ***low-spin*** complex would be formed, while the more straightforward situation of parallel spin coupling is referred to as the ***high-spin*** case (Figure 2.10).

Obviously high- or low-spin complexes could also be formed for the d^5, d^6 and d^7 cases, while in octahedral symmetry the d^8 and d^9 configurations must be associated with high-spin ground states (for d^9 only one spin-state is possible, analogous to the d^1 case).

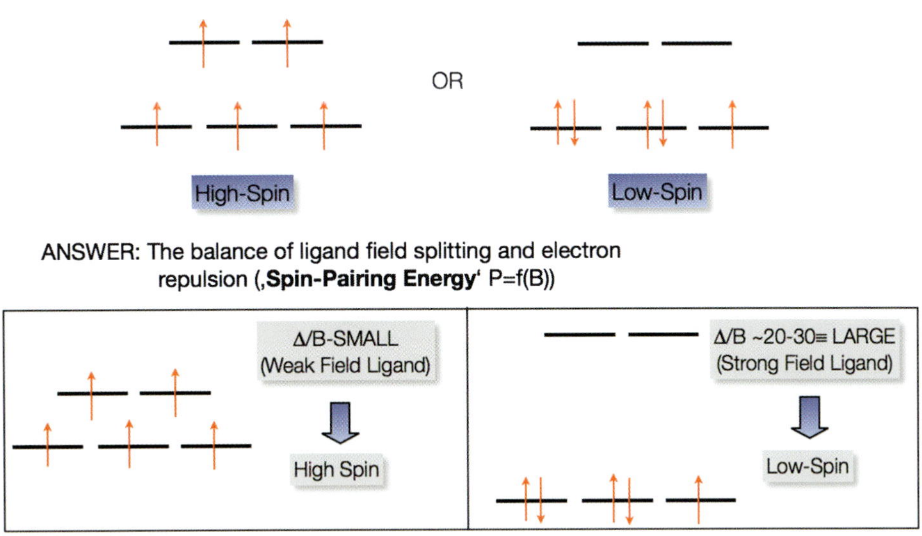

FIGURE 2.10 Competition between ligand field splitting and spin pairing energy in determining whether the ground state of a given ion is of high- or low-spin type.

The question of how to put these relationships into more quantitative terms has been solved by the founding fathers of CFT. Taking inspiration from atomic spectroscopy, it became evident that within the crystal field model the interelectronic repulsion can be represented by a single semi-empirical parameter, the so-called ***Racah B-parameter***. This parameter measures the interelectronic repulsion strength. There actually is a second parameter C that is typically set to ~4B. A third Racah parameter, A, is of no physical concern as it enters identically into the energy expression of each multiplet and hence cancels out upon taking chemically meaningful energy differences. The value of B varies roughly from 400 to 1200 cm^{-1} over the first transition series with energy differences being in the order of 1−10B.

The occurrence of high- and low-spin states of transition metal ions has major impact on their magnetic properties. The magnetic moment associated with a given transition ion is directly related to the spin-multiplicity of the ground state multiplet. Hence, the physical properties, the electronic structure and consequently also the reactivity of the transition ion differ fundamentally between the high- and low-spin states. By understanding that the cross-over between high- and low-spin states is a function of the ligand field strength, chemists have been able to synthesise complexes where the transition between spin-states can be triggered by an external stimulus, such as light. This has led to important advances in technological devices. This is one example how LFT impacts chemical thinking and guides chemical intuition. In addition, realizing that the spin state reflects the ligand field of a transition ion, biochemists have made good use of the spin state as a marker for changes at a transition metal active site. For example, the accessibility of a given active site can be probed by using CO or NO which are known to be strong field ligands and induce a change from high-spin to low-spin states upon binding to the transition metal centre.

Tanabe-Sugano Diagrams

Everything that has been described so far can be summarised in a highly compact set of diagrams, the so-called Tanabe-Sugano diagrams (Figure 2.11). For each d^N configuration, there is a separate Tanabe-Sugano diagram. On the x-axis of the diagrams, there is a measure of the ligand field strength (in fact 10Dq divided by the Racah parameter B). The origin corresponds to the case of the free atom or ion when the ligand field strength is zero.

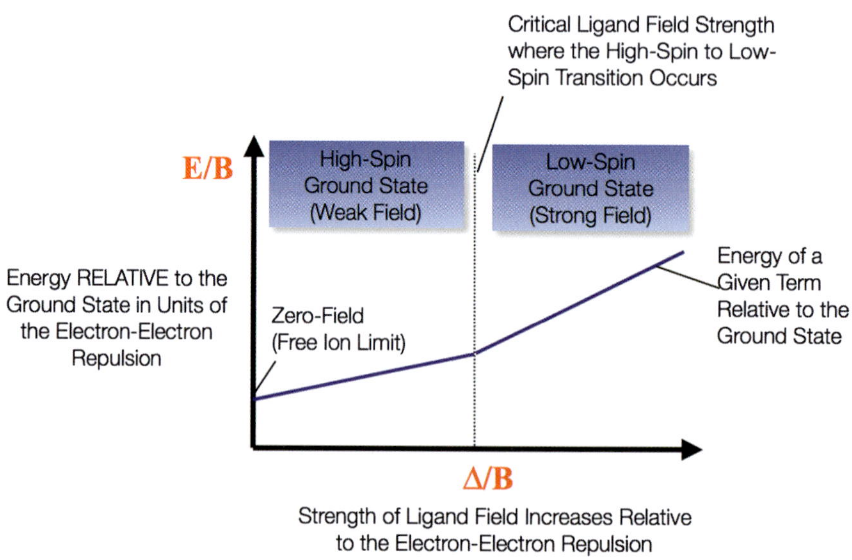

FIGURE 2.11 The anatomy of a Tanabe-Sugano diagram.

On the y-axis of the diagrams, the energy of each term is plotted *relative to the electronic ground state* (again in units of B). This is important especially for the case of the d^4–d^7 configurations where there is an apparent discontinuity. At the discontinuity point, the system passes from a high-spin ground state to a low-spin ground state. The part between the origin of the x-axis and the critical ligand field strength is called the ***weak-field regime***, the right part of the diagram is called the ***strong-field regime***.

The Tanabe-Sugano diagrams nicely summarise which terms exist for a given d^N configuration and also provide a reasonable idea of where they can be expected to show up in the electronic spectrum. In fact, in order to assign a given d-d spectrum all that it takes is a ruler and a bit of patience to extract semi-qualitative estimates for the ligand field parameters 10Dq and B (Figure 2.12).

In dealing with the optical absorption spectra it must, of course, be recognised that optical transitions do not only have an energetic position but also an intensity. It is generally found that the intensities of d-d transitions are fairly low (the extinction coefficient at the band maximum is typically not more than a few 100 M^{-1} cm^{-1}). This behaviour was understood early on the basis of group theory. With each spectroscopic phenomenon, there is a ***selection rule*** that determines whether a process is allowed on the basis of quantum mechanics or not. In the case of d-d transitions, the group theoretical selection rules state that in complexes that feature a centre of inversion (as is the case for O_h) d-d transitions are forbidden as the initial and final states are both "gerade" under inversion while the selection rule states that one of the states must be "ungerade". This is known as the ***Laporté rule***. Thus, d-d transitions are typically more intense in transition metal complex that do not feature a centre of inversion (for example tetrahedral complexes). A second important selection rule concerns the total spin of the initial and final

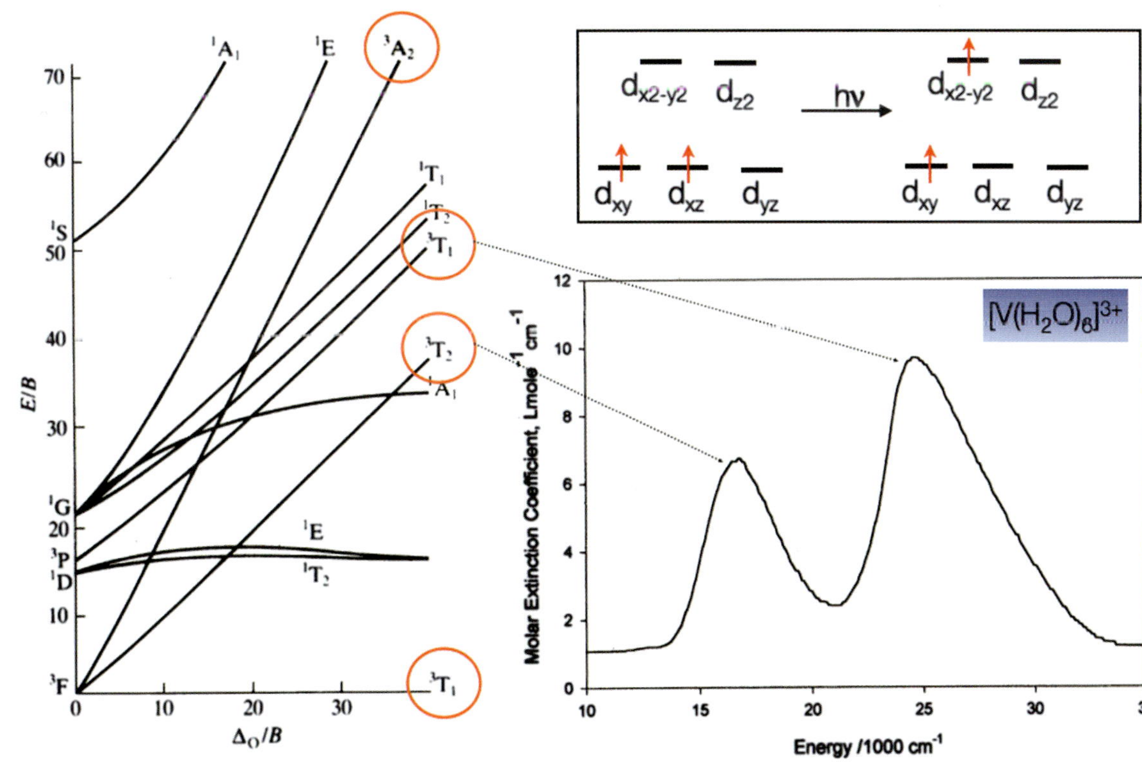

FIGURE 2.12 The Tanabe-Sugano diagram for the d^2 configuration and its application to interpret d-d spectra as exemplified for the case of the hexaquo-vanadium^{2+} complex.

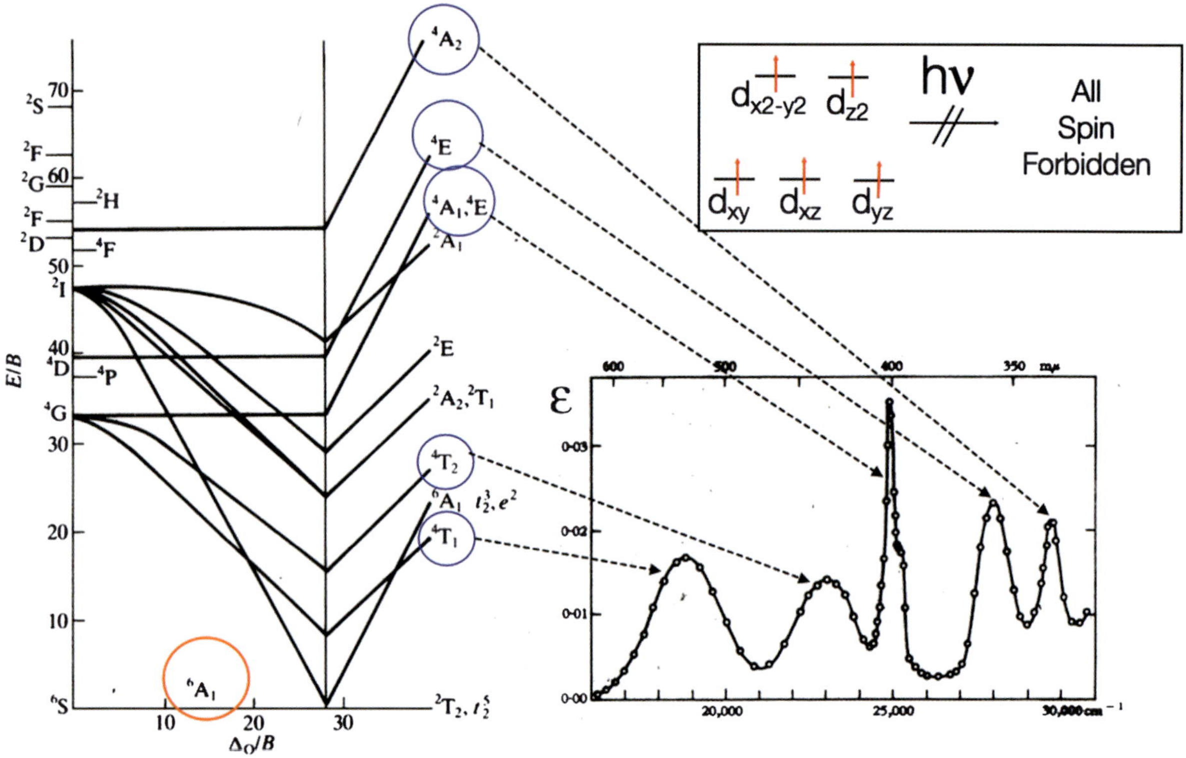

FIGURE 2.13 The Tanabe-Sugano diagram for the d^5 configuration and its application to interpret d-d spectra as exemplified for the case of the hexaquo-manganese^{2+} complex.

states. According to the quantum mechanical selection rules, the total spin quantum number in the two states must be identical for the transition to be allowed.

No selection rule rigorously holds and in reality even forbidden transitions can often be observed, albeit with low intensities. In transition metal spectroscopy, the ***spin-selection rule*** is stronger than the Laporté rule. Spin forbidden transitions are typically observed with extinction coefficients below 1 M^{-1} cm^{-1} (Figure 2.13). The spin selection rule becomes weaker for the second- and third-row transition metals due to the phenomenon of ***spin-orbit coupling***. This is a relativistic effect the treatment of which is outside the scope of this chapter.

Ligand Field Spectra for Alternative Coordination Geometries

So far the discussion has been centred around octahedral transition metal complexes. From what has been said it is, however, evident that the d-orbital splitting pattern is sensitive to the actual coordination geometry with the "two e_g orbitals over three t_{2g} orbitals" being a peculiarity of the octahedral group. The splitting patterns that occur in alternative common coordination geometries are summarised in Figure 2.14. The general form of the splitting patterns can be intuitively understood.

Removing a ligand from the z-axis to pass from octahedral to square pyramidal coordination will lead to a stabilization of the d_{z2} orbital relative to the $d_{x2\text{-}y2}$ orbital and likewise a stabilization of the d_{xz}/d_{yz} orbitals over the d_{xy} orbital. The resulting point group is C_{4v}.

Removing a second ligand from the −z direction leads to a square planar four-coordinate complex that belongs to the point group D_{4h}. The changes in the orbital splitting pattern follow the same trend with further stabilization

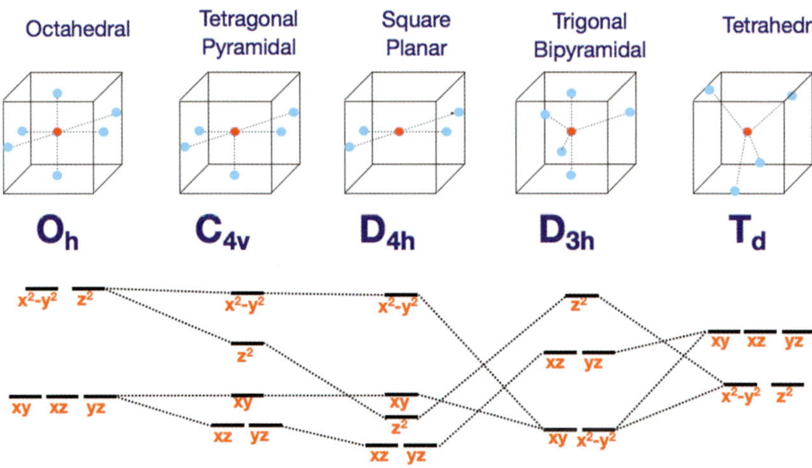

FIGURE 2.14 d-orbital splittings in different ligand fields.

of the d_{z2} and d_{xz}/d_{yz} orbitals. In fact, the d_{z2} orbital might be so strongly stabilised that it falls energetically below the d_{xy} orbital in this coordination geometry.

A completely different splitting pattern is observed in trigonal symmetry. The prototypical coordination geometry is represented by a trigonal bipyramid which belongs to the point group D_{3d}. The characteristic splitting pattern is a high lying d_{z2} orbital followed by two sets of double degenerate orbitals. The $d_{x2\text{-}y2}/d_{xy}$ pair of orbitals is the higher lying of the two sets because of the interaction with the three equatorial ligands.

Finally, the alternative four-fold coordination pattern is represented by a tetrahedral complex which belongs to the point group T_d. Interestingly, the analysis shows that the splitting pattern is exactly opposite to the one met in octahedral symmetry. Thus, there is a higher lying t_2 set of orbitals which contain the d_{xy},d_{xz} and d_{yz} orbitals and a lower lying e set which is comprised of the $d_{x2\text{-}y2}$ and d_{z2} orbitals. One very interesting relationship that has been derived early on is that the ligand field splitting in the tetrahedral case — all other factors being identical — is $-4/9$ times the splitting in octahedral symmetry, e.g. $10Dq(T_d) = -4/9\ 10Dq(O_h)$. This means that the splittings are always much smaller in the tetrahedral case. This has two immediate and important consequences: (1) d-d absorption bands of tetrahedral complexes typically fall into the near- or far infrared regions of the spectrum and (2) there are no known truly tetrahedral low-spin complexes.

Using LFT in Bioinorganic Chemistry

At this point, the exposition of CFT has progressed to a stage where we can consider how the theory can be used in biochemistry. As an example the case of a mononuclear non-heme iron enzyme will be considered. These enzymes have been studied intensely by Solomon and co-workers and we will draw from their work in this section.

The group of nonheme iron enzymes is a large family of enzymes that carries out a wide variety of biological transformations. Typically, these enzymes have the iron in the Fe(II) (ferrous) state in the absence of substrate. They then react with dioxygen to form highly reactive intermediates that can attack organic molecules. Using a combination of CFT and optical spectroscopy, one can obtain insight into the events that occur at the active site during catalysis.

To this end, it is important to have "benchmark" spectroscopy available. The ferrous ion is difficult to study experimentally since it has only very weak absorption bands that typically fall into the near-infrared region of the spectrum where overtones of water vibrations strongly complicate absorption measurements. In addition, the ferrous ion has a $^5T_{2g}$ ground state. Such integer spin ($S = 2$) ions are difficult to study with magnetic resonance

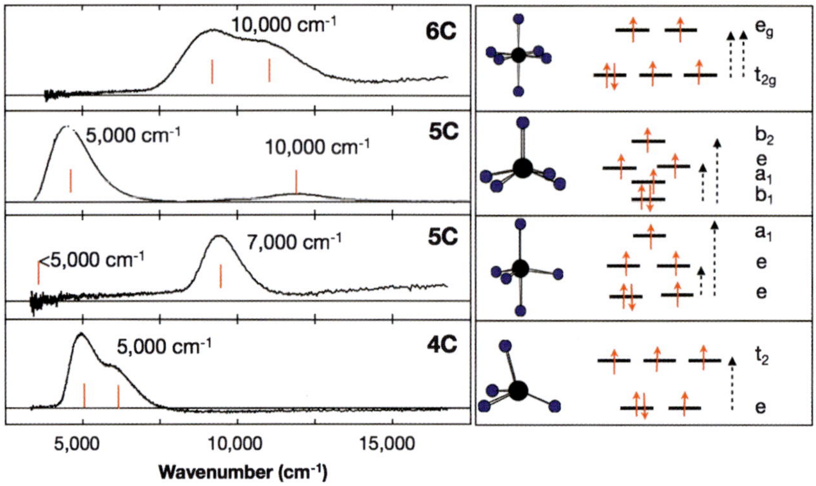

FIGURE 2.15 Correlation between d-orbital splittings and d-d spectra in high-spin Fe(II) complexes as measured with near-infrared MCD spectroscopy (see Chem. Rev. Solomon et al. in the cited literature).

techniques. Thus, only a few methods, like X-ray absorption and Mössbauer spectroscopy can be applied. However, a powerful technique to investigate Fe(II) is offered by magnetic circular dichroism (MCD) spectroscopy. An introduction to MCD spectroscopy is provided elsewhere in this volume. For the purposes of the present section, it suffices to know that using MCD one can detect the weak ligand field transitions all the way out to about 2000 nm (5000 cm^{-1}). Examples of a series of Fe(II) complexes is shown in Figure 2.15. Quite obviously the spectra change drastically with coordination geometry. Roughly octahedral complexes are characterised by two bands that are centred around 10,000 cm^{-1} and split by about 2000 cm^{-1}. Five-fold coordinated complexes with tetragonal symmetry are characterised by a band around 10,000 cm^{-1} and a band around 5000 cm^{-1}. The origin of the enormous downshift of one of the bands is readily understood through CFT and is related to the stabilization of the d_{z2} orbital upon removal of one axial ligand. Trigonal complexes typically feature one band around 7000–8000 cm^{-1} while tetrahedral complexes only show features in the 5000 cm^{-1} region. All these features are readily assigned on the basis of CFT as indicated in Figure 2.15. Thus, these model studies of Solomon and co-workers demonstrate that coordination geometry can be reliably deduced from MCD measurements.

As an application consider the case of a Rieske dioxygenase, a nonheme iron enzyme. In addition to a mononuclear nonheme iron active site, the enzyme also features an iron sulfur cluster of the Rieske type that serves as an electron transfer site. The enzyme hydroxylates aromatic rings as indicated in Figure 2.16. The MCD spectra of this enzyme in the resting state are superpositions of the contributions from the active site and the Rieske centre. In order to remove the contributions from the Rieske site, a form of the enzyme can be prepared in which only the Rieske centre is retained and the active site is removed (Figure 2.17, middle). Subtraction of the spectra of the holoenzyme and the Rieske only form provides the MCD spectrum of the active site.

In Figure 2.17, spectra are presented in the absence of dioxygen and without and without the substrate present. As shown in Figure 2.17, the active site spectra change drastically in the presence of substrate. Importantly, there is a new band appearing around 5000 cm^{-1}. Using the spectra/structure relationship described above, the appearance of this band indicates that the active site assumes a five coordinate form upon substrate binding. This is of very great mechanistic importance.

The picture that emerges from this (and many related) study is that nonheme iron enzymes "protect" their active sites against harmful Fenton chemistry (the production of hydroxyl radicals through reaction of Fe(II) with O_2) by keeping the sites coordinately saturated as long as the substrate is not around (Figure 2.18). Once substrate binds, the enzymes open up a coordination site to which dioxygen can bind. What follows is a chemical reaction

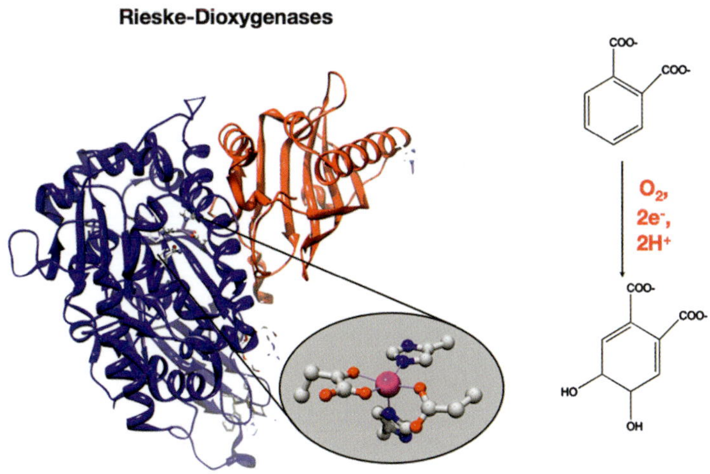

FIGURE 2.16 The Rieske-dioxygenases.

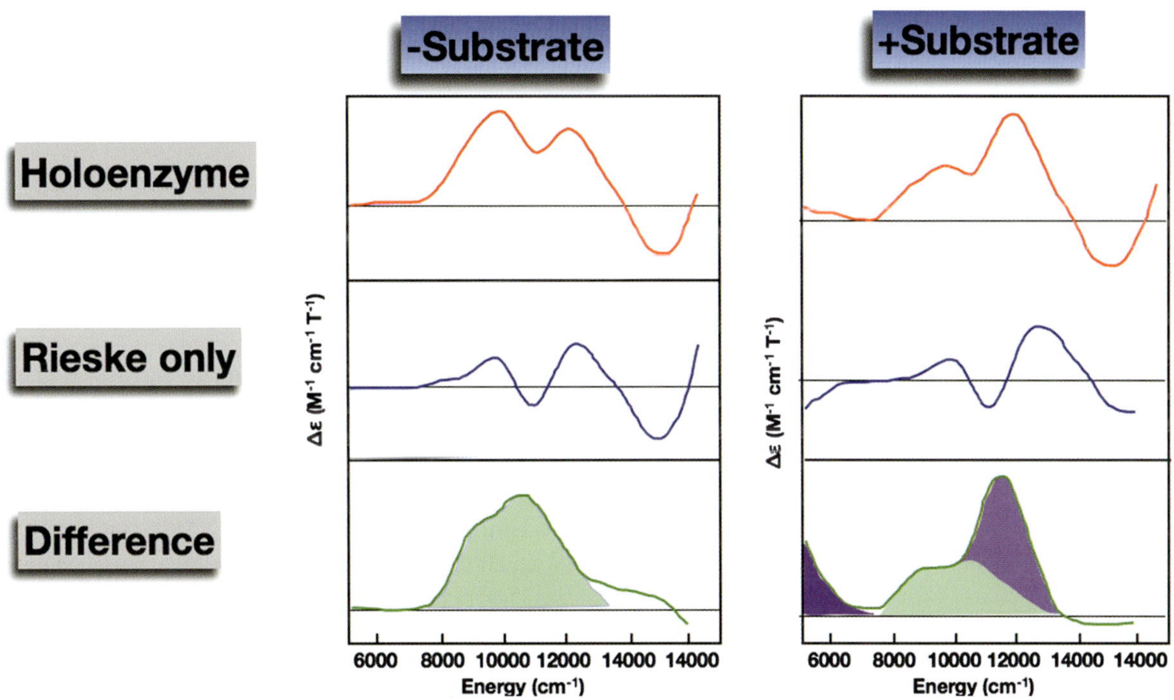

FIGURE 2.17 MCD spectra of Rieske dioxygenases (see text for explanation).

between dioxygen and ferrous ion that lead to a highly reactive high valent iron species that can then attack rather stable C-H bonds in substrates. Obviously, the production of such highly reactive species in the absence of dioxygen is potentially dangerous as it could lead to protein self-destruction. It is important that the crucial six- to five-coordinate transition can be experimentally observed by MCD spectroscopy. The correct interpretation of the observed spectral features was made possible by CFT.

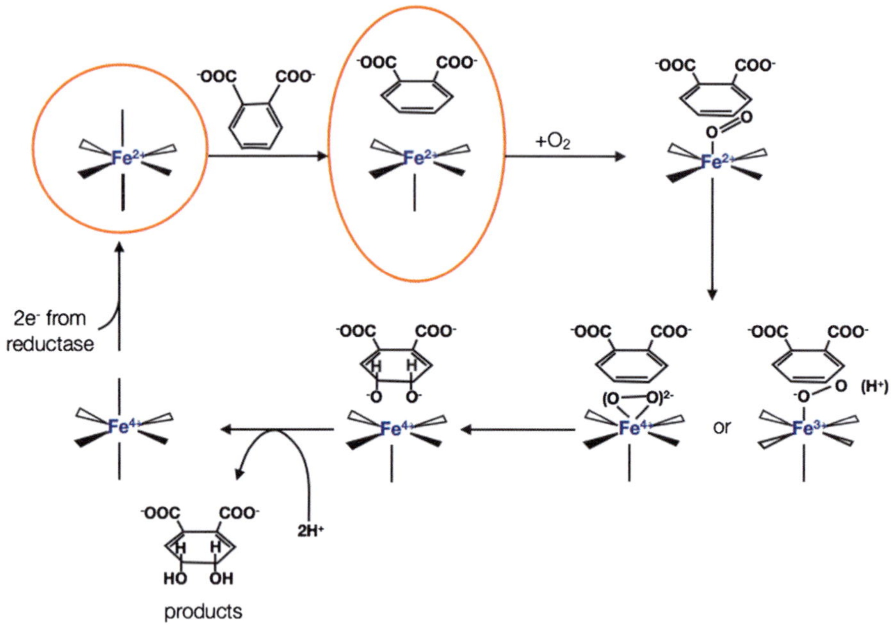

FIGURE 2.18 Mechanistic implications for the mechanism of Rieske dioxygenases arising from the combination of MCD spectroscopy and ligand field theory.

Further Aspects of LFT

In this section, some additional aspects of CFT and LFT will be briefly mentioned in order to provide incentives for further study.

The Jahn–Teller effect

As discussed above, some d-orbitals remain energetically degenerate in certain coordination geometries. If these orbitals are only partially filled an orbitally degenerate state may arise. For example, a (e^1) or a (e^3) configuration both give rise to orbitally degenerate ^2E states while for (e^2) configuration with triplet-coupled electrons an orbitally non-degenerate $^3A_2$ state results. The Jahn–Teller theorem has at its heart the remarkable insight that such orbitally degenerate states are always unstable in the sense that there always exists a nuclear motion that lifts the degeneracy and leads to an orbitally non-degenerate state of lower energy (Figure 2.19). Thus, Jahn–Teller active ions will distort into a less symmetric point group in which they can assume an orbitally non-degenerate state.

Classical examples of Jahn–Teller active configurations are the ground states of octahedral high-spin d^4 (^5E) or d^6 ($^5T_{2g}$) ions as well as octahedral d^9 (^2E) systems. Consider the case of Mn^{3+} ion, a d^4 system in octahedral geometry. In the electronic ground state, the e_g ($d_{x2\text{-}y2}$, d_{z2}) carries a single electron. Moving the axial ligands along the $\pm z$ directions outwards will stabilise the d_{z2} orbital. Hence, the electronic ground state will be $^5A_1$ in tetragonal symmetry. For the opposite motion, in which the ligands move inward, the d_{z2} orbital would be destabilised and hence the single e_g-based electron will energetically prefer to occupy the $d_{x2\text{-}y2}$ orbital which leads to a $^5B_2$ state in tetragonal symmetry. Importantly, the Jahn–Teller theorem makes no statement about in which direction the distortion will occur or how large it is. It simply states that some distortion will take place. In practice, the overwhelming majority of Mn^{3+} complexes are axially elongated.

The theory of the Jahn–Teller effects quickly becomes fairly complicated upon consideration of the detailed shapes of the potential energy curves as a function of nuclear coordinates. These potential energy surfaces may

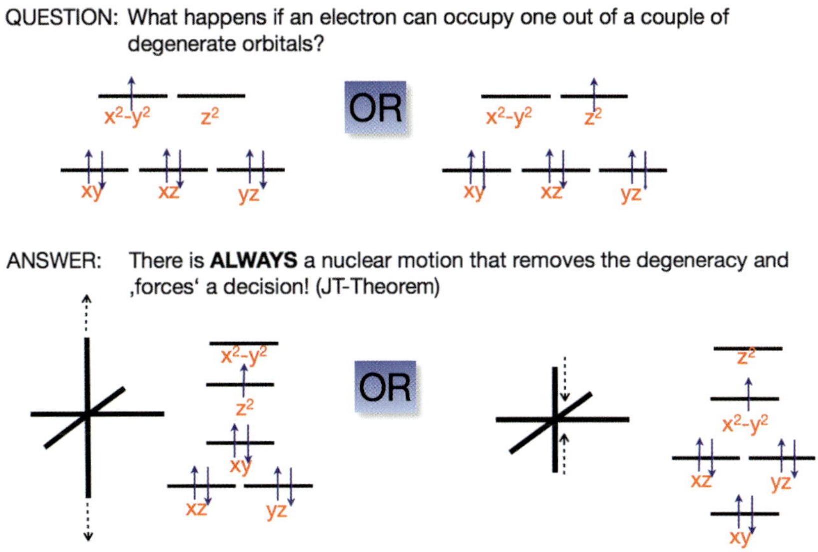

FIGURE 2.19 Illustration of the Jahn–Teller effect for the low-spin d^7 configuration.

show several minima that are separated by low energetic barriers such that the system may thermally "jump" from one minimum (one geometry) to another minimum (another geometry). If such jumping occurs one speaks of a "dynamic" Jahn–Teller effect. In the more common case that the barriers between different minima are low one has a "static" Jahn–Teller effect.

In biology, the coordination environment of the metallic active sites is always of low symmetry. Hence, no orbital degeneracies exist in the strict sense and the Jahn–Teller effect is not highly prominent. Nevertheless, near orbital degeneracies *do* occur even in biology and then there are recognisable Jahn–Teller distortions that can be observed in high-resolution crystal structures. A very prominently discussed example is the blue copper site in electron transfer proteins such as plastocyanin.

The Nephelauxetic Effect, Covalency and LFT

Up to this point, the treatment is based on a purely electrostatic picture in which the metal orbitals are pure d-orbitals and the ligand are pure point charges. This is, of course, a drastic oversimplification. That this must be the case has already been realised by the founding fathers of crystal and LFT. Upon analyzing optical absorption spectra of coordination complexes, they found that appropriate values for the Racah B-parameter are always lower than what is deduced from atomic spectroscopy. One way to explain this is to assume that the ions "expand" upon complex formation. In this case, the orbitals would be larger than in the free ion. Consequently, the average interelectronic distance would be larger and the interelectronic repulsion energy would go down which would in turn be reflected in a reduced Racah B-parameter. This was referred to as the "nephelauxetic" (cloud expanding) effect. While the nephelauxetic effect was deduced from optical spectroscopy, there is an accompanying "relativistic nephelauxetic effect" derived from magnetic spectroscopy that was interpreted in a similar manner.

As appealing as the cloud expansion idea might appear, it is inconsistent with the modern theoretical point of view. In fact, MO theory, to be sketched below, offers a more satisfying and equally straightforward explanation for both, the nephelauxetic and relativistic nephelauxetic effects.

In fact, the assumption of the existence of pure metal d-orbitals after complex formation has been realised to be much too strong. Consequently, refined theories that acknowledge the mixing of metal- and ligand orbitals have been devised and in a broad and collective sense these are referred to as "LFT" while the name "CFT" is reserved for the original purely electrostatic theory.

Magnetism

As will become evident throughout this volume, magnetic properties of ions are extremely helpful in studying their structure and their reactivity. Of course, all physical properties of molecules are intimately related to their electronic structures. Hence, it is not surprising that there also is a ligand field theoretical explanation for the magnetic properties of molecules. However, its detailed description would require more space than is available in the context of this chapter and would also require some familiarity with the mathematical methods of quantum chemistry. The interested student is encouraged to consult the pertinent literature on the subject.

MO THEORY OF TRANSITION METAL COMPLEXES

Where CFT Breaks Down

The fact that CFT cannot be completely correct has been realised relatively early on. An experimental demonstration is shown in Figure 2.20, which shows the EPR spectrum of $[Cu(imidazole)_4]^{2+}$. This is a regular square planar d^9 complex. According to CFT, the unpaired electron occupies the $d_{x2\text{-}y2}$ orbital which has its lobes pointing towards the coordinating imidazole nitrogen ligands. What is striking about this spectrum is the series of sharp lines around 3400 Gauss. If the same experiment is repeated with ^{15}N labelled imidazole (not shown) this pattern is changing thus proving that the structure of the signal in that region is related to the magnetism of the nitrogen nuclei ($S = 1$ for ^{14}N and $S = ½$ for ^{15}N). Without going into any great detail about the interpretation of the EPR spectrum, it is worth mentioning that the spacing between the sharp lines is a measure of the probability to find the unpaired electron at one of the nitrogen nuclei (or in a different picture – a measure of the fractional time the unpaired electron spends on any of the nitrogen nuclei).

The important point is that this experiment proves that there is a finite probability of finding the unpaired electron on the nitrogen nuclei. In the purely electrostatic picture implied by CFT, this could never have happened. Hence, the experiment proves that the singly occupied orbital is more "spread out" than implied by conventional CFT. Hence, an extension is necessary that takes the ligand orbitals into account to at least some extent. Obviously, this takes away some of the appealing simplicity of the crystal field model. It is, however, necessary to go beyond CFT if the experiments are to be interpreted properly.

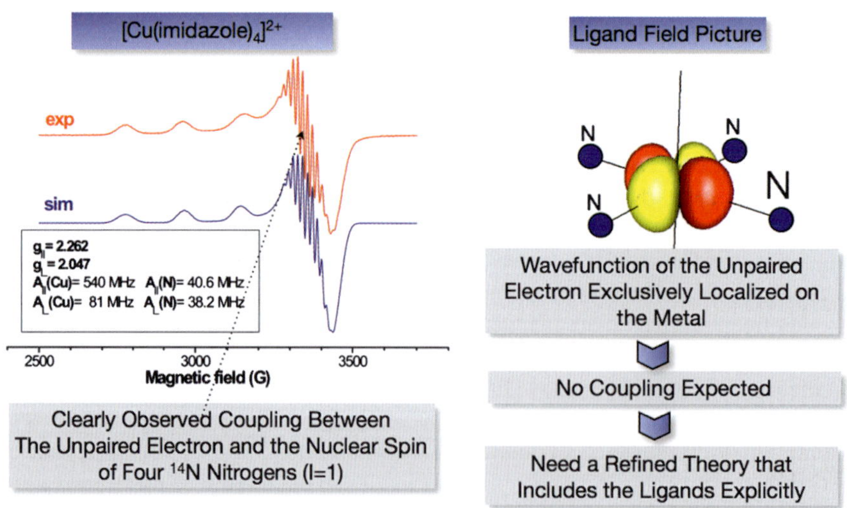

FIGURE 2.20 The limitations of CFT as revealed by the EPR spectrum of $[Cu(imidazole)_4]^{2+}$.

This refined theory is known as MO theory. MO theory is very powerful and very general. Using MO theory, the behaviour of all molecules can be predicted and interpreted. MO theory can be made quantitative and predictive. However, this comes at the price of complexity. It is difficult to work through MO theory and its various extensions with paper and pencil. Quite frequently, computers must be relied upon to perform MO calculations. LFT, on the other hand, might be thought of as a simplification of MO theory that retains the conceptual simplicity of CFT but incorporates ideas of MO theory into its formulation.

In this chapter, a fully exposition of MO theory cannot be provided. However, it is possible to grasp a few key ideas.

General MO Theory

In general MO theory, MOs are formed which, in principle, extend over the entire molecule under investigation. These MOs are formed from constituent atomic orbitals (AOs; in the quantitative form of MO theory these are only atomic orbitals in a very loose sense) of the atoms involved in the system. From this perspective, no special role is assigned to the metal.

Whenever two AOs 'interact', new orbitals (MOs) are formed that are mixtures of the constituent AOs (Figure 2.21). When the two AOs are equivalent (e.g. two pz-orbitals on adjacent carbon atoms), they have the same energy and the resulting MOs are equal mixtures of the two AOs. One of the MOs will be found at lower energy than the constituent AOs and will be called a "***bonding***" MO. It is characterised by a constructive overlap of the constituent AOs. This means that in the bonding region between the atoms the AOs enter such that lobes of the same sign combine. To higher energy, there will be an "***antibonding***" MO that features destructive overlap between the constituent AOs. It is characteristic that the antibonding MO is more strongly destabilised than the bonding one is stabilised (this accounts for electron-pair repulsion). In general, occupation of the bonding MO by an electron will lead to the formation of a chemical bond. It is maximally strong, if the bonding component is occupied by two electrons. If the antibonding component is occupied by one electron, the bond is weakened. Full occupation of both the bonding and the antibonding component destroys the bond and leads to a repulsive interaction between the constituent atoms.

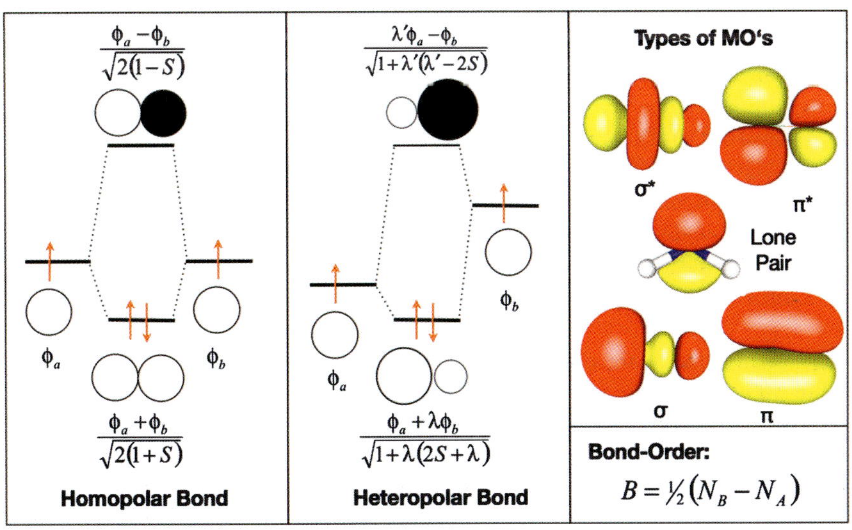

FIGURE 2.21 Basic principles of bond formation in molecular orbital theory. The bond order B is defined as one-half the number of bonding minus the number of antibonding electrons in a given bond.

The more general case is met when two AOs interact that are not equivalent. In this case, one will be at lower and one will be at higher energy. In this case, there will still be two MOs formed. However, the lower energy bonding MO will dominantly reflect the lower lying AO while the higher lying antibonding MO will dominantly reflect the higher lying AO. The extent of mixing depends on the energetic separation as well as the strength of interaction (in the most elementary form simply related to the overlap of the two AOs).

After forming all possible combinations (in general carried out by a computer), one ends up with a set of MOs. These MOs broadly fall into roughly three categories: (a) bonding MOs, (b) antibonding MOs and (c) nonbonding MOs. The latter are basically AOs that do not interact with other AOs and are usually referred to as ***lone pairs***.

Whether an interaction can take place is dictated by symmetry – only if two AOs can overlap will there be an interaction. Of course, MOs are not restricted to just two constituent atoms but can, in principle, involve all atoms of the molecule. In fact, group theory can, again be very helpful, in finding so called ***symmetry adapted orbitals*** that transform under the irreps of the point group. The bottom line of this statement is that there can be a symmetry label associated with each MO in a similar way as it can be associated with a given state.

The next important concept is the so-called ***Aufbau principle***. The Aufbau principle states that MOs will be filled with electrons according to their energetic order. Thus, starting from a set of atoms and their associated atomic orbitals one can form a set of MOs, order them according to increasing energy and then start filling electrons in that will occupy these orbitals. The result will be an approximate representation of the electronic ground state of the system (cases exist in nature where this concept does not work but this would lead too far astray here). Excited states of the system are (approximately) formed by redistributing electrons among the available orbitals in just the same way it was discussed above for atomic spectroscopy and CFT. The symmetry of the resulting states can be determined from the symmetries of the partially occupied MOs, again in a similar way as was discussed for the case of CFT.

MOs for Octahedral Complexes

It is illuminating to look at the MOs of an octahedral complex as in Figure 2.22. While this diagram is fairly complex it has a number of striking features that allow for a connection to be made between MO theory and CFT. It should first of all be appreciated that metal atoms are harder to ionize than ligand atoms. Thus, while ionisation potentials of typical donor atoms vary from 10 to 20 eV (generally increasing for more electronegative elements), the ionisation potentials of the first row transition metals fall into the region 6–10 eV. Thus, quite typically, the metal valence orbitals are higher in energy than the ligand orbitals. Thus, upon forming MOs, the lower energy binding components will be dominantly the filled ligand orbitals while the higher lying antibonding components will be the metal d-based MOs. This expectation is confirmed by the actual MOs (resulting form a so-called density functional calculation).

It is clear that the set of ligand-based orbitals is fairly complex and difficult to memorise. However, more importantly, the highest occupied MOs are based on the metal d-orbitals that correspond precisely to the orbitals predicted by CFT. Thus, there is a lower lying set of t_{2g} orbitals that contain the metal d_{xy}, d_{xz} and d_{yz} orbitals as main constituents and a higher lying set of e_g orbitals with the metal d_{x2-y2} and d_{z2} AOs as the main components.

Thus, the orbitals assumed to be the ones to be partially occupied by crystal field are also the ones seen to be partially occupied in MO theory. The main difference is that MO theory acknowledges that these are *not* pure d-orbitals but are, in fact, "diluted" by ligand orbitals (Figure 2.25). The amount of dilution depends on the relative energies of the metal and ligand orbitals as well as the strength of their interaction. This is what is referred to as the "***covalency***" of a metal ligand bond. A bond is perfectly covalent if the constituent AOs enter with identical weight and is completely ionic if it only consists of a given AO of one of the interacting partners. Thus, the fractional ligand character in the partially occupied MOs is a measure of covalency. Covalency typically increases as the metal orbitals go down in energy (higher oxidation states, late metals), or the ligand

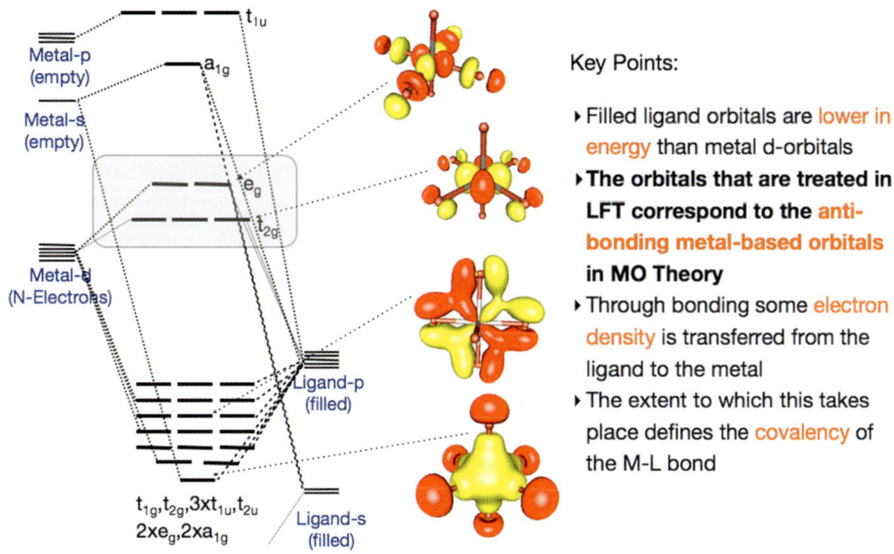

FIGURE 2.22 Partial molecular orbital diagram for an octahedral transition metal complex.

orbitals are high lying (anions) and the metal–ligand overlap is large ("soft") ligands. Obviously, the covalency will, in general be different in the various metal d-based orbitals. In this case, one speaks of ***anisotropic covalency***.

The covalency of a given metal–ligand bond is a very important concept that determines many of the physical properties of transition metal ion complexes. The "dilution" is obviously the origin of the nephelauxetic and relativistic nephelauxetic effects described in section The Nephelauxetic Effect, Covalency and LFT. Using MO theory, it can be made even somewhat quantitative. The higher the covalency, the more charge is donated from the ligand to the metal. This obviously has a bearing on the redox potential of the metal. More strongly charge donating ligands will drive the redox potential more negative.

A particularly interesting situation arises, if the ligand has low-lying empty orbitals of the correct symmetry to interact with the metal d-orbitals. In actual coordination complexes, these are invariably low-lying π^*-orbitals of ligands that posses π-orbitals. Such ligands are, for example, CO, NO or CN^- or hetero-aromatic systems such as imidazole, pyridine or porphyrin. By symmetry, these empty π^*-orbitals can only interact with the metal-based t_{2g} orbitals. Since the π^*-orbitals are higher in energy than the t_{2g}-based metal orbitals, the interaction will be *bonding* for the metal-based t_{2g} orbitals, which consequently will be *stabilised*. This special kind of bonding is called ***backbonding***. Through backbonding, electron density will be transferred back from the metal to the ligand. Since the ligand orbitals that receive the electron density are antibonding this back-transfer will weaken the intra-ligand bonds. The more normal case where a filled ligand orbital interacts with the metal-based t_{2g}-orbitals is simply called ***π-bonding***. It will destabilise the metal-based t_{2g}-orbitals and lead to electron transfer from the ligand to the metal. One also speaks of ***π-acids*** and ***π-bases*** in relation with ligands that allow for ***π-backbonding*** and ***π-bonding***, respectively (Figure 2.24).

Acknowledging covalency, one moves from CFT to LFT. In LFT, one typically accounts for covalency by using "diluted" forms of the metal d-orbitals, for example:

$$\psi_{x^2-y^2} \approx \alpha d_{x^2-y^2} - \sqrt{1-\alpha^2 L} \tag{2.2}$$

In this equation, $\psi_{x^2-y^2}$ is the d-based MO, $d_{x^2-y^2}$, a metal d-based AO and L a suitably constructed ligand orbital of the same symmetry. The parameter α is a measure of the admixture of the ligand orbital into the

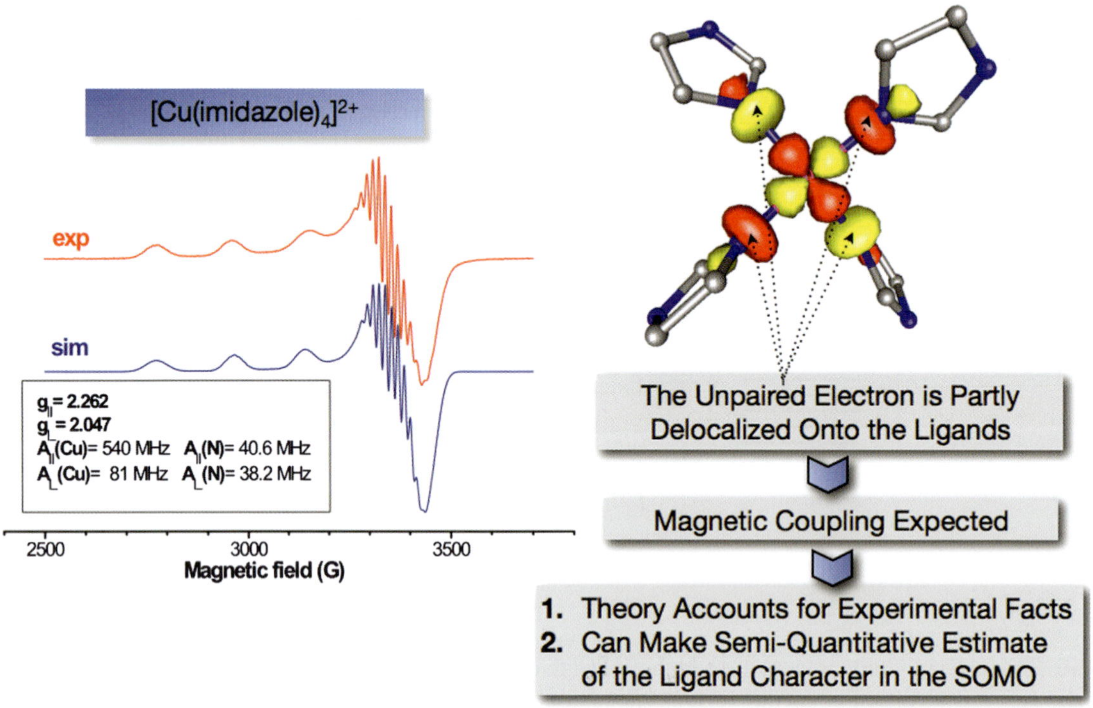

FIGURE 2.23 Explanation of the EPR spectrum of $[Cu(imidazole)_4]^{2+}$ in terms of molecular orbital theory.

metal d-based MO. It has a range of 0−1. The value $1 - \alpha^2$ is "the covalency". Using approximate expressions of this kind, one can derive expressions for physical observables that will contain atomic constants that are scaled down by α^2 or α^4. The value of α can be treated as an adjustable empirical parameter. While in general it will be different for each metal d-based orbital, most treatments only consider an average value and refer to it as the ***orbital reduction factor***.

In refining CFT, one also gives up on the electrostatic origin of the crystal field splitting but instead is inspired by MO theory. One of the most successful variants of such a LFT is the ***angular overlap model***. In the AOM model, one takes specific account of the number and nature of the ligand and derives for each ligand a set of transferable and additive semi-empirical parameters in terms of which the ligand field splitting pattern can be parameterised in the semi-quantitative way. Details can be found in the specialist literature.

In closing this section, it is important to return to the statements made in section States Versus Orbitals and stress that all observables are eventually related to the states a system can assume, not to any set of orbitals. The orbitals that are described here serve as a powerful aid in interpreting the physical properties of transition metal complexes. There is, however, no absolute truth in orbitals of whatever kind. They are simply helpful devices to guide our intuition and *can* (but do not even have to) serve as building blocks for constructing approximate states of molecules.

Optical Spectra of Coordination Complexes Revisited

Based on the discussion about the MOs of octahedral complexes in the preceding section, a fresh look can be taken at the spectrochemical series discussed in section The Spectrochemical Series. In fact, the apparent irregularities that arose within the attempt to interpret this series on the basis of the electrostatic model find a natural and appealing explanation in terms of MO theory (Figure 2.25).

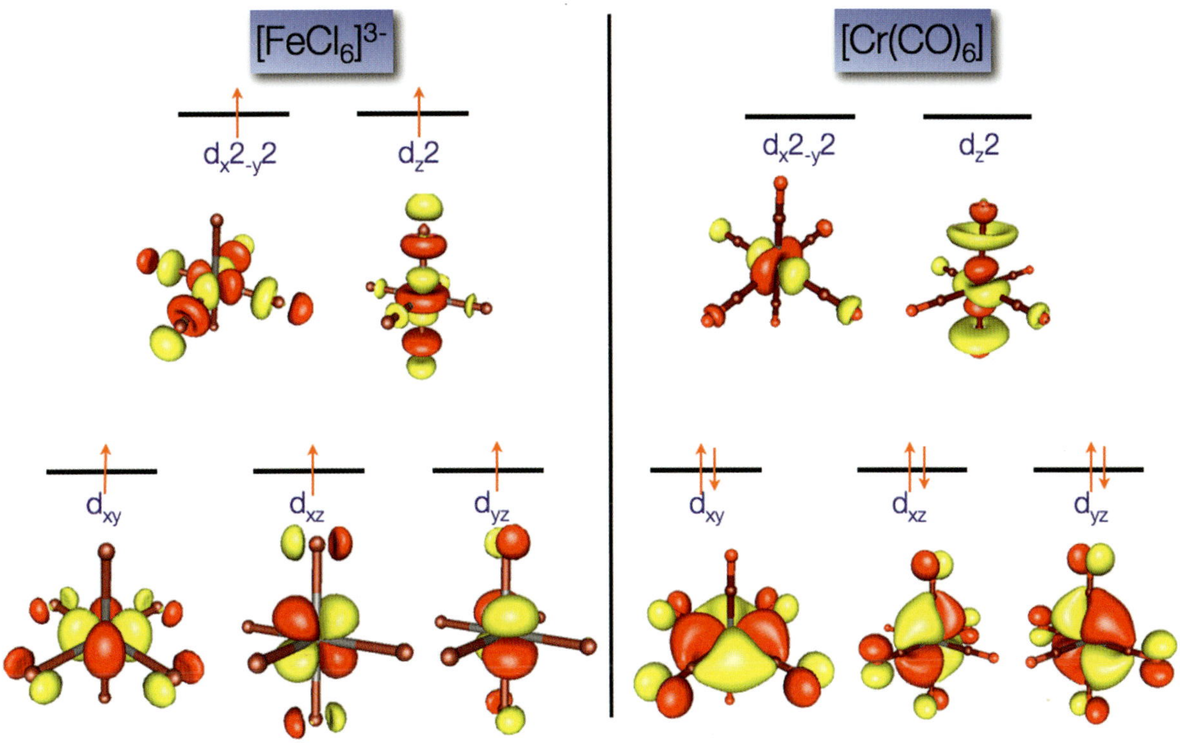

FIGURE 2.24 Illustration of π bonding (left) and π backbonding (right).

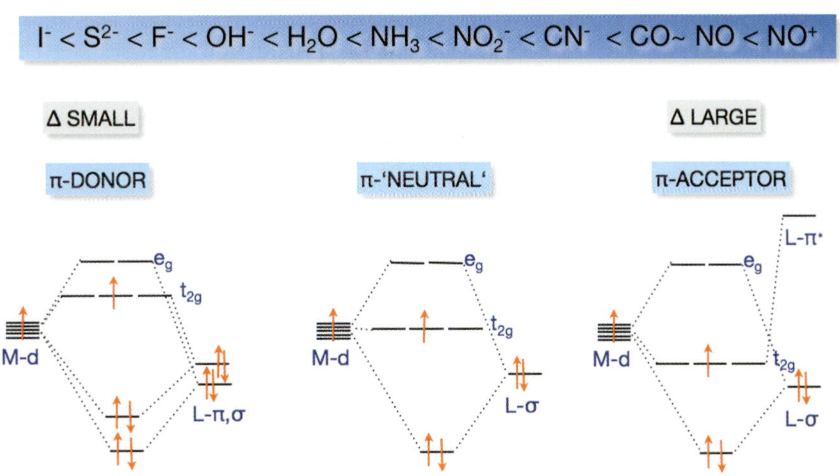

FIGURE 2.25 Interpretation of the spectrochemical series on the basis of molecular orbital theory.

If one takes a close look at the ligands that occur on the extreme strong-field side of the spectrochemical series (CO, NO, NO^+) one notices that these are ligands that allow for π-backbonding. On the very left (weak field) side of the spectrochemical series, one finds ligands like the halide anions that have appropriate orbitals to engage in π-bonding. In the middle of the spectrochemical series, one finds ligands like NH_3 that do not have the capability for any π-bonding or antibonding interaction with the metal. Assume for the sake of this discussion that the σ-donor strength of all ligands would be identical such that the e_g orbitals are all destabilised by σ-bonding to the same extent. In this situation, a π-backbonding ligand will *increase* the t_{2g}-e_g separation while π-bonding ligands will *decrease* the t_{2g}-e_g separation and hence the value of 10Dq. This also nicely explains why OH− occurs at apparently lower ligand field strength than H_2O − OH^- is a strongly π-bonding ligand whereas H_2O is essentially π-neutral. Thus, the apparent ligand field strength is lower for OH^- than for H_2O. It should become evident from this discussion that the apparent ligand field strength has no relation to the binding energy of a given ligand. Both, π-bonding as well as π-backbonding ligands can bond strongly to a given metal. Yet, their 10Dq values will be very different.

On the Meaning of the Oxidation State in Transition Metal Complexes

Before closing this chapter, a final remark should be made about how one understands the concept of an "oxidation state" in transition metal complexes. There appears to be little to no consensus among researchers and hence it seems appropriate to discuss this subject in the context of this introductory chapter.

The discussion started from an electrostatic picture. In this picture, one introduced a naked metal $n+$ ion (say, dication, trication, ...). This naked ion was associated with a number of d-electrons giving rise to a d^N configuration. Upon surrounding the metal with (partially) negatively charged ligands, the orbital and state degeneracies present in the free ion were resolved into a series of molecular multiplets. Yet, one still was arguing from the perspective of a naked metal $n+$ ion, implying that the ionic charge is a realistic representation of the actual charge of the ion in the complex. It is astonishing how much meaning and order can be brought into coordination chemistry on the basis of this picture. There are many properties (spectra, thermochemistry, reactivity) that are characteristic of a given d^N system with hundreds if not thousands of examples confirming the trends that are correctly predicted by CFT.

Despite this success, it became clear in MO theory that partially covalent bonding will lead to charge transfer from the ligand to the metal. This will reduce the effective ionic charge to some extend. In fact, charge equalisation is absolutely required by the laws of electrostatics that overall state that positive and negative charges attract each other. Hence, when one speaks about a metal oxidation state, say Mn(III), one should *not* imply that the manganese ion realistically has a positive charge of +3. Quite to the contrary, together with other channels of charge equalisation the net result is that pretty much any atom in any molecule is close to charge neutral − physics demands it this way. What one does understand that about Mn(III) is that it is a d^4 system and will magnetically, optically, thermochemically and reactivity-wise behave like a d^4 system with all those properties being modulated by the covalency of the metal−ligand bond (and obviously also steric effects imposed by the ligands).

It is likewise important to understand that the partial charge of an atom in a molecule is *not* a physical observable. There is no quantum mechanical operator that properly returns the value "partial charge" when operated upon with an accurate many particle wave function. The concept of a partial charge is, however, central to chemical thinking and it is hence important to find these elements of chemical thinking in a successful theoretical treatment. However, one should refrain from reading too much into actual values of partial charges. Just as there is no objective truth in MOs, there is no objective truth in partial charges. Both are invaluable interpretation aids − but not required by the fundamental laws of physics.

Obviously, the connection between crystal/LFT and MO theory rests upon the assumption that orbitals of mainly metal d-character can be identified that cleanly map onto the pure metal d-orbitals used in the electrostatic theory. There are actual cases where this is not possible. If the covalency becomes very strong, the high-lying "redox active" orbitals may become nearly equal mixtures of metal and ligand contributions. In this case, the meaning of an oxidation state is blurred and the system will behave in a non-classical way. Taken to the extreme, there might be "redox active" orbitals that are dominantly ligand in character and that are partially occupied. In this case, one has a so-called ***non-innocent*** ligand and enters the fascinating field of ***metal−radical interactions***.

Many exciting new properties can be found in such systems. Electronic structure theory is an invaluable aid in the interpretation of their physic-chemical behaviour.

CONCLUDING REMARKS

This chapter has provided an introduction to LFT and has highlighted its role as an interpretation aid for a wide range of properties of transition metal complexes. The mathematical elaboration of the theory will require further study. It is, however, dangerous to over-mathematise a simple theory like LFT whose main virtue are the intuitively appealing pictures that it creates.

In modern times, electronic structure calculations have become a commonplace in chemical research and education. Many program packages are available, commercial or free of charge, which can be used to obtain useful predictions about a wide range of molecular properties and chemical reactivity. However, it hopefully became evident in this chapter that such calculations are not a substitute for LFT but are complimentary. If one looks at electronic structure calculations through the eyes of LFT much insight can be obtained. Furthermore, the calculations can be guided and useless calculations can be avoided.

Using this combination of ligand field and ab initio methods, theory has become a very strong partner of experiment. It is very likely that this trend will continue and intensify for many years to come. Hence, studying LFT is as worthwhile as it was during the past four decades.

FURTHER READING

Atanasov, A., Comba, P., Daul, C. A., & Neese, F. (2008). The ligand field paradigm and new insight into the electronic properties of transition metal complexes based on modern electronic structure calculations. In J. Boeyens, & J. Ogilvie (Eds.), *Models, Mysteries and Magic of Molecules*. Berlin: Springer-Verlag. *This text describes a connection between ligand field theory and density function theory.*

Ballhausen, C. J. (1962). *Introduction to Ligand Field Theory*. New York, San Francisco, London: McGraw Hill Inc. *This textbook provides an excellent introduction to ligand field theory.*

Ballhausen, C. J. (1977). Approximate methods for the electronic structures of inorganic compounds. In G. Segal, (Ed.), *Semiempirical Methods of Electronic Structure Calculations* (pp. 129–162), New York: Plenum Press, *This textbook is a more modern exposition that eludes to molecular orbital theory in many places. The quantitative aspects of this text are outdated.*

Cotton, F. A. (1990). *Chemical Applications of Group Theory*. New York: Wiley Interscience. *This text is used in most courses on group theory in chemistry world-wide. It is very accessible.*

Griffith, J. S. (1964). *The Theory of Transition Metal Ions*. Cambridge: Cambridge University Press. *The classic text on the theory of transition metal ions. Written in an authoritative manner. A very difficult text – only recommended to the very serious student.*

Neese, F. (2009). Prediction of molecular spectra and molecular properties with density functional theory: from fundamental theory to exchange coupling. *Coord. Chem. Rev., 253*, 526–563. *This review describes the density functional approach in detail and its applications to transition metal spectroscopic properties.*

Neese, F., & Solomon, E. I. (2003). Calculation and interpretation of spin-hamiltonian parameters in transition metal complexes. Invited review. In J. S. Miller, & M. Drillon (Eds.), *Wiley series: Magnetoscience - From Molecules to Materials* (Vol. IV), pp. 345–466. *A detailed review about the magnetic properties of transition metal ions and the connection between first- principles and ligand field theory.*

Ray, K., Petrenko, T., Wieghardt, K., & Neese, F. (2007). Joint spectroscopic and theoretical investigations of transition metal complexes involving non-innocent ligands. *Dalton Trans.* 1552, *A review on non-innocent ligands and their investigation by a combination of spectroscopy and quantum chemistry.*

Schaeffer, C. E., & Jörgensen, C. K. (1965). The angular overlap model: an attempt to revive ligand field approaches. *Molec. Phys. 9*, 401. *The original paper that introduces the angular overlap model.*

Solomon, E. I., Brunold, T. C., Davis, M. I., Kemsley, J. N., Lee, S. K., Lehnert, N., et al. (2000). Geometric and electronic structure/function correlations in non-heme iron enzymes. *Chem. Rev., 100*, 235–349. *A detailed review dealing with nonheme iron enzymes and their properties and reactivity. Contains many applications of ligand field theory.*

Chapter 3

EPR Spectroscopy

W.R. Hagen
Delft University of Technology, Department of Biotechnology, Julianalaan 67, 2628BC Delft, The Netherlands

Chapter Outline

WHY EPR SPECTROSCOPY?

In biochemistry, spectrometers are used for two reasons: to determine a concentration or to determine a structure. Measured as a function of time, a change in concentration gives a reaction rate, and a change in structure affords information on a reaction mechanism. The reactions that we are interested in, are typically conversions of a substrate into a product under the influence of a biocatalyst. By far the easiest way to measure a concentration is by UV−vis spectroscopy. If a substrate and its product do not have a measurable absorption in the spectral range from near-UV to near-IR, we turn to other detection methods. EPR (electron paramagnetic resonance), = ESR (electron spin resonance) = EMR (electron magnetic resonance) spectroscopy, can only detect systems with unpaired electrons. Most metabolites in living cells are relatively stable organic molecules; these are so-called closed-shell systems, i.e. they have an even number of electrons arranged in a pairwise manner such that no electron remains unpaired, and no EPR is possible. Only in relatively rare cases does a substrate or product have an odd number of electrons with one electron unpaired. In the language of quantum mechanics (QM) these molecules are called spin-one-half systems, or $S = 1/2$. Some of these radicals (or: paramagnets) may have a colour and others do not, but they all have an EPR spectrum.

A more consequential application area of EPR spectroscopy pertains not to the substrate but to the biocatalyst, which is typically an enzyme with an active site that may encompass a radical, but much more frequently a transition metal ion, or a cluster of metal ions. By definition transition ions are open-shell systems, i.e. they have

Practical Approaches to Biological Inorganic Chemistry, 1st Edition. http://dx.doi.org/10.1016/B978-0-444-56351-4.00003-8.

partially filled d or f shells, and they have one ($S = 1/2$), or more ($S > 1/2$) unpaired electrons (i.e. they carry paramagnetism) in at least one of their common oxidation states. Thus, biological EPR spectroscopy is predominantly a means to study the structure and functioning of active sites of enzymes.

This chapter is an introduction to the subject of continuous-wave (CW) EPR of biomolecules and their models. EPR is a quantum mechanical phenomenon, and the theory of EPR likewise makes ample use of QM methods. Knowledge of QM is not required to read this chapter, or to do a range of useful EPR experiments on biological systems. Key equations are given here without derivation. Those interested in a more extensive treatment of the subject, including derivation of equations, are referred to the book *Biomolecular EPR Spectroscopy* (Hagen, 2009). Associated with that book a suite of self-explanatory computer programs for Windows is freely available for the analysis of CW-EPR spectra (www.bt.tudelft.nl/biomolecularEPRspectroscopy).

WHAT IS EPR SPECTROSCOPY?

EPR spectroscopy is the absorption of microwave radiation between energy levels of a molecules. In EPR one does not vary the frequency ν of the radiation (or the wavelength $\lambda = c/\nu$ in which $c = 299{,}792{,}458$ m/s is the speed of light), but one uses a monochromatic source at a single fixed frequency. The most commonly used frequency is in the range 9–10 GHz (the X-band). The unit of energy used in EPR is the 'wavenumber' or 'reciprocal cm'. A frequency of $\nu = 9.5$ GHz means an energy $h\nu \cong 0.3\ \mathrm{cm}^{-1}$. This is a very small energy quantum. Compare an optical transition at 500 nm, which represents an energy quantum of $500^{-1}\ \mathrm{nm}^{-1}$ or $20{,}000\ \mathrm{cm}^{-1}$. The small molecular energy splittings required for EPR are due to quantisation of the electron spin S and require an external dipole magnet: when the paramagnetic molecule is placed in an axial magnetic field B (i.e. a vector $\vec{B}$) the unpaired electron behaves like a little bar magnet but with the quantum mechanical property that its orientation in the field can assume only two values 'parallel' or 'anti-parallel' to the external field. These orientations correspond to a higher and a lower energy state of the molecule (in QM language: $m_S = +1/2$ and $m_S = -1/2$) between which microwave energy can be absorbed. The two states are 'degenerate' (i.e. they have identical energies) when there is no external field (when the magnet is switched off).

Since the frequency ν is fixed we have to vary the field B to create a spectrum. So in contrast to optical spectroscopy were we submit an invariant molecule to radiation of continuously varying energy, in EPR we throw radiation of invariant energy on a molecule whose paramagnetism is continuously varied. Absorption occurs when the splitting of the two spin energy levels, caused by the magnetic field, happens to be exactly equal to the energy of the microwave. This is called the resonance condition, $h\nu = g\beta B$, in which $h = 6.62606957 \times 10^{-34}$ J·s is Planck's constant and $\beta = 9.27400915 \times 10^{-24}$ J/T (T is tesla; 1 tesla $\equiv$ 10,000 gauss) is the Bohr magneton. The two electron spin energy levels are $E = \pm g\beta B/2$. The proportionality constant g is what is determined in an EPR experiment: the g value is specific for the molecule under study; it contains (bio)chemical information. Therefore, we rewrite the resonance condition in the practical form:

$$g = 0.714477\frac{\nu(\mathrm{MHz})}{B(\mathrm{gauss})} \tag{3.1}$$

As an estimate of practical magnetic field values we can use the (theoretical) g value for a free electron in vacuo, $g_e = 2.00232$, and a microwave frequency of $\nu = 9500$ MHz. This requires a field of $B \cong 3390$ gauss, or 0.339 T, a field that is readily produced with an electromagnet. Since radicals are 'almost-free' electrons, i.e. molecules with a reactive, delocalised, loosely bound unpaired electron, they typically exhibit g values close to g_e, say 2.002 ± 0.005 for carbon-, nitrogen-, and/or oxygen-based radicals. This does not hold for transition metals for which the g value can take any value. The interaction between the electron spin S and a magnetic field B, which causes the splitting of otherwise degenerate spin states, is called the electronic Zeeman interaction.

If Eqn 3.1 would be a sufficient description of EPR, then all spectra would consist of a single line, and the frequency of the microwave source would be an irrelevant choice. In practice EPR spectra can be fairly complex with many details, and they usually change with the frequency (note: a change in microwave frequency typically means the use of two or more spectrometers that operate with different source frequencies). There are four main reasons why this complexity (therefore: increased chemical information content) occurs.

Firstly, the electronic Zeeman interaction between unpaired electrons and the magnet ($S \leftrightarrow B$ interaction) is essential for EPR to occur, but it is not the only magnetic interaction to determine the spectrum. Also nuclei can have a spin, I, and this nuclear magnet can interact not only with the external magnet ($I \leftrightarrow B$ or nuclear Zeeman interaction) but also with the electron spin ($S \leftrightarrow I$ or hyperfine interaction). If a paramagnet has more than one unpaired electron (i.e. a high-spin system), these electron spins can mutually interact ($S \leftrightarrow S$ or zero-field interaction). A similar effect occurs for nuclear high-spin systems ($\mathrm{I} \leftrightarrow \mathrm{I}$ or quadrupole interaction). When a system has more than one paramagnet, either within one molecule or between molecules, this can lead to dipolar interaction between spins (another form of $S \leftrightarrow S$ interaction). Finally, many metalloproteins contain clusters of metals (i.e. metal ions at a mutual distance of one or two chemical bonds) leading to exchange interaction. Since only the Zeeman interactions are linear in the field (therefore linear in the frequency) and all other interactions are independent of the field, changing the microwave frequency (and therefore the field) changes the relative weight of different interactions and thus changes the EPR spectrum. Note that the quadrupole interaction and, especially, the nuclear Zeeman interaction are very often too weak to be resolved in regular EPR, however, they are observed in more complicated double-resonance experiments like electron nuclear double resonance (ENDOR) or electron spin echo envelope modulation (ESEEM).

Secondly, as a consequence of the non-spherical structure of molecules the above mentioned interactions are all dependent on the orientation of the dipole magnet (the vector $\vec{B}$) with respect to the molecule (or with respect to a cartesian coordinate system defined by the molecular structure of the paramagnet). In other words, the EPR spectrum depends on the direction of orientation of the molecule in the magnetic field, and even if the electronic Zeeman interaction is the only relevant spectral determinant, the spectrum is almost never a single line.

Thirdly, if we use a microwave to excite an $S = 1/2$ molecule from its lower spin energy level, or ground state, to its higher level, or excited state, then the molecule subsequently has to return to its ground state again by some mechanism if only to be ready to absorb a next microwave quantum and thus to maintain the spectrum in time. This so-called 'spin relaxation' influences the shape and, particularly, the width of EPR spectra, and it makes the spectral shape dependent on the temperature T and on the intensity P (for power) of the microwaves.

Fourthly, all molecules are subject to conformational distribution: the exact relative atomic coordinates slightly (or not so slightly) vary in a real sample from one molecule to the other. This means that each molecule has a slightly different g value (and values for the other interactions), which in turn leads to EPR broadening and spectral changes. This phenomenon is called 'g-stain'.

Understanding EPR means to understand the nature and the practical consequences of these phenomena, and this is the subject of the remainder of this chapter.

ANISOTROPY

The unpaired electron of a paramagnetic molecule does not only experience the external magnetic field, but it also 'sees' the molecular structure as it is part of the molecule's electronic structure. Electrons as moving charges represent a magnetic field, and this internal field adds to, or subtracts from, the external field in the Zeeman interaction: $h\nu = g_e\beta(B_{ext} + B_{int})$. By convention, this is usually written as an observed shift in the g value of the free electron: $h\nu = (g_e + \Delta g)\beta B$, or simply $h\nu = g\beta B$, where g, or rather the deviation from g_e now contains electronic (and therefore structural) information on the molecule.

Magnetic fields are vectors in 3D space, for example in the space defined by the coordinates of the paramagnetic molecule. The external field is the dominant one, so it determines the direction along which the EPR spectroscopist 'looks at' the molecule. In other words, the orientation of the molecular structure in the external field is of the essence: rotation of the molecule in the field (or, alternatively, rotation of the field around the magnet) will result in a change of the EPR spectrum. Each orientation has its own spectrum, and so a single molecule can give rise to an infinite number of different spectra. If the spectra are determined by the electronic Zeeman interaction only, then they all consist of a single absorption line. Single-molecule EPR spectroscopy does not exist, because the signal of a single molecule is too weak to detect. However, a single crystal consists of many identical molecules all with the same orientation in space. A single crystal of $S = 1/2$ molecules gives single-line Zeeman EPR spectra.

Our biomolecular or model compound samples will usually not be single crystals; they are homogeneous solutions, or frozen solutions, or powder samples. Each molecule has a different orientation and the EPR of such a sample is the sum of many different single-line spectra. All these spectra have different g values, because the internal field seen by the unpaired electron depends on the orientation of the molecule in the external monitoring field. The result, illustrated in Figure 3.1 is a specific spectral shape called the 'powder pattern'; it covers a defined field range (or g value range) between two extreme values. Most EPR spectrometers produce the first derivative of this powder absorption spectrum, and since the slope of the powder pattern rapidly changes around the three g values: g_x, g_y, g_z (together also called: the g tensor or – by mathematics purists – the g matrix), corresponding to the molecular x, y, z-axes, one gets the impression that the EPR derivative spectrum consists of three 'peaks'.

This three-featured form (peak, derivative, negative peak), called the 'rhombic powder pattern', is the general fingerprint of an $S = 1/2$ system without any specific symmetry properties ($g_x \neq g_y \neq g_z$). Note that the labelling

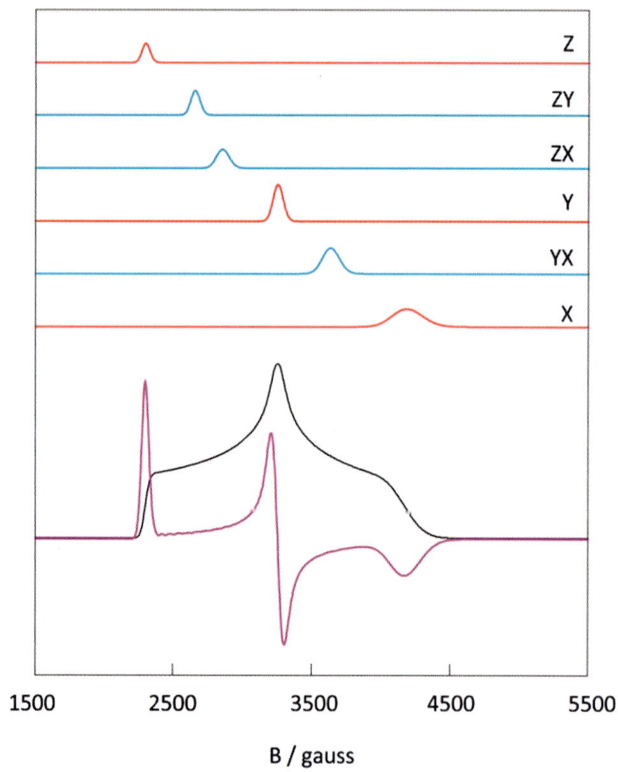

FIGURE 3.1 The construction of an X-band ($\nu = 9.5$ GHz) EPR powder pattern (the black trace) by summation of single-orientation spectra from individual molecules in the XYZ molecular axis system with respect to the magnetic field vector B. Red traces are for B along one of the molecular axes, and blue traces are examples of spectra from intermediate orientations. The label ZY, for example, means that B is oriented halfway between the molecular Z- and X-axis. The shown orientations are only a few examples of an infinite number of possibilities. The magenta trace is the observable first derivative of the EPR powder absorption spectrum with g values: $g_x = 1.62$, $g_y = 2.09$, $g_z = 2.95$.

with (x, y, z) is arbitrary; one might just as well use (z, y, x) or (x, z, y) or (a, b, c) or (1, 2, 3), etc. Symmetric properties of the coordination complex may simplify the EPR pattern. A metal ion at the centre of a perfect octahedron with six identical ligands with identical metal-to-ligand bond lengths will give a single (derivative) line spectrum, called an isotropic pattern, with $g_x = g_y = g_z$. Such a highly symmetrical structure is not likely to occur in biology. However, for metalloproteins we do frequently find (near) 'axial' spectra with $g_z \neq g_y \cong g_x$. An axial pattern can occur when a perfect octahedron is elongated (or compressed) along one of the axes (which is then defined as the z-axis). More generally, if one, or two of the ligands along the z-axis are different from the others (as in many tetrapyrrole complexes, for example in hemoproteins) the structure and the spectrum are (near) axial. Strictly speaking axiality of the EPR spectrum reflects axiality in the electronic structure, i.e. in the wave function(s) occupied by the unpaired electron, and may not necessarily be retraceable to an exact axial geometry. Several so-called blue copper proteins, such as plastocyanin with His, His, Met, Cys ligands, exhibit axial EPR, although the copper coordination is a very strongly deformed NNSS tetrahedron. In summary, the basic EPR spectrum is a one-, two-, or three-featured pattern, but it is not always easy to link this to structural symmetry.

A COMPARISON OF EPR VERSUS NMR

The EPR and the NMR effect were both discovered in the mid-1940s, and their subsequent technical and theoretical developments initially took a parallel course. It is instructive to compare the two spectroscopies in terms of present-day similarities and differences. The simplest system giving EPR is one with a single unpaired electron: $S = 1/2$. Most nuclear magnetic resonance spectroscopy, e.g. ^{1}H-NMR, is done on systems with a nuclear spin $I = 1/2$. There is a strong analogy between $S = 1/2$ and $I = 1/2$: they both give rise to two quantised energy states in an external magnetic field by means of a Zeeman interaction, although in EPR, due to the negative sign of the electron charge, the orientation antiparallel to the field is the lowest in energy, while in NMR the parallel orientation is the ground state. Both the nuclear Zeeman interaction and the electron Zeeman interaction are measured in the form of a spectroscopic shift. The electron shielding induced chemical shift away from the resonance frequency of a standard compound such as tetramethylsilan is the NMR equivalent of the internal field induced shift Δg from the free-electron value g_e in EPR. Of course for the same external field strength, nuclear spin energy level differences are much smaller than the separations of electron spin energy levels, and the radiation to induce nuclear resonance is in the radio frequency range (MHz instead of GHz), which is one of the reasons why NMR spectroscopy typically has a lower concentration sensitivity than EPR spectroscopy.

Arguably the most important historical divergence between NMR and EPR was the divulgence of commercial pulsed NMR spectrometers starting in the late 1960s. From then on it became routine to apply a broad spectrum of frequencies (i.e. a short single-frequency pulse turning into a frequency range by virtue of Heisenberg's uncertainty principle) to a sample in a single-valued, invariant external magnetic field. In this approach the recording of a single NMR spectrum (a 'shot') takes a short instance, and thus extensive averaging becomes practical. Even today such an experiment is technically impossible in EPR spectroscopy because one cannot make a sufficiently short gigahertz pulse of sufficient intensity and homogeneity to cover all the frequencies of an EPR spectrum at constant field. Therefore, EPR still uses monochromatic CW radiation in combination with a varying magnetic field, and the recording of a single EPR spectrum typically takes a few minutes, if only to avoid hysteresis in the field scan.

Pulsed versions of EPR have also been developed based on nanosecond electronics, but there is a paradigmatic difference in the application of pulsed EPR versus that of pulsed NMR. EPR studies always start with the CW experiment; pulsed EPR is an optional 'next step' or 'advanced' follow-up experiment providing information additional to the CW data and obtained at additional cost and effort. Because the EPR pulse cannot cover the whole spectrum, it is used to measure a small range of the spectrum at increased resolution. Technically, this is typically set up in the form of a double-resonance experiment. With the CW spectrum known, the magnetic field is fixed at a value corresponding to a single point of the powder spectrum, and then a pulse of radiation is applied to probe a small frequency spectral range around the fixed powder point. Today a variety of pulsed EPR methods is at one's disposition such as pulsed ENDOR, ESEEM, pulsed ELDOR (electron–electron double resonance) also called PELDOR or DEER (double electron–electron resonance). These techniques are usually only available in specialised laboratories, and each one has its own possibilities and limitations. Their details are beyond the scope of this chapter.

From the perspective of the biochemist there is also a major conceptual difference in the application of NMR versus EPR spectroscopy. A high-resolution solution NMR experiment affords a global picture of all protons (and/or ^{13}C, ^{15}N, etc.), i.e. of all parts of a biomacromolecule. On the contrary, EPR only looks at paramagnets and thus typically provides a picture of a local spot of the macromolecule, e.g. a metal coordination complex in a metalloprotein. In other words, the EPR result is spatially limited and spectroscopically more simple. On the other hand, the focus point is usually on the most important part of the molecule: the active site. And here the two methods show considerable complementarity, because the NMR resonances are extensively broadened (frequently beyond detection) for the nuclei near the paramagnetic centre.

Both $S = 1/2$ EPR and $I = 1/2$ NMR exhibit anisotropy, however, in liquid-state NMR tumbling of molecules affords complete averaging of this pattern to single-line spectra for each nucleus even in large biomacromolecules. In EPR this only works for relatively small molecules: proteins are so big that their tumbling on the EPR timescale is too slow to average anisotropy away, and therefore, e.g. metalloproteins always give powder-pattern spectra even when in solution. In solid-state NMR one does observe powder patterns, very similar to the EPR ones, as spectra for individual nuclei.

Finally, 'increasing the frequency' has a rather different connotation in NMR as compared to EPR. Over the years the proton NMR frequency has been steadily increased from tens of megahertz to around 1 GHz at present, concomitant with improving technology for the generation of stable, homogeneous static magnetic fields first with electromagnets and subsequently with superconducting solenoids. Each frequency-field increase has led to an increase in sensitivity and an increase in resolution in these nuclear Zeeman interaction-dominated spectra. On the contrary, concentration sensitivity of the EPR detection system typically decreases above circa 10 GHz due to several reasons such as increasingly noisy detection diodes or increased power losses in high-frequency microwave components.

EPR SPECTROMETER

The vast majority of EPR spectrometers operate at a frequency in the range 9–10 GHz, which is part of the X-band of circa 8–12 GHz. The most efficient, or least lossy, way to transfer microwaves at X-band frequencies is the waveguide, typically a rectangular tube made of brass filled with air. The waves move through the inner 'skin' of the waveguide, i.e. through a layer of a few micrometers thick on the inside of the guide. The second best choice for transport in X-band is the coaxial cable, an inner and outer conductor separated by a dielectric such as teflon. Thus, the microwave part of the spectrometer is a spaghetti of components connected by waveguide and coax, most of which is, however, invisible to the operator, because it is built in a box called 'the bridge'. This design gives the spectrometer a somewhat austere look, in which only a single piece of waveguide sticks out of the bridge to terminate in the heart of the machine called the cavity, centred in between the poles of the magnet (Figure 3.2).

The cavity is also called the resonator, but note that this name has nothing to do with the EPR of the sample. The bridge-coupled cavity can be described as a resonator circuit with a quality factor Q. In brief this means that the inner dimensions of the cavity and the material properties of its inner walls are such that a unique frequency in X-band affords the sustaining of a standing wave pattern with an energy density that is Q-times greater than in a setup in which the cavity would be absent. Typical Q-factors for X-band cavities are of the order of 5000, and the sensitivity of the spectrometer is approximately increased by this factor compared to that of a simple transmission or reflection instrument without a cavity.

The 'mode pattern', or the field lines in the cavity, are such that the magnetic component of the microwave (i.e. the field required to make EPR transitions) is maximal along a vertical line through the middle, where the electric component of the microwave (i.e. the field that only disturbs the EPR measurement by non-resonant

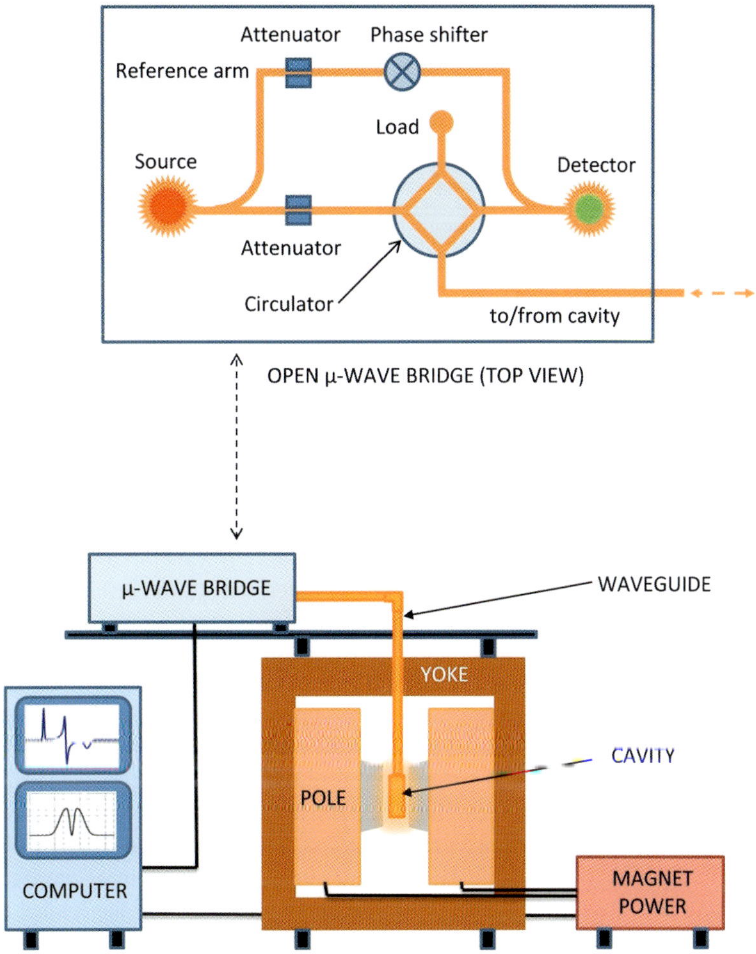

FIGURE 3.2 Drawing of a conventional CW-EPR spectrometer consisting of a dipolar electromagnet, a microwave bridge (with waveguide schematic) with a reflection resonator or cavity, and a computer console whose screens exhibit an EPR spectrum and a tuning mode pattern.

absorption) is minimal. The shape of an EPR sample in an X-band cavity is a vertical cylinder with a length of circa 15 mm and a diameter of circa 1 mm for aqueous samples and a diameter of circa 4 mm for all other samples including frozen aqueous solutions. This gives approximate volumes of 15 and 200 μl held in quartz tubes; the reduced diameter for aqueous samples is to compensate for the increased non-resonant absorption of microwaves due to the high dielectric constant of water.

Microwave intensities are expressed in decibels attenuation with respect to a reference value in watts. The decibels scale is a logarithmic one:

$$n(\text{dB}) \leftrightarrow 10^{0.n} \tag{3.2}$$

so for a source of 200 milliwatt the attenuated power in decibels versus watt is as in Table 3.1 (in which m = milliwatt, μ = microwatt, n = nanowatt).

The source of X-band microwaves is either a klystron (vacuum tube technology) or a Gunn diode with a typical initial output power of circa 400 milliwatt slowly decreasing over the source's lifetime (>10,000 h of operation).

TABLE 3.1 Conversion of Decibels Attenuation to Power in Watt for a Source Levelled at 200 milliwatt

Decibels	0	−2	−4	−6	−8
0	200 m	126 m	79.6 m	50.2 m	31.7 m
−10	20.0 m	12.6 m	7.96 m	5.02 m	3.17 m
−20	2.00 m	1.26 m	796 μ	502 μ	317 μ
−30	200 μ	126 μ	79.6 μ	50.2 μ	31.7 μ
−40	20.0 μ	12.6 μ	7.96 μ	5.02 μ	3.17 μ
−50	2.00 μ	1.26 μ	796 n	502 n	317 n
−60	200 n				

This power is 'levelled' (i.e. reduced) to a fixed value of 200 milliwatt to assure constant output over time. The 'road map' of the microwaves is then as follows (cf. Figure 3.2): a wave of 200 milliwatt leaves the source. A small amount of this intensity (circa 1% or −20 dB) is 'coupled out' to the reference arm by a device called a directional coupler. The remaining 99% intensity in the main arm can be reduced to the required level (see Saturation section) by means of an attenuator, with the maximal attenuation for a good spectrometer being 60 dB (i.e. 1,000,000 times attenuation to 200 nanowatt). After the attenuator the main wave enters a circulator, a device that can be thought of as a right-hand roundabout, and is forced to go into the waveguide that ends with the cavity. Waves reflected back from the cavity pass the circulator to go to the third arm with the detector diode, and any wave reflected from there is caught in the dead end of the fourth arm called 'load' to be converted into heat waste.

As an essential prelude to actual measurement the spectrometer has to be 'tuned', which means that we have to (1) adjust the frequency, (2) make the spectrometer reflectionless, and (3) adjust the reference arm. The microwave frequency of the source in the bridge is tuneable over a small range (9−10 GHz in X-band) to make it correspond to the unique 'Eigen frequency' (or the frequency of the fundamental mode) of the loaded cavity, i.e. the cavity with a sample tube and (if necessary) a cooling system in place. This is accomplished by making the oscilloscope tuning mode pattern symmetric with the 'dip' centred (cf. the lower screen on the computer in Figure 3.2). Then the spectrometer is made 'reflectionless', which means that one adjusts a teflon screw at the back of the cavity with a metal end plate in front of a little aperture called the iris between the waveguide and the cavity until the dip in the tuning pattern is maximal (touches the baseline) and no current is measured at the detection diode. Reflectionless means that a standing wave is set up in the cavity whose energy dissipates only through the cavity's side walls with no loss by means of back-reflection into the waveguide. In the language of electronics the impedance of the cavity is now 'matched' to that of the rest of the system. This delicately balanced system will be detuned when, in an actual experiment, some of the radiation is absorbed by the sample, and as a result some radiation will be reflected out to cause a voltage change over the detection diode. However, for proper operation of the diode the voltage change should not be around zero but around a finite value called the voltage bias. One can create this bias by slightly detuning the cavity with the iris screw, but the disadvantage of this approach is that the spectrometer has to be re-adjusted every time the power, i.e. the intensity of the microwaves, is changed by adjustment of the attenuator. To avoid this complication we use the reference arm through which a constant fraction of the unattenuated microwave is directly by-passed to the detection diode. The reference arm has its own attenuator, usually called 'bias', to optimise its output power to the characteristics of the diode (i.e. typically a diode current output of 200 μA). The reference arm also has a device called 'phase shifter' to make the phase of the microwave equal to that through the main arm, which is accomplished by adjusting the phase shifter until the tuning mode pattern is perfectly symmetric.

The apparatus is now ready to run a spectrum, which means that we scan the magnetic field (the x-axis of the spectrum) over the required range over a period of typically a few minutes. The field is produced by a water-cooled electromagnet, or in high-frequency EPR (circa $\nu \geq 90$ GHz) by a superconducting solenoid. To further increase spectrometer sensitivity we use a technique called phase-sensitive detection: the very slowly varying magnetic field is modulated with a very rapidly varying (100 kHz) small field of the order of ± 1 gauss, and the EPR signal from the detection diode is measured with a lock-in amplifier, i.e. a device that takes an in-phase 100 kHz 'look' at the signal. The result of this technique is twofold: (i) the EPR signal-to-noise ratio increases because electronic system noise at frequencies other than 100 kHz is not amplified, and (ii) we obtain the first derivative of the EPR absorption spectrum.

WHAT (BIO)MOLECULES GIVE EPR?

All compounds with one or more unpaired electrons give an EPR spectrum. The very vast majority of biologically relevant radicals have one single unpaired electron: a doublet with $S = 1/2$ (e.g. flavin radicals, nitric oxide, NO, superoxide, $O_2^{\bullet -}$). In rare cases one finds two unpaired electrons: a triplet with $S = 1$, the most notorious example being molecular oxygen, O_2. Radicals with more than two unpaired electrons exist, but none has yet been reported for a biological system. Occasionally, it may be possible to excite a diamagnet (no unpaired electrons) by continuous illumination with UV light into the cavity to an excited triplet state with two unpaired electrons. Transition ions can have up to five unpaired d electrons: a quintet state with $S = 5/2$ (e.g. Fe^{3+}), or up to seven unpaired f electrons: a heptet state with $S = 7/2$ (Gd^{3+}), and exchange-coupled clusters of transition ions can have many unpaired electrons (e.g. the [4Fe-2O-2S] cluster in the hybrid-cluster protein in one of its oxidation states has nine unpaired electrons: a nonet state with $S = 9/2$).

From an EPR spectroscopist's point of view it is practical to divide all molecular systems into four groups: diamagnets ($S = 0$), doublet systems ($S = 1/2$), half-integer high-spin systems ($S = n/2$), and integer high-spin systems ($S = n$). Diamagnets have a ground state that is not split by a magnetic field, i.e. a singlet system, and they cannot have an EPR spectrum. Doublet systems are 'easy' not only because they have only a single electron Zeeman transition, but also because electron spin relaxation (see Saturation section) is usually slow, which means that their EPR can be measured at relatively high temperatures (i.e. ambient temperature or nitrogen-flow temperatures). Half-integer high-spin systems are 'not so easy' because they have more than one electron Zeeman transition, and because their relaxation is usually so fast that cooling with cryogenic helium gas is required to obtain sharp lines. Integer high-spin systems are 'difficult' because usually they additionally have very broad and asymmetric EPR features. Some $S = n/2$ systems behave as 'effective' $S = 1/2$ systems, which means that only one of the several Zeeman transitions is detectable, but the relaxation can still be fast. Some $S = n$ systems behave as effective $S = 0$ systems, which means that none of the several possible Zeeman transitions is detectable, and so these systems are called 'EPR silent'. An overview of these classes each with an iron–protein example is given in Table 3.2.

TABLE 3.2 Overview of Classes of Spin Systems

A	Diamagnets	$S = 0$	Fe(II) in myoglobin sulfide
B	Low spin	$S = 1/2$	Fe(III) in ferrimyoglobin sulfide
C	High spin	$S = n/2$	Fe(III) in rubredoxin
C′	Single-transition high spin	Effective $S = 1/2$	Fe(III) in ferrimyoglobin
D	Integer spin	$S = n$	Fe(II) in myoglobin
D′	EPR silent integer spin	Effective $S = 0$	Fe(II) in rubredoxin

And here follows a more extensive list of biologically relevant systems and the spins of their electronic ground states.

$S=0$ systems: most organic molecules, complexes of all main group elements, low-spin Fe(II), Co(III), square planar Ni(II), Cu(I), Zn(II), Mo(VI), W(VI), $[2Fe\text{-}2S]^{2+}$, $[4Fe\text{-}4S]^{2+}$;

$S=1/2$ systems: most organic radicals, e.g. flavin radicals, quinone radicals and most inorganic radicals, e.g. nitric oxide NO, superoxide $O_2^{\bullet -}$, low-spin Fe(III), low-spin Co(II), Ni(III), Cu(II), Mo(V), W(V), $[2Fe\text{-}2S]^{1+}$, $[3Fe\text{-}4S]^{1+}$, $[4Fe\text{-}4S]^{3+}$, $[4Fe\text{-}4S]^{1+}$, Fe(II)-O-Fe(III);

$S=n/2$ systems: Mn(II), Mn(IV), high-spin Fe(III), high-spin Co(II), linear $[3Fe\text{-}4S]^{1+}$, some $[4Fe\text{-}4S]^{1+}$;

$S=n$ systems: bi-radicals (i.e. triplet-state molecules), e.g. light-excited reaction centres, molecular oxygen, Mn(III), high-spin Fe(II), Fe(IV), high-spin Ni(II), $[3Fe\text{-}4S]^0$, the Cu(II)-heme Fe(III) dimer in cytochrome oxidase.

BASIC THEORY AND SIMULATION OF EPR

EPR spectroscopy of single crystals from biomolecules is rare because, e.g. protein crystals are much smaller than the X-band sample size of 200 μl, and they do not give sufficient signal intensity. EPR samples are almost always (frozen) solutions of biomolecules, which means that they contain very many molecules (circa 10^{17} for a millimolar solution), each one with a different orientation with respect to the external axial magnetic field. To quantitatively understand EPR spectra we must carry out what is called a 'walk over the unit sphere', which means that we conceptually place the molecule of interest at the origin of an xyz cartesian axes system (i.e. the molecular axes system), and then let a vector from the origin, of unit length, parallel to the magnetic field B, sample 'all' possible orientations with respect to the molecule. This is done by defining a sphere around the origin with a radius of unity, dividing up the surface of this sphere in a large number of little areas of equal size and shape, and letting the unit vector subsequently point towards each one of these little areas. The 'walking' of the unit vector is conveniently defined in polar coordinates $(1, \theta, \varphi)$ as shown in Figure 3.3, and we must make steps in $\cos\theta$, and in φ to keep the little areas constant in size.

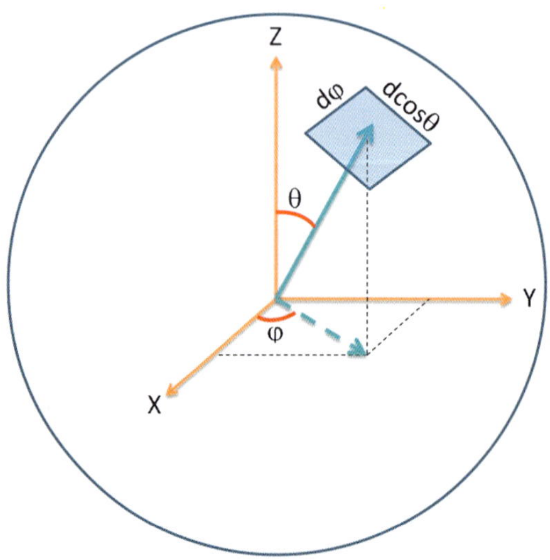

FIGURE 3.3 Sketch of a walk over the unit sphere. The solid blue angle is a unit vector along the magnetic field B with polar angles θ and ϕ in the cartesian XYZ molecular axis system. The broken blue arrow is the projection of the unit vector onto the XY plane. The paramagnet is at the origin; the surface of a surrounding unit sphere is divided into fragments of equal area defined in terms of the polar angles as $d\cos\theta \times d\phi$. The unit vector along B samples each of these fragments once.

For a rhombic spectrum as in Figure 3.1, this means that we must solve the equation for the field position of the EPR absorption, $B = 0.714477\ \nu/g$ with

$$g(\theta,\phi) = \sqrt{g_x^2 l_x^2 + g_y^2 l_y^2 + g_z^2 l_z^2} \tag{3.3}$$

in which we used the definition of the so-called direction cosines

$$l_x = \sin\theta\cos\phi;\ l_y = \sin\theta\sin\phi;\ l_z = \cos\theta \tag{3.4}$$

and the equation for the intensity (or transition probability)

$$I(\theta,\phi) = g^{-1}\sum_{i=x,y,z}\left(g_i^2 - g^{-2} l_i^2 g_i^4\right) \tag{3.5}$$

by making equidistant steps in $\cos\theta$ and in θ. For example, to generate a simulation such as the spectrum in Figure 3.1, we write a computer program that has four nested FOR loops, two for the walk over the unit sphere, one for a scan of the magnetic field range, and one for a scan through a line shape function, e.g. a Gaussian distribution:

```
FOR cosθ = 1 to 0                                   !step from z to xy plane
    FOR ϕ = 0 to π/2                                !step from x to y
        g_r = g(θ, ϕ)                               ! compute g-value
        I_r = I(θ, ϕ)                               ! compute intensity
        B_r = 0.714484 * v/g_r                      ! compute resonance field
                FOR B = B_start to B_end            ! step in field
                FOR b = B − 3σ to B + σ             ! step in Gaussian line shape
                    S = S + I_r exp[−ln2(b − B)^2/σ^2]   !compute amplitude
```

This example program is written in 'pseudo code' and should be rewritten in your favourite programming language in order to work. Alternatively, you can of course use existing programs, such as described in (Hagen, 2009) and available from www.bt.tudelft.nl/biomolecularEPRspectroscopy. Note also that the walk over the unit sphere in the example is actually over only one octant, because the other seven octants give identical $g(\theta, \varphi)$ and $I(\theta, \varphi)$ values. Also, we do not really have to calculate 10^{17} different molecular orientations; typically some 100 steps in $\cos\theta$ and in φ, i.e. circa 10^4 orientations, is enough, which means that an increase in the number of steps will not change the shape of the final simulation. For a digital spectrum of say 1024 points this means calculating circa 10^7 amplitudes, which your garden variety laptop can easily accomplish in less than a second.

SATURATION

How much microwave power should we use; how far should we open the attenuator of the main arm in the bridge of Figure 3.2? There is no danger of destroying the sample: the maximum output power is typically 200 milliwatt, which is some 5000 times less than what one uses in a household microwave oven. In X-band an aqueous sample at room temperature subjected to full power may warm up by a few degrees, which is why most work on aqueous samples is done at reduced power of −10 dB (i.e. 20 milliwatt) or less. The temperature of a frozen sample in a cryogenic flow of nitrogen or helium will not even noticeably change at full power. From a statistical viewpoint higher power means more transitions per unit time and therefore higher EPR amplitude. However, the phenomenon of saturation limits the maximum power that we may use.

This is readily illustrated on the energy level scheme of a simple $S = 1/2$ system. At resonance the two levels are separated by an energy difference $\Delta E = h\nu$, and this means a Boltzmann population distribution $n_1 = n_0 e^{\Delta E/kT}$, in which $k = 0.69503476\ \mathrm{cm^{-1}/K}$ is the Boltzmann constant. In other words, for a total number of $n_0 + n_1$

molecules we will have n_0 in the ground state and n_1 in the excited state. For a sample in X-band ($\nu = 9.5$ GHz; $\lambda = 3.156$ cm) at room temperature ($T = 295$ K) we find $n_1/n_0 = 0.9985$, i.e. a very small difference in population indeed. Inducing transitions by microwave absorption will further reduce this small difference, the more so at higher power levels. This reduction is counteracted by relaxation, i.e. the falling back to the ground state of excited molecules by dissipation of the energy difference ΔE to the surroundings. If the relaxation cannot keep up with the microwave power input, then there will be a net decrease in population difference. At high microwave power the difference will eventually become zero and the possibility for net microwave absorption will be abolished: the spin system is completely saturated and the EPR signal has disappeared. In summary, with increasing microwave power the EPR amplitude increases linearly with low power, then levels off at higher power, then decreases at even higher power, and eventually disappears at very high power.

This amplitude versus power relation is usually measured and plotted for normalised amplitudes, i.e. corrected for the power. Experimentally, this is done as follows. The power expressed in decibels is a logarithmic scale; the gain, or electronic amplification, on EPR spectrometers is also expressed on a logarithmic scale either in hardware on older spectrometers or in software on newer spectrometers. In other words, one can only choose from a limited number of gain values, and these are 1.25, 1.6, 2.0, 2.5, 3.2, 4.0, 5.0, 6.3, 8.0, 10 times a power of 10. These numbers are equal to 10^x where $x = 0.1, 0.2, \ldots, 1.0$, i.e. a logarithmic scale. In practice this means that when a signal is *not* saturated, increasing the power by $m \times 2$ dB and at the same time decreasing the gain by m steps should afford the same EPR amplitude. So we can make a saturation graph (or 'power plot') by running a spectrum at a power of, say, −30 dB and a gain of, say, 4.0×10^3; then we take the next spectrum at power = −28 dB and gain = 3.2×10^3, and so on until we reach a power/gain combination at which the EPR amplitude starts to decrease. We have then reached the onset of power saturation, and we should go back one step (2 dB) to get to the optimal measuring conditions in terms of signal-to-noise ratio.

The experiment is schematically illustrated by the green trace in Figure 3.4: increasing the power from the very low value of −60 dB initially leaves the normalised amplitude unaffected. When we reach a power of circa −30 dB the signal starts to decrease, and at the highest power of 0 dB we face some 90% saturation. The optimal power value for this case would be approximately −32 dB: the spectrum has maximal signal-to-noise ratio without being significantly saturated. There are two good reasons why saturation is to be avoided. First, under (partial) saturation the signal amplitude is no longer linear in the applied power (expressed in decibels) and so determination of the spin concentration (cf. next section) versus an external standard is no longer possible. Secondly, relaxation rate, and therefore saturation, is anisotropic: its extent with increasing power is different for

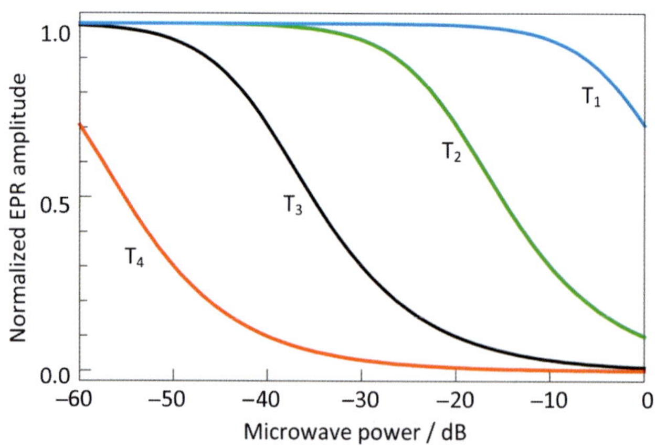

FIGURE 3.4 Theoretical power plots showing saturation as a collapse of normalised EPR amplitude with increasing microwave power. Traces are shown for four different temperatures with $T_1 > T_2 > T_3 > T_4$. At lower temperature the onset of saturation occurs at lower power. At T_1 the signal is only slightly saturable; at T_4 the signal cannot be measured under non-saturating conditions.

different parts of the spectrum. Therefore, under partial saturation a spectrum will change shape in a complex manner for the analysis of which no theory is available to date.

Figure 3.4 gives the power plots for one single EPR signal taken at four different sample temperatures with $T_1 > T_2 > T_3 > T_4$. The values of the T's are not specified because the quantitative temperature dependence of relaxation rates in EPR is usually rather complex. Qualitatively, however, we can take the relaxation rate to always decrease with decreasing temperature: is it easier to saturate a signal at low T than at high T. This has the very important practical implication that for a given EPR signal we have to experimentally determine the optimal microwave power for each temperature at which we want to measure. In literature one frequently finds reports of the temperature dependence of an EPR spectrum taken at a single, intermediate power level. It should be obvious from Figure 3.4 that such an approach is far from optimal. Suppose, for example, that we take spectra at power $= -30$ dB for all four T's. At T_1 the signal is unsaturated, but the signal-to-noise ratio is suboptimal; at T_2 we have a near-optimal situation with high signal-to-noise and hardly any saturation; at T_3 however the signal is seriously saturated (and therefore deformed) at this power; and at T_4 there is hardly any intensity left. In fact for the example of Figure 3.4 it is impossible to measure an EPR spectrum at T_4 with any available power level without serious saturation problems. T_4 is simply too cold for this sample. This situation is commonly encountered for $S = 1/2$ systems, such as $[2Fe\text{-}2S]^{1+}$ proteins, recorded in X-band at the boiling point of liquid helium, i.e. $T = 4.2$ K.

CONCENTRATION DETERMINATION

Arguably one of the most useful applications of EPR in biochemistry is 'spin counting', i.e. the determination of the concentration of spin systems. Picture a complex protein containing several paramagnets, for example the enzyme succinate dehydrogenase that contains a [2Fe-2S] cluster, a [3Fe-4S] cluster, a [4Fe-4S] cluster, a heme, and an FAD radical. Determination of their stoichiometry (at a given redox potential) would be a very difficult task indeed if one would not have quantitative EPR spectroscopy available. Since the intensity of a spectrum is defined by Eqn 3.5, one can simply take the area under the EPR absorption envelope (cf. the black trace in Figure 3.1), corrected for the intensity expression, as a measure for the concentration of a paramagnet. An absolute concentration is obtained by comparison with the area under the EPR absorption envelope from a standard compound of known concentration. A commonly used standard is the hydrated Cu^{2+} ion, e.g. 10 mM $CuSO_4$ + 10 mM HCl + 2 M $NaClO_4$ with g values 2.404, 2.076, 2.076 (Hagen, 2006). The anisotropic intensity expression in Eqn 3.5 is usually replaced with an approximate average scalar according to Aasa and Vänngård (1975):

$$I = (2/3)\sqrt{(g_x^2 + g_y^2 + g_z^2)/3} + (g_x + g_x + g_x)/9 \tag{3.6}$$

In practice, the procedure is as follows: run a spectrum of the paramagnet with unknown concentration C_U; read the approximate g values from the spectrum (cf. Figure 3.1); take the second integral $\int\int_U$ (i.e. the area under the EPR absorption spectrum) and calculate the intensity factor I in Eqn 3.6 for normalisation. Do the same with the standard compound of known concentration C_K, and then correct the final number $\int\int_K$ for any difference in experimental measuring condition between the unknown and the standard. When the two spectra are taken under identical experimental conditions, we have

$$\frac{C_U}{C_K} = \frac{\iint_U I_K}{\iint_K I_U} \tag{3.7}$$

Most of the experimental differences between unknown and standard require linear corrections, e.g. the modulation amplitude (M), the electronic amplification (G), the sample tube diameter (d), the sample temperature (T). However, if the two spectra have different x-axis dimensions, i.e. different magnetic field scan widths (W), then this correction should be taken squared, because the EPR derivative spectrum is integrated once to obtain the absorption and then once more to obtain the area under this absorption. Note also that a difference in the microwavepower (P) used in decibels requires a logarithmic correction. Altogether, the concentration C_U is now obtained as

$$\frac{C_U}{C_K} = \frac{\iint_U I_K\, M_K\, G_K\, d_K\, T_U}{\iint_K I_U\, M_U\, G_U\, d_U\, T_K}\left(\frac{W_K}{W_U}\right)^2 10^{\frac{(P_U - P_K)}{20}} \quad (3.8)$$

Sometimes it may be preferable to replace the experimental spectrum with its simulation, for example, when integration of the experimental spectrum is unreliable because the baseline is of poor quality, or because the spectrum overlaps with other spectra, or because the spectrum extends over a field range that is not covered by the used electromagnet. Internal stoichiometries follow directly from the weighing factors used in the simulation.

An example is given in Figure 3.5: the spectrum of oxidised bovine cytochrome oxidase (EXP) is simulated (SIM) with four different spectral components, Cu_A, cyt *a*, cyt a_3, dirtyFe, with relative stoichiometries 1:1:0.04:0.0005. These ratios make biochemical sense: Cu_A is the mixed-valence [Cu(II)-Cu(I)] cluster with $S = 1/2$, cyt *a* is the low-spin heme *a* with $S = 1/2$, and they occur once per cytochrome oxidase molecule. The other two metal centres, Cu_B ($S = 1/2$) and cytochrome a_3 ($S = 5/2$) are antiferromagnetically coupled to a cluster spin $S = 2$, which is not detectable under the conditions used. A small fraction of the Cu–Fe cluster is uncoupled (presumably by reduction of Cu_B to Cu(I)), and we observe a high-spin heme a_3 spectrum of low intensity.

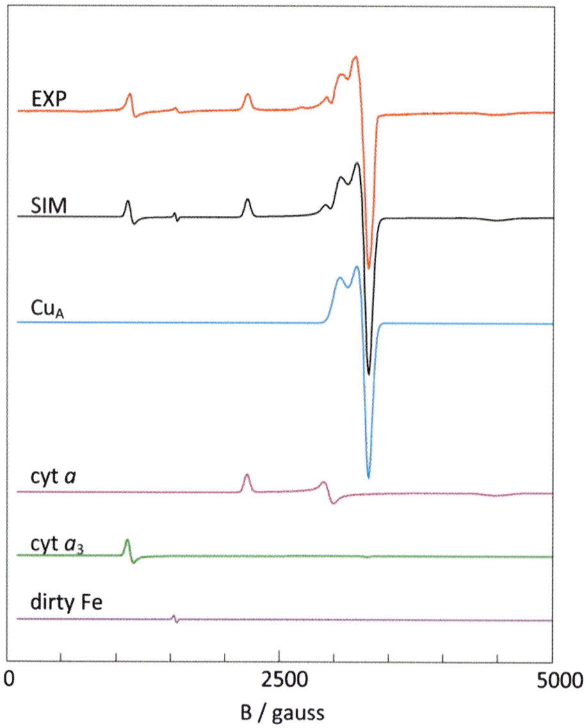

FIGURE 3.5 Spin counting in a complex EPR spectrum. The experimental X-band ($\nu = 9.25$ GHz) spectrum of oxidised beef heart cytochrome *c* oxidase is simulated as a sum of four signals with intensity ratios 1:1:0.04:0.0005. The relative intensities of the simulated spectral components show that the dinuclear mixed-valence copper-A centre and the cytochrome *a* are stoichiometric, that the EPR-silent dimer of copper-B and cytochrome a_3 is slightly uncoupled affording a small amount of high-spin heme Fe(III), and that the preparation is very slightly contaminated with 'dirty' iron.

The fourth component is a contamination with aspecifically bound high-spin Fe(III) of very low intensity. Note that this analysis could not have been based on inspection of the amplitudes of the four spectral components.

HYPERFINE INTERACTIONS

Many nuclei have a nuclear spin I $\neq$ 0 and they are experienced by unpaired electrons in EPR as extra magnets, affording $S \leftrightarrow I$ or hyperfine interactions. Nuclei studied with NMR are mainly limited to those with $I = 1/2$, with two energy states in an external field labelled $m_I = 1/2$ and $m_I = -1/2$, because the complexity of NMR spectroscopy with $I \geq 1/2$ rapidly increases. This does not usually hold for EPR because the $S \leftrightarrow I$ interaction is a perturbation to the electronic Zeeman interaction ($S \leftrightarrow I << S \leftrightarrow B$) and, therefore, allowed transitions are limited by the selection rule $\Delta m_I = 0$. For example, for the copper ion Cu^{2+}, with $S = 1/2$ and $I = 3/2$, one finds only four EPR transitions: $|m_S(\mathrm{i}); m_I(\mathrm{k})> \leftrightarrow |m_S(\mathrm{j}); m_I(\mathrm{k})>$, i.e. the copper EPR line is split by a magnitude A (the hyperfine splitting expressed in gauss) into four lines centred around the field corresponding to the g value. So for an $S = 1/2$ system with a single nuclear spin the number of EPR transitions is equal to 2I + 1. This description holds under the condition that the hyperfine interaction is indeed significantly less than the Zeeman interaction, which is a reasonable assumption at X-band and higher frequencies. At low frequencies (circa 1 GHz for copper) the assumption no longer holds, and the spectra become more complex and their analysis more complicated.

Just like the Zeeman interaction, the hyperfine interaction is also generally anisotropic, i.e. different for different molecular orientations in the external magnetic field. The magnitude of the splitting in terms of the polar angles θ and φ of Figure 3.3, used to define the direction cosines of Eqn 3.4, is

$$A(l_i g_i) = g^{-2}\sqrt{l_x^2 g_x^4 A_x^2 + l_y^2 g_y^4 A_y^2 + l_z^2 g_z^4 A_z^2} \tag{3.9}$$

from which it can be seen that the three features in a powder pattern (cf. the black trace in Figure 3.1), namely the low field absorption-like peak around g_z, the intermediate field derivative-like line around g_y, and the high field negative absorption-like peak around g_x, will each be split into 2I + 1 features, but with different splitting magnitudes A_i (together also called the A-tensor or the hyperfine tensor) for the different g values. Spectra with hyperfine structure can now be analysed by simulation using Eqn 3.3 for the anisotropic g value and Eqn 3.9 for the anisotropic hyperfine splitting with the resonance field defined as

$$B_{\mathrm{res}} = \frac{h\nu}{g\beta} - \sum_{m_I} A m_I \tag{3.10}$$

A simple, but practically important, example is given in Figure 3.6. Spin traps are diamagnetic compounds that readily react with unstable radicals to from stable paramagnetic adducts. The spin trap DMPO (5,5-dimethyl-pyrroline N-oxide) reacted with superoxide radical, $O_2^{\bullet -}$, affords the spectrum given in the red trace (exp). The unpaired electron of the superoxide delocalises over the DMPO molecule, and consequently has hyperfine interaction with a nitrogen (^{14}N with $I = 1$) and a hydrogen (^{1}H with $I = 1/2$) nucleus. Because the adduct is a small molecule in water, all anisotropy is averaged away, and one observes a single g value and single hyperfine splitting A values according to the resonance field expression

$$B_{\mathrm{res}} = \frac{h\nu}{g_{\mathrm{iso}}\beta} - A_{\mathrm{iso}}^N m_I^N - A_{\mathrm{iso}}^H m_I^H \tag{3.11}$$

which has been used to simulate the spectrum in the black trace (sim). To understand how such a spectrum comes about, the figure also gives simulations for the theoretical cases of interaction only with a single nitrogen (blue) or only with a single proton (green). The experimental spectrum can be thought of as a single line split into three by the nitrogen, with subsequent splitting of each of these three into two by the hydrogen. In general this should result in a total of six lines, however, since the splittings from the nitrogen and from the hydrogen by chance happen to be identical ($A^N = A^H = 15$ gauss), some lines overlap and the result is a four-line spectrum with 1:2:2:1 intensity pattern.

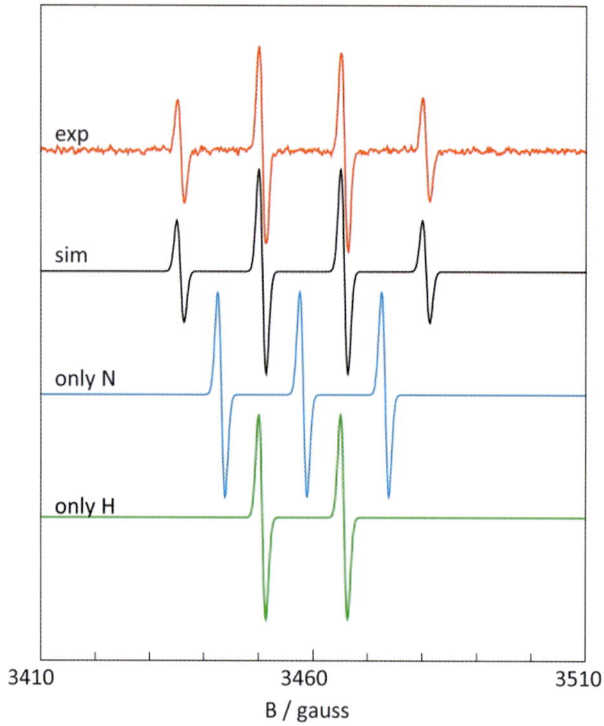

FIGURE 3.6 Analysis of isotropic hyperfine interaction. The spin trap DMPO has been mixed with superoxide and the room temperature aqueous solution EPR spectrum of the adduct has been simulated as a single peak split by hyperfine interaction with one nitrogen and one proton with $A_N \approx A_H$.

A more complicated example, now involving anisotropy, of the hydrated vanadyl ion ($VOSO_4$ dissolved in acidified water) is analysed in Figure 3.7. Vanadium proteins with the vanadium reduced to V(IV) exhibit similar spectra. The spin system is $S = 1/2$ and $I = 7/2$, so we expect a powder pattern whose features are split into $2 \times 7/2 + 1 = 8$ lines. The structure of the hydrated VO^{2+} ion (five H_2O ligands) is quasi octahedral with the molecular z-axis defined along the V=O bond. The x- and y-axis are equivalent, so the symmetry is axial, and this implies that $g_x = g_y$ and also $A_x = A_y$. The red trace in Figure 3.7 is the experimental X-band spectrum taken with the sample immersed in liquid nitrogen (i.e. $T = 77$ K). The black trace is an axial simulation. The blue trace is a simulation of what the spectrum would look like if there would not be any hyperfine interaction. The green trace is a simulation of the EPR absorption spectrum for a single molecular orientation (the 'parallel' orientation), namely, for the external magnetic field B along the molecular z-axis; similarly, the magenta trace is for a single molecular orientation (perpendicular), namely, for B along the molecular x-axis (or, for that matter, anywhere else in the xy plane). Note that the hyperfine splitting in the z-direction, A_z, affords positive peaks on the low-field side but negative peaks on the high-field side of the powder pattern. A good estimate of the value A_z is obtained by taking the difference in field position between the highest-field negative peak and the lowest-field positive peak divided by seven (i.e. $2 \times 7/2$). Also, the average field position affords a good estimate of the g_z value.

It is more difficult to estimate the g_x and A_x values from the 'messy' middle part of the spectrum. One of the reasons can be appreciated by inspection of the single-orientation spectrum of the magenta trace: the distance between the eight hyperfine peaks is not constant, and the width of the peaks is also not constant. The first phenomenon is called a 'second-order effect', which means that Eqn 3.9 (derived using perturbation theory) is not exactly correct for these large hyperfine interactions and requires a significant correction in X-band. The details

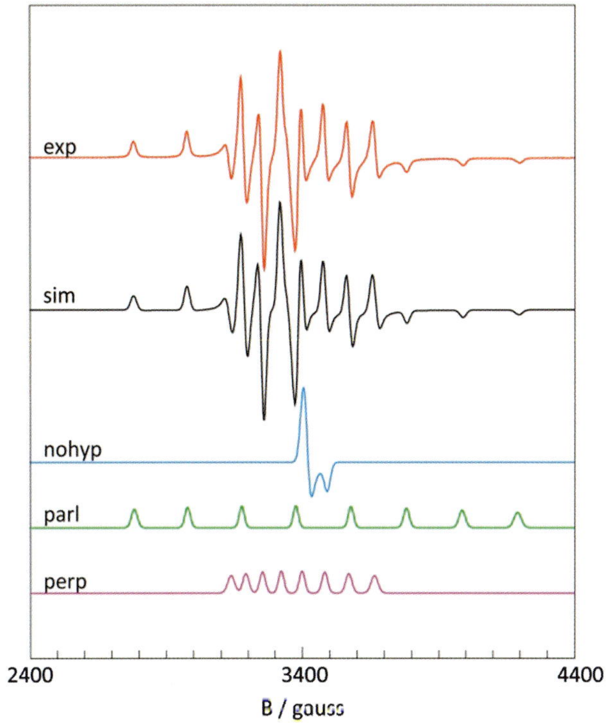

FIGURE 3.7 Analysis of anisotropic hyperfine interaction. The 9.42 GHz frozen solution spectrum ($T = 77$ K) of the $S = 1/2$ system V^{4+} in vanadyl sulfate in acidic water is split by hyperfine interaction with the ^{51}V nucleus (I = 7/2). The hydrated vanadyl ion $VO(H_2O)_5^{2+}$ has axial symmetry with $g_{\|} = 1.927$, $g_{\perp} = 1.972$, $A_{\|} = 201$ gauss, $A_{\perp} = 75$ gauss. The blue trace 'nohyp' is a simulation with $I = 0$; the green trace 'parl' simulates the absorption EPR for $\theta = 0$ (i.e. along the Z-axis), and the magenta trace 'perp' is for $\theta = 90°$ (i.e. in the XY plane).

are beyond the scope of this chapter but the required equations can be found in Hagen (2009). The variable line width is a reflection of conformational distribution of the $VO(H_2O)_5^{2+}$ structure, which leads to an m_I dependence of the width of the form

$$W(m_I) = W_0 + c_1 m_I + c_2 m_I^2 \tag{3.12}$$

The bottom line is that approximate g- and A values may be estimated directly from the spectrum, however, accurate analysis of spectra of the type shown in Figure 3.7 requires computer simulation affording high-quality fits (www.bt.tudelft.nl/biomolecularEPRspectroscopy).

Even without this sophisticated analysis inspection of hyperfine structure can be very useful, because simple 'line counting' provides direct information on the chemical elements involved in the spin system. For example, in the spectrum of Figure 3.7 with two A_z lines on the low-field side resolved and three A_z lines on the high-field side, it is straightforwardly established that there are eight lines and, therefore, that $I = 7/2$. In a biochemical setting this would define the spectrum to be either from a low-spin Co(II) complex or from a V(IV) complex. Since crystal field theory dictates that Co(II) has $g > g_e$ and V(IV) has $g < g_e$ ($g_e = 2.0023$) the distinction is readily made. There are no other splittings in addition to those due to the vanadium nucleus, which is consistent with all ligands being oxygen (^{16}O has $I = 0$). The protons (^1H has $I = 1/2$) of the water ligands are apparently too far away from the unpaired electron to afford resolved splittings, but they could be studied with special techniques, e.g. ENDOR or ESEEM.

The following metal isotopes are relevant in the frame of hyperfine structure in biological EPR (with percentage of natural abundance in parentheses): ^{51}V, I = 7/2 (99.8%); ^{55}Mn, $I = 5/2$ (100%); ^{57}Fe, $I = 1/2$ (2.1%); ^{59}Co, $I = 7/2$ (100%); ^{61}Ni, $I = 3/2$ (1.1%); ^{63,65}Cu, $I = 3/2$ (100%); ^{95,97}Mo, $I = 5/2$ (25.5%), ^{183}W, $I = 1/2$ (14.3%). Note that when the percentage is less than 100, the remainder is made up of isotopes with $I = 0$, and the EPR spectrum is the sum of hyperfine split and unsplit spectra. For example, natural molybdenum consists of the isotopes with mass number 92, 94, 95, 96, 97, 98, 100, and the EPR spectrum is 25.5% split into six lines and 74.5% unsplit. Since the splitting into six reduces the amplitude by a factor of six, the amplitudes of the split spectrum and the unsplit spectrum relate as 5.4: 94.6.

Biologically relevant ligand isotopes are: ^{1}H, $I = 1/2$ (99.99%); ^{2}H, $I = 1$ (0.01%); ^{13}C, $I = 1/2$ (1.1%); ^{14}N, $I = 1$ (99.6%); ^{15}N, $I = 1/2$ (0.4%); ^{17}O, $I = 5/2$ (0.04%); ^{19}F, $I = 1/2$ (100%); ^{31}P, $I = 1/2$ (100%); ^{33}S, $I = 3/2$ (0.8%); ^{35,37}Cl, $I = 3/2$ (100%); ^{75}As, $I = 3/2$ (100%); ^{77}Se, $I = 1/2$ (7.6%), ^{79,81}Br, $I = 3/2$ (100%), ^{127}I, $I = 5/2$ (100%).

Sometimes the symmetry of (biological) complexes is so low that it is not clear how to define a cartesian axis system, with related direction cosines (cf. Eqn 3.4), in terms of the molecular structure. Analysis of the EPR may then reveal that the axis system for the g values (defined by Eqn 3.3) differs from the axis system for the hyperfine A values (defined by Eqn 3.9). Mathematically this means that Eqn 3.9 has to be rotated in 3D space with respect to Eqn 3.3, and the EPR analysis becomes very involved. In practice this so-called 'axes non-colinearity' from low molecular symmetry can be qualitatively recognised in the EPR spectrum as small extra peaks and shoulders in between the hyperfine lines (Hagen, 2009).

HIGH-SPIN SYSTEMS

High-spin systems are intrinsically more complex than $S = 1/2$ systems because they have more than two (namely: 2S + 1) magnetic sublevels, and this has two important consequences: (i) the number of possible EPR transitions is greater than one, and (ii) the observed spectral features define 'effective' g values, i.e. the real g values from the Zeeman interaction are dramatically shifted by the $S \leftrightarrow S$ interaction, or zero-field interaction (because it is always present, also without a magnetic field), between unpaired spins. We use the example of $S = 5/2$ (e.g. high-spin Fe^{3+}) to illustrate these matters.

There are six sublevels labelled $m_S = +5/2, +3/2, +1/2, -1/2, -3/2, -5/2$, which could in principle afford up to 15 EPR transitions. In practice, however, this number is usually seriously restricted by the selection rule $\Delta m_S = 1$ and by the fact that some zero-field interlevel splittings are greater than the microwave quantum $h\nu$. The $\Delta m_S = 1$ selection rule is a quantum mechanical shorthand notation to state that some of the possible transitions have low probability, and are practically undetectable. In many high-spin biological systems and model compounds at X-band frequencies the zero-field $S \leftrightarrow S$ interaction is much stronger than the Zeeman $S \leftrightarrow B$ interaction, and this results in a grouping of sublevels in so-called Kramers pairs (or doublets) with $m_S = \pm n/2$ separated by an energy $\Delta > h\nu$. For $S = 5/2$ we find the pairs $m_S = \pm 1/2$, $m_S = \pm 3/2$, and $m_S = \pm 5/2$, which in hemoproteins (e.g. ferrimyoglobin) are separated by $\Delta_1 = 2D$ (see below) and $\Delta_2 = 4D$ with typically $D \approx 10\ \text{cm}^{-1}$, i.e. energy differences much greater than the X-band quantum of circa 0.3 cm^{-1}, cf. the left-hand panel of Figure 3.8.

Together with the $\Delta m_S = 1$ selection rule only a single transition is possible, namely the intra-doublet transition within the $m_S = \pm 1/2$ pair. This would make the system very similar to a low-spin $S = 1/2$ one; indeed, it is therefore called an 'effective' spin 1/2 system. There are two important differences with real $S = 1/2$ systems. First, the real g values for a system with a half-filled outer electron shell (Fe^{3+} is d^5) are predicted to be very close to g_e, however, for this effective $S = 1/2$ system we find two of the effective g values close to $g = 6$. Secondly, although the $m_S = \pm 3/2$ and $m_S = \pm 5/2$ doublets may not contribute to the EPR spectrum, they *are* populated by molecules, and so the temperature dependence of the EPR intensity is more complex than that of a real $S = 1/2$ system. At high temperature only one-third of all molecules is in the $m_S = \pm 1/2$ doublet, so when counting spins we have to multiply the double integral value by three. At very low temperature, e.g. 4.2 K, only the $m_S = \pm 1/2$

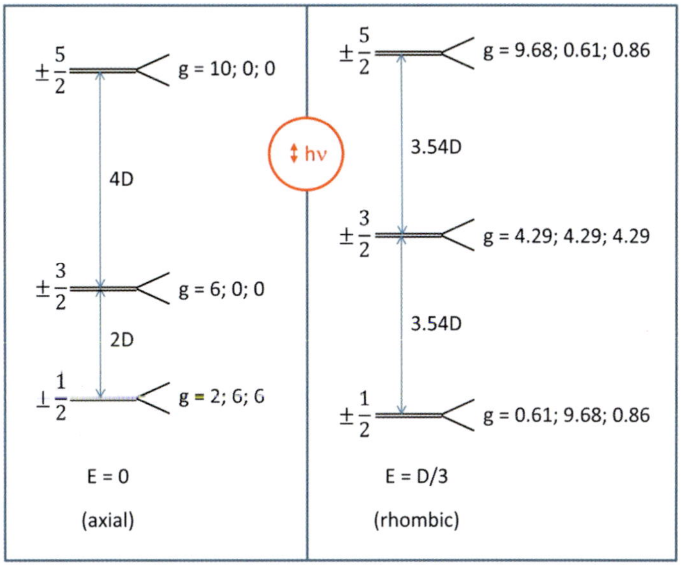

FIGURE 3.8 $S = 5/2$ spin manifold for axial symmetry ($E = 0$) and for maximal rhombicity ($E/D = 1/3$). Effective g values are given for all three intradoublet spectra assuming a real isotropic $g = 2.00$.

doublet may be populated, and so no correction is required. At intermediate temperatures a correction has to be made according to the Boltzmann distribution over the three doublets (Hagen, 2006).

In 3D space the anisotropic Zeeman interaction $S \leftrightarrow B$ and the anisotropic hyperfine interaction $S \leftrightarrow I$ are characterised by a set of three parameters: g_x, g_y, g_z (the g tensor), and A_x, A_y, A_z (the A tensor). Similarly, the anisotropic zero-field interaction $S \leftrightarrow S$ between electrons is characterised by the set of parameters D_x, D_y, D_z (the D tensor). However, in this case it can be shown that the three parameters are not independent: $D_x^2 + D_y^2 + D_z^2 = 0$, and so the set can be reduced to two independent parameters $D \equiv 3D_z/2$ and $E \equiv (D_x - D_y)/2$. In case of axial symmetry we have $g_x = g_y$ and $A_x = A_y$, but $D_x = D_y$ implies that $E = 0$, so an axial zero-field interaction is described by a single parameter D. In X-band the hyperfine interaction is a perturbation to the Zeeman interaction ($S \leftrightarrow I << S \leftrightarrow B$) leading to characteristic splittings of the EPR spectrum as described in the previous Hyperfine Interactions section. Similarly, *if* the zero-field interaction is a perturbation to the Zeeman interaction ($S \leftrightarrow S << S \leftrightarrow B$), it leads to splittings of the EPR spectrum. For example the $S = 1$ EPR of organic bi-radicals is split by $D/g_e\beta$ and $(D \pm 3E)/g_e\beta$ gauss (cf. Wasserman et al., 1964). However, for metal complexes we noted, above, that in X-band the zero-field interaction is usually dominant ($S \leftrightarrow S >> S \leftrightarrow B$) and, therefore, the spectrum of a high-spin transition ion complex does not show easily readable spectral splittings from $S \leftrightarrow S$ interactions. And while we have generally valid simple analytical expressions for the resonance field as a function of g- and A values (cf. Eqs 3.3 and 3.9) it is not possible to deduce such an expression for the effect of D and E on B_{res}. In fact, in most cases we simply cannot determine D from analysis of an X-band EPR spectrum, although we can sometimes estimate D from fitting the temperature dependence of the EPR intensity (i.e. from spectra taken at several temperatures) to a Boltzmann distribution over the sublevels (e.g. Hagen, 2006). To understand high-spin EPR spectra, generally requires making simulations based on energy matrix diagonalisation techniques. This complex subject is beyond the scope of this chapter; it is addressed in detail, e.g. in Hagen (2009).

Remarkably, however, it turns out that solutions in terms of effective g values (and thus: B_{res} values) from such an approach for all half-integer (i.e. $S = n/2$) systems form a simple set dominated by a single parameter: $\eta = E/D$. This result is illustrated for Fe(III) in the form of a 'rhombogram' in Figure 3.9: the rhombicity E/D is on the x-axis and it runs from zero to its theoretical maximum 1/3. On the y-axis are effective g values, and the graph defines

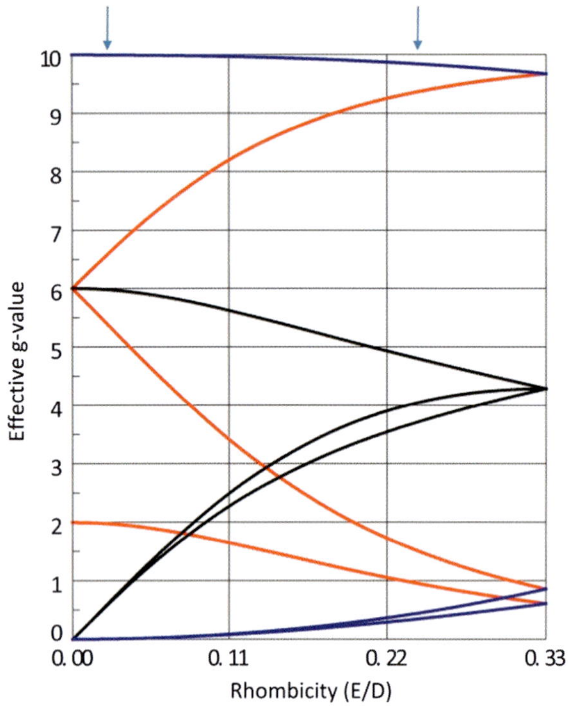

FIGURE 3.9 Rhombogram for $S = 5/2$. Effective g values as a function of rhombicity are given for all three intradoublet spectra: the red traces are for the $\pm 1/2$ doublet, the black traces are for the $\pm 3/2$ doublet, and the blue traces are for the $\pm 5/2$ doublet. The real g tensor is assumed to be isotropic: $g = 2.00$. The arrows indicate rhombicities for the real spectra in Figure 3.10.

g_x^{eff}, g_y^{eff}, and g_z^{eff} for the intradoublet spectra from the three doublets $m_S = \pm 1/2$ (red), $m_S = \pm 3/2$ (black), and $m_S = \pm 5/2$ (blue). So a given $S = 5/2$ compound will have its unique rhombicity value $\eta = E/D$, related to the molecular symmetry at the paramagnetic centre, which defines, according to Figure 3.9, three sets of g values for three overlapping spectra. The graph also gives an indication of the relative intensity of these three spectra: the closer the three g values to each other the narrower the spectrum and the higher its amplitude will be. Note, however, that spectra are taken on a linear magnetic field scale, i.e. on a reciprocal g value scale. So moving g values towards zero will rapidly increase the field range of the spectrum and thus rapidly decrease its amplitude. In fact, $g_{eff} = 0$ means a resonance at infinite field and thus an infinitely wide scan range and an infinitely small amplitude. In other words: the transition is forbidden. This is what we noted, above, to be the case for ferrimyoglobin, for which the energy level of the left-hand panel of Figure 3.8 holds: two of the three possible spectra have $g_{eff} = 0$, and, therefore, are totally unobservable. On the other hand, the Fe(III) in the protein rubredoxin, coordinated by four cysteinato sulfurs in a strongly deformed tetrahedron, has $\eta = E/D \approx 1/3$, for which the right-hand panel of Figure 3.8 holds: all three spectra are allowed, but the one from the middle doublet will have the highest amplitude by far, because it incidentally has an isotropic g_{eff} value ($g_x^{eff} = g_y^{eff} = g_z^{eff} = 4.29$).

Figure 3.10 gives examples of two experimental spectra for cases of intermediate rhombicity: the slightly rhombic ($\eta = 0.023$) heme iron site in a catalase and the rather rhombic ($\eta = 0.238$) iron site in a superoxide dismutase. Their effective g values are indicated by arrows on the top of the rhombogram of Figure 3.9. Hagen (2009) gives rhombograms and examples of spectra for $S = 3/2 - 9/2$ and www.bt.tudelft.nl/biomolecularEPRspectroscopy links to a simulation program to generate these type of spectra as a function of rhombicity.

All metalloproteins and models are subject to conformational distributions, which in turn lead to distributions in EPR parameters, i.e. g strain. Where a distribution in g values leads to broadening and skewing of the main

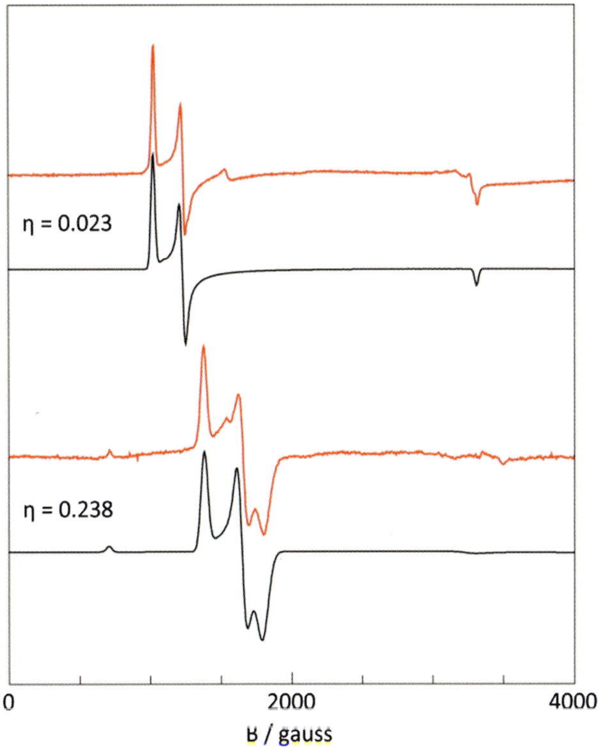

FIGURE 3.10 Examples of $S = 5/2$ spectra (red traces) and their simulations (black traces) based on effective g values from rhombograms with rhombicity $\eta = E/D$. The top spectrum ($\nu = 9.18$ GHz; $T = 15$ K) is from high-spin heme Fe(III) in PAC (*Penicillium simplicissimum* atypical catalase), and the bottom spectrum ($\nu = 9.30$ GHz; $T = 90$ K) is from high-spin Fe(III) in *Escherichia coli* iron superoxide dismutase.

features in the powder pattern, and a distribution in A values leads to nuclear orientation-dependent line width (cf. Eqn 3.10), a distribution in D values leads to broadening of all features in high-spin spectra, however, with the broadening increasing over the field scan. In other words, the low-field features of spectra from $S = n/2$ systems with large zero-field splittings always exhibit the best resolution (Hagen, 2007).

Thus far, we have limited the discussion to half-integer spin system. Integer spin systems with $S = 1, 2, 3$, etc., form a special class from the viewpoint of EPR spectroscopy. Many of these systems are 'EPR silent', which means that no signal is found in X-band EPR. This situation is very different from that of the half-integer spins, which should always afford at least one signal at any frequency. The reason is that the equivalence of Kramers pair doublets in half-integer spins is *non*-Kramers pair doublets in integer spin. According to QM Kramers pairs are always degenerate (i.e. they have the same energy) in zero field; non-Kramers pairs do not have this restriction and, in fact, they are always split in zero field (with the exception of axial $S = 1$ systems) (Hagen, 2009). If this splitting happens to be greater than the microwave quantum $h\nu$, then no transitions are possible at frequency ν, and one has to turn to a spectrometer with (much) higher frequency to detect EPR. When the zero-field intradoublet splitting is small ($\Delta < h\nu$) then EPR is possible, but the transition probability is low for normal spectrometers in which the external field B is perpendicular to the magnetic component B' of the microwave ($B \perp B'$). One frequently has to use a special cavity, the parallel-mode resonator, in which $B \parallel B'$, to get reasonably sharp spectra of reasonable intensity. See Hagen (2006, 2009) for further theoretical and experimental details. A practical summary of integer-spin X-band EPR of metalloproteins and models is as follows. $S = 2$ is the most commonly observed system. $S = 1$ systems are rarely detectable. $S = 2$ systems are either EPR silent or give a single, broad line at low field with $g_{eff} \geq 8$ (e.g. the $[3Fe\text{-}4S]^0$ cluster). $S > 2$ systems in biology are restricted to metal clusters (e.g. the P-cluster in nitrogenase enzymes), and they may exhibit a single, relatively sharp line in parallel-mode EPR with $g_{eff} \approx 4S$ (Hagen, 2009).

APPLICATIONS OVERVIEW

EPR is a 'something for everyone' spectroscopy: practical and useful EPR applications on biomolecules and models can range from very simple to very involved experiments and analyses. "Wow, this protein contains a metal cofactor 'X'!" could well be the verbal synopsis of a breakthrough result from a 5-minute, first-trial EPR measurement. Such an identification does not necessarily require any knowledge of EPR theory, as it can be based on fingerprint correlation, i.e. comparing the EPR spectrum (e.g. shape, peak positions, broadening with increasing temperature) with literature data on characterised systems. Rather more often, however, some qualitative or (semi-) quantitative understanding of EPR may support assignments. For example, "This signal is split into four lines, so it must be from a copper complex, and since the A_z-value is in the range of 30−100 gauss we must have a type-I, or 'blue' copper protein". Or, "The $g_z > g_\perp$ so EPR combined with ligand-field theory stipulates that this Cu^{2+} probably has an axially elongated octahedral coordination". In other words, the element identification usually also affords a qualitative conclusion about the coordination environment. And, of course observing a copper EPR signal from a protein also immediately defines the oxidation state of the metal to be 2+.

Input of slightly more effort can make the experiment quantitative, and thus considerably increase its biological relevance. Establishing non-saturating measurement conditions and then counting spins from the doubly-integrated EPR with respect to an external standard, provides a concentration of the metal, which, when compared to protein concentration, allows for important interpretational discriminations such as: 'The Cu/protein ratio is approximately unity, so this must be a copper protein and 'The Cu/protein ratio is circa 0.05, so this is either a heavily demetallated copper protein, or, more likely, it is a non-copper protein that in the course of the purification procedure became contaminated with extraneous copper'.

Observation and qualitative understanding of super-hyperfine splittings (the 'super-' means: from ligands, not from the central metal) can identify ligands such as nitrogens in metal coordination. For example, observation of a line that is split into five lines with an intensity ratio of 1:2:3:2:1, would be consistent with coordination by two equivalent ^{14}N's (i.e. with the same A value). Alternatively, a line that is split into nine lines of equal intensity, would be consistent with coordination by two non-equivalent N's (i.e. with significantly different A values). A chemically more sophisticated form of this type of experiment is to isotopically label an enzyme's substrate (or better: a non-convertible substrate analogue), and then, after reaction of substrate and enzyme, to look for hyperfine splittings, e.g. in the ENDOR of the metal in the catalytic centre of the enzyme. Identification of such splittings from, e.g. ^2H, ^{13}C, or ^{15}N, not only proves that the substrate directly binds to the metal (which is also shown by a change in the EPR upon incubation with substrate), but it may also reveal which part of the substrate is coordinated by the metal. By the way, observation of superhyperfine interaction in EPR is the most direct way a (bio)chemist has available to show that a coordination bond possesses covalent character: it proves that the unpaired electron of the metal must be delocalised and spend some of its time on the ligand.

Clustering of metal ions, especially in iron−sulfur clusters, is another phenomenon that is readily identified in EPR experiments. Since iron−sulfur clusters in proteins are relatively labile entities (they contain acid-labile sulfur, so the cluster disintegrates upon acidification), iron content is readily determined colourimetrically. Suppose a protein is found to contain circa four irons. If the EPR of the protein only shows a single spin system, then the straightforward conclusion is that all irons are electronically coupled by exchange interaction (more precisely: superexchange, because the coupling is not directly between irons, but rather over two chemical bonds via the sulfurs).

Interaction of electrons by exchange is significant only over a distance of one or two chemical bonds, but through-space interaction between electrons (i.e. dipolar interaction) is significant over longer distances. This particular form of $S \leftrightarrow S$ interaction gives small splittings and shoulders in X-band EPR for spin systems at mutual distances of the order of 5−15 Å. Its observation is a qualitative indicator for the fact that two centres 'see each other', i.e. are separated by some 10 Å, and therefore are located in the same protein complex. Quantitative interpretation in terms of accurate distance and mutual orientation is possible but requires involved analyses usually based on data taken at several microwave frequencies.

Finally, if the identification of a signal with a particular structure is unequivocal, then one can also forget about EPR theory and use the spectrometer as a black box. In these types of experiments the amplitude of a signal is usually measured as a function of some external parameter, e.g. redox potential (to determine reduction potentials), concentration (to determine binding affinities), reaction time (to determine rate constants), acidity (to determine pK's) etc. The experimental challenges are now not so much in the EPR spectroscopy, but rather in the setups to chemically prepare the EPR samples, e.g. anaerobicity, rapid mixing, rapid freezing, etc.

REFERENCES

Aasa, R., & Vänngård, T. (1975). EPR signal intensity and powder shapes: a reexamination. *J. Magn. Reson., 19*, 308–315.

Hagen, W. R. (2006). EPR spectroscopy as a probe of metal centers in biological systems. *Dalton Trans., 2006*, 4415–4434.

Hagen, W. R. (2007). Wide zero field interaction distributions in the high-spin EPR of metalloproteins. *Mol. Phys., 105*, 2031–2039.

Hagen, W. R. (2009). *Biomolecular EPR Spectroscopy*. Boca Raton, Fl: CRC Press Taylor & Francis Group. pp. 249.

Wasserman, E., Snyder, I. C., & Yager, W. A. (1964). ESR of the triplet states of randomly oriented molecules. *J. Chem. Phys., 41*, 1763–1772.

TEST QUESTIONS

[q1] (A) The g value of a radical measured at $\nu = 9$ GHz is $g = 2.006$; what is its g value at $\nu = 250$ GHz? (B) Why does an $S = 1/2$ EPR spectrum generally have three 'peaks'?

[q2] (A) Why do EPR spectrometers have a resonator/cavity? (B) Why does the microwave bridge of an EPR spectrometer have a reference arm?

[q3] (A) What is the EPR equivalent of the extinction coefficient in optical spectroscopy? (B) The detection limit for X-band EPR of a reduced [2Fe-2S] cluster is circa 5 μM. How many milligrams of a 11-kDa ferredoxin do you have to prepare minimally to measure a spectrum?

[q4] In a small coordination complex in water an $S = 1/2$ metal ion is coordinated by four nitrogens. (A) How many EPR lines do we observe when the hyperfine interaction with all ^{14}N's is of the same strength? (B) And how many lines do we get if each N has a very different A-value?

[q5] With reference to the rhombogram in Figure 3.9, what approximately would be the relative intensities of the three sub-spectra of an $S = 5/2$ system with rhombicity $E/D = 0.11$?

ANSWERS TO TEST QUESTIONS

[a1] (A) Real g value is frequency invariant: if ν goes up, then B_{res} goes up too, and their ratio is constant, therefore, also $g = h\nu/\beta B$ is constant; (B) The derivative of the EPR absorption spectrum emphasises the turning points of the powder pattern.

[a2] (A) Creating a standing microwave increases the spectrometer's sensitivity by $Q \approx 5000$; (B) Directing a fixed, small amount of power to the detection diode makes it possible to change the power to the sample without need to re-tune the spectrometer.

[a3] (A) The EPR extinction coefficient is unity, and the anisotropic EPR intensity is given by Eqn 3.5; (B) 1 mM is 11 mg/ml, so 5 μM in 0.2 ml is 0.011 mg.

[a4] (A) 9; (B) 81.

[a5] The anisotropy in the effective g tensor of the $\pm 1/2$ and the $\pm 3/2$ doublet are similar, and, therefore, the two sub-spectra will have comparable intensity. Anisotropy for the $\pm 5/2$ intradoublet transition is very extensive, so the intensity of this subspectrum will be extremely low. Use the software from www.bt.tudelft.nl/biomolecularEPRspectroscopy to try this out.

Chapter 4

Introduction to Biomolecular NMR and Metals

Ricardo O. Louro
Instituto de Tecnologia Química e Biológica, Universidade Nova de Lisboa, Oeiras, Portugal

Chapter Outline

INTRODUCTION

Like any other kind of spectroscopy, Nuclear Magnetic Resonance (NMR) spectroscopy uses the outcome of the interaction between electromagnetic radiation and matter to obtain information on the sample being analysed. The sample is placed in a strong and homogeneous magnetic field, and is subjected to electromagnetic radiation. Some nuclei will absorb radiation at a frequency determined by three factors: the chemical nature of the nucleus, the chemical environment surrounding the nucleus and the strength of the magnetic field. These three aspects can be manipulated experimentally giving rise to the myriads of applications of NMR. These applications, in addition to the biomolecular research that will be the focus of the present chapter, extend to studies in materials sciences and medical imaging, two examples of contemporary applications of great societal importance. It should however be realised that the present day applications are the product of continuous support for nearly a century of fundamental physical research.

Practical Approaches to Biological Inorganic Chemistry, 1st Edition. http://dx.doi.org/10.1016/B978-0-444-56351-4.00011-7.

The first suggestion that atomic nuclei could have magnetic properties was made by Pauli in the 1920s at a time when Bohr's atomic model was being developed. In the 1930s, Rabi (Nobel Prize in Physics in 1944) and collaborators observed absorption of energy by a beam of molecular hydrogen subject to a homogeneous magnetic field. Observation of the NMR phenomenon in 'ordinary mater' was made by the end of the Second World War by Purcel and Bloch (Nobel Prizes in Physics in 1951) using water and paraffin, respectively. Chemists began to take note of the NMR technique in the beginning of the 1950s following the observation of three signals in a sample of ethanol, therefore revealing the existence of chemical shifts. However, the low sensitivity of the technique remained a challenge for biomolecular applications. This was solved only in the late 1960s and early 1970s with the convergence of three independent developments: The increase in computing and memory storage power enabled the widespread transition to pulsed NMR techniques associated with Fourier transform. Additionally, the introduction of superconducting magnets lead to a dramatic increase in the field strength. The concomitant spectral sensitivity and dispersion supported the literal explosion in the development of applications of NMR spectroscopy. The award of the 1991 Nobel Prize in Chemistry to Richard Ernst, for his pioneering contributions to high-resolution NMR, highlights the importance of these developments for chemistry and biology. Great efforts were dedicated to applications that explored strategies to obtain structural information from biological molecules; using either distance-geometry constraints from paramagnetic probes or from nuclear Overhauser effect. These were recognised in 2002 by the award of the Nobel Prize in Chemistry to Kurt Wüthrich for pioneering the strategies to determine the structure of biological macromolecules by NMR, establishing this as *a bona fide* structural biology method. Also during the 1970s, Paul Lauterbur introduced the concept of projection reconstruction using field gradients to encode spatial information on the recorded data. Together with the speeding up of the data collection by the echo-planar method developed by Sir Peter Mansfield, these two methods are the bases for the Magnetic Resonance Imaging technique for which both received the Nobel Prize of Physiology and Medicine in 2003.

PROPERTIES OF THE MATTER RELEVANT TO NMR

NMR spectroscopy can be performed on nuclei that are sensitive to an external magnetic field, B_0, generated by the spectrometer. Both neutrons and protons have the quantum mechanical property of spin (I), which assumes values of + or −1/2. Just like in the case for electrons in orbitals, nuclear particles occupy defined energy levels. Within each of these levels, the most stable configuration for a pair of protons or a pair of neutrons is established when these spins are antiparallel. As a consequence, atomic nuclei with an even number of neutrons and protons possess zero or integer net spin. Nuclei with and odd number of neutrons and of protons have an integer spin, whereas nuclei with either an odd number of neutrons or an odd number of protons have a non-integer spin. Only nuclei with non-zero spin are sensitive to the external magnetic field.

Table 4.1 shows that some nuclei of biomolecular relevance that are NMR-active occur at low abundance in Nature. This often forces the need to enrich the sample using chemical or biochemical methods. This is now routinely performed to enrich samples in ^{13}C or ^{15}N for biomolecular NMR. Alternatively, isomorphous replacement can substitute one nucleus by another with similar chemical properties and more convenient NMR characteristics. This is sometimes done for metalloproteins to facilitate the characterisation of the active site.

The spin of subatomic particles like the neutron, the proton, and the electron is a quantum mechanical attribute intrinsic to these particles. However, some of its properties relevant for NMR can be described using analogies with classical electromagnetism. In the context of classical electromagnetism, spin is described as arising from the rotation of a charged particle. A rotating particle has mass and possesses an angular momentum **J**. This is a vectorial quantity pointing perpendicular to the direction of rotation according to the right-hand rule. This rule states that the thumb indicates the direction of the angular momentum, and the other fingers curl in the direction of rotation. A classical rotating particle with charge also has a magnetic moment $\boldsymbol{\mu}$, which is also a vectorial quantity. These two vectorial quantities are related to each other by a scalar quantity called the gyromagnetic ratio γ (Eqn 4.1 and Figure 4.1). The gyromagnetic ratio is a constant for each nucleus and depends on an adimensional number, called the nuclear g factor, on the charge of the nucleus and on its mass. As seen on Table 4.1 the

TABLE 4.1 Examples of Nuclei Relevant for Biomolecular NMR and their Spin Properties

Nucleus	Spin	Natural abundance (%)	Gyromagnetic ratio ($\times 10^6$) rad T^{-1} s^{-1}	Sensitivity (relative to proton)	Larmor frequency at 2.3488 T (Mhz)
^1H	½	99.98	267.52	1	100.00
^2H	1	1.5×10^{-2}	41.06	1.45×10^{-6}	15.35
^{13}C	½	1.108	67.28	1.76×10^{-4}	25.15
^{14}N	1	99.63	19.34	1.01×10^{-3}	7.23
^{15}N	½	0.37	−27.13	3.85×10^{-6}	10.14
^{23}Na	3/2	100	70.80	9.27×10^{-2}	26.47
^{31}P	½	100	108.39	6.65×10^{-2}	40.52

gyromagnetic ratio can assume positive or negative values, which indicate whether the angular momentum and the magnetic moment are parallel or antiparallel, respectively.

$$\mu = \gamma \cdot \mathbf{J} \quad (4.1)$$

When nuclei with non-zero spin are placed in an external magnetic field, the magnetic moment of the nuclei will not align with the direction of the magnetic field but will precess about the direction of the field. This precessional motion occurs with a frequency ω, which is proportional to the intensity of the effective magnetic field and to the gyromagnetic ratio of the nucleus.

$$\omega = \gamma \mathbf{B} \left(\text{rad s}^{-1}\right) \quad (4.2)$$

Therefore, in the presence of an external magnetic field, nuclei with different gyromagnetic ratios can be discriminated because their precession frequencies are different. Equation (4.2) is at the basis of the common designation of NMR magnets by the precession frequency of protons by showing that this property is directly related to the field strength.

ENERGY OF NMR TRANSITIONS

The magnitude of the magnetic moment μ depends on the angular momentum, which is defined with respect to the nuclear spin quantum number I. Because this is a quantum mechanical property, a discrete number of states is possible and is given by $(2I+1)$. Quantum mechanical selection rules determine that only integer transitions between the discrete states are allowed between the $-I$ and the $+I$ states. For example, for $I = 1/2$, only one

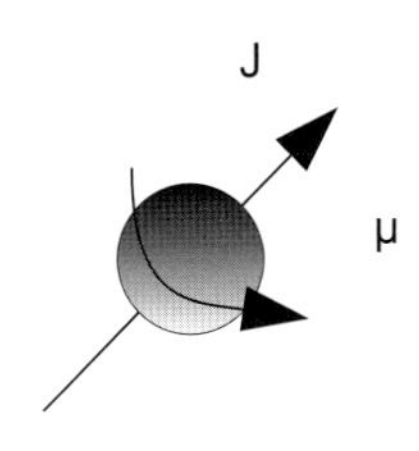

FIGURE 4.1 A classical view of spins. For particles of positive gyromagnetic ratio, the angular momentum and the magnetic moment point in the same direction.

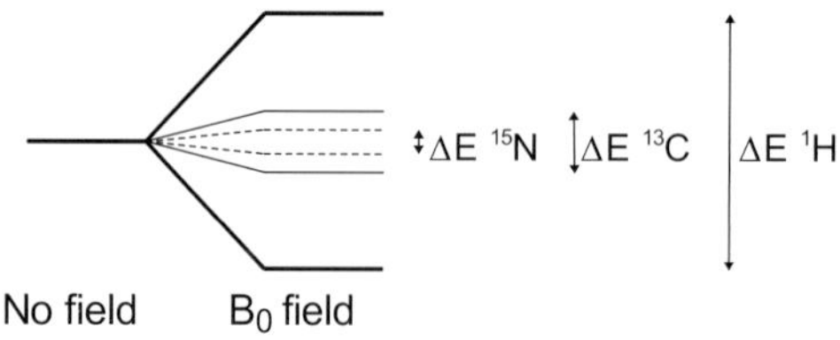

FIGURE 4.2 Nuclear Zeeman splitting for commonly used nuclei in biomolecular NMR given to scale to show in a pictorial way that the energy splitting of protons is much larger than for other nuclei.

transition exists from $I = -1/2$ to $I = +1/2$. These states are degenerate in the absence of a magnetic field, i.e. they have the same energy. It is only in the presence of a magnetic field that the degeneration is lifted, and the states assume different energies. This is called the Zeeman effect (Figure 4.2).

The potential energy of a magnetic dipole in a magnetic field is given by:

$$E = -\mu \mathbf{B} \tag{4.3}$$

Because μ is dependent on the angular momentum, which is quantised according to the nuclear spin quantum number (m), Eqn (4.3) can be rewritten as:

$$E = -g_I \mu_N \mathrm{B}\, m = \gamma\, h/2\pi \mathbf{B}\, m \tag{4.3a}$$

Only transitions between states that lead to unitary changes in the spin quantum number are allowed. The energy associated with such transitions is:

$$\Delta E = \gamma \frac{h}{2\pi} B \tag{4.4}$$

And according to Planck's equation:

$$hv = \gamma \frac{h}{2\pi} B \tag{4.5}$$

Eliminating h from both sides of the equation and converting the frequency from Hz to radians/s (multiplying by 2π), we obtain again Eqn (4.2), but in this case, it was derived from quantum mechanical considerations:

$$\omega = \gamma \mathbf{B}$$

This is called the Larmor equation. This is the fundamental equation in NMR because it relates the quantum mechanical and the classical description of the behaviour of magnetic particles in the presence of an external magnetic field. The Larmor equation defines both the frequency of the precession of the magnetic moment about the direction of the external field and the energy splitting associated with the transitions between quantised nuclear magnetic states.

MACROSCOPIC MAGNETISATION

Up to now we have dealt with an idealised isolated spin in order to obtain the fundamental expressions that define its behaviour in a magnetic field. However, a real life sample contains a large number of spins (of the order of 10^{17} for a micromole of material). When placed in a magnetic field, these spins will be found precessing predominantly in the Zeeman state of lower energy (Figure 4.3). The difference in the number of spins in the various Zeeman states depends on I and on the energy separation between the states. For two states with ΔE energy difference, the Boltzman equation defines the ratio between the respective populations as

$$N_\beta / N_\alpha = \mathrm{e}^{-\Delta E/\mathrm{kT}} \tag{4.6}$$

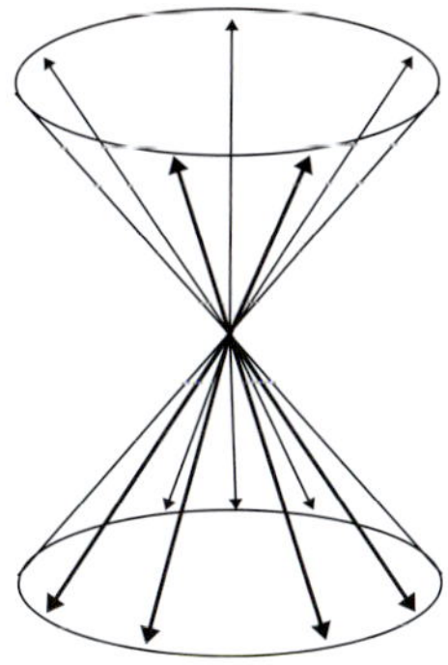

FIGURE 4.3 Representation of an ensemble of spin precessing in a external magnetic field showing the slight excess in the low energy state.

At contemporary operating field strength of NMR spectrometers, this difference is tiny, with a ratio of the order of 0.999999. This small number reveals two important aspects of NMR. First, the technique is inherently insensitive and therefore requires a considerable amount of material for good signal detection. Second, from Eqns (4.4) and (4.6) it is clear that the stronger the field the larger the population difference. This gives rise to a more intense signal, and therein lays one of the drivers for stronger fields (with some caveats to be seen later on).

When faced with a dilute sample and if a spectrometer with a more intense field is not available or cannot be used, the sensitivity can be improved by repeating the experiment several times and adding up the result. The signal to noise ratio will increase with the square root of the number of repetitions. These repetitions are called scans for historical reasons. The original continuous wave NMR machines would scan a range of field strengths to detect the signals. Nowadays, the field is fixed but the designation has survived the instrumental evolution.

The small excess of spins precessing in the state of low energy, and the fact that these spins are found randomly distributed throughout the surface of the precessional cone, generates a macroscopic magnetisation vector **M**, aligned with the magnetic field and with a strength given to a good approximation by Curie's Law:

$$M \approx \frac{\gamma^2 h^2 BI(I+1)}{3\mathrm{kT}} \tag{4.7}$$

This vectorial description of the NMR phenomenon in terms of a macroscopic magnetisation behaves in a classical fashion. Although it is applicable strictly to non-interacting spins, the pictorial representations that are derived from this description are often useful to interpret more complex situations. In this description, the NMR experiment involves manipulation of the orientation of the magnetisation vectors, and therefore, it is convenient to define an axes system. In this axes system, the B_0 field is aligned with the z direction, and the x and y directions are related as shown in Figure 4.4. This is called a right-hand axes system because the x and y axes are oriented in the same way as the fingers curl around the thumb in the right hand.

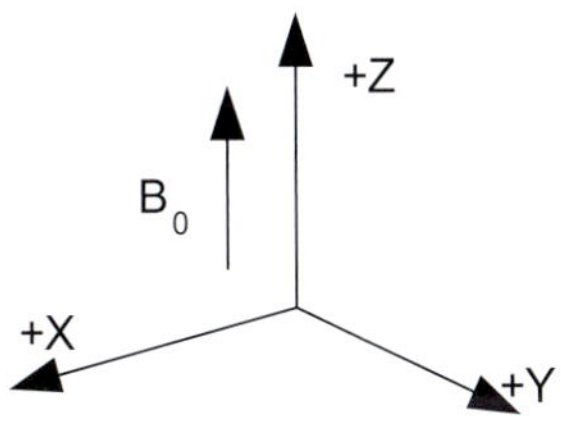

FIGURE 4.4 The right-hand axis system.

ACTING ON MAGNETISATION

Pulses

The introduction of pulsed NMR methods associated with Fast Fourier Transform for processing the data was a tipping point in the development of NMR, opening the possibility of numerous experimental developments.

In pulsed NMR, the macroscopic magnetisation is disturbed by applying a magnetic field $\mathbf{B_1}$ perpendicular to the static field generated by the magnet. This $\mathbf{B_1}$ field is applied as electromagnetic radiation during a short period, and from there came the designation pulse. This pulse is not monochromatic. It is a polychromatic emission of radio frequency, and to a good approximation can be considered to cover a bandwidth of frequencies given by the reciprocal of the pulse length. Therefore, the shorter the pulse the larger the bandwidth covered. When the $\mathbf{B_1}$ field matches the Larmor frequency, energy will be absorbed by the nuclei, and the difference between spins up and down will be reduced, reducing the macroscopic magnetisation in the z direction. However, magnetisation will not be eliminated. Instead, it will be flipped away from the direction of the static $\mathbf{B_0}$ field by an angle theta because the nuclear spins are no longer randomly distributed but will tend to point in the direction of the $\mathbf{B_1}$ field, i.e. they develop coherence. The flip angle achieved by the pulse depends on the nature of the nucleus, the strength of the $\mathbf{B_1}$ field, and on the duration of the pulse. It will become apparent below that the maximum signal is observed for a pulse with a flip angle of 90°, which is also typically called a $\pi/2$ pulse.

$$\theta = \gamma B_1 t \tag{4.8}$$

From Eqn (4.8), it is evident that multiple combinations of pulse power and length can be used to obtain the same flip angle.

Pulses also have phases. Most frequently, they are applied in the x, y, $-x$, or $-y$ direction. A 90° pulse about the x direction will leave the magnetisation in the y axis.

The Rotating Frame

The description of NMR often includes the implicit assumption that the axes system has been transformed from the static laboratory frame to the rotating frame. The description of the pulse represented in Figure 4.5 follows this frame. The rotating frame is a system of axes with the z axis aligned with the $\mathbf{B_0}$ field and x and y axes rotating. When the frequency of the rotating frame is equal to the Larmor frequency, a magnetisation vector that has been flipped by a $\mathbf{B_1}$ pulse will appear to be stationary, simplifying the description of the events. This is equivalent to the everyday situation of describing the trajectory of an object that is dropped as a vertical fall. Such a description ignores the fact that the Earth is rotating while the object is falling. The advantage of using the rotating frame, just like in the gravitational analogy, is that it simplifies the mathematical expressions used.

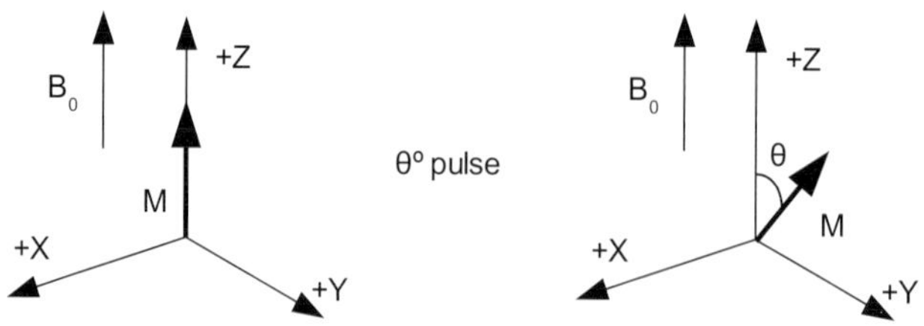

FIGURE 4.5 Effect of the application of a pulse about the x direction.

RELAXATION

After switching off the $\mathbf{B_1}$ field, the spins gradually lose the coherence, and the macroscopic magnetisation returns to the direction of the static $\mathbf{B_0}$ field. These two phenomena are called relaxation and are assumed to follow an exponential decay. This is the free induction decay or FID, and it is described by the Bloch equations:

$$M_x(t) = \left[M_x(0)\cos(\omega t) - M_y(0)\sin(\omega t)\right]e^{(\ t/T_2)} \quad (4.9a)$$

$$M_y(t) = \left[M_x(0)\sin(\omega t) + M_y(0)\cos(\omega t)\right]e^{(-t/T_2)} \quad (4.9b)$$

$$M_z(t) = M_{eq} + \left[M_z(0) - M_{eq}\right]e^{(-t/T_1)} \quad (4.9c)$$

Each of the equations defines how the macroscopic magnetisation evolves in the direction of each axis and combining them shows how the precessing macroscopic magnetisation returns to the direction of the static field (Figure 4.6).

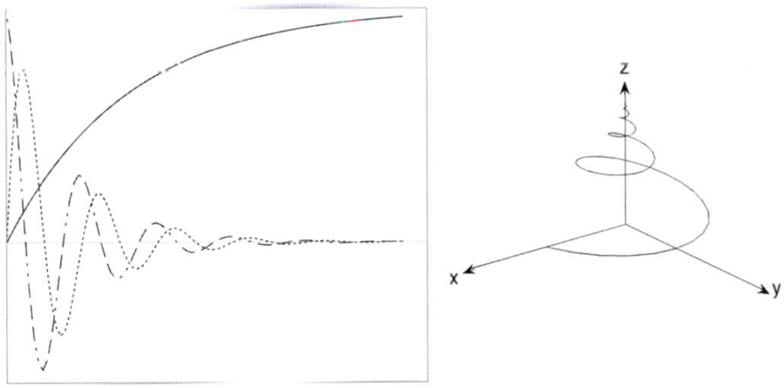

FIGURE 4.6 Time evolution of the macroscopic magnetisation during relaxation following a 90° pulse. M_z, continuous line; M_x, dotted line; M_y, dash–dot line. 3D plot of the trajectory followed by the magnetisation.

The Bloch equations show that the time it takes to recover the macroscopic magnetisation in the direction of the static field $\mathbf{B_0}$ depends on the Larmor frequency and on two new parameters called T_1 and T_2. Only M_x and M_y depend on the Larmor frequency, and therefore detection of NMR signals has to be made on the *xy* plane in order to be able to distinguish different nuclei in the sample. The maximum signal is obtained when the magnetisation is placed on the *xy* plane. This is what is achieved by a 90° pulse.

The T_1 and T_2 parameters introduced by the Bloch equations are time constants that describe the relaxation of the macroscopic magnetisation. T_1 is called longitudinal relaxation, because it affects solely M_z and defines how fast the macroscopic magnetisation returns to the direction of the static field. T_2 is called transverse magnetisation, because it affects M_x and M_y and defines how fast the magnetisation loses coherence after the $\mathbf{B_1}$ pulse has been switched off.

It is clear from Eqn (4.9) and Figure 4.7 that coherence can be lost without return of the magnetisation to the direction of the static field. Therefore, T_2 is always shorter or equal to T_1.

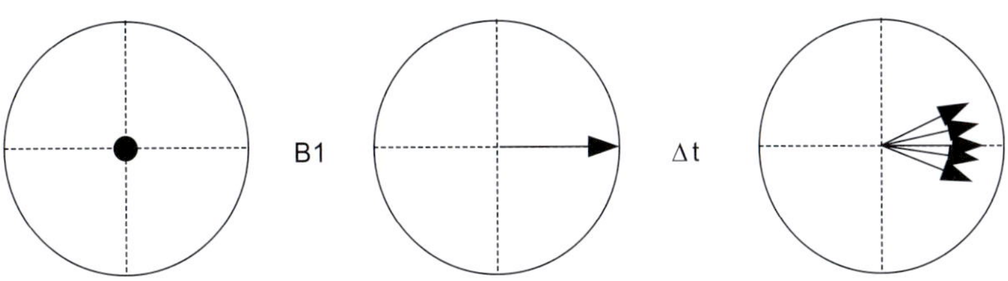

FIGURE 4.7 Representation of the magnetisation vectors looking down on the *xy* plane from the *z* axis. The rotating frame tracks the Larmor frequency, and therefore the macroscopic magnetisation is stationary. In equilibrium, the magnetisation is aligned with the *z* axis and pointing towards the observer (left panel). After the $\mathbf{B_1}$ pulse all vectors are aligned (middle panel). As time goes by (right panel), the projection on the *xy* plane becomes shorter due to T_1, and the vectors spread evenly about the Larmor frequency (loose coherence) due to T_2.

What are the Physical Mechanisms of Relaxation?

The difference in energy levels in an NMR experiment is very small. In these conditions, spontaneous emission has a low probability of occurring and is not an effective relaxation mechanism. Collisions are also not capable of promoting relaxation of nuclear dipoles, because nuclei are not naked but surrounded by electrons. Therefore, relaxation occurs because of local fluctuations on the effective magnetic field that affect the nuclei and stimulate them to switch between the energy levels. These can be dipolar interactions with other nuclei, paramagnetic interactions with unpaired electrons or quadrupolar interactions in the presence of nuclei with $I > ½$. For nuclei with $I = ½$, the dominant sources of relaxation are dipolar interactions modulated by molecular motion. However, the underlying physical mechanism is different in each case:

T_1: arises from energy transfer to the surrounding medium, and for this reason is also called spin-lattice relaxation. It is caused by rotations and translations of a frequency similar to the Larmor frequency so that spin transitions take place. This is the mechanism by which the nuclei dissipate the energy absorbed from the B_1 pulse, and it is an enthalpic process.

T_2: arises from energy transfer to other spins, and for this reason is called spin−spin relaxation. It is stimulated by molecular motions of frequencies similar to the Larmor frequency and also by molecular motions of very low frequency close to zero Hz. Because this is associated with the loss of coherence of the magnetisation, i.e. increased disorder, it is an entropic process.

As seen above, relaxation depends on molecular motions, therefore on changes with time. Changes of a parameter over time are described by a correlation function. For rotation of a rigid body under Brownian motion, this function is given by Eqn (4.10):

$$G(\tau) = G(0)e^{\frac{-\tau}{\tau_c}} \tag{4.10}$$

The rotation correlation time τ_c is the average time taken by a molecule to rotate by a radian. The correlation time is obviously a function of molecular size and shape, temperature and medium viscosity. It is evident that, all else being equal, the larger the molecule the slower the typical motions and consequently, the correlation time is longer. To put this in a more precise way, the spectral density function, which defines the distribution of frequencies of molecular motions, depends on τ_c according to:

$$J(\omega) = \frac{2\tau_c}{1+\omega^2\tau_c^2} \tag{4.11}$$

Figure 4.8 shows that as the τ_c decreases, i.e. the molecular size decreases, the range of motional frequencies sampled extends to higher values. The spectral density function shows that for shorter τ_c, slower motions are less probable.

Given that T_1 depends on motions of frequency similar to the Larmor frequency, it will be shorter when the molecular motions occur close to this frequency. On the other hand, because T_2 depends on motions with frequencies close to the Larmor frequency and motions with frequency close to zero Hz, T_2 continues to decrease as the correlation time increases, i.e. as the molecular size increases. This is seen in Figure 4.9.

Because T_2 is related with the way M_x and M_y lose coherence, i.e. with the increased spread of vectors in the *xy* plane, it is related to the signal linewidth.

$$\Delta\nu_{1/2} = \frac{1}{\pi T_2} \tag{4.12}$$

Therefore, as the molecular size increases the lines become broader. The T_2 estimated from the experimental linewidth includes the contributions from sample and field inhomogeneities.

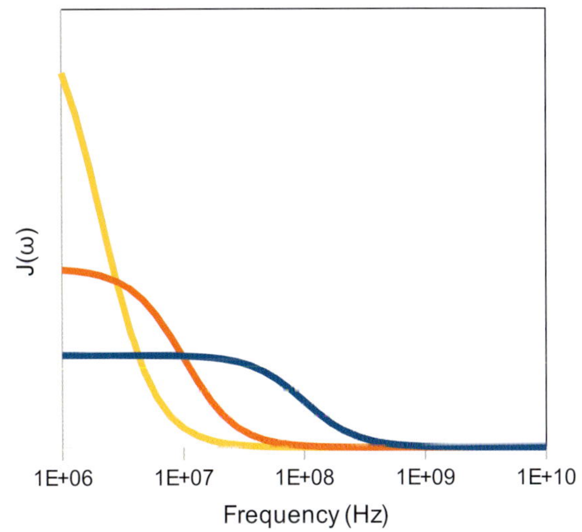

FIGURE 4.8 Normalized spectral density functions calculated for different values of τ_C. Green, red and blue lines were calculated for decreasing τ_C.

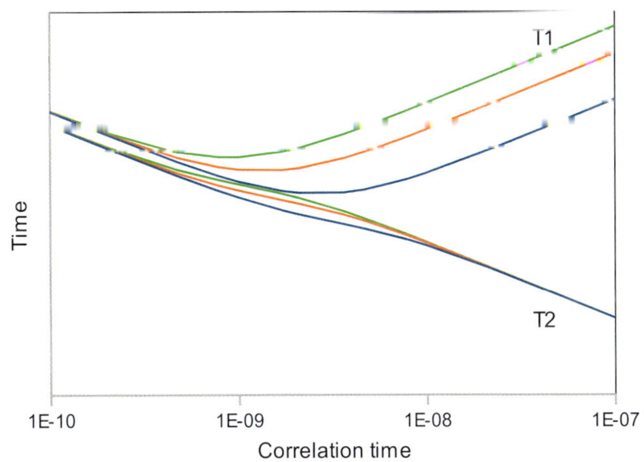

FIGURE 4.9 Dependence of T_1 and T_2 with correlation time. Blue, red and green are calculations made respectively at 300, 600 and 900 MHz field strength. Molecular size increases to the right.

AN NMR EXPERIMENT

All NMR experiments can be described as organized in two general parts:

- the preparation period, during which the spins are manipulated in order to generate the effect of interest.
- the detection period, where the time evolution of the transverse magnetisation is observed.

The time dependence of the magnetisation decay is measured at discrete intervals and subjected to a mathematical procedure called Fourier transform to give the familiar frequency domain spectrum.

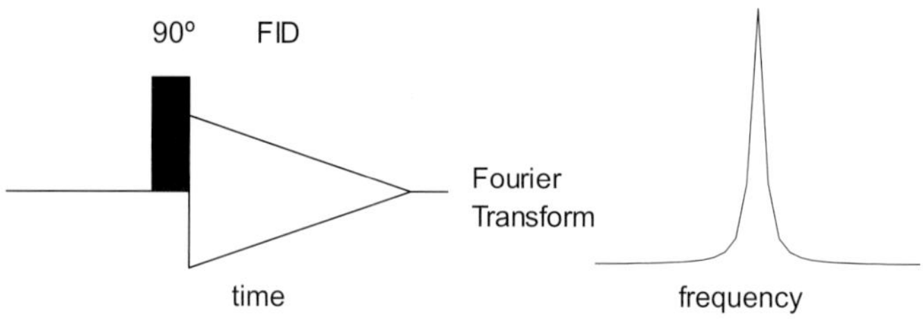

FIGURE 4.10 Diagram of a simple NMR experiment.

This very simple NMR experiment consists of a 90° pulse and detection of a single signal and gives two observables:

- The intensity of the signal. This is dependent on the concentration of the sample, the strength of the B_0 magnetic field and also on the value of the gyromagnetic ratios of the nuclei being excited and detected (which are the same for this particular experiment).
- The linewidth of the signal. As seen above, this is inversely proportional to the rate of transverse magnetisation decay (T_2).

Three consequences arise from these observables:

- The stronger the magnetic field, or the larger the gyromagnetic ratio of the nucleus, the stronger the signal.
- If the transverse relaxation is too fast, the signals will be too broad to be detected. Therefore, very large molecules will be difficult to study.
- if there was no transverse relaxation, the signal would be very sharp, and the linewidth would arise only from the Uncertainty Principle.

A spectrum such as that in Figure 4.10, with a single signal, is not of much use. Fortunately, nuclei of the same nature can have slightly different Larmor frequencies depending on the surrounding environment, therefore increasing the information content in a NMR experiment.

THE CHEMICAL SHIFT

Atomic nuclei are immersed in atomic or molecular electron clouds. Electrons, being charged particles that are not static, are sensitive to the presence of the magnetic field of the spectrometer and generate a very weak local magnetic field. Therefore, the nuclei sense an effective field that is different from the static field due to the influence of the surrounding electronic environment – the nuclear shielding (σ). Typically, the distribution of electrons around the nuclei is not spherical. The nuclear shielding tensor is anisotropic and has to be defined with respect to each of the directions of the system of axes (the three principal components). This is the origin of the phenomenon of chemical shift anisotropy. Fortunately, when the reorientation of the molecules is fast and isotropic, i.e. in solution, this effect averages out and a single value for the nuclear shielding can be used. The Larmor frequency can be expressed as:

$$\omega = -\gamma(1 - \sigma)B_0 \tag{4.13}$$

Throughout this chapter, isotropic conditions will be assumed, and each chemical environment gives rise to a characteristic shielding. Therefore, there are typical spectral regions where the signals of nuclei of a particular chemical nature are found.

As seen from the Larmor equation, the resonance frequency is dependent on the static field strength, and therefore the frequency difference between two signals increases with field strength. On the other hand, the chemical shift (δ) is defined as the ratio between the nuclear shielding and the static magnetic field and is reported in parts per million (ppm). The very useful outcome of defining the chemical shift this way is that signals measured in spectrometers of different field strength can be matched because they will have the same chemical shift (in ppm). Nonetheless, the spectral dispersion is larger in the experiment performed at the strongest field. In practice, chemical shifts are measured as the difference in ppm between the signal of interest and the signal of a reference substance added to the sample. Different compounds are used as reference depending on the solvent, applications and nuclei being observed. In water or deuterated water, the methyl signal of TMS (tetramethylsilane) is typically used as marker of 0 ppm for protons.

$$\delta = \frac{\nu_{\text{sample}} - \nu_{\text{ref}}}{\nu_{\text{ref}}} \cdot 10^6 \tag{4.14}$$

The chemical shift defined this way means that a positive δ indicates that the frequency of resonance of the nucleus under observation is higher than that of the reference. In the old CW spectrometers, where the spectrum was obtained by varying the intensity of the field, observation of such signal required that the field was lowered relative to that required to observe the reference. Therein lays the still often used designation of 'low field' for signals with positive chemical shift.

The phenomenon of chemical shift has a very important operational consequence in modern FT spectrometers of fixed B_0. The frequency of the B_1 pulse may need to be modified depending on the chemical shift of the nuclei to be observed.

Carrier Frequency

Differences in nuclear shielding place NMR signals in different spectral positions. Therefore, the choice of the carrier frequency is essential for a good performance of the NMR experiment. It was seen above that when the B_1 field matches the Larmor frequency, there is absorption of energy. By analogy with a relay race where the runners match their speed to transfer the baton, the carrier frequency should match as closely as possible the Larmor frequency of the nuclei being observed so that excitation occurs and is most effective.

Sampling Bandwidth and the Nyquist Theorem

In addition to correctly choosing the B_1 pulse power, length and carrier frequency to ensure that all interesting nuclei are excited, the existence of chemical shifts also forces the need to ensure that the measurement conditions are chosen so that all signals of interest are observed. NMR spectrometers record an FID that is Fourier transformed to obtain a spectrum. Because of this arrangement, the sampling bandwidth (usually referred to as spectral width) observed is dependent on the sampling frequency during the time of recording the FID. According to the Nyquist theorem, for any signal present in the FID, at least one point per period must be sampled in order to define correctly the frequency upon Fourier transformation.

$$F = 1/(2\Delta t) \tag{4.15}$$

In Eqn (4.14), F is the highest frequency that can be correctly defined when using the sampling interval Δt. Nonetheless, in some cases where the spectra are sparsely populated with signals, researchers may deliberately opt for sampling below the requirements of the Nyquist theorem. The advantage is that for the same experimental time and data size, better spectral resolution is obtained, and signals outside of the spectral window are folded into the spectrum. A signal with a frequency equal to $F + \text{f}$ will be aliased into a position that is equal to $F - \text{f}$. Therefore, these aliased signals can be identified by changing the sampling bandwidth, because unlike all other signals, they will change position.

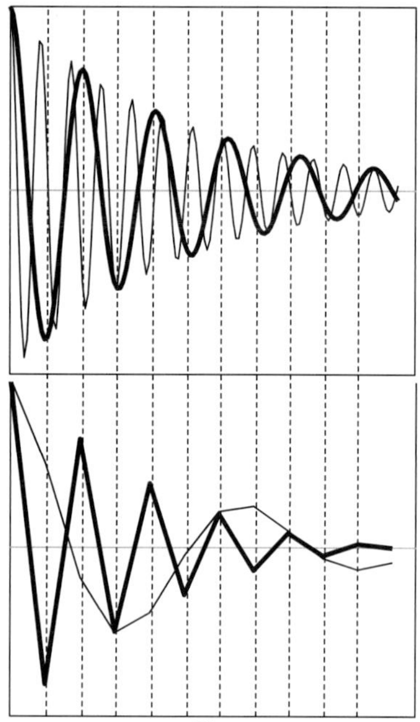

FIGURE 4.11 Effect of incorrectly chosen sampling frequency. The top panel shows the continuous decay of two signals of different frequency. The bottom panel shows the result of the discrete measurements at the sampling frequency indicated by the dotted lines. Clearly, for the signal with thick lines, this sampling is adequate, whereas for the thin line, the apparent recorded frequency is clearly wrong and of lower value that the real signal.

At current operating field strength, the Larmor frequency is of the order of hundreds of MHz, whereas in typical solution state, biomolecular applications the chemical shifts show differences at most of the order of kHz. Therefore, the importance of correctly selecting the carrier frequency becomes apparent. The most favourable condition occurs when the carrier frequency can be placed in the centre of the spectrum because of the following: (1) The excitation profile of the $\mathbf{B_1}$ pulse extends symmetrically from the carrier frequency covering a range inversely proportional to the pulse length. Therefore, at the centre of the spectrum we have the lowest demands on pulse duration and power. (2) Also, this is the condition where the minimum sampling is necessary, because the difference between the carrier frequency and any signal in the spectrum is, at most, half of the spectral width. The consequence is the minimisation of the experimental time. This is a consequence of the total experimental time, being the number of samplings made, multiplied by the sampling interval (Figure 4.12).

Until now, we have seen that NMR provides information on the number and relative abundance of chemically distinguishable nuclei in the sample. It also provides information on the transverse relaxation rate that is related with the molecular size. If there is need to perform multiple scans to improve the signal to noise of the spectrum, the T_1 must be determined in order to know how long one must wait until the magnetisation has returned to the z direction. T_1, unlike T_2, cannot be directly estimated from the spectrum, and its measurement requires that a specially designed experiment is performed.

Measuring T_1

Spin-lattice relaxation is measured using the inversion recovery experiment. The sample is subjected to a 180° pulse that inverts the magnetisation. The magnetisation will start relaxing back to the equilibrium position.

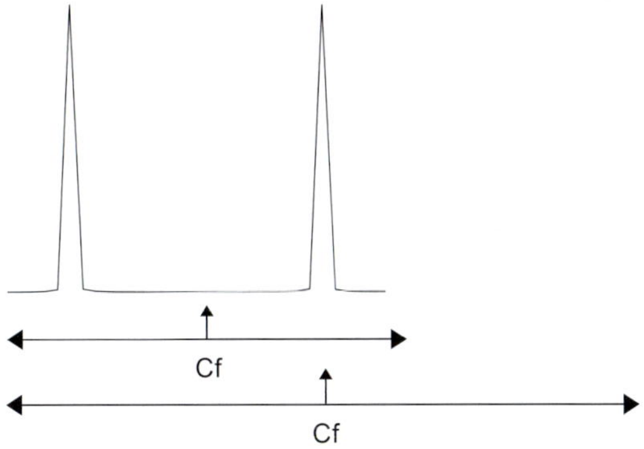

FIGURE 4.12 Effect of the choice of the carrier frequency (Cf) on the spectral window that has to be excited and measured. While the top option shows a good choice for the carrier frequency, the bottom option is clearly not the best choice for the two signals reported.

Waiting a variable length of time and applying a 90° pulse provides snapshots of the recovery process back to equilibrium, which is assumed to be first order. Therefore, the null point occurs at a time equal to $T_1 \cdot \ln 2$ (Figure 4.13).

With the above information, it would still be difficult to obtain the exquisitely detailed information that NMR can provide. However, magnetic nuclei in a molecule create a slight disturbance of the effective magnetic field felt by each other. This disturbance, called coupling, can be propagated through chemical bonds or through space, providing chemical or spatial information.

COUPLING: THE INTERACTION BETWEEN MAGNETIC NUCLEI

Two types of interactions between nuclei are known: dipolar couplings and scalar couplings.

Dipolar couplings depend on the relative orientation of the vector connecting the interacting nuclei with respect to the direction of the static magnetic field. Dipolar couplings are observed in solids and in liquid crystal samples. The interaction energy is given by

$$E = \frac{\mu_{Iz}\mu_{Sz}}{r^3}\left(3\cos^2\theta - 1\right) \qquad (4.16)$$

This equation applies to all kinds of dipoles, and the dipolar interaction disappears for the angle 54.74°, called the 'magic angle' for this reason. In amorphous solid samples, this effect is used to simplify the spectrum by rotating the sample holder tilted at this angle, and therefore remove the dipolar couplings.

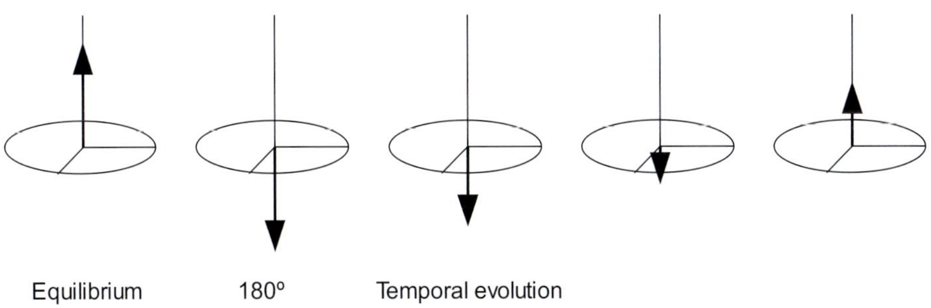

FIGURE 4.13 Schematic representation of the temporal evolution of magnetisation in an experiment designed to measure spin-lattice relaxation.

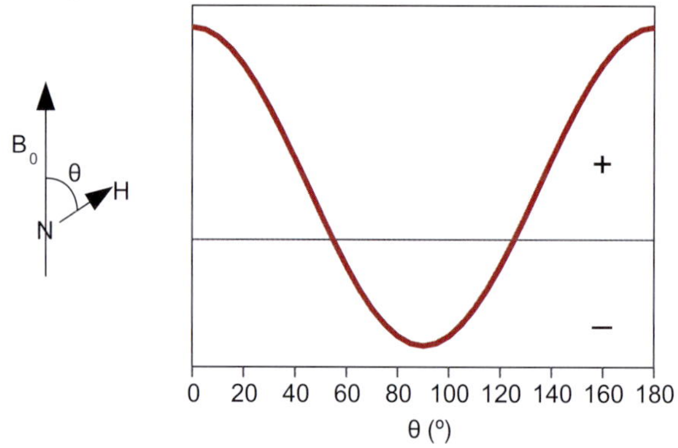

FIGURE 4.14 Schematic representation of the angle between the internuclear bond and the static field giving rise to the RDC, using an NH bond as an example (left), and angular dependence of the RDC (right). The actual values depend on characteristics of the alignment and on the internuclear distance, but the crossing point occurs always at the magic angle.

In solution, in contrast, the molecular motion is typically sufficiently fast and random for this mode of coupling to average to zero. However, it is possible to prepare samples with a preferred molecular orientation and measure the residual dipolar coupling (RDC) that is generated in these conditions. Also, macromolecules with metal centres that contain unpaired electrons can spontaneously assume a slight preference for a particular orientation in solution in the presence of the static magnetic field. This is a consequence of the interaction of the unpaired electron with the magnetic field, allowing for measurement of the resulting RDCs without further manipulation (Figure 4.14).

Scalar couplings are another consequence of the electron orbitals that envelope the magnetic nuclei. The scalar coupling is mediated by the Fermi contact interaction, which describes the coupling between nuclear and electronic spins. The consequence is that only electrons in orbitals with non-zero spin density in the nuclei will contribute to this effect, i.e. bonds with some sigma character (Figure 4.15).

Scalar couplings are mediated by the electrons that form the chemical bonds connecting the nuclei. Therefore, the effect fades as the number of bonds increases. According to the Pauli principle, the electrons in the bonding orbital will be antiparallel. The consequence is that parallel and antiparallel nuclear spins connected by such a bond will have a slightly different energy, and therefore the signal will be split $2I+1$ times. For example, a nucleus coupled with another nucleus of spin 1/2 will have its signal split in two. When several equivalent nuclei are the source of the coupling, some signals will overlap. For nuclei of spin 1/2, the multiplicity becomes $N+1$, and relative intensities are given by the Pascal triangle, as seen on Table 4.2.

Scalar couplings are named according to the number of intervening bonds between the coupled nuclei and designated ^{n}J:

^{2}J or geminal couplings are observed between nuclei of different chemical shifts bonded to one common nucleus. The value depends on the hybridisation of the molecular orbitals of the connecting atom.

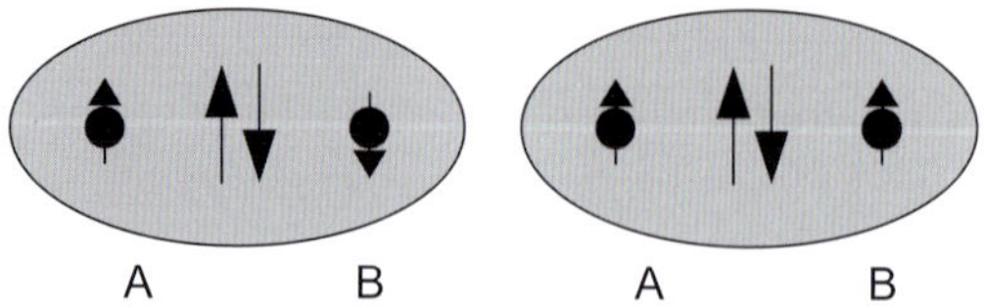

FIGURE 4.15 Diagram of two bonded nuclei A and B with spin 1/2, and the two options for the orientation of the nuclear spin B. The configuration on the left has slightly lower energy. The short arrows crossing the circles are the nuclear spins. The long arrows are the paired electron spins responsible for the bonding orbital depicted in grey. The nucleus A shows two signals, one of each of the two configurations and the same happens for nucleus B.

TABLE 4.2 Multiplicity and Relative Intensity for the Various Peaks of the Multiplets of a Signal Coupled to Nuclei with Spin 1/2

Number of $I = 1/2$ nuclei	Multiplicity	Relative intensity
0	1 (singlet)	1
1	2 (doublet)	1:1
2	3 (triplet)	1:2:1
3	4 (quadruplet)	1:3:3:1

^3J or vicinal couplings are observed between nuclei separated by two other nuclei. Vicinal couplings have a well-defined geometrical dependence on the dihedral angle between the coupled nuclei. This is easily visualised in diagrams called Newman projections that indicate the dihedral angle between two hydrogen atoms bonded to the front and back carbons.

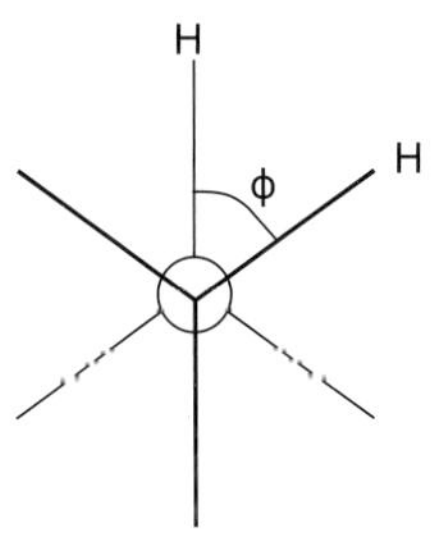

There are different mathematical expressions for different kinds of nuclei – all denominated Karplus equation. For vicinal protons $A = 7$ Hz, $B = -1$ Hz and $C = 5$ Hz. If rotation is fast, the coupling settles into an average value of the dominant configurations, but when rotation is prevented the value of the scalar coupling provides geometrical constraints for structural determination and definition of stereochemistry.

$$^3J = A + B\cos\varphi + C\cos 2\varphi \quad (4.17)$$

Up to now, this chapter showed that NMR can provide knowledge on:

- the number of different magnetically active nuclei, from the number of signals.
- how many equivalent nuclei contribute to each signal, from its intensity.
- how many nuclei are coupled to each nucleus, from the pattern of scalar splitting.
- the stereochemistry form the magnitude of the vicinal couplings.
- molecular motions, from the signal linewidths.

However, Figure 4.16 shows that this information is insufficient for the characterisation of biological molecules of even moderate size. Fortunately, more information can be obtained from NMR experiments.

Decoupling

We have seen that scalar coupling is of considerable use to identify the nature of neighbouring groups and to define molecular geometry. However, in a complex spectrum such as that shown in Figure 4.16, it may not be obvious

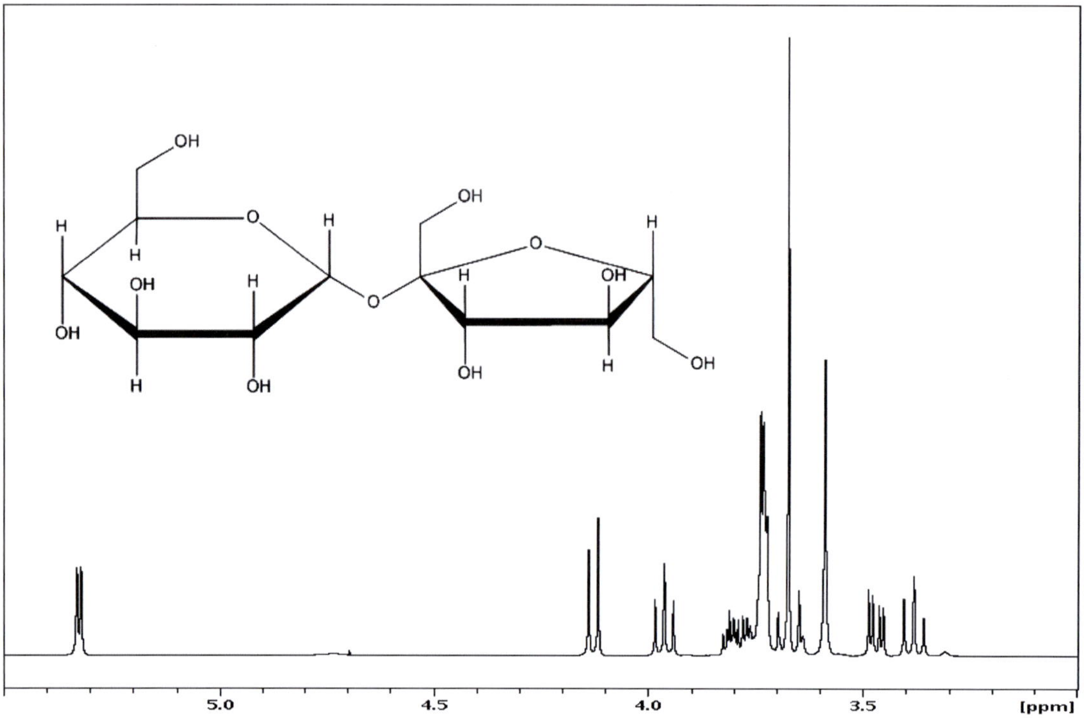

FIGURE 4.16 1D ^{1}H spectrum of sucrose and respective molecular structure represented with the Haworth projection.

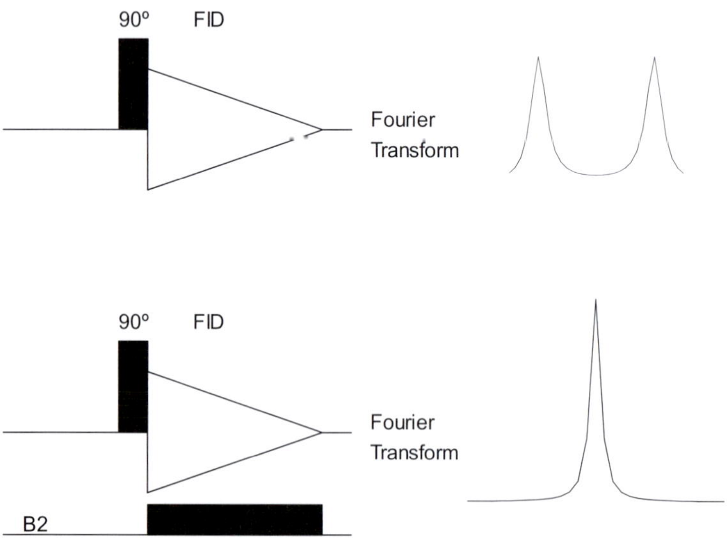

FIGURE 4.17 Comparison of the pulse sequences for a standard acquisition and an acquisition with a $\mathbf{B_2}$ decoupling pulse during acquisition. The effect on the spectrum is represented on the right side.

which pairs of signals are coupled and therefore belong to neighbouring nuclei. Saturation of the signals of one nucleus with a weak selective B_2 pulse during detection causes the coupled nuclei to feel the average orientation of the irradiated nucleus and lose the splitting, becoming a singlet. This way, the signals of nuclei separated by a few chemical bonds can be identified, and the *S*/*N* ratio is increased, improving the experimental sensitivity (Figure 4.17).

CHEMICAL EXCHANGE

The phenomenon of chemical exchange arises from the fact that nuclei may experience different chemical environments during the course of an NMR experiment. This is of great importance in the study of biological macromolecules. For example, the conformational interconversion of a flexible region of a protein, makes the chemical shift of a particular nucleus different in the various conformations. Equation (4.18) shows the simplest formulation for chemical exchange showing the forward and backwards rate constants for two states.

$$A \underset{K_{-1}}{\overset{K_1}{\rightleftharpoons}} B \tag{4.18}$$

If a nucleus experiences two chemical environments, it can have signals in two different spectral positions. Furthermore, Eqn (4.18) shows that for the simple case of a two-state exchange, kinetics will be of paramount importance. Indeed, the appearance of the spectrum will be determined by the ratio between the rates of the exchange phenomenon and the frequency difference between the signals in exchange. Therefore, we can modify the appearance of the spectrum by modifying the experimental conditions. This can be achieved by changing sample concentration, solvent viscosity, temperature or pH. Also, the spectral appearance can be modified by running the experiment at different field strengths, which changes the frequency difference between the signals in exchange without modifying the sample (Figure 4.18).

Three limiting regimes of exchange can be defined:

- Slow exchange: the rates of chemical exchange are much slower than the difference in the Larmor frequency of the signals. For the case of Eqn (4.18) two signals are observed, one each for chemical environment with intensities proportional to the relative occupancy of each chemical environment.

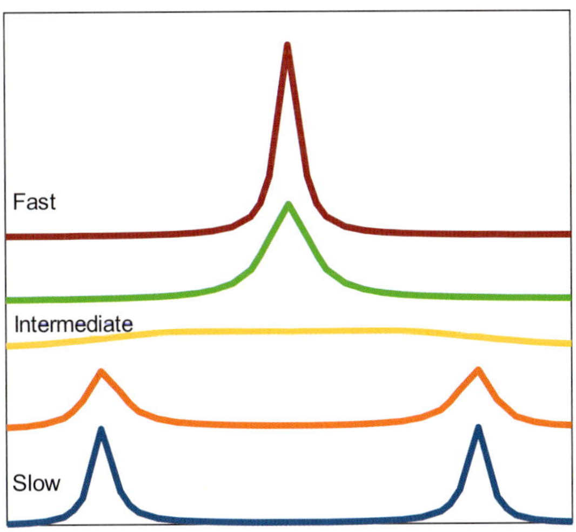

FIGURE 4.18 Calculated NMR spectra for exchange between two sites with equal occupancy. From top to bottom exchange becomes increasingly slow.

- Fast exchange: The rates of chemical exchange are much faster than the difference in the Larmor frequencies of the signals for the (two) chemical environments. They must also be faster than the sums of spin–spin and spin-lattice relaxation rates, i.e. $1/T_1$ and $1/T_2$, for the (two) chemical environments. In this case, a single signal is obtained in a position that is the average of the extreme positions weighted by the relative occupancy of each chemical environment.

$$v = v_A f_A + v_B f_B \tag{4.19}$$

- Intermediate exchange: also known as coalescence, which occurs when the exchange rates are of the order of magnitude of the differences in Larmor frequencies. For experiments run in this regime, a very broad peak between the two extreme positions of the slow exchange regime may be observed under favourable conditions.

THE NUCLEAR OVERHAUSER EFFECT

As explained above, dipolar couplings typically cannot be observed directly in solution due to fast and random reorientation of the molecules. However, their presence influences the spectra in ways that can be detected. The NMR signal arises because the various states of a nucleus (two in the case of nuclei with spin ½) have different energies. Nuclei of different chemical shifts have different energy separations. In these conditions, the Boltzman distribution determines that the populations will also be different. When nuclei are coupled, their resonance frequencies are modified depending on the spin state of the coupled nucleus. For the simplest case of two nuclei with spin ½, four states exist with a population slightly different depending on their energy separation according to the Boltzman distribution, as shown in Figure 4.19. In equilibrium, the states $\alpha_1\alpha_2$, $\beta_1\alpha_2$, $\alpha_1\beta_2$ and $\beta_1\beta_2$ will be successively less populated. It is possible to equalize experimentally the population of one of these two nuclei, for example nucleus 1, by applying a selective pulse at the Larmor frequency of that nucleus – this is called a saturating pulse.

The consequence is that the states $\alpha_1\alpha_2$ and $\beta_1\alpha_2$ will be equally populated. The same happens to the states $\alpha_1\beta_2$ and $\beta_1\beta_2$. This means that the coupled system is out of equilibrium, and as soon as the saturating signal is switched off, re-equilibration will occur. This re-equilibration can occur through any of three pathways: along the edges of the diagram (the single quantum transitions, W1), sideways across the diagram (the zero quantum transitions, W0) or longitudinal between top and bottom (the double quantum transitions, W2). As the name of the transitions implies, zero quantum transitions occur at frequencies close to zero Hz, single quantum transitions

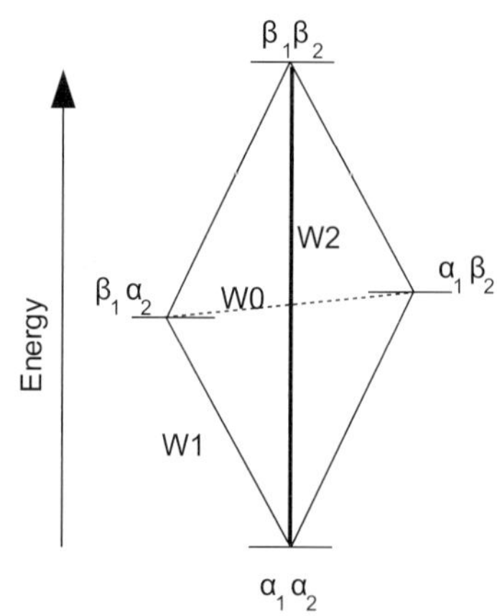

FIGURE 4.19 Energy diagram for a pair of coupled spins. Zero quantum (W0), single quantum (W1) and double quantum (W2) transitions are represented by dotted, thin and thick lines, respectively. The height difference between the transitions provides a visual cue for the kind of transition connecting each pair of states.

have frequencies close to the Larmor frequency, whereas double quantum transitions have frequencies close to double the Larmor frequency. The consequence of the re-equilibration is that the intensity of the signal of the coupled nucleus (in this case nucleus 2) will be modified; it is this change in intensity that reveals the **nuclear Overhauser effect (NOE)**.

$$\eta = (I - I_0)/I_0 \tag{4.20}$$

The change in intensity is designated the NOE enhancement. It depends on the distance and on the molecular motion and therefore is related with molecular size.

$$\eta = \mathrm{f}\left(\tau_c \mathrm{r}^{-6}\right) \tag{4.21}$$

It is the dependence with distance that makes the NOE effect, one of the most important tools for the characterisation of molecular structures. The dependence with molecular motion, and therefore size, arises from the fact that zero, single and double quantum transitions, like all NMR transitions are stimulated by fluctuating magnetic fields. For smaller molecules, molecular motion is fast and therefore transitions of higher frequency (double quantum) are enhanced. For large molecules, motion is slow, and transitions of low frequency (zero quantum) are enhanced. This has two consequences: (1) for small molecules, the coupled signal is larger than in the absence of saturation, whereas for large molecules, the coupled signal is smaller; (2) for molecular sizes where the correlation time is similar to the reciprocal of the Larmor frequency there is no NOE enhancement.

A simple rule of thumb provides a rough estimate of the correlation time for proteins in water at room temperature: the molecular weight in kDa is approximately twice the correlation time in ns. For the conditions of Figure 4.20 at a molecular weight of approximately 700 Da, there is no NOE enhancement. It also shows that the maximum enhancement for macromolecules (−1) is double and of opposite sign to what can be obtained for small molecules (0.5).

Therefore, NMR experiments also allow

- the identification of nuclei separated by a short number of chemical bonds using decoupling.
- the determination of bond geometry from the magnitude of scalar couplings.
- the observation of exchange phenomena.
- the determination of distance constraints between pairs of nuclei from the magnitude of the NOE effect.

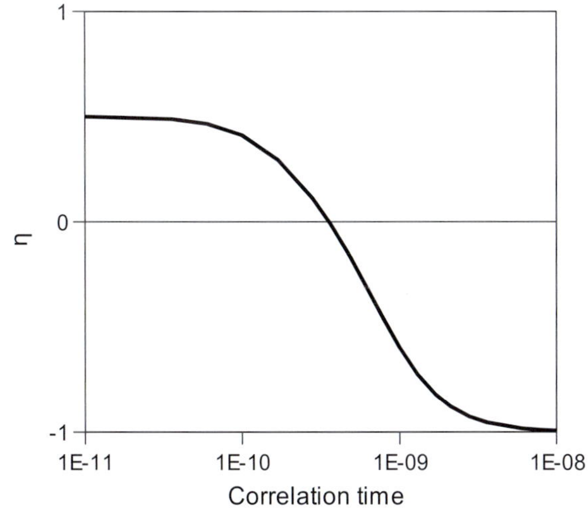

FIGURE 4.20 Dependence of the NOE enhancement on correlation time calculated at 500 MHz for a pair of protons separated by 2 Å.

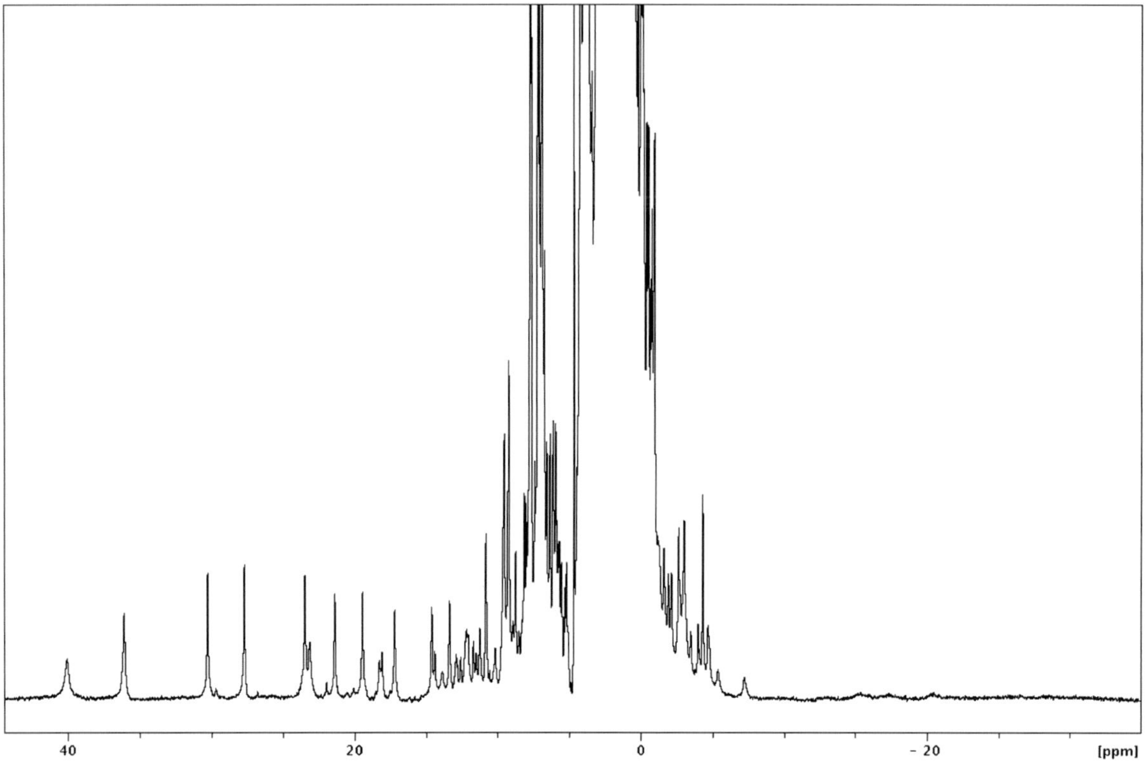

FIGURE 4.21 1D ^{1}H NMR spectrum of a protein containing low-spin paramagnetic centres.

However, spectra of biological macromolecules are extremely complex with hundreds to thousands of signals, as is evident in Figure 4.21. Spectral overlap of the signals is severe, and the techniques described up to now are not up to the challenge. Two strategies can be used to address this problem; spectral editing and expansion into multiple dimensions. Spectral editing, which is typically performed to distinguish the hybridisation of carbons in biomolecules, requires labelling of the sample to achieve convenient sensitivity and will not be described here.

MULTIDIMENSIONAL NMR

To obtain a 1D-NMR spectrum, the nuclei are excited with a pulse of radio frequency, the decay is recorded in the time domain and Fourier transformed to obtain the spectrum in the frequency domain. In order to obtain further frequency dimensions, it is necessary to introduce further time domains that can be Fourier transformed. Without loss of generality, we can consider the case of a 2D spectrum. A two-dimensional spectrum is obtained by recording a series of 1D spectra. The experiment is initiated by a preparation procedure that can have different degrees of complexity depending on the specific experiment, but includes a 90° pulse. There is a waiting period, called t_1, during which magnetisation evolves. The magnetisation is again manipulated by procedures that include another 90° pulse, and finally signals are recorded during a period called t_2. For each subsequent spectrum, the t_1 period is incremented so that the signal recorded is a function of the t_1 period. At the end, we have a series of FIDs that depend on the t_1 period and therefore can be Fourier transformed relative to this t_1 time and to the normal acquisition time t_2. The results are reported as a 2D-contour map with signals located at the correct frequency for both evolution periods. Multidimensional techniques allow the dispersion of the signals in various dimensions, and the observation of correlations between different signals. Depending on what is done during the experiment, different correlations are generated.

How Do the Correlations Arise, Generating the Cross-Peaks?

The vector model is not suitable to describe the process in detail but an energy diagram will be helpful. Without loss of generality, consider again the energy diagram of a pair of coupled nuclei. For the present case, it is not relevant whether the coupling is scalar or dipolar because the preparation period is designed to select which kind of coupling will be observed during the experiment.

After the first $\mathbf{B_1}$ pulse, the spins of these two nuclei were flipped to the xy plane, and they begin to precess during t_1 at their characteristic Larmor frequencies. During the second manipulation period, in particular the 90° pulse, some magnetisation will be flipped to the z direction (becoming invisible), and some remains in the xy plane to be detected. This is the trivial consequence of pulsing twice. However, because the energy of the $\mathbf{B_1}$ pulse is of the correct frequency to excite the nuclei, some of it will be used to promote the single quantum transitions depicted by thick lines in Figure 4.22. Following this, the magnetisation decay is detected in the t_2 period. The consequence of this is that some signals will precess with a particular frequency during t_1 but will then precess with a different frequency during t_2. Therein lies the origin of the cross-peak. The coupling between nuclei therefore provides a path for magnetisation to be transferred during the second pulse.

The enhanced discrimination provided by the multiple dimensions, together with the knowledge on the physical bases of the couplings that give rise to the cross-peaks, allows the characterisation of biological macromolecules such as proteins and nucleic acids.

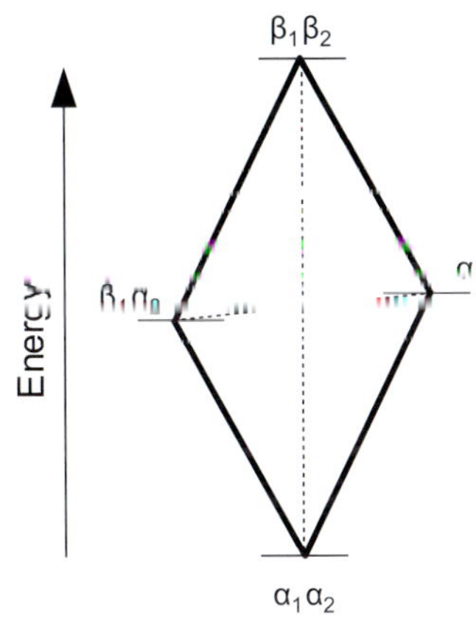

FIGURE 4.22 Energy level diagram for a pair of coupled spins. Single quantum transitions between pairs of states are depicted as thick lines.

Multidimensional NMR experiments can be **homonuclear** when they correlate signals of nuclei of the same nature in all dimensions, or **heteronuclear** when nuclei of different nature are correlated in the various dimensions.

Increasing the number of dimensions requires that relaxation is slow (long T_1 and T_2) so that there is still sufficient magnetisation in the xy plane to be observed after the various waiting periods.

Despite the myriads of experiments developed over the years, a few basic types are of widespread usage and important for biomolecular NMR.

The COSY

The name stands for COrrelation SpectroscopY and this is the simplest 2D experiment with respect to implementation. It detects scalar couplings (Figure 4.23).

The first 90° pulse generates transverse magnetisation that evolves during t_1. The second 90° pulse flips part of that magnetisation into the z axis but leaves part in the xy plane. Very importantly, it also promotes transitions

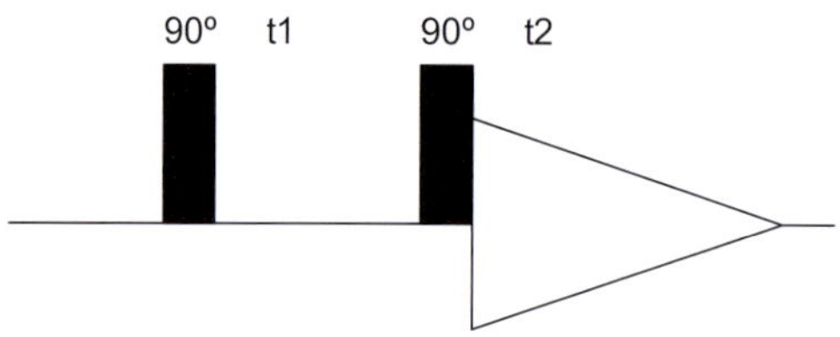

FIGURE 4.23 Diagram of the pulse sequence of a COSY experiment.

among the coupled spins. The signals are detected during t_2. For those spins that were stimulated to change state by the second pulse, the precessional frequency during t_2 is different from that during $t1$ and therefore will show a cross-peak. The cross-peaks in this experiment indicate the presence of scalar coupling between a pair of nuclei.

The NOESY

The name stands for Nuclear Overhauser Enhancement SpectroscopY and is the 2D implementation of the detection of transient NOE effects. For crowded spectra, it is more convenient because the selective saturation in 1D NOE experiments cannot be strictly achieved. As a consequence, some spill over of the saturation to the neighbouring signals cannot be avoided leading to uncertainty in assigning the effect observed. The NOESY experiment can also be used to observe chemical exchange. Chemical exchange provides an easier visualisation when describing what happens in the NOESY experiment (Figure 4.24).

The first 90° pulse generates transverse magnetisation that evolves during t_1. The second 90° pulse flips part of the magnetisation into the z axis and leaves part in the xy plane. The longitudinal magnetisation is unobservable but is evolving. In case of chemical exchange during the mixing time (tm), the nucleus can change chemical environment. The third 90° pulse brings this longitudinal magnetisation back into the xy plane to be detected during t_2. For those nuclei that experienced exchange during tm, the precession frequencies during t_1 and during t_2 are different giving rise to a cross-peak.

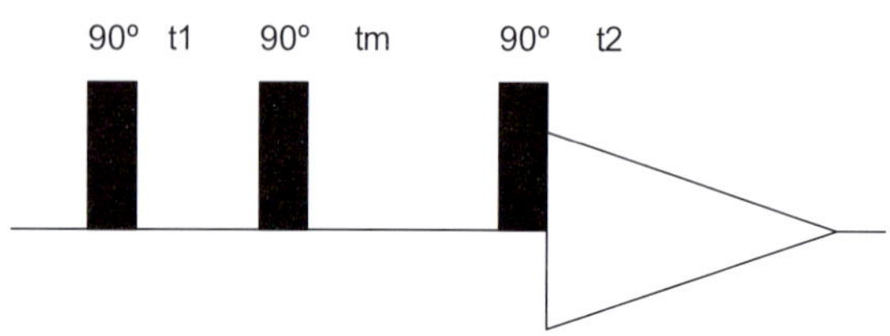

FIGURE 4.24 Diagram of the pulse sequence of a NOESY experiment.

The HSQC

The name stands for Heteronuclear Single Quantum Coherence, and this experiment generates a signal for each pair of different NMR active nuclei that are bonded together. When developing heteronuclear correlation experiments, two aspects must be considered: (1) the larger difference in Boltzmann population difference for protons (the nucleus with the highest gyromagnetic ratio) favours the transfer of magnetisation from protons to the heteronuclei; (2) the higher detection sensitivity of protons favours the transfer of magnetisation for the heteronuclei to the protons.

Since the S/N is approximately $\gamma_{\text{excited}}(\gamma_{\text{detected}})^{3/2}$, experiments that start and finish in proton with a passage by the heteronucleus of interest have a much better S/N. A diagram of the pulse program for an HSQC organised this way is depicted in Figure 4.25. A further advantage is that the spin-lattice relaxation of protons is typically faster than that of the other nuclei in a biomolecular sample. This allows for a shorter acquisition time for each scan.

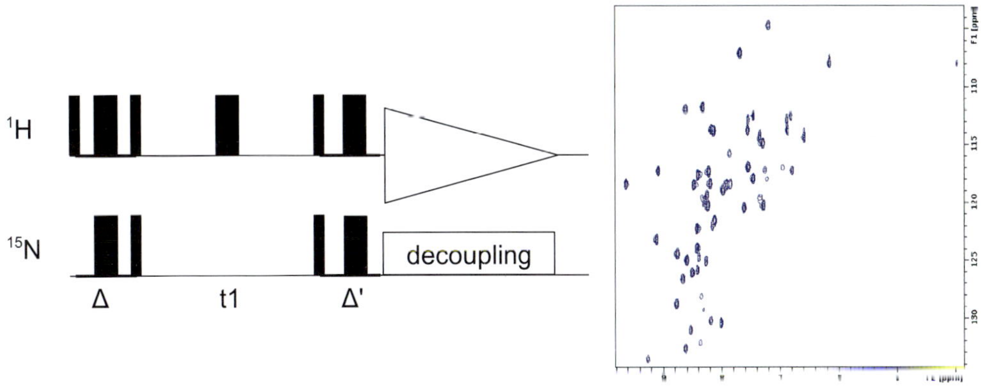

FIGURE 4.25 Diagram of a $^{1}H^{15}N$ HSQC pulse sequence. Thin bars represent 90° pulses and thick bars represent 180° pulses. A $^{1}H^{15}N$ HSQC of a protein showing the region of the NH groups with the good spectral dispersion characteristic of a properly folded protein. This has made this experiment very popular in protein NMR studies because each signal is associated with a single amino-acid backbone NH or a side chain.

The experiment is initiated by using the protons coupled to the heteronuclei to generate a population difference in the heteronuclei that is larger than what could be achieved by a 90° pulse applied to the heteronuclei. This is achieved by making the Δ delay equal to the reciprocal of the coupling constant between protons and the heteronuclei. The enhancement is approximately $\gamma_{proton}/\gamma_{heteronucleus}$. The magnetisation of the heteronucleus evolves during t_1. In order to avoid decoupling of the signals of the heteronucleus by the protons, a 180° pulse is applied to protons in the middle of the t_1 period. 90° pulses applied simultaneously to the proton and heteronucleous promote magnetisation transfer. Detection is made in the proton frequency during t_2. If decoupling of the ^{1}H signals by the heteronucleus is to be prevented, a broadband decoupling pulse has to be applied in the heteronucleus channel after waiting for a Δ′ delay. This is necessary to prevent complete loss of the coupling that would result in an empty spectrum. With the sequence represented in Figure 4.25, a seemingly impossible feat is achieved. A spectrum is obtained that results from scalar coupling between two nuclei, and decoupling is achieved in both dimensions.

METALS IN BIOMOLECULAR NMR SPECTRA

Metals are an essential component of the biological systems. They participate in the catalytic activity of numerous enzymes, in the capture of small molecules, inter-cellular signalling, in the structural organisation of numerous proteins, or as extracellular structural support of cells and tissues. The presence of metals in an NMR sample can be used to directly observe the metal when it is NMR active, or to observe the effect of the metal or its electrons on other NMR active nuclei.

Direct Observation of Metals

Of those metals that can be directly observed in biological samples and tissues, Na^{+} and K^{+} are sufficiently abundant and display an effective sensitivity that is compatible with NMR studies. These monovalent cations play key roles in cellular transport, maintenance of transmembrane electrical potential, and cell signalling. However, both nuclei present $I = 3/2$. This means that these nuclei have three allowed single quantum transitions. Nuclei of $I > 1/2$ have an electric quadrupole moment, which can be interpreted as arising from a non-spherical charge distribution on the nucleus. The presence of the electric quadrupole moment gives rise to greater spectral complexity. In a uniform electric field, all three transitions have the same energy, and a single signal is observed. However, when the nucleus is in an electric field gradient, the three transitions can have different energies, with the splitting of the three signals proportional to the electric field gradient. This can be observed in single crystals or in liquid crystalline samples. An important consequence is that when analysing un-oriented samples or samples where the nuclei are subject to rapidly fluctuating electric fields, only the central resonance may be visible. This leads to an apparent reduction of the signal intensity to approximately 40% relative to the expected value, a phenomenon known as 'invisibility'.

Transition Metals and Interaction with the Unpaired Electron(s)

Transition metals are characterised by having orbital shells that are not fully occupied. Depending on the bonding environment of the metal, some orbitals may be partially filled leading to electronic spin states that can range from 1/2 to 9/2. Electrons, like protons and neutrons, are subatomic particles with the quantum mechanical property of spin. Being a spin, typically denoted *S*, the electron has a magnetic moment that is quantised. When placed in a magnetic field, the energy of this magnetic moment is expressed in a fashion similar to that of the nuclear magnetic moments seen in Eqn (4.3).

$$E = g\mu_B B_0 S \tag{4.22}$$

Because the electrons are located in orbitals with defined shapes, *g* is not a number. When the orbital is not spherically symmetrical, *g* becomes a matrix that accounts for the differences in energy that may arise from the orientation of the molecule relative to the magnetic field. Unpaired electrons are spins and interact with nuclear spins via the same physical mechanisms that were presented above. This is called the hyperfine interaction.

Scalar Coupling

Scalar, or contact, coupling arises from non-zero density probability of the unpaired electron at the nucleus and is independent of the molecular orientation in the magnetic field. For transition metals, these electrons are located in the valence *d* orbitals that participate in bonds with the remainder of the molecule. The electron has a magnetic moment and generates a local magnetic field that modifies the effective field felt by the nuclei in its proximity. The Fermi contact shift generated by the unpaired electron is described by Eqn (4.23).

$$\delta^{\text{con}} = \frac{A}{h}\frac{g_e \mu_B S(S+1)}{3\gamma kT} \tag{4.23}$$

A is the coupling constant between electron and nucleus. It has the same physical meaning as *J* between nuclei.

Dipolar Coupling

As seen above, dipolar interactions are propagated through space. In this case, they depend on the distance between the unpaired electron and the nucleus. They also depend on the orientation of the vector connecting the electron and the nucleus with respect to the external magnetic field. When *g* is isotropic and under fast rotation, the dipolar interaction energy averages to zero. In these conditions, the chemical shift of the signal is not perturbed, and the dipolar effect arising from the presence of the unpaired electron is only felt at the level of enhanced relaxation, i.e. broader signals (see below). When the unpaired electron is found in an orbital of non-spherical symmetry, *g* is orientation dependent, and the dipolar contribution does not average to zero. This gives rise to the pseudocontact coupling. The dipolar coupling is very complex, but when describing its effect in protons, a simplification can typically be applied. This simplification considers that the unpaired electron orbital is centred in the metal coordinates and is axially symmetrical. The axial symmetry means that the *x* and *y* directions cannot be distinguished. With this simplification, the pseudocontact shift is defined by Eqn (4.24).

$$\delta^{\text{pc}} = \frac{\mu_0}{4\pi}\frac{\mu_B^2 S(S+1)}{9kT}\left(g_\parallel^2 - g_\perp^2\right)\frac{1}{r^3}\left(3\cos^2\theta - 1\right) \tag{4.24}$$

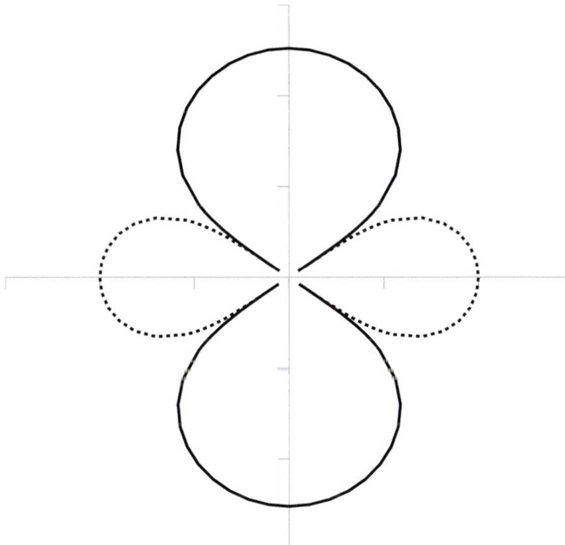

FIGURE 4.26 Contour of pseudocontact shifts of equal magnitude for an axial system calculated using Eqn (4.24). Solid and dotted lines represent shifts of opposite sign.

The g values can be measured in an EPR experiment (see chapter on EPR). Equation (4.24) reveals several important aspects of the pseudocontact shift. First of all, it shows that isotropic electrons do not generate a pseudocontact shift because $g_{\parallel}^2$ and $g_{\perp}^2$ are equal. The shift has a different sign depending on the position of the atom vs the magic angle. For the same position, the pseudocontact shifts change signs depending on whether the $g_{\parallel}$ or $g_{\perp}$ is larger.

Figure 4.26 shows how the pseudocontact shift extends around a paramagnetic centre of axial symmetry, clearly revealing the existence of the magic angle. It also shows that the effect extends more in the direction of the z axis than in the direction of the xy plane. Therefore, for nuclei not bonded to the metal and its ligands, the magnitude of the pseudocontact shift provides information on the distance and geometry relative to the magnetic axes defined by the unpaired electron. This is the basis for using paramagnetic shift constraints to improve the structure of proteins containing unpaired electrons.

It can be seen from Eqns (4.23) and (4.24) that both contact and pseudocontact effects are dependent on the reciprocal temperature. Therefore, paramagnetic shifts decrease with increasing temperature as long as there are no changes in coordination or transitions to excited electronic states. In solution, the molecule under observation is typically in conditions of fast reorientation, and therefore the contact and pseudocontact terms of the hyperfine interaction are measured together.

RELAXATION

When paramagnetic molecules are placed in a magnetic field B_0, the unpaired electron(s), which are magnetic spins 1/2, will distribute among the two Zeeman energy levels. Spontaneous transitions between the two Zeeman energy levels are accompanied by changes in the electron magnetic moment. Therefore, these transitions generate local fluctuations in the magnetic field capable of stimulating nuclear relaxation.

Contact Relaxation

This mechanism arises from the delocalisation of the electron onto the nucleus and is proportional to the electronic spin, and the correlation function. In this case, there is no contribution of the molecular motion to relaxation, which is restricted to electronic relaxation and (eventual) chemical exchange. It is significant for nuclei directly bonded to the metal through bonds displaying significant covalence.

Although the unpaired electron(s) occupies an orbital with a defined shape, relaxation can be analysed considering that the unpaired electron(s) is localised on the metal nuclear coordinates. This approximation ignores the contact contribution towards relaxation and simplifies the calculation of the dipolar contribution.

Dipolar Relaxation

Within this simplification, and in the fast motion limit, i.e. the transition frequencies for the nuclei and the electron are slower than the reciprocal of the correlation time for the electron-nuclear interaction, the simplest formulation of the Solomon equation is obtained:

$$R_{(1,2)} = \frac{4}{3}\frac{\gamma^2 g_e^2 B_e^2 S(S+1)}{r^6}\tau_C \tag{4.25}$$

Even in this simplified form, it is easy to see from Eqn (4.25) that relaxation is proportional to the square of the gyromagnetic ratio of the nucleus under observation, and therefore much more dramatic for protons than for other nuclei. Therefore, and given the strong dependence of the effect with distance, the radius around the paramagnetic centre, where broadening is so excessive that signals cannot be observed, is much larger for protons than for nuclei with low gyromagnetic ratio such as ^{15}N or ^{13}C. This is the driving force for the development of direct heteronuclear detection (protonless) NMR experiments for the characterisation of paramagnetic proteins.

The dipolar relaxation of the nucleus under observation caused by the unpaired electron can be partitioned into three components:

- transition between the Zeeman levels inverts the induced field generated by the unpaired electron that is felt by the observed nucleus.
- rotation of the molecule changes the angle subtended by the vector connecting the unpaired electron and the observed nucleus vs the direction of the static field.
- chemical or conformational exchange modifies the relative position (distance and/or angle) between the unpaired electron and the observed nucleus.

All of these components are of dynamic nature. We have seen above that dynamic effects caused by random and unpredictable changes in a system that is in equilibrium are described by a correlation function characterised by a correlation time. In the case of the electron-nuclear dipolar relaxation, three mechanisms contribute towards the correlation function as seen above. Given their nature, it is reasonable to assume that they are independent, and therefore the correlation function decays as the product of the three exponential components. As a consequence, the overall correlation time relates to the correlation times for each mechanism according to the following expression:

$$\frac{1}{\tau_C} = \frac{1}{\tau_S} + \frac{1}{\tau_R} + \frac{1}{\tau_M} \tag{4.26}$$

where τ_S is the correlation time for electron relaxation, τ_R is the correlation time for molecular tumbling and τ_M is the correlation time for chemical exchange.

Electronic relaxation times range from 10^{-13} to 10^{-7} s. Rotational correlation times at room temperature range from 10^{-11} s for aquated metal complexes to 10^{-6} s for large macromolecules, and exchange correlation times are slower than 10^{-10} s. It is clear from these ranges that depending on the molecule under analysis, one of the components may dominate the overall correlation time. If the overall correlation time is dominated by electronic relaxation, a situation that occurs for metals with short τ_S, nuclear relaxation is slow and relatively sharp signals can be observed. This is the case for low-spin Fe(III), high-spin Fe(II) and some lanthanides historically known as shift agents such as Pr(III) and Eu(III), which typically have electronic relaxation correlation times in the range of 10^{-12} to 10^{-11} s. However, if electronic relaxation is slower and the rotational correlation time dominates the overall correlation time, nuclear relaxation is enhanced leading to broad signals that may be undetectable in the vicinity of the metal. This is the case for Cu(II), high-spin Fe(III), and lanthanides such as Gd(III) which are known as relaxation agents. In the case of Gd(III), this property is used in Magnetic Resonance Imaging to improve the contrast of the image, by administering to the patient compounds with the metal.

Curie Relaxation

In the description of dipolar relaxation made so far, it was implicitly considered that electronic spins were evenly distributed among the two states. In fact, this is not true, and there is a slight excess of electronic spins in the low-energy state. As expected, the difference in population increases with increasing field strength. Curie relaxation arises from this difference in electron population in the presence of the external magnetic field. It causes a small electronic magnetic moment aligned with the field, the Curie Spin. Molecular motion modulates the effect of this spin on the nuclei, which gives rise to a dependence with r^{-6}. As the molecular size increases, the contribution of Curie relaxation towards the total nuclear spin relaxation increases and can become dominant. Curie relaxation is also proportional to B_0^2 which means that for paramagnetic molecules, the use of a stronger field may not be the most favourable for observing the signals, in particular for large paramagnetic macromolecules (>40 kDa). If a significant contribution of this relaxation mechanism is suspected, it can be identified by observing dramatic increases in the linewidth of the signals with increased field.

AN NMR SPECTROMETER AND HOW THE MACROSCOPIC MAGNETISATION AND RELAXATION ARE MEASURED

Considering the separation between the energy levels of magnetic nuclei in contemporary magnetic fields, the electromagnetic radiation that is absorbed during an NMR experiment has a wavelength of the order of a metre. This is the range of radio waves and therefore, apart from the magnet, the NMR spectrometer can be viewed as a very sophisticated radio that generates music for the nuclei and then observes the resulting choreography.

An NMR spectrometer is composed of:

- a magnet.
- probeheads.
- the lock system.
- the shim system.
- signal generators.
- analogue to digital converter.

All of these are now operated via a computer interface that also serves to store and process the experimental data.

The magnet is the core of the spectrometer and defines it. Nowadays, it is composed of a coil of wire that becomes superconducting when immersed in a liquid helium bath. This is isolated from room temperature by a liquid nitrogen bath and vacuum seals. The coil is superconductive so that a strong electrical current can flow through it indefinitely without generating heat that would boil off the helium bath. It is this electrical current flowing in the coil that generates the static field. Among the NMR community, the strength of the magnetic field is typically indicated by the Larmor frequency that the field imposes on hydrogen nuclei instead of the SI unit Tesla. Therefore, a spectrometer of 21.2 T is referred to as a 900 MHz. In the centre of the coil and of the liquid helium and nitrogen baths, there is a hole called bore, where the probehead is inserted.

The probeheads serve to hold the sample in the correct place to run the experiment. They are also like radio antennas that emit the electromagnetic pulses and detect the magnetisation decay of the sample. They contain coils of wire of specific design that do both tasks. An RF pulse is generated by passing current in the coils, and subsequently the signal is detected by measuring the current induced in the coils by the macroscopic magnetisation of the sample. When the sample contains very strong signals, the current induced in the detecting coils interacts with the nuclei in the sample. This phenomenon causes extra broadening of the signals and is called radiation damping. It is relevant for biomolecular samples given that protons in water can have a concentration of up to 110M.

The probes must have at least two coils. One for generating and detecting the signals of interest, and the other for locking the sample (see below). Because many modern NMR experiments used in biomolecular NMR excite and detect more than one kind of nucleus, the probes to execute such experiments must have more coils and are called multichannel.

The lock system is used to track the natural drift in the static magnetic field with time. Because field stability is essential for NMR experiments, the lock system measures the position of deuterium nuclei added to the sample (for instance deuterated water in biomolecular samples) and makes sure that it does not change by applying electrical current in a coil in the magnet. Because of the need to follow the deuterium signal, all probes have a dedicated coil to perform this task.

The shim system is used to correct small imperfections in the static magnetic field. This is done by passing electrical current in coils of specific shapes. The shims can be on-axis (in the zz direction) or off-axis (other directions). The names of shims have superscript numbers that define the degree of the polynomial function that describes their effect on the static field. Therefore, a Z shim has a linear effect in the z direction, whereas a x^2 shim has a parabolic effect in the x direction.

Signal generators produce the radiation that is used to manipulate the nuclei such as the $\mathbf{B_1}$ pulse. They produce sinusoidal waves of the desired frequency, phase and shape. Typically, there is more than one signal generator, each connected to a channel dedicated to a specific nucleus or frequency range.

The analogue to digital converter measures the analogue signal induced in the detection coil by the macroscopic magnetisation of the sample at discrete time points. The consecutive measurements (see Figure 4.11) are stored in a digital form that can subsequently be Fourier transformed. The period separating consecutive measurements is called sampling interval. Contemporary ADCs contain 16 bits, which means that the dynamic range is 2^{16}. The consequence is that the smallest signal that can be distinguished from noise need to be larger than 1/65536 the size of the largest signal.

A computer is used for controlling all components and for recording and processing the experimental data. As seen above, the NMR signal is a digitised periodic variation of electrical current induced in a coil that decays with time. For a simple case, inspection of the signal may reveal the frequency. When several nuclei with different chemical shift exist, the signal becomes too complex for direct inspection of the FID. In these conditions, a Fourier transform converts the time domain data into the frequency domain spectrum.

CARE IN OBTAINING NMR SPECTRA OF PARAMAGNETIC SAMPLES

NMR samples of biomolecules are in the vast majority of cases liquid. Therefore, the target molecules should be soluble at concentrations preferably above 1 mM due to the low sensitivity of the NMR phenomenon. For diamagnetic samples, molecular oxygen must be removed because it is paramagnetic and enhances the relaxation of the nuclei. Very often, paramagnetic samples are oxygen sensitive and therefore, oxygen also has to be excluded to avoid degradation or changes in the redox state. This is typically done by subjecting the sample to repeated cycles of vacuum and inert gas.

Given the dependence of the paramagnetic shift with temperature, a very good control of the sample temperature is necessary in order to obtain good quality spectra. Furthermore, careful temperature calibration is necessary to allow the matching of signals measured in different experiments, for instance, for structure determination of paramagnetic proteins by NMR. Because of the enhanced spectral dispersion caused by the paramagnetic shifts, a larger spectral bandwidth has to be excited, which requires short strong pulses to achieve the required flip angles. NMR pulses are made of electromagnetic radiation, and therefore samples should contain the minimum amount of salt. This minimises sample heating, resulting from movement of the ions stimulated by the radiation.

When really demanding paramagnetic samples are to be analysed, containing for instance signals with very large spectral dispersion, sampling of the FID may need to be so fast that special digitisers are required to handle data collection if folded peaks are to be avoided. Alternatively, a weaker field strength can be used, where the spectral window in frequency units is narrower, therefore reducing the demands on the digitiser. This is also advantageous for large paramagnetic proteins due to the reduced contribution of Curie relaxation.

Paramagnetically relaxed signals can be very broad and require a large number of scans to be observed. In order to optimise the signal strength for the available experimental time, the recycle delay, i.e. the time between successive scans, may need to be reduced from the typically recommended value of 5 T_1. With recycle delays between T_1 and 1.5 T_1, a better sensitivity per unit time is obtained but the $\mathbf{B_1}$ pulse should not aim to achieve a 90°-flip angle. For each recycle time and T_1, there is an optimum flip angle known as the Ernst angle:

$$\cos\alpha_E = \exp\left(\frac{-T}{T_1}\right) \tag{4.27}$$

Finally, it was shown above that spectral calibration is typically done by addition of standards to the sample. When these are not available or are not convenient because of spectral overlap, the residual HOD signal can be used to calibrate the spectra in samples prepared in deuterated water as is typical in biomolecular NMR.

$$\delta = 5.051 - 0.0111\,T \tag{4.28}$$

Equation (4.28) shows that the chemical shift of HOD is temperature dependent. This dependence is approximately linear in the range of 0–50 °C for which Eqn (4.28) is valid.

CONCLUSIONS

The contents of this chapter were the result of a choice of subjects that would touch various key aspects of NMR relevant for biomolecular samples and in particular those containing metals. Given the intended target audience and the available space, depth was traded for the vast ground covered. One glaring omission is the lack of reference to solid-state NMR methods, although these are gathering increased relevance for biomolecular characterisation. Considering the introductory nature of this chapter, an effort was made, to the extent possible, to describe NMR spectroscopy in ways that are pictorial rather than mathematical. However, it should be clear that true mastery of NMR will require a good command of the mathematical tools that support the underlying physical phenomena. The interested reader should explore more detailed literature, and some of it is listed below.

REFERENCES

Bertini, I., Luchinat, C., & Parigi, G. (2001). *Solution NMR of Paramagnetic Molecules*. Amsterdam: Elsevier.

Cavanagh, J., Fairbrother, W. J., Palmer, A. G., III, Skelton, N. J., & Rance, M. (2007). *Protein NMR Spectroscopy* (2nd ed.). San Diego: Academic Press. Principles and Practice.

Derome, A. E. (1993). *Modern NMR Techniques for Chemistry Research*. Oxford: Pergamon Press.

Freeman, R. (1997). *Spin Choreography, Basic Steps in High Resolution NMR*. Oxford: Spektrum Academic Publishers.

Gil, V. S. M., & Geraldes, C. F. G. C. (1987). *Ressonância Magnética Nuclear, Fundamentos Métodos e Aplicações*. Lisboa: Fundação Calouste Gulbenkian. (in Portuguese).

Levitt, M. H. (2008). *Spin Dynamics, Basics of Nuclear Magnetic Resonance* (2nd ed.). John Wiley & Sons.

USEFUL PHYSICAL CONSTANTS

Name	Symbol	Value
Free-spin electron g factor	g_e	2.0023
Electron Bohr magneton	μ_B	9.2740×10^{-24} JT^{-1}
Nuclear Bohr magneton	μ_N	5.0508×10^{-27} JT^{-1}
Plank's constant	h	6.6261×10^{-34} J
Boltzmann constant	k	1.3807×10^{-23} JK^{-1}
	μ_0	

EXERCISES

1. Calculate the resonance frequency of protons in a spectrometer that has a magnet of 18.79 T.
2. Calculate the energy difference for nuclear spin transitions of proton, ^{13}C and ^{15}N at 9.39 T.
3. Calculate the ratio for the spin-up spin-down populations for the nuclei used in the previous exercise at 25 °C.
4. Calculate the ratio between the macroscopic magnetisation of protons and ^{15}N at 9.39 T.
5. Calculate the strength for the B_1 field necessary to generate a 90° pulse in proton and ^{13}C with a pulse length of 8 μs.
6. calculate the difference in frequencies between two signals separated by 0.1 ppm when measured in a 500-MHz and a 900-MHz spectrometer.
7. A spectrometer has the choice of a 16-bit ADC with a minimum sampling interval 7 μs, or a 12-bit ADC with a sampling interval of 1 μs. Determine the maximum frequency that can be sampled by each ADC.
8. Calculate the vicinal coupling for a pair of protons when the dihedral angle is 0° and 60°.
9. Assuming that the coupling constant between the unpaired electrons and the nucelus does not change, what is the relationship between the contact shift for a low-spin paramagnetic molecule ($S = 1/2$) and a high-spin molecule ($S = 5/2$)?
10. A proton in a paramagnetic protein exchanges between two positions that have the same angle with respect to the orbital of the unpaired electron. One of these positions (A) is located 2 Å from the centre of coordinates and gives a pseudocontact shift of 1 ppm. What is the pseudocontact shift for the other position (B) that is 4 Å away from the centre?
11. Considering the approximation that the electron is found in the coordinates of the metal nucleus that gives rise to Eqn (4.25), how is the relaxation between the two positions related, considering that the metal nucleus is located at the centre of coordinates of the previous exercise?
12. Copper II has an electronic relaxation correlation time of 1×10^{-9} s. Consider that there are no exchange phenomena in a small molecule of 0.5 kDa and a protein of 40 kDa. Estimate the rotational correlation time for each case and indicate what mechanism dominates the correlation function for the electron-nuclear dipolar interaction in both cases? Do you expect sharp or broad lines for protons?

ANSWERS

1. 800 MHz.
2. 2.65×10^{-25} J, 0.67×10^{-25} J, 0.27×10^{-25} J.
3. 0.999936, 0.999984, 0.999993.
4. 97.23.
5. 0.734×10^{-3} T, 2.92×10^{-3} T.

6. 50 Hz and 90 Hz.
7. 71.4 kHz and 500 kHz.
8. 11 Hz and 4 Hz.
9. The magnitude of the contact shift for the low spin case is 8.6% of that observed for the high-spin case.
10. 0.125 ppm.
11. Relaxation in position A is 64 times the relaxation in position B.
12. The correlation times are approximately 0.25 ns and 20 ns. For the small molecule, it is the rotational relaxation time that dominates, whereas for the macromolecule, it is the electronic relaxation time that dominates.

Chapter 5

^{57}Fe-Mössbauer Spectroscopy and Basic Interpretation of Mössbauer Parameters

Eckhard Bill
Max-Planck Institute for Chemical Energy Conversion, Stiftstrasse 34-36, D-45470 Mülheim/Ruhr, Germany

Chapter Outline

INTRODUCTION

More than 5 decades after the discovery of *nuclear gamma resonance absorption* by R. L. Mößbauer in 1958 (Mössbauer, 1958a, 1958b), the spectroscopy based on the so-called Mössbauer effect is an established tool within bioinorganic spectroscopy, that is useful for the investigation of all types of iron-containing materials including metallo-enzymes. This exotic nuclear method owes its success to the fact that the ^{57}Fe nucleus has the most favourable physical properties among the family of Mössbauer isotopes, and iron in general is one of the most relevant transition metal ions in chemistry. Mössbauer spectroscopy detects iron in all oxidation and spin states, provided the samples are solids, powders or frozen solutions. The spectra can be readily assigned to local properties of the iron sites in the sample, and effects from other constituents cannot easily obscure the comprised chemical information. Moreover, the isotope is not particularly expensive, and small molecules up to molar masses of a few kDa can be studied with natural ^{57}Fe abundance (2.2%). Introductions to the method at various levels of sophistication can be found in a number of reviews and books (Debrunner, 1989; Münck, 1978; Trautwein et al., 1991; Münck, 2000; Martinho and Münck, 2010; Schünemann and Winkler, 2000; Paulsen et al., 2005; Gütlich et al., 2011), some of which address particular applications in transition metal chemistry (Gütlich et al., 2011; Greenwood and Gibb, 1971; Long and Grandjean, 1989; Gütlich and Schröder, 2010); also applications in biochemistry are covered in great breadth and depth by a number of the articles. The purpose of this contribution is to provide a concise overview of what Mössbauer spectroscopy eventually can do for scientific problems in (bio)inorganic chemistry and how to look at Mössbauer data published in literature. The

Practical Approaches to Biological Inorganic Chemistry, 1st Edition. http://dx.doi.org/10.1016/B978-0-444-56351-4.00004-X.

article is mostly focused on the interpretation of isomer shift and electric quadrupole splitting obtained from zero-field measurements, although magnetically perturbed spectra paramagnetic interactions are also briefly mentioned.

PRINCIPLES

Mössbauer spectroscopy is based on resonant absorption of γ-radiation by atomic nuclei. Chemists are familiar with resonant absorption of electromagnetic radiation from the phenomena of light-induced electronic transitions. Visible light from a white incident beam is absorbed by coordination compounds at exactly the energies of *d*-electron transitions or metal-to-ligand charge transfer transitions; this is the main cause of the colours of inorganic complexes. Only when the quantum energy of the light (photons) matches the energy gap between the electronic states does such resonant absorption occur.

An analogous process is possible for γ-radiation, for which nuclear states are involved as emitters and absorbers (Figure 5.1). Observation of this resonance is interesting because nuclear levels are extremely sharp such that hyperfine interactions between the nucleus and the charge and spin in the atomic shell can be observed. This is the concept of Mössbauer spectroscopy. Electric and magnetic fields from the electron shell affect the energies of the nuclear levels, which is seen as shifts and splittings of the resonance lines in the spectra. The data can be interpreted in terms of electronic parameters that provide chemical information about the valence and bonds of the Mössbauer atom.

The Mössbauer Light Source

The γ-photons used in the experiments usually originate from nuclear decay of a radioactive mother isotope (Figure 5.2). The decay cascade in the Mössbauer source passes through the excited nuclear state of the corresponding Mössbauer isotope. In the case of ^{57}Fe spectroscopy, the source isotope is ^{57}Co, which decays by so-called *K*-capture with a lifetime of 270 days. In these events, an electron from the *K*-shell reacts with a proton of the nucleus, yielding ^{57}Fe in a meta-stable excited state. The decay of this state yields a cascade of γ-transitions, the lowest of which has quantum energy $E_0 = 14.4$ keV. For reasons given below, the radiation coming from this transition is selected for the Mössbauer experiment. The lifetime τ of the Mössbauer level is about 100 ns.

According to Heisenberg's uncertainty principle, the lifetime of 100 ns is equivalent to a natural line width ΔE of the Mössbauer emission line of only 4.55×10^{-9} eV, which corresponds to an energy resolution $\Delta E/E_0$ of the order 3×10^{-13}. This is about 100,000 times higher than what is known for typical optical transitions in the electronic shell. For instance, the D-line of sodium atoms has an energy resolution of $\Delta E/E = 2.1 \times 10^{-8}$. The narrow width of the γ-transition leads to complete selectivity for resonant γ-spectroscopy; photons originating from a nuclear transition of a certain Mössbauer isotope cannot be resonantly re-absorbed by other nuclei than those of the same isotope. This means that a Mössbauer spectrometer equipped with ^{57}Co as a 'light source' can and will detect only ^{57}Fe in the Mössbauer sample, which is not radioactive since ^{57}Fe is a stable isotope.

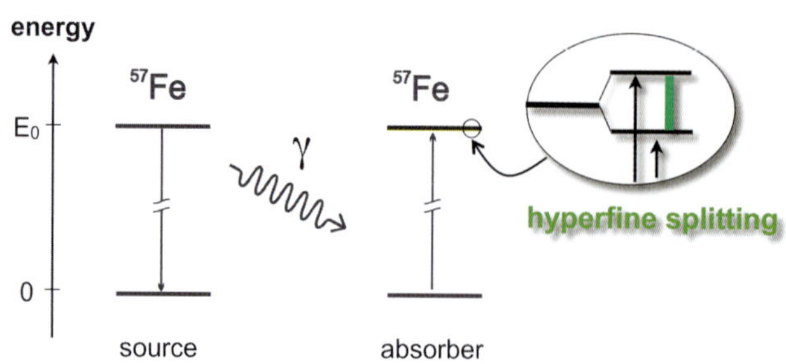

FIGURE 5.1 Emission of γ-rays from the excited state of ^{57}Fe in the Mössbauer source and resonant absorption in the absorber.

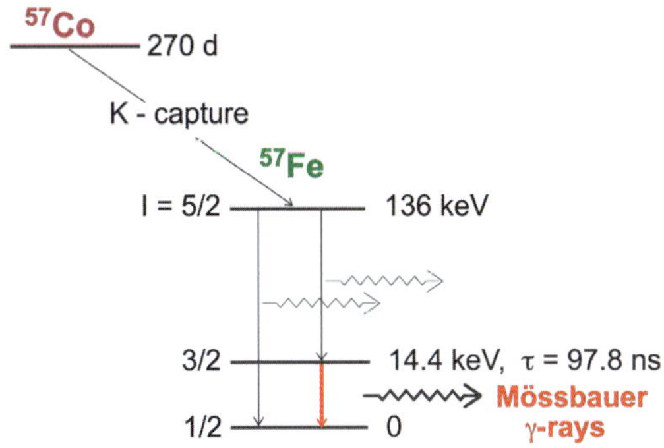

FIGURE 5.2 The light source for ^{57}Fe-Mössbauer spectroscopy. Meta-stable ^{57}Fe nuclei are generated by *K*-capture decay of radioactive ^{57}Co. The nuclear reaction yields energy to populate the excited state at 136 keV, the decay γ-cascade of which passes through the Mössbauer level at 14.4 keV.

Recoil Energy is a Problem

Unfortunately, the narrow width of γ-lines in principle hampers the observation of nuclear γ-resonance, because photons have a momentum and γ-transitions are affected by a loss of nuclear energy due to the recoil imparted to the nucleus during emission and absorption. This is a general effect that occurs also for optical transitions, but there it is usually insignificant because the recoil energy E_R is negligible compared to the experimental line widths. The recoil energy depends on the square of the photon energy, $E_R \propto E_0^2$, and for the 14 keV radiation of ^{57}Fe, E_R exceeds the width of the Mössbauer line by a factor of 3,000,000! As a consequence, nuclear γ-resonance absorption is not possible when the Mössbauer atoms are free to move (Figure 5.3). (In practice, Doppler broadening of the emission and absorption profiles due to thermal motion of the atom and molecules in gases and liquids restores some minor overlap of the profiles. However, the absorption effect is so small that it cannot be used for practical applications.)

Recoilless Emission and Absorption – The Mössbauer Effect

Rudolf L. Mößbauer discovered in his Ph.D. work (Mössbauer, 1958a, 1958b) that photon emission and absorption in solids is not necessarily affected by recoil energy loss. There is instead a certain probability *f* that a γ-transition does not pass all of its energy to the 'crystal lattice', which would mean exciting a vibration of the Mössbauer atom as a part of a phonon excitation; the elastic bonds prevent translational motions anyhow. Such zero-phonon processes yield sharp and unaffected lines with natural line width in the spectrum of a γ-source, that are necessary for resonant absorption by nuclei of the same kind. The discovery and understanding of the effect (Mössbauer, 1958a, 1958b; Mössbauer, 1959) merited a Nobel prize. Since then nuclear resonance absorption of γ-rays observed in solid materials, powders and frozen solutions has been called the *Mössbauer effect*.

The probability factor *f* for zero-phonon processes is obtained from a quantum mechanical treatment of the nucleus and its vibrational environment. The values are inversely proportional to the recoil energy, or in terms of the γ-energy: $f \propto 1/E_0^2$. The *f* factor directly determines the possible intensity of the Mössbauer spectrum. ^{57}Fe is a good Mössbauer isotope, because it has high *f* factors due to the low-energy transition at 14.4 KeV (but the radiation is still 'hard' enough to penetrate real samples). The *f* factor, also known as the Lamb–Mössbauer factor, is virtually equivalent to the Debye–Waller factor for elastic X-ray scattering by atoms. Characteristic *f*-values for the 14.4 keV transition of ^{57}Fe can be as low as 0.1 when the Mössbauer atom is in a 'soft' environment like a protein sample at 270 K, but the values can increase to 0.6–0.9 for bulk solids or upon cooling to 77 K or lower. Therefore, freezing of solution samples is essential, and cooling to cryogenic temperature is required for good absorption effects.

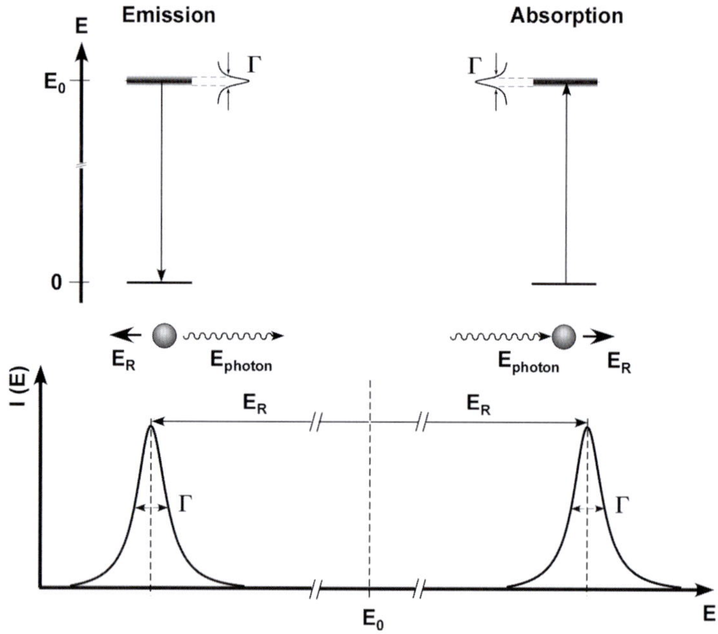

FIGURE 5.3 The natural line width of the excited state of the Mössbauer nucleus is many orders of magnitude smaller than the loss of the photon energy E_{photon} due to recoil energy E_R imparted to nuclei when it is free to move during the nuclear emission and absorption. Since the resulting energy separation of the Lorentzian emission and absorption profiles by $2E_R$ prevents any significant overlap of the two lines, resonant nuclear absorption of γ-photons is not possible for free atoms.

An overview of the isotopes for which Mössbauer effect has been demonstrated is given in Figure 5.4. Summarizing the above arguments about nuclear states, vibrations and γ-rays, one may note again that ^{57}Fe is virtually the best Mössbauer isotope, not only because it has an appropriate low-lying excited nuclear state, affording high f factors, but also the lifetime is suitable, yielding a (not too) sharp γ-line. Moreover, a convenient mother isotope (^{57}Co) is available, and last but not least, the nuclear spin and hyperfine interactions are reasonable and sensitive, providing valuable chemical information (which is described below).

The Mössbauer Experiment

According to the decay scheme of Figure 5.2, the radioactive Mössbauer source emits monochromatic γ-rays that are perfectly tuned for the observation of Mössbauer effect when the Mössbauer isotope has the same chemical environment as in the source and, thus, has exactly the same nuclear energies. However, for recording the full absorption *spectrum* of a general Mössbauer absorber, in order to obtain the energy and shape of the absorption line(s), the energy of the incident photons arriving at the absorber has to be varied to probe the absorption probability as a function of photon energy, – just like in any spectroscopy.

Since the nuclear levels in the Mössbauer source cannot be tuned experimentally, the energy modulation has to be done after γ-emission by moving the source during emission relative to the absorber, so that the Doppler effect adds or subtracts energy increments to the energy of the photons in the laboratory system. Doppler modulation depends on the ratio of the source velocity v and the speed of light, v/c (Figure 5.5). Nevertheless, the observation of usual hyperfine splitting in ^{57}Fe-Mössbauer spectra does not require higher Doppler velocities than ca. ±10 mm/s; for ^{57}Fe, 1 mm/s corresponds to 4.8×10^{-8} eV.

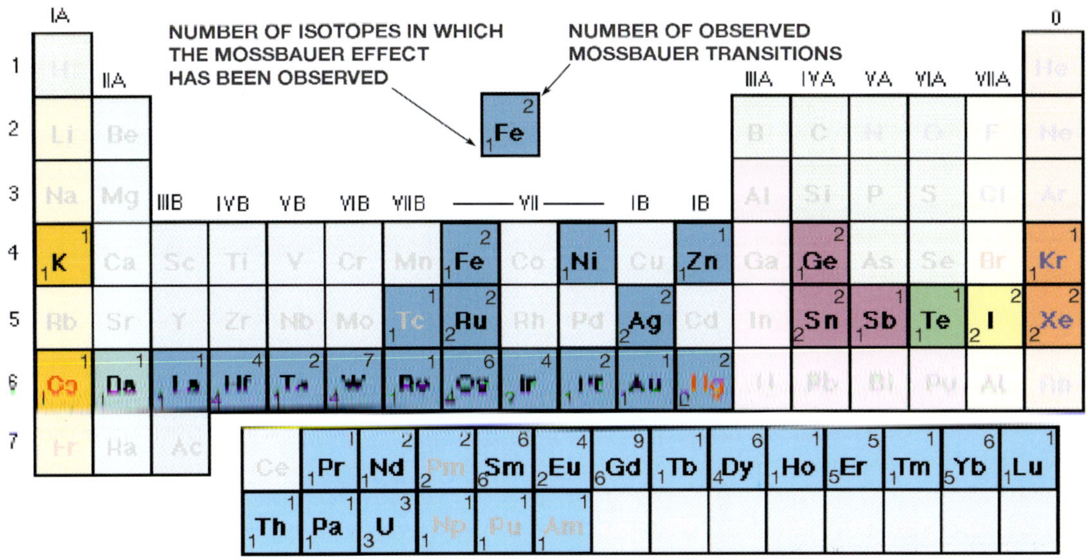

FIGURE 5.4 Table of Mössbauer active elements. The most suitable isotope for applications and most easy to use is ^{57}Fe. *(Table by courtesy of Prof. J.G. Stevens, Mössbauer Effect Data Center, previously UNC North Carolina, since 2010 Dalian Institute of Chemical Physics within Chinese Academy of Sciences. For on-line database of Mössbauer isotopes and more see: http://www.medc.dicp.ac.cn/Resources.php).*

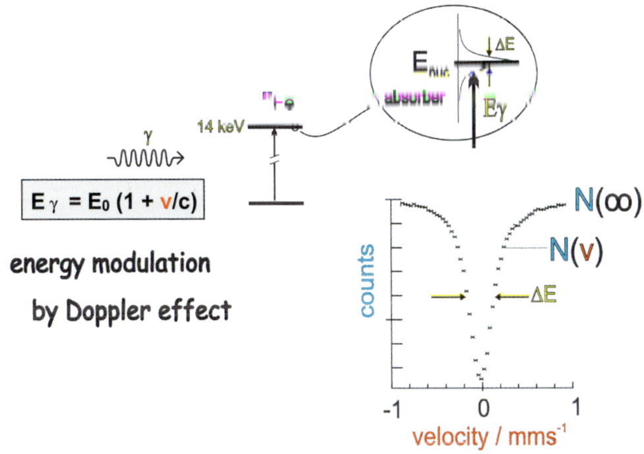

FIGURE 5.5 Recording the line shape of the nuclear energy level(s) in an absorption experiment requires that the energy of the photons emitted by the Mössbauer source gets modulated by the Doppler effect. This is achieved by moving the source relative to the absorber with velocity v. The energy modulation is $E_\gamma = E_0\ (1 + v/c)$, where E_0 is the energy of the nuclear transitions and c is the speed of light. The transmission spectrum, presented as numbers of photons passing through the absorber as a function of the Doppler velocity, shows a 'dip' at the Doppler velocity for which resonant absorption occurs. The shape of the absorption profile reveals the Lorentzian line shape and natural line width of the nuclear energy levels.

The Mössbauer Spectrometer

Most Mössbauer spectra are recorded with a transmission setup as shown in Figure 5.6. The light is coming from the Mössbauer source and passes through the absorber, which is the sample under investigation. The absorber is not radioactive and rests in a cryostat or another compartment. The source is mounted on the moving part of a Mössbauer drive system. For recording the spectrum, the number of γ-rays passing the absorber is counted with

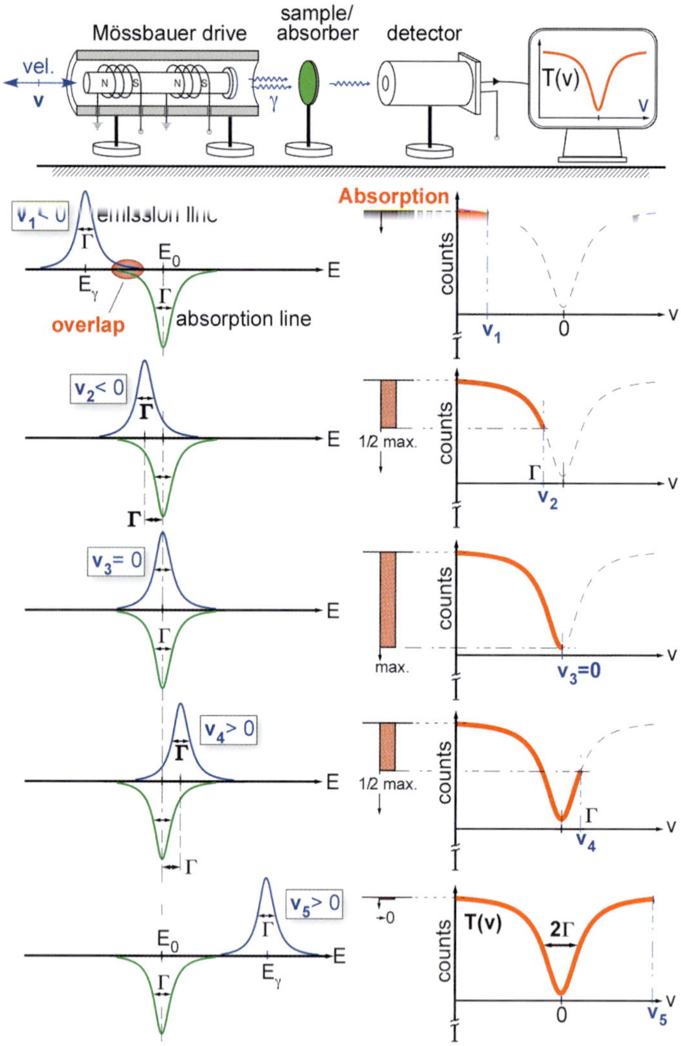

FIGURE 5.6 Mössbauer transmission experiment in five steps. The absorption probability depends on the energy shift of the emission line due to Doppler modulation ($E_\gamma = E_0(1 + v/c)$), affording a certain 'overlap' of emission and absorption lines. The transmission spectrum $T(v)$ is usually normalised to the count rate $C(\infty)$ at high velocity, $v \rightarrow \infty$. In the so-called thin absorber approximation, the absorption spectrum of a single-line absorber is a Lorentzian line given by $\frac{C(\infty) - C(v)}{C(\infty)} = f_s \frac{t}{2} \frac{\Gamma^2}{[E_0(v/c) - \Delta E]^2 + \Gamma^2}$, where f_S is the Lamb–Mössbauer factor for the source, and t is the so-called effective thickness of the absorber (Gütlich et al., 2011), Γ is the natural line width of the Mössbauer transition, and ΔE denotes a hyperfine shift of the absorption line. Note that the experimental line width is 2Γ because an emission line scans an absorption line of the same width.

a γ-counter and detection system that is tuned to discriminate the 14.4 KeV Mössbauer photons from other radiation, which get electronically rejected from counting. During the measurement, the radioactive source is periodically moved with controlled Doppler velocities, $+v$ towards and $-v$ away from the absorber. The counting system is synchronised with the velocity control of the driving system such that the full velocity range $\pm v$ is divided into a number of discrete velocity increments, v_i, for storing the detected data (so-called channels). The numbers of counts, $C(v)$, arriving at the detector and detected in each channel represent the Mössbauer spectrum.

In this experiment, the probability for resonant nuclear γ-absorption at each velocity is determined by the amount of overlap of the Doppler-shifted Lorentzian emission line and the Lorentzian absorption line, as indicated on the left side of Figure 5.6. Stronger overlap yields more absorption probability and, thus, less transmission. Maximum resonance occurs at complete overlap when the energies of emission and absorption lines fully match.

^{57}Fe HYPERFINE INTERACTIONS

The electrons in the vicinity of the Mössbauer nucleus affect the nuclear energy levels involved in the Mössbauer experiment due to the electric and magnetic fields that they generate. There are essentially three different types of resulting hyperfine interactions that shift and split the nuclear states. The effects are seen in the Mössbauer spectra as isomer shift, electric quadrupole splitting and nuclear magnetic Zeeman splitting (Figure 5.7). Hyperfine interactions can be formally described in terms of the nuclear spin of the Mössbauer isotope, which is $I = 1/2$ for the ground state of ^{57}Fe and $I = 3/2$ for the excited state.

The material of the radioactive source in a Mössbauer experiment is selected such that no hyperfine splitting occurs for the nuclear levels of the Mössbauer isotope, obviously in order to get emission from a non-split single-line (Figure 5.7, top left). In the Mössbauer sample (absorber), *magnetic dipole interaction* may completely lift the degeneracy of ground and excited state (Figure 5.7, top right). However, only six transitions are allowed in this case, because of the selection rules $\Delta m_I = 0, \pm 1$ for the Mössbauer radiation. As a consequence, a characteristic six-line pattern is observed. In contrast, *electric quadrupole interaction* affords partial splitting of the excited state levels into a doublet, whereas the ground state remains non-split (middle panel) because the state with $I = 1/2$ does not have a quadrupole moment. A line doublet is observed in the spectrum, the splitting of which is called electric quadrupole splitting ΔE_Q. (For powders and frozen solutions, the intensity distribution of the quadrupole doublet is symmetric). In addition to magnetic Zeeman and electric quadrupole interaction, the nuclear levels may be subjected to common energy shifts, arising from Coulomb (monopole) interaction of the nucleus with the surrounding electrons. Since the effect is different for ground and excited state of the Mössbauer isotope, and it is different for the sample and the source (Figure 5.7, top left and middle), this gives rise to a uniform *isomer shift* of all Mössbauer lines in the spectra. It should be noted that isomer shifts can be quoted only relative to a reference,

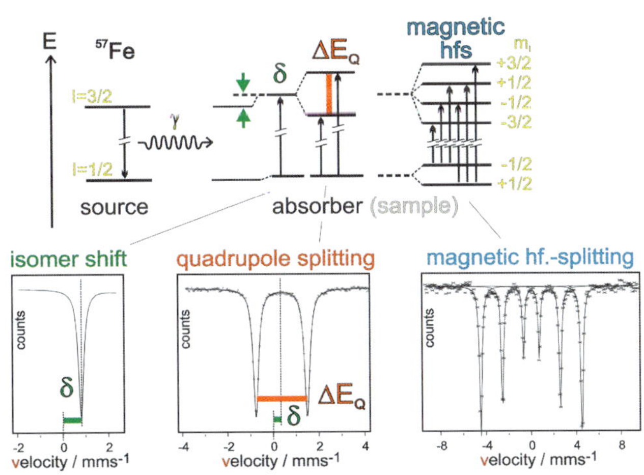

FIGURE 5.7 Overview of ^{57}Fe hyperfine interactions. Electric monopole interaction in source and absorber causes the isomer shift δ of the spectra (left spectra panel). Electric quadrupole interaction splits the excited state in the absorber and yields a symmetric quadrupole doublet in the spectrum with quadrupole splitting ΔE_Q (middle panel). The centre of the quadrupole doublet may be additionally subjected to isomer shift δ. Magnetic hyperfine interaction totally lifts the degeneracy of the nuclear spin states, such that six transition lines are observed due to the selection rule $\Delta m_I = 0, \pm 1$ (right panel). The centre of the magnetic spectrum also may be displaced by an isomer shift δ.

which can be the material of the Mössbauer source (e.g. ^{57}Fe in rhodium metal) or a reference absorber. The usual reference nowadays is metallic (α-)iron, which is mostly used also as a standard absorber for velocity calibration of Mössbauer spectrometers. In old literature, other reference material like nitroprusside etc. may be found.

ISOMER SHIFT AS INFORMATIVE HYPERFINE INTERACTION

The phenomenon of isomer shift is related to the fact that the nucleus has a finite size, which changes upon the Mössbauer transition. Moreover, it is immersed in a cloud of negative electron charge density that depends on the chemical properties and environment of the Mössbauer atom. The Coulomb energies of this system apparently depend on the nuclear properties and the chemical environment of the Mössbauer atom. The resulting isomer shift δ is proportional to the electron density $e|\psi(0)|^2$ at the Mössbauer nucleus in the absorber, which can be expressed as $\delta = \alpha\{\psi|(0)|_A^S - C\}$, where $\psi(0)$ is the radial part of the electron wave function at distance $r=0$, α is an isomer shift calibration constant, and C summarises the relevant properties of the source material or a reference absorber (Gütlich et al., 2011).

The electron density at the nucleus, $e|\psi(0)|^2$, originates primarily from the ability of s-electrons to penetrate the nucleus, whereas p- and d-valence electrons essentially do not contribute to $|\psi(0)|^2$ because their orbitals have nodes at $r=0$. Nevertheless, experimental isomer shifts depend on the electronic structure and bond of the Mössbauer atom, particularly the oxidation state (Figure 5.8). In an (over)-simplified approach, derived from the situation for hypothetically free atoms and ions, this effect is often explained by the expansion of inner s-orbitals

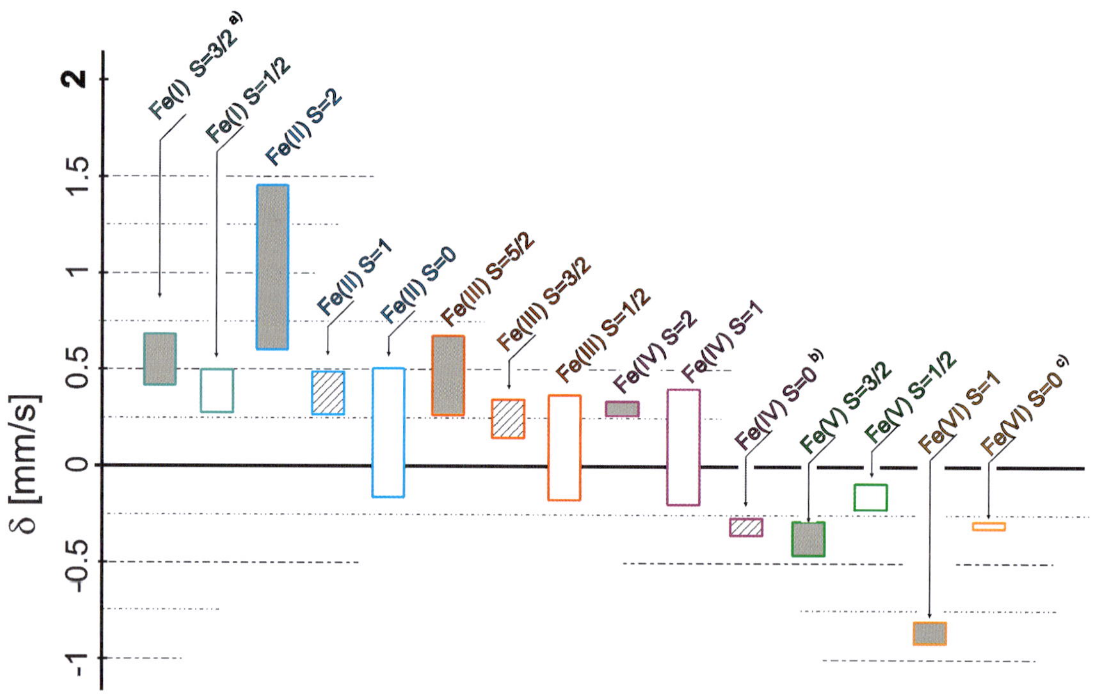

FIGURE 5.8 Isomer shifts observed for ^{57}Fe compounds relative to metallic iron at room temperature. Filled boxes indicate high-spin compounds, hashed boxes mark intermediate spin, and empty boxes are used for low-spin complexes. [a)]Only examples of unambiguous monovalent character are mentioned; these are structurally characterised molecular Fe(I)-diketiminate and tris(phosphino)borate complexes with *three-coordination*, which explains the low δ values as resulting from short bonds. [b)]The usual low-spin state of the $3d^4$ configuration for Fe(IV) in quasi-octahedral or tetrahedral symmetry has $S=1$. The 'low-low-spin' state with $S=0$ is found for distorted trigonal-prismatic sites with strong ligands. [c)]Diamagnetic Fe(VI) with spin $S=0$ occurs only in ferrates, and there is only one example of a *molecular* iron(VI) complex, which is six-coordinate and has spin $S=0$ (Berry et al., 2006).

due to shielding of the nuclear potential by electrons in the valence shell. For instance, six *d*-electrons of iron(II) ions afford more shielding and more *s*-expansion than five *d*-electrons of iron(III). As a consequence, $|\psi(0)|^2$ should be lower for iron(II) than for iron(III), and the isomer shift should be higher, because the calibration parameter α is negative. The explanation is qualitatively correct and in accord with experiment, but in reality it describes only a minor contribution to the isomer shift.

In addition to the indirect shielding effect, a direct contribution to the isomer shift arises from partial population of valence 4*s*-orbitals due to their participation in molecular orbitals (MOs). Quantum chemical calculations show that this effect plays a decisive role, particularly because the 4*s*-contribution to $|\psi(0)|^2$ shows significant variation in relation with iron-ligand bond lengths and coordination numbers. Chemical coordination of the Mössbauer atom apparently causes a compression of the radial distribution of the 3*s*- and 4*s*-iron orbitals and thus alters their density at the nucleus. The response of this effect on different bonding situations determines the major part of the variations of the isomer shift found for iron complexes. Based on these theoretical perceptions, it was suggested that iron-ligand *bond lengths* are the key parameters for understanding isomer shifts (Neese, 2002) (Neese and Petrenko, 2011).

The most obvious correlation of the isomer shift with a chemical parameter, and the most useful for practical applications, is a general relationship with the formal oxidation and the spin state of iron as shown in Figure 5.8 (which can also be explained by effects of the bond lengths involved!). The trends in the diagram are more obvious for high-spin complexes than for low-spin and low-valent states, for which the dependence of δ on the oxidation state increasingly fades. However, such overall comparisons for unrestricted types and numbers of compounds are not generally conclusive. Only isomer shifts above ca. 1 mm/s reveal a distinct electron configuration, namely Fe(II) high-spin; in all other cases, various interpretations are possible. The ambiguity is due to the fact that isomer shifts, among other factors, depend on the number of ligands, and via the bond length relation on the σ-donor and the π-acceptor strengths of the ligands. Rather conclusive isomer shift correlation diagrams are found for systematic series of related compounds. An example for synthetic iron centres is presented later in Figure 5.14, which was used to assign high-valent Fe(IV), Fe(V) and Fe(VI), modelling intermediates of the reaction cycle of oxygen- and nitrogen-activating enzymes. A macrocyclic ligand, cyclam, with four nitrogen donor atoms were used throughout the series of compounds and the number of valence electrons was 'titrated' only within the range of t_{2g} orbitals, i.e. all compounds have low-spin iron centres. The minimal number of chemical variations in that series yields a remarkably straight correspondence.

Tight isomer shift correlations have also been established also for *iron-sulfur clusters*, most of which are oligonuclear and include one to four tetrahedral high-spin [FeS_4]-sites. The iron centres are redox-active and show a variety of charge-delocalised mixed-valence states, as has been demonstrated by Mössbauer spectroscopy (Schünemann and Winkler, 2000; Beinert et al., 1997; Mouesca and Lamotte, 1998; Holm et al., 1996; Solomon et al., 2008; Rao and Holm, 2003). Isomer shifts have been empirically related to the average oxidation number x of the iron centres according to the expression $\delta(x) = [1.43 - 0.40\ x]$ mm/s, with x ranging from 2.0 to 3.0, depending on how many iron sites and valence electrons are available (Hoggins and Steinfink, 1976).

In summary, isomer shifts are experimentally correlated to bond lengths of the Mössbauer atom as follows:

- Higher oxidation states show shorter bond lengths and lower (more negative) isomer shifts.
- High-spin states have longer bond lengths than low-spin states and higher isomer shifts.
- Four-coordination affords shorter bond lengths than six-coordination and lower isomer shifts.
- More covalent bonds are shorter and induce lower isomer shifts. Complexes with 'soft' ligands like sulfur have lower isomer shifts than complexes with 'hard' ligands like oxygen or nitrogen.

ELECTRIC QUADRUPOLE SPLITTING

Electric quadrupole interaction is related to the orientation of non-spherical nuclei in the inhomogeneous electric field generated by an asymmetric charge distribution of the surrounding electrons. The resulting hyperfine coupling energy depends on the nuclear quadrupole moment and the *electric field gradient* (*efg*) generated by the electrons. Only nuclei with spin $I > 1/2$ are subjected to quadrupole interaction, because others do not have a quadrupole moment. Thus, the ground state of ^{57}Fe is not affected, but the excited state is split into a two Kramers doublets with magnetic quantum numbers $m_I = \pm 1/2$ and $\pm 3/2$. (Figure 5.7, top middle; the reader may note the similarity of nuclear quadrupole splitting and zero-field splitting in the electric shell for $S = 3/2$ systems). Therefore, two γ-transitions are possible for ^{57}Fe with electric quadrupole interaction, the energy difference of which is seen as quadrupole splitting ΔE_Q of the resulting spectrum.

The *efg* is a traceless tensor a ($\Sigma V_{ii} = 0$, see Figure 5.9) with only two independent components in the principal axes system. Usually, the so-called main component V_{zz} and the asymmetry parameter $\eta = (V_{xx} - V_{yy})/V_{zz}$ are selected as independent parameters. The range of η can be restricted to $0 \leq \eta \leq 1$, since the tensor axes are selected such that V_{zz} is the largest component. For ^{57}Fe, the quadrupole splitting is given by the relation $\Delta E_Q = 1/2 \cdot eQV_{zz}\sqrt{1 + \eta^2/3}$, where eQV_{zz} is called the quadrupole coupling constant. The main component of the *efg* may be positive or negative, depending on the type of asymmetry of the electronic charge distribution that causes the *efg* (see Figure 5.9, right). Next, we will discuss some archetypical charge distributions in the valence shell that cause an *efg* at the Mössbauer nucleus.

The quadrupole splitting $|\Delta E_Q|$ of ^{57}Fe in bioinorganic compounds is usually found in the range zero to about 4 mm/s; whereas some particular synthetic nitrido complexes related to nitrogen fixation reactions show up to 6.23 mm/s (Hendrich et al., 2006; Vogel et al., 2008; Scepaniak et al., 2011). A finite *efg* indicates always the presence of a non-spherical or non-cubic distribution of electronic charge around the nucleus. Axial elongated distributions cause an *efg* with negative V_{zz} (negative ΔE_Q), and axial compressed or oblate charge distributions cause positive V_{zz}. Due to a $1/r^3$ – dependency of the *efg* – components on the distance of the nucleus to the

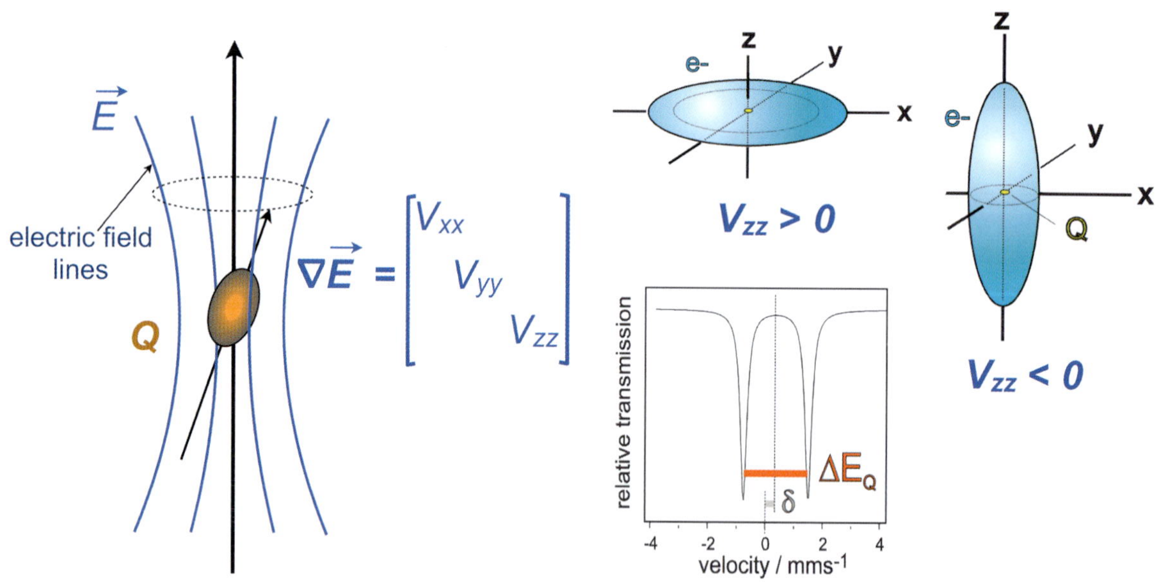

FIGURE 5.9 Nuclei with $I > 1/2$ (like ^{57}Fe in the excited state) have an electric quadrupole moment Q, which has rotational energy in an electric field gradient (*efg*) ∇E (left). The *efg* is mathematically described by a 3×3 matrix V_{ij}, with $i,j = x,y,z$. Precession of the electric quadrupole moment Q in the electric field gradient ∇E is quantised, which gives rise to discrete quadrupole coupling energies of the nucleus and the electron shell. This is seen as quadrupole splitting ΔE_Q of the Mössbauer spectrum (middle). Oblate axial electronic charge distributions give rise to a positive *efg* at the nucleus ($V_{zz} > 0$, $\eta = 0$), whereas elongate, cigar-shaped charge distributions cause a negative *efg* ($V_{zz} < 0$, $\eta = 0$) (right). The sign of V_{zz} can be determined only from magnetically perturbed spectra discussed below, not from pure quadrupole doublets as shown here.

originating charges, the electrons close to the nucleus have the most important effect, i.e. the valence electrons of the Mössbauer atom. The corresponding *valence contributions* to the *efg* parameters, $(V_{zz})_{\text{val}}$ and η_{val}, are determined by the iron-centred MOs formed between the central Mössbauer atom and the ligands. Strong values for $(efg)_{\text{val}}$ arise when the oxidation state of the metal ion and the chemical bonds afford significant differences in the population of the local atomic valence orbitals. This is often the case for open-shell complexes. As long as covalence effects are not too strong, first 'hand-waving' interpretations of iron quadrupole splittings can be derived from the values estimated for $(efg)_{\text{val}}$ from the $3d^n$ electron configuration, given in a simple crystal field model. In this approximation, $(efg)_{\text{val}}$ is incrementally built up by summarizing the contribution of all *d*-electrons, using the expectation values given in Table 5.1. (One may note the different signs of V_{zz} for the different 'elongate' and 'oblate' orbitals d_{z2} or $d_{x2\text{-}y2}$ for instance).

Addition of the tensor elements for, e.g. the five *d*-orbitals in the $3d^5$ configuration of Fe(III) high-spin, yields zero (one may focus on V_{zz} and sum up the values in that column). The result corresponds to the experimental observation of small $|\Delta E_Q|$ in the range 0–0.5 mm/s found for most 'ionic' ferric high-spin compounds. In contrast, the contribution of six electrons for the $3d^6$ configuration of Fe(II) high-spin yields a finite value for $(V_{zz})_{\text{val}}$. If we assume that the sixth electron is found in a low-lying doubly occupied d_{z2} orbital, the expectation value for $(V_{zz})_{\text{val}}$ is $-4/7e<r^{-3}>$. This can be converted into a quadrupole splitting of about −4 mm/s by using the parameters given in the caption of Table 5.1. This value is considered to be large. Alternatively, if $\{d_{xz}, d_{yz}\}$ would be ground state orbitals and doubly occupied, the expectation for $(V_{zz})_{\text{val}}$ would be only half the value, $+2/7e<r^{-3}>$. Correspondingly, most ferrous high-spin compounds show in fact large quadrupole splitting of the order 2–4 mm/s. In similar manner, one can predict large quadrupole splitting for ferric low-spin compounds with $e_g^0 t_{2g}{}^5$ configuration, having a 'hole' in the t_{2g} orbital set, or vanishing ΔE_Q for ferrous low-spin compounds due to the filled t_{2g} orbital set. More practical examples will be presented below.

Crystal-field theory dealing with pure metal *d*-orbitals can often provide a reasonable insight into the origin of 'large' or 'small' quadrupole splitting, but quantitative estimates of *efgs* are not possible, even for ionic compounds. In reality, ligand orbitals participate substantially in the valence orbitals forming covalent chemical

TABLE 5.1 Expectation Values of $(efg)_{\text{val}}$ Tensor Elements for *d*-Electrons in Quasi-Octahedral Symmetry. The Values are Given in Units of e $<r^{-3}>$, Where e is the Proton Charge and $<r^{-3}>$ Is the Expectation Value for the Cube Orbital Radius r^{-3}. The Expectation Values Can Be Converted into Quadrupole Coupling Energies (in mm/s) by Inserting the Actual Values for the Quadrupole Moment and for $<r^{-3}>$ of the Orbitals into the Appropriate Equations (Gütlich et al., 2011). An Estimate of $(\Delta E_Q)_{\text{val}}$ May be Obtained with the Conversion Factor 4.5 mm/s per 4/7 e $<r^{-3}>$ (Obtained for $<r^{-3}> = 5a_0^{-3}$ and $Q = 0.16$ b). Accordingly, a Single Electron in a Hypothetically Pure d_{x2-y2}-Orbital, e.g. Is Expected to Yield a Quadrupole Splitting of +4.5 mm/s When Covalence Effects and Lattice Contributions Are Neglected

Orbital	$\frac{(V_{xx})_{\text{val}}}{e<r^{-3}>}$	$\frac{(V_{yy})_{\text{val}}}{e<r^{-3}>}$	$\frac{(V_{zz})_{\text{val}}}{e<r^{-3}>}$
d_{xy}	−2/7	−2/7	+4/7
d_{xz}	−2/7	+4/7	−2/7
d_{yz}	+4/7	−2/7	−2/7
d_{x2-y2}	−2/7	−2/7	+4/7
d_{z2}	+2/7	+2/7	−4/7

bonds so that the electron configuration cannot be realistically described by integer numbers $3d^n$. Resulting non-equivalences of the *d*-population numbers reflect charge asymmetries around the nucleus that may cause substantial covalence contributions to the *efg*. In addition, ligand orbitals can also contribute to this effect. Therefore, quantum chemical calculations are necessary for quantitative interpretation of Mössbauer quadrupole splitting (Neese and Petrenko, 2011).

The most prominent examples of large *covalence contribution* to the *efg* in bioinorganic chemistry are found for ferric high-spin ions in μ-oxo dimers with an Fe(III)-O-Fe(III) core. The centres are typically encountered in non-heme iron proteins like ribonucleotide reductase. They show quadrupole splitting as large as −2.45 mm/s ($\delta = 0.53$ mm/s) (Atkin et al., 1973; Vincent et al., 1990), in spite of the fact that iron has formally a half-filled *d*-shell, which in a crystal field model affords vanishing *efg*, $(\Delta E_Q)_{val} = 0$. However, there is considerable charge anisotropy as the five or six covalent bonds around the iron centres are very different. Particularly, the bridging oxo group contributes a uniquely short Fe-O bond ($d \approx 1.64-1.68$ Å) (Vincent et al., 1990) whereas the other O- and N- bonds are 'normal' and less covalent. Thus, one (or two) of the iron *d*-electrons will be much more affected by the covalence of the Fe−O bond than the others, namely those that have the right orientation for strong σ- and π-interactions. Their population numbers deviate significantly from those for a half-filled shell in a purely ionic compound, signifying distinct charge anisotropy. The (negative) sign of the resulting *efg*, indicating an *elongate* electron distribution, is consistent with strong σ-donation from the oxygen atom into the half-filled d_{z^2} orbital, if the iron-oxo bond is in *z*-direction.

MAGNETIC HYPERFINE SPLITTING

The energy of the magnetic Zeeman interaction is given by the product of the nuclear magnetic moment μ_N and the field B at the nucleus, $E = -\mu_N B$. Large magnetic hyperfine splitting as shown in the right panel of Figure 5.7 is encountered in bioinorganic chemistry, e.g. for iron(III) high-spin paramagnetic compounds at liquid helium temperatures with external fields applied to polarise the spin system. Then, strong spin magnetic moments from unpaired valence electrons can yield internal fields of more than 50 T. The resulting magnetic splitting of the Mössbauer spectra are 16 mm/s or more. Although applied-field Mössbauer spectroscopy of paramagnetic iron centres is an interesting topic, an introduction would be too long to include here and the topic is covered in various reviews and book articles (Debrunner, 1989; Münck, 1978; Trautwein et al., 1991; Münck, 2000; Martinho and Münck, 2010; Schünemann and Winkler, 2000; Paulsen et al., 2005; Gütlich et al., 2011).

For first, basic interpretations of magnetic spectra the overall magnetic splitting between the outermost lines can be converted into units of the internal field at the nucleus by using the factor 0.323 mm/s per Tesla.

COMBINED HYPERFINE SPLITTING

For molecular iron compounds, the combined effect of magnetic Zeeman interaction and electric quadrupole interaction is quite common when paramagnetic compounds are measured and magnetic fields are applied to the Mössbauer samples. This is expected as most iron sites are somehow asymmetric and have an *efg*, affording quadrupole interaction; cubic symmetry is extremely rare except for solid-state materials with regular crystal lattices. The general description of combined hyperfine interaction is difficult and requires numerical procedures, particularly when both interactions are of about the same strength (Gütlich et al., 2011). However, when one of the two interactions is weaker than the other, the resulting nuclear level splitting can be rather easily predicted from perturbation treatments (Figure 5.10).

The result in both cases of combined hyperfine interaction, ($\mu_N B >> eQV_{zz}$) or ($eQV_{zz} >> \mu_N B$), a perturbed spectrum is observed, from the asymmetry of which the sign of the *efg* can be determined. The use of the latter case, i.e. magnetically perturbed quadrupole spectra, will be demonstrated below for some applications.

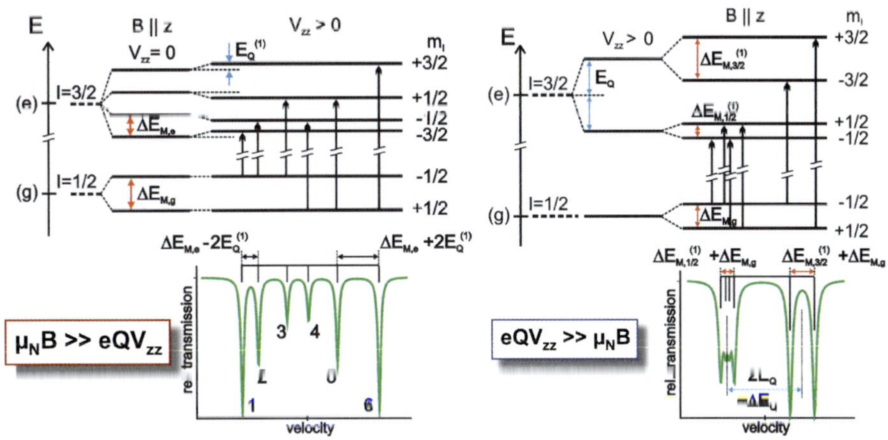

FIGURE 5.10 Combined magnetic dipole (Zeeman) and electric quadrupole interaction for the cases of weak quadrupole perturbation of a magnetic Zeeman interaction (left), and weak magnetic perturbation of a quadrupole spectrum (right). In the first case ($\mu_N B >> eQV_{zz}$), the inner four lines of the six-line pattern are shifted relative to the outer two. The direction of the shift depends on the sign of *efg* component V_{ij} along the magnetic field (here the field B is parallel $V_{zz} > 0$). The corresponding quadrupole splitting in a non-magnetic situation would be twice the quadrupole shift E_Q seen in the perturbed magnetic spectrum, $\Delta E_Q = 2E_Q = 0.5[(L_6 - L_5) - (L_2 - L_1)]$, where L_i indicates the Doppler energies of lines *i*. In the second case ($eQV_{zz} >> \mu_N B$), the manifold of absorption transition from the ground state levels to the excited state levels underlying an unperturbed quadrupole doublet is split in characteristic manner, providing a 3:2 splitting pattern of the generic quadrupole lines. The asymmetry of the spectrum reveals the sign of the *efg*. Here, the situation for $V_{zz} >$ is shown; for negative values, the perturbed quadrupole spectrum would be inverted on the velocity axis.

APPLICATIONS – SELECTED EXAMPLES

Oxidation and Spin States in a Non-Heme Di-Iron Centre

Most of the enzymes involved in the activation of molecular oxygen have one or more iron atoms in their active site. Our first example for the characterisation of oxidation and spin states of biological iron centres deals with the enzyme component called BoxB, which catalyses the de-aromatisation key step in a particular pathway for metabolic degradation of aromatic compounds (Rather et al., 2011). The enzyme has a two-iron centre, the structure of which is shown on the left side of Figure 5.11. The enzyme occurs in diferric, diferrous, and semi-reduced redox states as was shown by Mössbauer and EPR spectroscopy (Rather et al., 2011). Productive O_2 binding starts from the semi-reduced state in presence of benzoyl-CoA.

Oxidised BoxB shows a quadrupole doublet with isomer shift $\delta = 0.49$ mm/s, and electric quadrupole splitting $\Delta E_Q = 0.69$ mm/s (Figure 5.11, right top). The values are typical of high spin Fe^{III} with five or six 'hard' oxygen or nitrogen ligands, similar to what was found for other non-heme diiron enzymes in oxidised state (Kurtz, 1990). The values reveal, in particular, that there cannot be a bridging oxo-ligand, since that would induce significantly higher quadrupole splitting, as mentioned already above in the discussion of quadrupole splitting. A recent example of an enzyme with a di-ferric centre occasionally having either an oxo- or an hydroxo bridging group, is found in stearoyl-acyl carrier protein Δ9D desaturase. Mössbauer studies of the resting enzyme revealed $\Delta E_Q = 1.53$ mm/s, $\delta = 0.54$ mm/s for the oxo-bridged dimer, and $\Delta E_Q = 0.72$ mm/s, $\delta = 0.49$ mm/s for the hydroxo-bridged centre (Fox et al., 2004). The difference owes its origin to the pronounced charge asymmetry that is induced by the short iron-oxo bond, in contrast to the 'normally' elongated bonds expected for hydroxo groups.

Upon reduction of BoxB, the Mössbauer spectrum changes significantly and shows a new, slightly asymmetric quadrupole doublet (Figure 5.11, right-bottom) that could be simulated with two Lorentzian doublets (in green) with parameters $\delta(1) = 1.20$ mm/s, $\Delta E_Q(1) = 2.39$ mm/s, and $\delta(2) = 1.31$ mm/s, $\Delta E_Q(2) = 2.77$ mm/s, using equal intensities but different line width $\Gamma(1) = 0.32$ mm/s and $\Gamma(2) = 0.50$ mm/s. The simulated values clearly

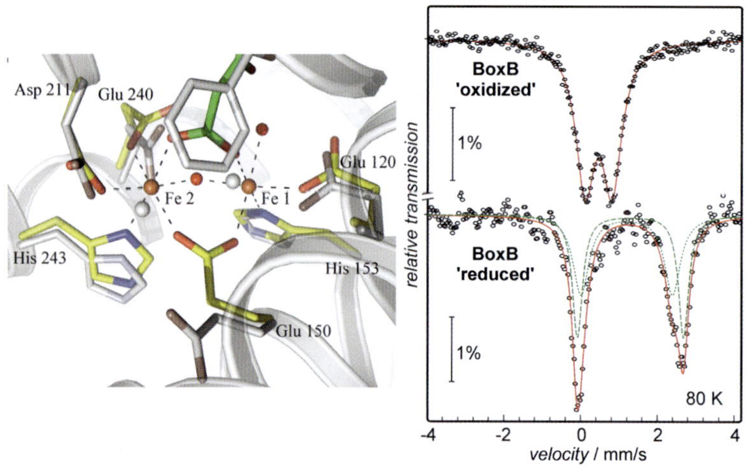

FIGURE 5.11 Structure of the diiron centre of the enzyme BoxB (Rather et al., 2011) in oxidised state and, superimposed in grey, a so-called half-reduced state (left). The carboxylate oxygen atoms of Glu150 bridge the two iron atoms in the oxidised but essentially not in the semi-reduced state. Right: Zero-field Mössbauer spectra of oxidised (top) and reduced (bottom) BoxB recorded at 80 K. The red lines represent fits with Lorentzian quadrupole doublets.

indicate two non-equivalents high-spin ferrous ions ($S_{Fe} = 2$). Particularly, the high isomer shifts, which are similar to those of other di-ferrous centres in bioinorganic chemistry (Merkx et al., 2001), are typical for an octahedral coordination shell with hard ligands such as oxygen from carboxylates or water.

Reaction Intermediates and Low- and High-Valent Iron Complexes

The most common oxidation states of iron in metalloproteins are Fe(II) and Fe(III). Higher oxidation states, Fe(IV) and Fe(V), are also known or have been proposed for short-lived intermediates in the reaction cycle of oxygen-activating enzymes and model systems (Merkx et al., 2001; Costas et al., 2004; Groves, 2006; Bell and Groves, 2009; Tinberg and Lippard, 2011). This situation has fuelled an interest in the chemistry of heme and non-heme iron complexes supporting high valence states, for which Mössbauer spectroscopy proved to be most valuable for the study of the electronic structure of the reaction intermediates captured by freeze-quench procedures (Krebs and Bollinger, 2009), as well as of stable model compounds.

The Heme Enzyme Horseradish Peroxidase

A particular problem encountered with heme complexes in high oxidations states is that not only iron but also the porphyrin ligand can be oxidised, forming a π-cation radical. This was first demonstrated for horseradish-peroxidase (HRP), a heme enzyme that catalyses degradation of peroxides. In the first step of the reaction cycle, a mononuclear iron-oxo intermediate called 'Compound-I' (HRP-I) is formed, which formally is two equivalents more oxidised than the ferric starting compound. Moreover, a second intermediate (HRP-II) could also be trapped after subsequent one-electron reduction. Mössbauer spectroscopy revealed "a fundamental change in iron configuration in going from HRP to HRP-II, and it is compatible with a net loss of electrons from the *d* shell. However, the Fe(V) state postulated for compound I probably does not exist: the electronic configurations of the iron in HRP-I and HRP-II appear essentially the same", as Moss stated already in 1969 (Moss et al., 1969). The conclusion was inferred from a comparison of the isomer shifts and quadrupole splittings seen in the Mössbauer spectra of HRP and its peroxide derivative (Figure 5.12).

The moderately low isomer shift $\delta = 0.25$ mm/s found for native HRP at 77 K, in conjunction with large quadrupole, $\Delta E_Q = 1.96$ mm/s (Moss et al., 1969), is typical of ferric low-spin heme iron complexes (Debrunner,

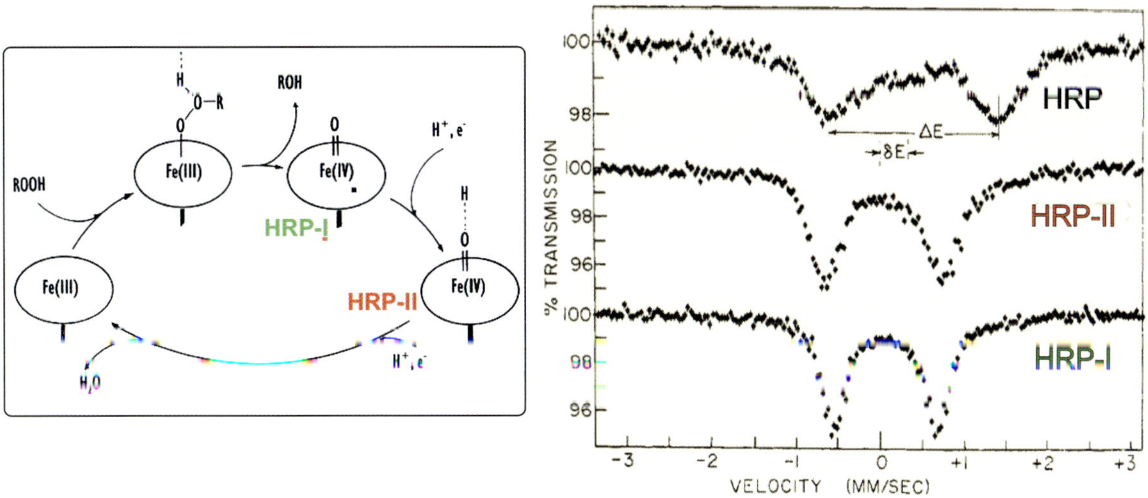

FIGURE 5.12 Reaction cycle of horseradish peroxidase (HRP) (left), and zero-field Mössbauer spectra of HRP (Moss et al., 1969) and its peroxide reaction derivatives (right).

1989). The quadrupole splitting arises mainly from valence electrons due to the asymmetry in valence charge distribution expected for the $3d(t_{2g})^5$ configuration. After oxidation, the distinctly lower isomer shift for the second intermediate HRP-II ($\delta = 0.03$ mm/s at 77 K, $\Delta E_Q = 1.36$ mm/s) reveals a loss of d-charge density due to a metal-based oxidation, which means that a ferryl species, Fe(IV)=O was formed. However, the second oxidation equivalent that distinguishes HRP-I from HRP-II must be ligand centred, since HRP-I has essentially the same Mössbauer parameters ($\delta = 0.0$ mm/s at 77 K, $\Delta E_Q = 1.20$ mm/s) as HRP-II. Later, applied-field Mössbauer studies (Schulz et al., 1979; Schulz et al., 1984) yielded detailed insight into the electronic structure of the ferryl heme iron(IV) oxo complex (low-spin $3d^4$ configuration, $S = 1$ with large zero-field splitting) and the interaction and spin coupling with the porphyrin π-cation radical species ($S' = 1/2$) (Groves, 2006; Jayaraj et al., 1997; Mandon et al., 1992; Rittle and Green, 2010; Shaik et al., 2009).

Non-Heme Model Compounds – Synthetic Iron(III) Complexes with the Macrocyclic Ligand Cyclam

In non-heme iron systems, the ligands bound to iron are generally considered to be redox-innocent, and high-valent intermediates containing Fe(IV) or Fe(V) have been found and studied by Mössbauer spectroscopy (Price et al., 2003; Pestovsky et al., 2005; de Oliveira et al., 2007; Hont et al., 2010). A robust supporting ligand for the synthesis of high-valent iron model complexes is the macrocycle 1,4,8,11-tetraazacyclotetradecane (cyclam), which coordinates iron with four equatorial nitrogen donors akin to the coordination mode of a porphyrin. In contrast to the latter, cyclam is 'redox-innocent', i.e. it cannot be oxidised or reduced. There are several cyclam derivatives, including some with a pendant acetate-arm that can bind as a fifth ligand (see Scheme).

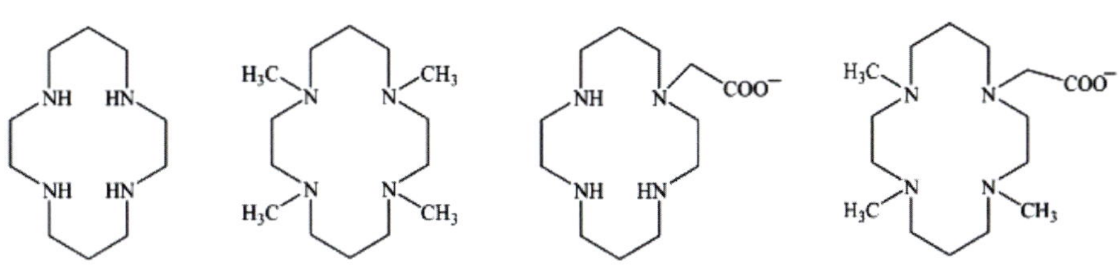

Cyclam Me$_4$Cyclam Cyclam-acetate Me$_3$Cyclam-acetate

Cyclam and its methylated and pendant-arm derivatives have been successfully used to generate high-valent transition metal complexes, particularly mononuclear complexes with high-valent Fe(IV), Fe(V), and Fe(IV) (Berry et al., 2006; Meyer et al., 1999; Grapperhaus et al., 2000; Rohde et al., 2003; Aliaga-Alcalde et al., 2005; Petrenko et al., 2007). While the reactive high-valent iron species in biological systems are mostly stabilised by a strong electron-donating oxo group, iron-nitrido systems are also interesting and have been used in synthetic systems, not only because of their relevance for enzymatic reactions in the biogeochemical nitrogen cycle, but also for theoretical and technical reasons. An interesting preparation method for high-valent complexes is based on photolytic cleavage of coordinated azide (N_3^-) ions, yielding dinitrogen (N_2) and a high-valent Fe=N species (Berry et al., 2006; Meyer et al., 1999; Grapperhaus et al., 2000; Rohde et al., 2003; Aliaga-Alcalde et al., 2005; Petrenko et al., 2007; Betley and Peters, 2004; Brown and Peters, 2005). In early studies, the starting material was a bis-azide iron(III)cyclam complex, $[(cyclam)Fe^{III}(N_3)_2]^+$. This can exist in two different conformations with the two azide ligands either in *cis*- or in *trans*-positions (see insets of Figure 5.13). The difference has major impact on the electronic structure, as can be inferred from the Mössbauer spectra (Figure 5.13, A + B). The *trans*-complex is a typical ferric low-spin compound with low isomer shift and large quadrupole splitting ($\delta = 0.29$ mm/s, $\Delta E_Q = 2.26$ mm/s), whereas the more open arrangement of the macrocyclic ligand with *cis*-arrangement of azides allows high-spin configuration with moderately large isomer shift and small quadrupole splitting, due to the quasi cubic symmetry of the half filled *d*-shell ($\delta = 0.46$ mm/s, $\Delta E_Q = 0.29$ mm/s) (Meyer et al., 1999).

Iron(II) Complexes

Electro-chemical reduction of the ferric compounds yields different ferrous products (Figure 5.13, C + D) (Meyer et al., 1999). The Mössbauer parameters of the reduced *cis*-complex unambiguously reveal formation of high-spin Fe(II) ($\delta = 1.11$ mm/s, $\Delta E_Q = 2.84$ mm/s); whereas the major part of the reduced *trans*-complex is in a low-spin

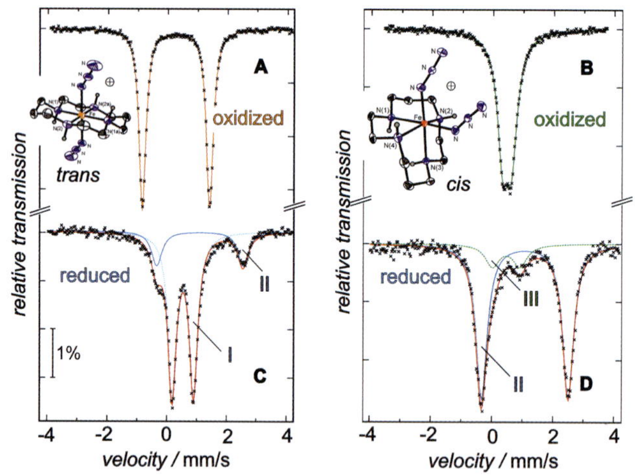

FIGURE 5.13 Zero-field Mössbauer spectra of *trans*- and *cis*-$[(cyclam)Fe^{III}(N_3)_2]^+$ complexes (35% ^{57}Fe-enriched, 1.5 mM in CH_3CN, panels **A + B**) (Meyer et al., 1999) and their electrochemically one-electron reduced CH_3CN solutions (**C + D**) recorded at 80 K. The spectra are simulated with Lorentzian doublets. The oxidised complex with *trans*-orientation of the azide (N_3^-) ligands is a typical low-spin Fe(III) compound (**A**), whereas the chemically identical *cis*-arrangement shown in (**B**) affords high-spin Fe(III). After coulometric reduction at −30 °C, the reduced *trans*-complex (**C**) shows a superposition low-spin Fe(II) (subspectrum I) and high-spin Fe(II) (subspectrum IL); prolonged incubation in solution at −30 °C provides strong enhancement of subspectrum II and concomitant attenuation of I. Reduction of the *cis*-complex shown in (**D**) yields only high-spin Fe(II). Subspectrum III in D, having parameters $\delta = 0.49$ mm/s, $\Delta E_Q = 0.90$ mm/s, is most probably from a contamination with a $[Fe^{III}\text{-O-}Fe^{III}]^{4+}$ μ-oxo dimer species due to oxidation with traces of air. This spectrum was exclusively obtained when a reduced sample was deliberately oxidised with air.

Fe(II) state which has a low isomer shift and low quadrupole splitting, due to the high symmetry of the filled t_{2g}^6 subshell (subspectrum I, $\delta = 0.55$ mm/s, $\Delta E_Q = 0.72$ mm/s). However, the *trans*-conformation is not stable for the reduced compound and tends to convert in solution to the *cis*-arrangement, as can be seen from the presence of the same high-spin component as found for the *cis*-complex (subspectrum II in panel C). The intensity of this component is found to increase with incubation time in solution.

Similar Mössbauer spectra and parameters as for *trans*-$[(\text{cyclam})\text{Fe}^{\text{III}}(N_3)_2]^+$ (Figure 5.13, A + C) have been observed for $[(\text{cyclam-acetato})\text{Fe}^{\text{III}}N_3]^+$ (not shown, $\delta = 0.27$ mm/s, $\Delta E_Q = 2.53$ mm/s). Coordination of the tethered acetate arm allows only *trans*-configuration for this compound. Apparently, the replacement of an (N_3^-) ligand by the acetato-oxygen group did not have much influence on the electronic structure of the metal ion. Similar trends were also observed for the corresponding reduction and photooxidation products, as discussed in the following.

Mixed-Valence Iron(III)-Iron(IV) Dimers and Iron(IV) Monomers

Iron(III)cyclam-azide complexes are light sensitive, and illumination of *trans*- and *cis*-$[(\text{cyclam})\text{Fe}^{\text{III}}(N_3)_2]^+$ complexes in fluid solution at cryogenic temperatures (−35 °C) with light of about 420 nm wavelength yields two new compounds which could be identified as mixed valence dimers with μ-nitrodo bridged di-iron cores of the type $[\text{Fe(III)-N-Fe(IV)}]^{4+}$. One has *cis*-symmetry of nitrido- and terminal azide ligands at the Fe(IV) site, the other *trans*-symmetry (Meyer et al., 1999). The formation of these products indicates that photooxidation as well as a photoreductions must have occurred as primary reactions, such that the corresponding Fe(IV) and Fe(II) products could form the observed Fe(III)/Fe(IV) mixed-valent dimers via bi-molecular reactions. Here, only the properties of the high-valent Fe(IV) sites of these dimers shall be considered. As revealed by applied-field Mössbauer and EPR measurements, they are hexa-coordinated Fe(IV) low spin centres with spin $S = 1$ due to $3d(t_{2g})^4$ electron configuration (four electrons in three orbitals). The electric Mössbauer parameters are $\delta = 0.14$ mm/s, $\Delta E_Q = 0.79$ mm/s for the *cis*-compound, and $\delta = 0.11$ mm/s, $\Delta E_Q = 0.97$ mm/s for the corresponding *trans*-complex (at 80 K in frozen solution). The values are similar to what is known for the Fe(IV)=O species found with porphyrins (Debrunner, 1989; Rittle and Green, 2010; Shaik et al., 2009; Jung, 2011).

Monomeric iron(IV) complexes without terminal or bridging oxo- or nitrido-groups could be obtained from electrochemical oxidation of a series of halide and azide complexes of the type $[(\text{Me}_3\text{cyclam-acetate})\text{FeX}]\text{PF}_6$, where X was Cl^-, F^-, and N_3^-, and Me_3cyclam-acetate is the methylated cyclam derivative with a pendant acetate arm shown in the scheme above (Berry et al., 2006). The Mössbauer parameters for these compounds are rather similar to those of the Fe(IV) sites in the dimer molecules: $\delta = 0.08/0.02/0.11$ mm/s, and $\Delta E_Q = 2.40/2.43/1.92$ mm/s for the Cl^-, F^-, and N_3^- ligands, respectively (Berry et al., 2006).

Iron(V) Complexes

Genuine iron(V) is a very rare oxidation state; only a single iron(V)-oxo compound has been reported in (bio) inorganic chemistry. That was synthesised with the redox-innocent tetranionic macrocyclic TAML ligand (de Oliveira et al., 2007), which provides four exceptionally strong amide-N σ-donor groups. These are capable of stabilizing iron(V) when an iron(III) precursor complex is treated with an oxygen-transfer agent. The presence of the expected $3d^3$ configuration was verified by Mössbauer spectroscopy ($\delta = -0.42$ mm/s at 4.2 K, $\Delta E_Q = 4.25$ mm/s), EPR spectroscopy ($S = 1/2$, $g = [1.99, 1.97, 1.74]$), and X-ray absorption spectroscopy (XAS and EXAFS).

Genuine iron(V) nitrido complexes could be generated by low-temperature photolysis of the cyclam iron(III) complexes described above (Berry et al., 2006; Meyer et al., 1999; Grapperhaus et al., 2000; Aliaga-Alcalde et al., 2005; Petrenko et al., 2007). In frozen solution, where bi-molecular reactions are prevented, the corresponding monomeric Fe(V)≡N species could be generated in high yield (Figure 5.14, left). Apparently, photoreduction of the immobilised molecules does not contribute to the result, presumably because in contrast to photooxidation

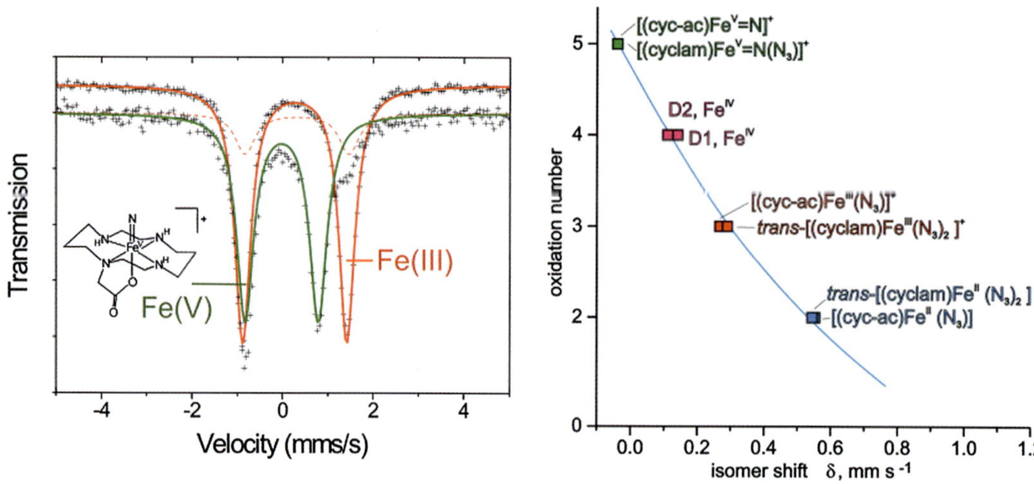

FIGURE 5.14 Left: Mössbauer spectra of $[(\text{cyclam-acetato})\text{Fe}^{\text{III}}(\text{N}_3)]^+$ (in red) and its low-temperature iron(V) photolysis product (in green) (Grapperhaus et al., 2000), measured at 80 K. Right: Isomer shift correlation diagram for low-spin iron complexes with cyclam and cyclam-acetato supporting ligands. The iron valences range from Fe(II) to Fe(V). The blue line is shown to guide the eye; it is not based on a theoretical concept beyond the general arguments about d^n-electron configurations and isomer shifts Sometimes also straight lines have been used here.

it is reversible, and azide radicals eventually generated by photoreduction can re-bind to the immobilised Fe(II) product molecules, restoring the starting material.

Comprehensive studies on $[(\text{cyclam-acetato})\text{Fe}^{\text{V}}{\equiv}\text{N}]^+$ using XAS and EXAFS measurements, magnetic susceptibility, Mössbauer spectroscopy and nuclear inelastic scattering (NIS) in conjunction with DFT calculations (Meyer et al., 1999; Grapperhaus et al., 2000; Aliaga-Alcalde et al., 2005; Petrenko et al., 2007; Berry, 2009), characterised the compound as a low-spin $(d_{xy})^2(d_{xz}, d_{yz})^1$ system. The Mössbauer parameters of $[(\text{cyclam-acetato})\text{Fe}^{\text{V}}{\equiv}\text{N}]^+$ ($\delta = -0.02$ mm/s, $\Delta E_\text{Q} = 1.60$ mm/s) are consistent with those for $[(\text{cyclam})\text{Fe}^{\text{V}}{\equiv}\text{N}]^+$ in CH_3CN solution which presumably provides a sixth ligand to the iron ($\delta = -0.04$ mm/s, $\Delta E_\text{Q} = 1.67$ mm/s).

The conclusions about the correct electronic structure have been substantially based on the close correlation of the isomer shifts of the cyclam complexes with the formal oxidation state of the iron, as shown in the right panel of Figure 5.14. The variation of δ appears to follow the expected decrease in bond lengths of the apical ligands with increasing oxidation state. A tight, conclusive correlation is obtained here because in contrast to the general scheme shown in Figure 5.8, a consistent series of quasi-isomorphous complexes with invariant equatorial ligands has been compared. The compounds are all six-coordinate and variations in the valence electron configuration are restricted to the t_{2g} subshell (low-spin) with increasing covalency of the axial ligand bonds. A similar correlation may be found for related complexes with axial oxygen or oxo ligands, but the slope of the corresponding plot for isomer shifts versus oxidation state will be different.

Four-Coordinated Iron(IV) and Iron(V) Compounds

An amazing *four-coordinate* iron(V) nitrido complex with a tripodal N-heterocyclic carbene ligand was reported recently (Figure 5.15, left). The high-valent compound was generated electrochemically by oxidation of the corresponding iron(IV)-nitrodo precursor, which had been synthesised previously by photolysis of a corresponding iron(II)-azide starting complex (Scepaniak et al., 2008). The iron(IV) precursor is a diamagnetic molecule ($S = 0$) with low Mössbauer isomer shift, $\delta = -0.28$ mm/s. The low-spin state is most unusual for four-coordination; the strong ligand field results here from short bonds and strong trigonal-axial distortion of the pseudo-tetrahedral symmetry. The quadrupole splitting is $\Delta E_\text{Q} = 6.23$ mm/s, which is the largest value ever

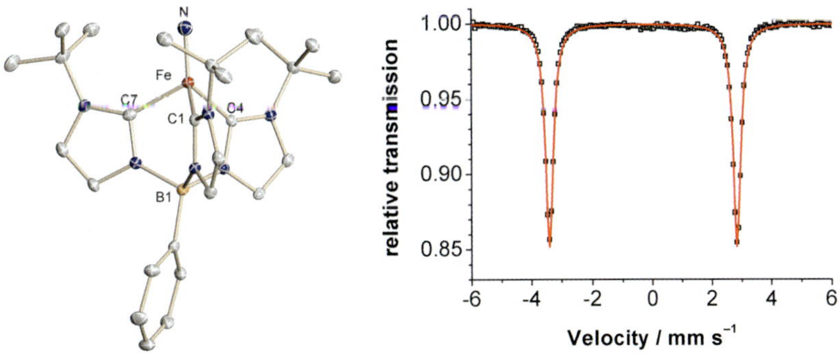

FIGURE 5.15 Molecular structure of the cation $[PhB(^tBuIm)_3Fe^V{=}N]^+$, where $PhB(^tBuIm)^{3-}$ is the phenyltris(3-tert-butylimidazol-2-ylidene)borato ligand (Scepaniak et al., 2011) (left) and zero field Mössbauer spectrum of a microcrystalline sample of the iron(IV) precursor (Scepaniak et al., 2008) $PhB(^tBuIm)_3Fe^{IV}{\equiv}N$ recorded at 77 K (right). The red line represents the best fit with a Lorentzian doublet obtained with $\delta = -0.28$ mm/s, $|\Delta E_Q| = 6.23$ mm/s.

observed for an iron compound. The value slightly exceeds the splittings of 6.01 mm/s and 6.04 mm/s found for two other, electronically similar four-coordinate compounds, $[PhBP^{iPr}{}_3]Fe{\equiv}N$ and $[(TIMEN^{mes})Fe{\equiv}N]^+$ (Vogel et al., 2008), respectively (isomer shifts of −0.31 mm/s (at 140 K) and −0.27 mm/s (at 77 K)).

The strong (trigonal) distortion of the pseudo-tetrahedral symmetry of these complexes affords unusually strong orbital splitting with a 'three-over-two' pattern. Two non-bonding *e*-orbitals are low lying, rendering linear combinations of d_{xy} and $d_{x2\text{-}y2}$ orbitals. Since they are energetically well isolated the four valence electrons are accommodated in the two orbitals (Scepaniak et al., 2008), affording diamagnetism ($S = 0$). The state may be called the 'low-low-spin' state of the d^4 configuration, since the usual $(t_{2g})^4$ low-spin configuration in octahedral symmetry has $S = 1$, arising from four electrons in three orbitals. Moreover, the quasi-planar charge distribution expected for the discussed $(d_{xy}, d_{x2\text{-}y2})^4$ configuration of the pseudo-tetrahedral nitrido complexes explains the unprecedentedly large *efg* at the Mössbauer nucleus (see V_{zz} expectation values for d_{xy} and $d_{x2\text{-}y2}$ in Table 5.1). Although not determined, one may presume a positive sign for V_{zz} because of the huge (positive) valence contribution to the *efg*.

The oxidised cation $[PhB(^tBuIm)_3Fe^V{\equiv}N]^+$ is also low-spin, $S = 1/2$, according to the corresponding $(d_{xy}, d_{x2\text{-}y2})^3$ configuration (Scepaniak et al., 2011). Its quadrupole splitting, $\Delta E_Q = 4.73$ mm/s, is less than that of the Fe(IV) precursor, but it is still remarkably large and exceeds the values usually reported for octahedral or tetrahedral iron compounds. The isomer shift, $\delta = -0.49$ mm/s, is below the range observed for the cyclam complexes with Fe(V)-nitrido groups, due to the overall shorter metal-ligands bonds. This demonstrates again that a single isomer shift value may be not conclusive if it cannot be compared with reference systems.

The First Molecular Iron(VI) Compound

Basic ligand-field considerations predict relatively high stability for iron complexes with d^2 (or d^1) configuration, due to the reduced number of anti-bonding *d*-electrons compared to $3d^3$ configuration. Nevertheless, iron(VII) ($3d^1$) is elusive and the first and only known molecular iron(VI) ($3d^2$) compound was discovered only recently. That is an Fe(VI)-nitrido complex with the bulky ligand (Me_3cyclam-acetate) (Berry et al., 2006). The high-valent $[(Me_3\text{cyclam-acetate})Fe^{VI}{\equiv}N]^{2+}$ complex (inset of Figure 5.16) was electrochemically generated from the iron(IV)-azide precursor $[(Me_3\text{cyclam-acetate})Fe^{IV}N_3]^{2+}$ by low-temperature photolysis in the sample cup of the Mössbauer spectrometer. The product is diamagnetic, $S = 0$, and the Fe≡N bond (157 pm obtained from EXAFS) is significantly shorter than that for the corresponding iron(V) compound (161 pm) (Aliaga-Alcalde et al., 2005). The isomer shift, $\delta = -0.29$ mm/s, is 0.40 mm/s lower than that of the corresponding iron(III)-azide starting material and 0.19 mm/s lower than that of the corresponding iron(V)-nitrido compound. The value fits perfectly to

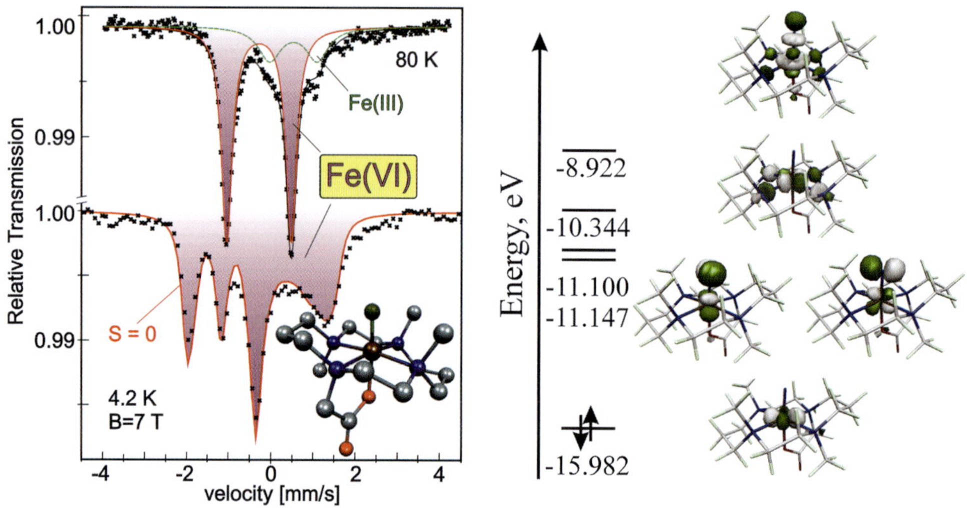

FIGURE 5.16 Left: Mössbauer spectra of a photolysed acetonitrile solution of [(Me$_3$cyclam-acetate)Fe^{IV}N$_3$](PF$_6$)$_2$ measured at 80 K (top) and at 4.2 K with an applied magnetic field of 7 T (bottom) (Berry et al., 2006; Berry et al., 2008). The major subspectrum marked in red represents the high-valent photoproduct [(Me$_3$cyclam-acetate)Fe^{VI}≡N]$^{2+}$. The red lines results from a fit with Lorentzian doublets for the 80-K measurement and from a spin-Hamiltonian simulation with $S = 0$ for the applied-field spectrum (bottom). The subspectrum shown in green was detectable only at 80 K and presumably represents a ferric contamination, which was split at 4.2 K/7 T and broadened beyond recognition due to intrinsic magnetic moments. The inset on the left panel shows the molecular structure predicted by density functional theory (DFT). Right: Orbital scheme for [(Me$_3$cyclam-acetate)Fe^{VI}≡N]$^{2+}$ from DFT.

the isomer shift correlations shown in Figure 5.14. This feature together with the short Fe≡N bond, and the spin $S = 0$ obtained from an applied-field Mössbauer measurement (Figure 5.16, left) supports the presence of genuine iron(VI) with an electronic structure that is well described as $3d^2$ configuration (Berry et al., 2006) (Figure 5.16, right).

The quadrupole splitting of the iron(VI) compound is $\Delta E_Q = +1.53$ mm/s, with asymmetry parameter $\eta = 0.3$. The sign of V_{zz} and η have been obtained from the simulation of the magnetically perturbed spectrum with spin $S = 0$ shown in Figure 5.16, left bottom. Apparently, the *efg* is significantly weaker than what would be expected for an ionic d^2 low-spin configuration with two valence electrons in one orbital (compare Table 5.1), which reveals competition of valence and covalence contributions attenuating the actual values. The various spectroscopic and structural parameters have been found consistent with the overall geometry and electronic structure predicted by quantum chemical calculations on [(Me$_3$cyclam-acetate)Fe^{VI}≡N]$^{2+}$ (calculated Mössbauer parameters: $\delta = -0.31$ mm/s , $\Delta E_Q = +0.75$ mm/s, $\eta = 0.7$, which is excellent for δ and satisfactory for ΔE_Q). The $3d$ orbital scheme obtained from DFT calculations as shown on the right side of Figure 5.16 nicely supports the interpretation as $3d^2$ configuration. The highest occupied MO is best described as a doubly occupied d_{xy} orbital. The xz and yz orbitals are strongly destabilised in energy because of considerable contributions (45%) from p orbitals of the terminal nitrido ligand and are best considered to be Fe—N p* antibonding orbitals. Occupancy of these orbitals reduces the Fe-N bond order, and is a major source of the *efg* (Berry et al., 2006; Berry et al., 2008).

PERSPECTIVES

The most spectacular application of Mössbauer spectroscopy in recent years has been the search for water-dependent or water-carrying minerals on Mars by the Mössbauer spectrometers of the NASA twin Mars Exploration Rovers, which landed in 2004 and are still going (Gütlich and Schröder, 2010; Klingelhöfer et al., 2003; Klingelhöfer and Fleischer, 2011). The tremendous amount of data harvested during this most successful mission has provided a world of information on extraterrestrial chemistry, which has certainly prompted further extraterrestrial research missions. The miniaturised spectrometers are now also used for terrestrial applications, in geo-sciences for instance (Klingelhöfer and Fleischer, 2011). Although less extravagant, but also very innovative are Mössbauer experiments with synchrotron radiation (SR); an introduction of which is given in Chapter 9 of Gütlich et al. (2011). It is easy to predict that applications of SR-based techniques in bioinorganic chemistry will develop vividly. The aim might be measurements of hyperfine interactions with several non-iron isotopes, with microscopically small samples with minimal volume, or NIS as a variant of vibrational spectroscopy, particularly when more intense synchrotron facilities will become available. But there will also be continuing applications for conventional Mössbauer spectroscopy with radioactive sources, although already since long in its 'mature' middle age. For instance, there is a revival of research on iron-sulfur proteins due to their role in catalysis as for example in hydrogenases (Pandelia et al., 2011), or the role of Fe:S cluster formation and enzyme maturation for gene expression and genetic disorder has captured much attention, not least due to the medical implications (Lill, 2009, Sheftel et al., 2010). In such fields there should be a promising future for interesting Mössbauer spectroscopy.

REFERENCES

Aliaga-Alcalde, M., George, S. D., Mienert, B., Bill, E., Wieghardt, K., & Neese, F. (2005). *Angew. Chem. Int. Ed., 44*, 2908.

Atkin, C. I., Thelander, L., Reichhard, P., & Lang, G. (1973). *J. Biol. Chem., 248*, 7464.

Beinert, H., Holm, R. H., & Münck, E. (1997). *Science, 277*, 653.

Bell, S. R., & Groves, J. T. (2009). *J. Am. Chem. Soc., 131*, 9640.

Berry, J. F., Bill, E., Bothe, E., George, S. D., Mienert, B., Neese, F., et al. (2006). *Science, 312*, 1937.

Berry, J. F., Bill, E., Bothe, E., Neese, F., & Wieghardt, K. (2006). *J. Am. Chem. Soc., 128*, 13515.

Berry, J. F., George, S. D., & Neese, F. (2008). *Phys. Chem. Chem. Phys., 10*, 4361.

Berry, J.F., 2009. *Comments on inorganic chemistry*. 30, 28.

Betley, T. A., & Peters, J. C. (2004). *J. Am. Chem. Soc., 126*, 6252.

Brown, S. D., & Peters, J. C. (2005). *J. Am. Chem. Soc., 127*, 1913.

Costas, M., Mehn, M. P., Jensen, M. P., & Que, L. (2004). *Chem. Rev., 104*, 939.

De Hont, R. F., Xue, G., Hendrich, M. P., Que, L., Bominaar, E. L., & MuÃànck, E. (2010). *Inorg. Chem., 49*, 8310.

de Oliveira, F. T., Chanda, A., Banerjee, D., Shan, X., Mondal, S., Que, L., Jr., et al. (2007). *Science, 315*, 835.

Debrunner, P. G. (1989). In A. B. P. Lever, & H. B. Gray (Eds.), *Iron Porphyrins Part III*, vol. III (p. 137). Weinheim: VCH.

Fox, B. G., Lyle, K. S., & Rogge, C. E. (2004). *Acc. Chem. Res., 37*, 421.

Grapperhaus, C. A., Mienert, B., Bill, E., Weyhermüller, T., & Wieghardt, K. (2000). *Inorg. Chem., 39*, 5306.

Greenwood, N. N., & Gibb, T. C. (1971). *Mössbauer Spectroscopy*. London: Chapman and Hall Ltd.

Groves, J. T. (2006). *J. Inorg. Biochem., 100*, 434.

Gütlich, P. G., & Schröder, C. (2010). *Bunsenmagazin, 12*, 1.

Gütlich, P., Bill, E., & Trautwein, A. X. (2011). *Mössbauer Spectroscopy and Transition Metal Chemistry*. Berlin Heidelberg: Springer Verlag.

Hendrich, M. P., Gunderson, W., Behan, R. K., Green, M. T., Mehn, M. P., Betley, T. A., et al. (2006). *Proc. Natl. Acad. Sci. USA, 103*, 17107.

Hoggins, J. T., & Steinfink, H. (1976). *Inorg. Chem., 15*, 1682.

Holm, R. H., Kennepohl, P., & Solomon, E. I. (1996). *Chem. Rev., 96*, 2239.

Jayaraj, K., Gold, A., Austin, R. N., Ball, L. M., Terner, J., Mandon, D., et al. (1997). *Inorg. Chem., 36*, 4555.

Jung, C. (2011). *Biochim. Biophys. Acta (BBA) - Proteins + Proteomics, 1814*, 46.

Klingelhöfer, G., & Fleischer, I. (2011). In P. Gütlich, E. Bill, & A. X. Trautwein (Eds.), *Mössbauer Spectroscopy and Transition Metal Chemistry* (p. 447). Berlin Heidelberg: Springer.

Klingelhöfer, G., Morris, R. V., Bernhardt, B., Rodionov, D., de Souza, P. A., Jr., Squyres, S. W., et al. (2003). *J. Geophys. Res., 108*, 8067.

Krebs, C., & Bollinger, J. M., Jr. (2009). *Photosynth. Res., 102*, 295.

Kurtz, D. M. (1990). *Chem. Rev., 90*, 585.

Lill, R. (2009). *Nature, 460*, 831.

Long, G. J., & Grandjean, F. (1989). *Mössbauer Spectroscopy Applied to Inorganic Chemistry*, vol. 1−3. New York: Plenum Press.

Mandon, D., Weiss, R., Jayaraj, K., Gold, A., Terner, J., Bill, E., et al. (1992). *Inorg. Chem., 31*, 4404.

Martinho, M., & Münck, E. (2010). In A. Bakac (Ed.), *Physical Inorganic Chemistry* (p. 39). Hoboken, New Jersey: Wiley & Sons.

Merkx, M., Kopp, D. A., Sazinsky, M. H., Blazyk, J. L., Müller, J., & Lippard, S. J. (2001). *Angew. Chem. Int. Ed., 40*, 2782.

Meyer, K., Bill, E., Mienert, B., Weyhermüller, T., & Wieghardt, K. (1999). *J. Am. Chem. Soc., 121*, 4859.
Moss, T., Ehrenberg, A., & Bearden, A. J. (1969). *Biochemistry, 8*, 4159.
Mössbauer, R. L. (1958a). *Naturwissenschaften, 45*, 538.
Mössbauer, R. L. (1958b). *Z. Physik., 151*, 124.
Mössbauer, R. L. (1959). *Z. Naturforsch.* 14a.
Mouesca, J.-M., & Lamotte, B. (1998). *Coord. Chem. Rev., 178–180*, 1573.
Münck, E. (1978). In S. Fleischer, & L. Parker (Eds.), *Methods in Enzymology*, vol. 54 (p. 346). New York: Academic Press.
Münck, E. (2000). In L. Que (Ed.), *Physical Methods in Bioinorganic Chemistry* (p. 287). Sausalito: University Science Books.
Neese, F., & Petrenko, T. (2011). In P. Gütlich, E. Bill, & A. X. Trautwein (Eds.), *Mössbauer Spectroscopy and Transition Metal Chemistry* (p. 137). Berlin Heidelberg: Springer.
Neese, F. (2002). *Inorganica Chim. Acta, 337*, 181.
Pandelia, M.-E., Nitschke, W., Infossi, P., Giudici-Orticoni, M.-T., Bill, E., & Lubitz, W. (2011). *Proc. Natl Acad. Sci, USA, 108*, 6097.
Paulsen, H., Schünemann, V., Trautwein, A. X., & Winkler, H. (2005). *Coord. Chem. Rev., 249*, 255.
Pestovsky, O., Stoian, S., Bominaar, E. L., Shan, X., Münck, E., Que, L., & Bakac, A. (2005). *Angew. Chem. Int. Ed., 44*, 6871.
Petrenko, T., George, S. D., Aliaga-Alcalde, N., Bill, E., Mienert, B., Xiao, Y., et al. (2007). *J. Am. Chem. Soc., 129*, 11053.
Price, J. C., Barr, E. W., Tirupati, B., Bollinger, J. M., & Krebs, C. (2003). *Biochemistry, 42*, 7497.
Rao, P. V., & Holm, R. H. (2003). *Chem. Rev., 104*, 527.
Rather, L. J., Weinert, T., Demmer, U., Bill, E., Ismail, W., Fuchs, G., et al. (2011). *J. Biol. Chem., 286*, 29241.
Rittle, J., & Green, M. T. (2010). *Science, 330*, 933.
Rohde, J.-U., In, J.-H., Lim, M. H., Brennessel, W. W., Bukowski, M. R., Stubna, A., et al. (2003). *Science, 299*, 1037.
Scepaniak, J. J., Fulton, M. D., Bontchev, R. P., Duesler, E. N., Kirk, M. L., & Smith, J. M. (2008). *J. Am. Chem. Soc., 130*, 10515.
Scepaniak, J. J., Vogel, C. S., Khusniyarov, M. M., Heinemann, F. W., Meyer, K., & Smith, J. M. (2011). *Science, 331*, 1049.
Schulz, C. E., Devaney, P. W., Winkler, H., Debrunner, P. G., Doan, N., Chiang, R., et al. (1979). *FEBS Lett., 103*, 102.
Schulz, C. E., Rutter, R., Sage, J. T., Debrunner, P. G., & Hager, L. P. (1984). *Biochemistry, 23*, 4743.
Schünemann, V., & Winkler, H. (2000). *Rep. Prog. Phys., 63*, 263.
Shaik, S., Cohen, S., Wang, Y., Chen, H., Kumar, D., & Thiel, W. (2009). *Chem. Rev., 110*, 949.
Sheftel, A., Stehling, O., & Lill, R. (2010). *Trends Endocrinol. Metab., 21*, 302.
Solomon, E. I., Xie, X., & Dey, A. (2008). *Chem. Soc. Rev., 37*, 623.
Tinberg, C. E., & Lippard, S. J. (2011). *Acc. Chem. Res., 44*, 280.
Trautwein, A. X., Bill, E., Bominaar, E. L., & Winkler, H. (1991). Structure and bonding, 78, 1.
Vincent, J. B., Olivier-Lilley, G. L., & Averill, B. A. (1990). *Chem. Rev., 90*, 1447.
Vogel, C., Heinemann, F. W., Sutter, J., Anthon, C., & Meyer, K. (2008). *Angew. Chem. Int. Ed., 47*, 2681.

Chapter 6

X-ray Absorption Spectroscopy in Biology (BioXAS)

Martin C. Feiters[a] and Wolfram Meyer-Klaucke[b]

[a] *Department of Organic Chemistry, Institute for Molecules and Materials, Faculty of Science, Radboud University Nijmegen, Heyendaalseweg 135, Nijmegen, The Netherlands,* [b] *Faculty of Science, Department of Chemistry, University of Paderborn, Warburger Straße 100, Paderborn, Germany*

Chapter Outline

INTRODUCTION TO BIOLOGICAL X-RAY ABSORPTION SPECTROSCOPY (BIOXAS)

X-ray absorption spectroscopy (XAS) is established as one of an armory of physical techniques that life scientists apply to the study of metal ions or other trace elements in whole biological systems or isolated components thereof. Many metal ions are now recognised as essential to life in addition to the 'bulk' elements (C, H, N, O, S, and P). The majority of biological systems studied by XAS are metalloproteins, but there is an increasing number of studies on other biomolecules and/or other trace elements. XAS can be applied regardless of the physical state of a sample and is also used for the characterisation of other non-crystalline systems such as metal ions in solution or on catalyst supports. At the so-called absorption edge of an element of choice, the XAS shows fine structures, both at the absorption edge itself as well as above the edge, which are related to the chemical structure by physical processes that will be described below. From the so-called X-ray Absorption Near Edge Structure (XANES) at and

Practical Approaches to Biological Inorganic Chemistry, 1st Edition. http://dx.doi.org/10.1016/B978-0-444-56351-4.00005-1.

near the absorption edge, information on the valence state and coordination geometry of the element involved can be derived; the Extended X-ray Absorption Structure (EXAFS) beyond the edge is interpreted to give information on the type, distance, and number of the surrounding ligands.

Using monochromatised synchrotron radiation (see Box 6.1 for experimental aspects), it is possible to select an element and scan the X-ray energy around the so-called *edge*, that is around the energy that is required to liberate an electron from that atom. Figure 6.1 gives an overview on the element specificity of these electron binding energies; it summarises the edge energies for the most important elements in biology. For most elements, in particular the 3*d* transition metals, the K edges (blue numbers in Figure 6.1), where the 1*s* electron is excited, are readily accessible; for elements higher up in the periodic table, it is more convenient to use the L edges where 2*s* and 2*p* electrons are excited (red numbers give L_3 edges in Figure 6.1).

BOX 6.1 The BioXAS Experiment

X-ray absorption spectroscopy of samples as dilute in the element of interest as biological systems requires an intense source of X-rays continuous in wavelengths around the absorption edge. This requirement was fulfilled in the 1970s at synchrotrons. In principle, the spectrometer consists of the following parts (Figure 6A): X-rays of all wavelengths (white beam) originating from the bending magnet (or insertion device like wiggler or undulator) in the synchrotron enter the spectrometer from the left and pass through an entrance slit into a monochromator. According to Bragg's law

$$n\lambda = 2d \sin \theta \tag{6.1A}$$

where n is a natural number (1, 2, 3, etc.), λ the wavelength, d the lattice parameter of the Si, and θ the angle between monochromator and beam, X-rays of a certain wavelength are diffracted by two parallel Si crystals into the ionisation chambers. The wavelength of the monochromatised beam is varied throughout the experiment by changing the monochromator angle with respect to the incident beam. Diffraction of the X-rays occurs for $n=1$ and for the harmonic contaminations ($n>1$), which have shorter wavelengths, corresponding to higher energies. These harmonics are selectively rejected by an order-sorting monochromator in which the positions of the two crystals are controlled so that they are slightly non-parallel. This selection is based on the acceptance angles that are by far smaller for higher harmonics.

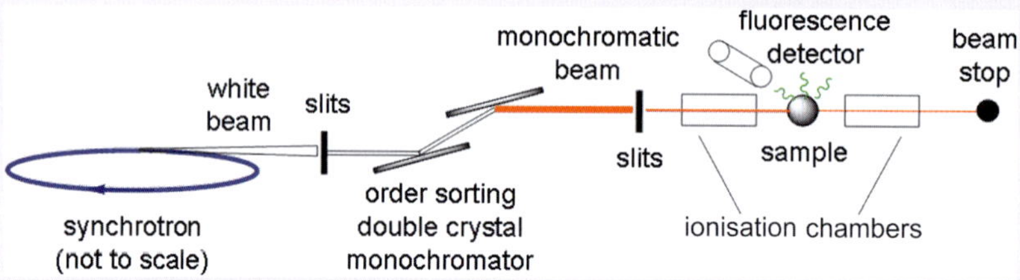

FIGURE 6A Schematic representation of an X-ray absorption spectrometer.

In order to determine the absorption of the sample according to the Beer–Lambert law (Eqn (6.1A)), the intensities of the X-rays before and after passage through the sample are measured by ionisation chambers, in which the X-rays ionise gas molecules; the resulting ions give a current that is proportional to the X-ray intensity. Absorption of X-rays causes fluorescence radiation (Figure 6B, left) which has lower energies than incident and scattered radiation. Using detectors of the solid state multi-element type, the fluorescence (I_f) can be separated from scattered radiation, which represents the background, by its energy; provided that the sample is thick, i.e. sufficiently absorbing the X-rays used in the experiment, but dilute in the element of interest, the energy-dependent absorption coefficient is given as $\mu_f = I_f/I_0$, which ultimately results in an extracted fine structure (see Box 6.3) equivalent to that measured in transmission mode. Biological systems are usually weakly absorbing, except for the trace elements, that they are relatively dilute in, resulting in hardly any edge step against a large background in the transmission experiment; therefore fluorescence is the preferred mode of detection for these systems. The fluorescence yield depends on both the atomic number and the absorption edge (Figure 6B, right). Even if fluorescence detection is applied, it is still important to accumulate multiple scans of a sample with a concentration in the millimolar range in the element of interest.

BOX 6.1—cont'd

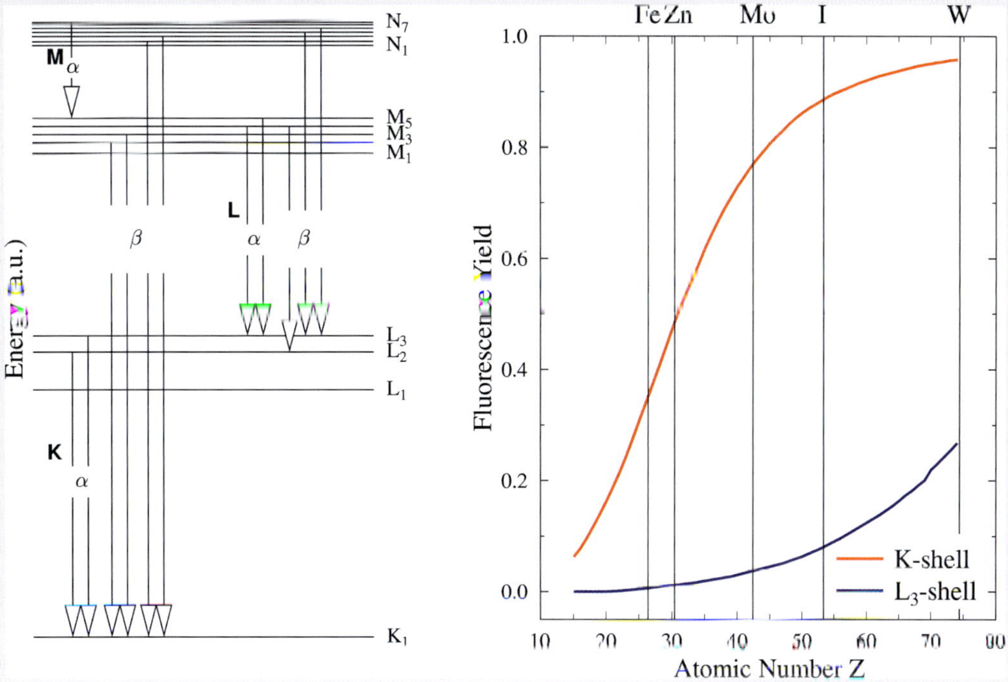

FIGURE 6B (left) In the X-ray absorption process a hole in one of the orbitals (e.g. 1*s* for K edge) is created, and the excited electron leaves the atom as a photoelectron and takes the remaining energy (= photon energy − binding energy) with it. X-ray fluorescence arises when electrons from higher orbitals fill up the resulting hole, giving off the excess energy as fluorescence, which has energy also in the X-ray range. (right) Dependence of K and L_3 edge fluorescence yields on the atomic number Z. The probability for filling the hole with an electron from an orbital with higher energy is called fluorescence yield. Note that this is rather high for K edges, whereas for L edges it requires rather high concentrations to detect a sufficient signal.

Because of the susceptibility of the biological matrix to radiation damage, measurements are usually carried out on frozen samples at cryogenic temperatures in order to reduce the mobility of any radicals generated by X-ray irradiation; this has also an added beneficial effect of decreasing the thermal component of the so-called Debye–Waller factor (see Box 6.5), enhancing the sensitivity to weak long-range contributions. Cryoprotectants such as glycerol may be added if it is suspected that freezing damages the biological sample, but it is necessary to check that they do not interfere with the biological activity of the biomolecule. A special kind of radiation damage is that resulting in a change of the valence state of the element under investigation, e.g. photoreduction, as its progress can be probed by comparing the edge structures (the so-called XANES part of the spectrum, see below) of consecutive scans during irradiation and data accumulation. Typically the requirement to have efficient data collection on a sample as small and dilute as possible using a high flux beamline (3rd generation source with insertion device) needs to be balanced against the risks of general and more specific radiation damage (Ascone et al., 2003). A typical solution is to spread the photon flux over a rather large area and/or replace the sample as soon as photoreduction is detected.

Besides the high intensity and the energy range, synchrotron radiation has another property that is of interest for biological samples, viz. it is polarised in the plane of the synchrotron. This means that if one wants to measure typical anisotropic XAS of a crystalline sample it is advisable to use a slurry of crystals rather than a single crystal, in order to avoid effects of preferential orientation with respect to the plane of the synchrotron beam. On the other hand, advantage may be taken of the polarised beam to study the *linear dichroism* in anisotropic biological samples, such as protein single crystals, or the metal centres in proteins in stacked layers of membranes, such as the organelles responsible for photosynthesis in plants, the thylakoids.

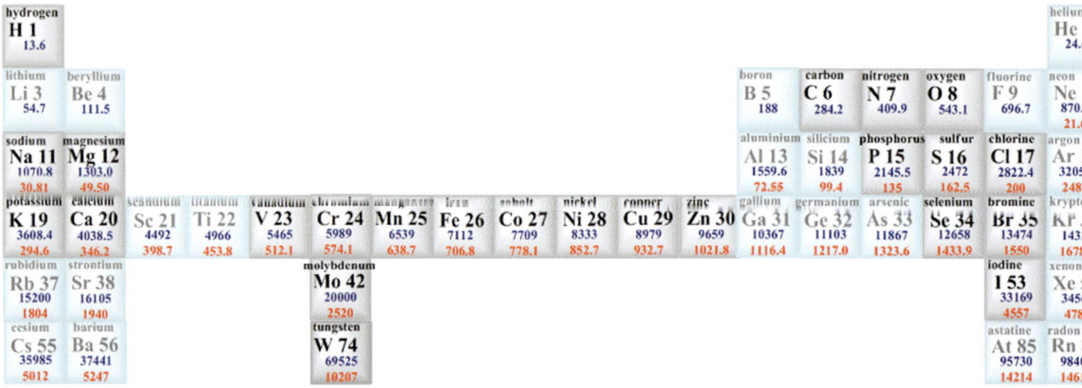

FIGURE 6.1 Name, symbol, atomic number, K (blue) and L_3 (red) edges in eV of biologically important elements (black) and some others (grey) in the periodic table.

Absorption is described by the Beer–Lambert law, with the transmitted intensity I_t depending on the incident intensity I_0, the sample thickness x and the energy-dependent absorption coefficient $\mu(E)$

$$I_t = I_0 \exp(-\mu x) \tag{6.1a}$$

Thus, the X-ray absorption spectrum is represented in the dimension of the X-ray absorption coefficient μ:

$$\ln(I_0/I_t) = \mu x \tag{6.1b}$$

with I_0 and I_t the X-ray intensities before and after the sample, respectively; it can also be measured, as in most examples discussed here, in fluorescence as μ_f (see Box 6.1).

AN INTRODUCTORY EXAMPLE: Mo, Cu, AND Se IN CO-DEHYDROGENASE FROM *OLIGOTROPHA CARBOXIDOVORANS*

Figure 6.2 shows XANES and EXAFS spectra of examples that were selected to demonstrate the power of XAS to probe chemical information that is relevant to biological structure–function relationships. In the top panel, we follow molybdenum uptake and utilisation by an enzyme and show the XANES of the K edge of molybdenum (absorption of X-rays by the 1*s* electron of Mo) of bioavailable, aqueous molybdate (MoO_4^{2-}), molybdate in one of the Mo transport proteins, here ModG, and finally Mo in the enzyme CO-dehydrogenase in its oxidised and reduced forms. The reduction of Mo in CO-dehydrogenase (from Mo^{6+} to Mo^{4+}) leads to a shift of the absorption edge (the part of the spectrum where the absorption increases most, 20,000–20,030 eV) to lower energy; this is a general observation in XANES. It is in agreement with the intuitive notion that it should require a little more energy to excite an electron from a metal ion in a relatively higher oxidation state, when it bears more positive charge, than in a lower oxidation state. There is also an interesting pre-edge structure at approximately 20,005 eV, which is very strong in the highly symmetric (tetrahedral) molybdate and much weaker in the low-symmetry site of the enzyme. Its intensity appears to be proportional to the number of oxo (O^{2-}) ligands around Mo, which is 4 for molybdate in aqueous solution and in the transport protein ModG vs. 2 for the enzyme.

When the enzyme CO-dehydrogenase was initially characterised, selenium was identified in various protein preparations. This led people to believe that in these preparations a dinuclear Mo–Se centre catalyses the dehydrogenation, possibly with the Se in the position of the question mark in the structure in the inset of

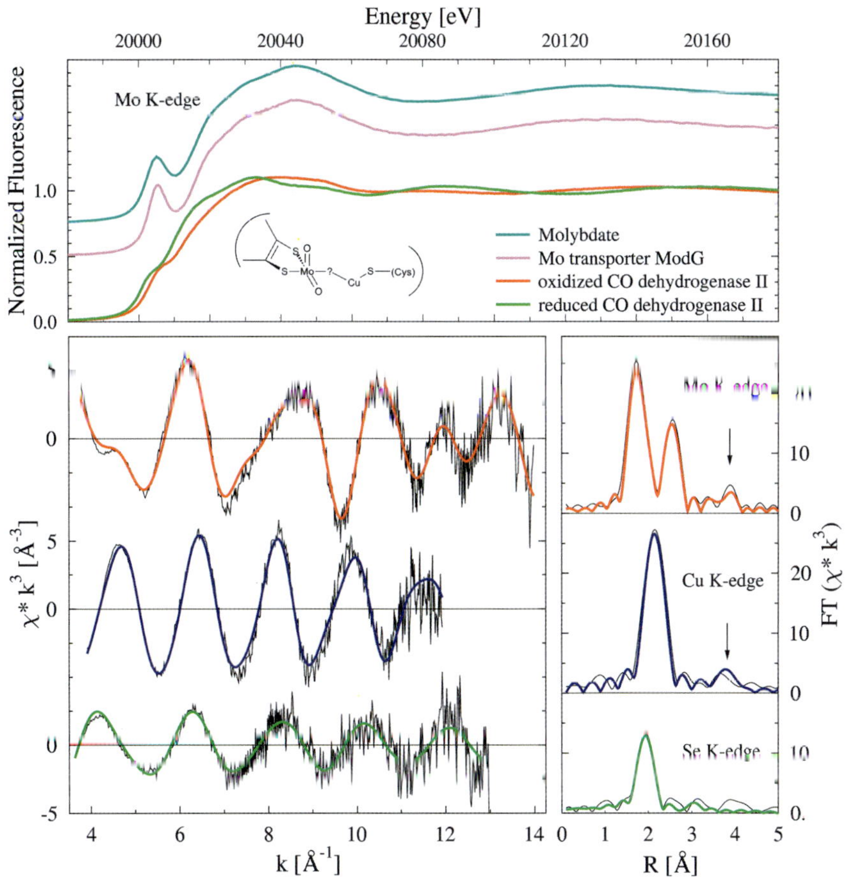

FIGURE 6.2 Top panel, Mo K edge XANES of (top to bottom) aqueous molybdate (blue), the Mo transport protein ModG (pink, (Duhme et al., 1999)), and oxidised CO-dehydrogenase (red, (Gnida et al., 2003)); inset, putative active site of CO-dehydrogenase. Bottom panel, experimental (black) and simulated EXAFS (left) and k^3-weighted phase shift-corrected Fourier transform (right) at (top to bottom) Mo (red), Cu (blue) and Se (green) K edge; metal–metal interactions highlighted with arrows.

Figure 6.2. This was inconsistent with the Mo K edge EXAFS analysis, however. The other panels of Figure 6.2 highlight the element specificity of X-ray absorption spectroscopy and show experimental EXAFS spectra (middle panel) and their Fourier transforms (FT) (right panel), respectively, of the Mo, Cu, and Se edge of the oxidised form of the enzyme. As will be discussed in more detail below, the FT give a radial distribution of atoms around the element at the edge of which the EXAFS was measured, and the phase-relationship between Fourier transform and EXAFS, together with the characteristic backscattering in the latter, allow the ligand atom types to be identified. The analysis of the EXAFS of enzyme preparations that were fully catalytically competent revealed the presence of a dinuclear MoCu cluster, bridged by a sulfur ligand. In the final model, Mo is directly surrounded by S and O ligands, the Cu by S ligands. The Fourier transform of the Mo and Cu EXAFS both show a small peak just below 4 Å, which represents the distance between Mo and Cu, which are connected by a bridging ligand. The Fourier transform of the Se EXAFS reveals the presence of the C atoms of the methionine, but no metal contribution. Thus Se is not a constituent of the active site.

OUTLINE OF THE BIOXAS CHAPTER

In this chapter, we will introduce the most important aspects of X-ray absorption spectroscopy that are relevant to biological studies. The main line of the text will discuss mostly principles and examples, while the reader is referred to specific Boxes for details on specific aspects: Box 6.1 on the BioXAS experiment; Box 6.2 on L edge, ligand edge, and X-ray emission techniques, Box 6.3 on data reduction, Box 6.4 on phase shift calculations, and Box 6.5 on the Debye–Waller factor. The first section discusses how XANES can be interpreted in terms of electronic spectroscopy, in order to extract information on oxidation state and ligand geometry. Subsequently we introduce the X-ray-induced electron diffraction that underlies EXAFS. We then treat the theory in terms of a single scattering approximation, and discuss the phase shifts that need to be calculated in order to be able to extract information on the distance and approximate atom type and number of the ligands by simulation. The next topic is multiple scattering, which is found to be particularly important in biological systems. We then discuss a strategy for the interpretation of the EXAFS and how it can be automated, strengths and limitations of the technique and its relation with other spectroscopic techniques and crystallography. The chapter concludes with a number of special topics, Conclusions and self-test Exercises.

XANES

The near edge region of the K edge XAS of Cu in the oxygen-binding protein haemocyanin of the arthropod *Panulirus interruptus* in its deoxygenated and oxygenated forms is shown in Figure 6.3. Upon oxygenation of haemocyanin, the oxygen molecule is bound to a pair of Cu^{1+} ions, and both Cu ions transfer an electron to the molecular oxygen, so that they become Cu^{2+}, resulting in a shift of the edge to higher energy. Intuitively one would expect on the basis of electrostatic arguments alone that it should be easier to liberate an electron from a metal in a low oxidation state (in this case Cu^{1+}) than in a high oxidation state (Cu^{2+}). The position of the edge, taken as the excitation energy corresponding to 50% of the maximum edge intensity, depends on: i) valence state, so that the edge appears at higher energy for a higher valence, but also on: ii) ligand type, and iii) the average R, so that it appears at higher energy when R is shorter ('Natoli's rule'). Valence state and average R play a role in the small (a few eV) edge shift observed in the direct comparison of the deoxygenated and oxygenated forms of the oxygen-binding protein haemocyanin shown in Figure 6.3.

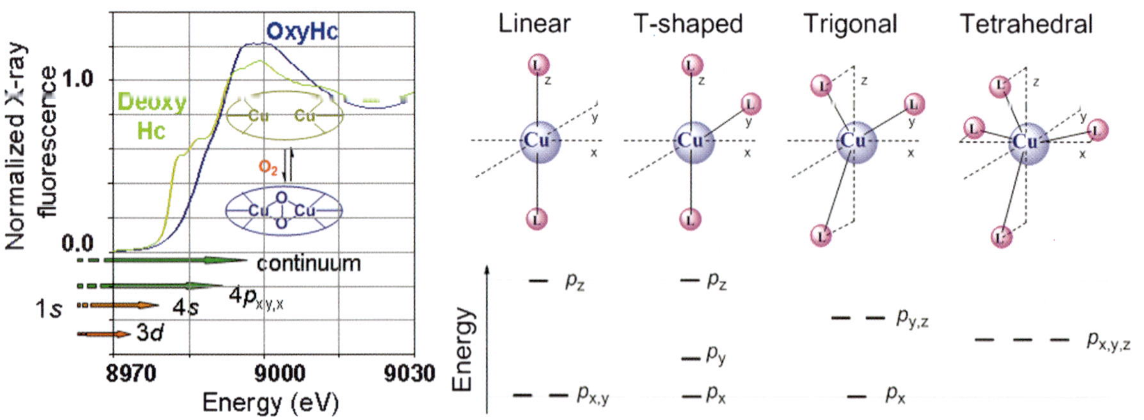

FIGURE 6.3 (left, top) X-ray Absorption Near Edge Spectra (XANES) of the Cu K edge of oxygenated (blue) and deoxygenated (green) haemocyanin from the spiny lobster *P. interruptus*; (left, bottom) approximate energies of excitations of the Cu 1*s* electron to non-occupied orbitals and to the continuum, with a colour code (red, formally forbidden; green, at least dipole- or spin-allowed) indicating the probability; (inset) oxygenation in the haemocyanin active site. (right) Effect of ligand geometry on the relative energies of the 4*p* orbitals of the Cu^{1+} ion.

In cases where the energy of the X-rays falls in the range 8980–8995 eV, where it is just not enough to liberate the $1s$ electron of Cu into the so-called continuum, it is probably still enough to excite the electron into one of the unoccupied higher orbitals of the absorber atom; the approximate energies of these transitions are schematically indicated by the arrows at the bottom of Figure 6.3, left. Such electronic transitions are governed by the same selection rules as those that apply to the UV-vis spectra of coordination compounds, and the possibilities for Cu are schematically indicated in Figure 6.3, left, with a colour code indicating the expected intensity. One rule is that they should be *spin-allowed*, i.e. they should not be accompanied by a change in the total spin S ($\Delta S = 0$), or, in other words, the *spin multiplicity* should be maintained. This means that some mechanism of relaxation of this rule must exist to allow transitions from a filled $1s$ orbital to an empty higher orbital, as it is spin-forbidden and its intensity is therefore at best relatively weak. Another rule, the Laporte rule, requires that the transitions should be *dipole-allowed*, by a change in symmetry between initial and final state, which is equivalent to the necessity of a change in secondary quantum number: $s \rightarrow p$ and $p \rightarrow d$ transitions are allowed (this is the most important reason that K and L_3 edges look differently, as they involve excitation of $1s$ and $2p$ electrons respectively, see Box 6.2), but $s \rightarrow d$ transitions are not. Like the $d \rightarrow d$ transitions in the UV-vis range of the electromagnetic spectrum, the transitions in systems with tetrahedral ligand geometry are more intense than for octahedral, as the rules are more likely to be relaxed in a system of lower symmetry.

Cu^{2+} ions have a vacancy in their $3d$ shell (d^9 system) which makes a $1s \rightarrow 3d$ transition possible at 8979 eV (Kau et al., 1987); this is spin-allowed but not dipole-allowed and therefore so weak that it is not visible in Figure 6.3. For some other transition metal ions, e.g. the d^6 and d^5 systems of Fe^{2+} and Fe^{3+}, respectively, there are various vacancies at different energies in the set of d orbitals, and the pattern of transitions is found to be sensitive to the spin state (high or low) of the ion (Westre et al., 1997). The energies and probabilities of the edge transitions are also influenced by the degree of overlap between the orbitals of metals and ligands, or in other words by the *covalency* of the bond between metal and ligand; this covalency can also be probed by measuring and interpreting the ligand XANES (see Box 6.2). For Cu^{2+} complexes, the K edge is at higher energy when the ligands are more electronegative, viz. increasing from approximately 8985 to 8988 eV going from S via Cl and N to O ligands. The intuitive explanation is analagous to that of the effect of a higher oxidation state, viz. that a more electronegative ligand would cause a decrease in electron density on the absorber atom and X-rays of higher energy would be required to liberate an electron. An analogous covalency shift is observed for octahedral Ni^{2+} complexes with varying amounts of S and N ligands (Colpas et al., 1991).

Cu^{1+} ions have a filled $3d$ shell (d^{10} system), but because of the lower coordination numbers they feature coordination geometries in which some or most of the $4p$ orbitals have no interaction with the ligands. For example, Cu^{1+} with linear 2-coordinate geometry has 2 ligands along the z-axis, which raises the energy of the $4p_z$ orbital above that of $4p_{x,y}$; this allows for 'pure' $1s \rightarrow 4p_{x,y}$ transitions, which are spin-forbidden but dipole-allowed, at relatively low energy (8983–8984 eV). The effects of coordination geometry for Cu^{1+} are illustrated in the right part of Figure 6.3, together with ligand-field orbital descriptions derived from the analysis of relevant model compounds (Kau et al., 1987). Another example of a pure transition is the $1s \rightarrow 4p_z$ transition in square planar Ni^{2+} (d^8 system with ligands in the xy plane (Colpas et al., 1991)).

In summary, the XANES of haemocyanin indicates that Cu^{1+} ions with a low (3) coordination number (various $1s \rightarrow 4p$ transitions observed) with low symmetry (no strong $1s \rightarrow 4p_{x,y}$ absorption indicative of linear geometry) are oxidised (edge shift) to a Cu^{2+} ion with some additional electronegative (O) ligands, consistent with the change in the chemical diagram in Figure 6.3, left (inset). We conclude that the XANES is very sensitive to biologically relevant changes in the metal environment; it can also be used to probe the stability of the metal ion towards non-biological processes like photoreduction and/or or radiation damage in other studies using X-rays, such as X-ray crystallography. Compared to EXAFS, which gives much more accurate information on ligand distances, it is much easier to record a XANES spectrum with high signal-to-noise ratio on a relatively small sample in a relatively short time. It is the only part of the spectrum that one can expect to measure satisfactorily in many time-resolved studies, and is much more suited than EXAFS for spatially resolved (imaging) studies where one aims at getting some more information (oxidation state, symmetry of coordination) on the trace element rather than just its distribution over the image.

BOX 6.2 Ligand Edge, L Edge, and X-ray Emission Spectroscopy (XES)

Ligand edge. An alternative for probing the covalency of metal–ligand bonds in proteins by metal K or L edge studies is to measure the ligand K edge XANES, e.g. for sulfur. The predominant transition in S K edge XANES is $1s \rightarrow 4p$, but the excitation of the S $1s$ to the empty metal $3d$ level can be stimulated by mixing of the latter with the S $3p$ level (Figure 6.2C, right panel). Thus the intensity of this pre-edge transition in the S K edge XANES is proportional to the degree of mixing, and thereby to the *covalency*. In this way the covalency of metal–S bonds in a number of electron transfer proteins has been probed, viz. for Cu–S bonds in the so-called blue copper proteins (Shadle et al., 1993), and for Fe-S in the [4Fe-4S] clusters in high-potential iron–sulfur protein and ferredoxins (Dey et al., 2007).

L edge. As mentioned in the text, the X-ray-induced transitions involving the $3d$ (or $4d$, $5d$) valence orbitals of transition metals can give information on the energies of these orbitals and thereby about the *spin state*, e.g. for Fe^{2+} and Fe^{3+} (Westre et al., 1997). On the one hand it is easier to probe the energy levels of the d orbitals, which gives information about the metal ion's spin state, from the transition metal's L_3 edge, since $2p \rightarrow 3d$ transition induced at this edge is dipole-allowed, whereas $1s \rightarrow 3d$ probed at the K edge is not; moreover, the L edge features are also sharper, and therefore better resolved, than the K edge features. From an experimental viewpoint the measurement at the softer L edge (see Figure 6.1 for typical excitation energies) is more demanding, since all parts of the spectrometer which can be in air in the instrument depicted in Figure 6A (Box 6.1) such as space between ion chambers, sample holder, and fluorescence detector, have to be in vacuum to avoid loss of intensity due to the X-ray absorption of the atmospheric gases; moreover, the fluorescence yield at the L edge is lower (Figure 6B, right). The higher resolution of the L edge features compared to the K edge makes the measurement with circularly polarised X-rays in the presence of a magnetic field (X-ray Magnetic Circular Dichroism, XMCD) very powerful. L edge spectroscopy has been applied to a number of Fe and Ni proteins (Cramer et al., 1997).

X-ray emission. Element-specific analyser crystals are required to resolve and measure the energies of the fluorescence emission lines arising from excitation of the $1s$ electron. The result of this Resonant Inelastic X-ray Scattering (RIXS) experiment is usually given as a two-dimensional plot of the excitation energy in the K edge energy range in one dimension, and the resolved fluorescence in the other (Glatzel and Bergmann, 2005). The final state of the combined excitation and emission is identical to that of the L edge excitation (Figure 6C, right part of left panel), but it has an experimental advantage of being probed by hard X-rays at the K edge rather than the soft L edge, and giving additional information.

In the X-ray emission spectrum of a first row transition metal (such as the Mn^{2+} represented in Figure 6C, left), the $K\alpha_1$ and $K\alpha_2$ lines are well resolved and more intense than the $K\beta_1$ and $K\beta_3$ lines, which are not resolved, by an order of magnitude; these are in turn more intense than the Kβ satellite lines $K\beta_{2,5}$ and $K\beta''$. Not indicated in Figure 6C, but usually present for transition metals which have a total electron spin $S \neq 0$ (such as Mn^{2+}), is the $K\beta'$ line at slightly lower energy than the $K\beta_{1,3}$ line. This results from emission from the metal $3p$ level combined with a spin flip of a $3d$ electron and is therefore sensitive to the *spin state* of the metal ion. Of the Kβ satellite lines, the cross-over emission line $K\beta''$ is extremely sensitive to the nature of the coordinating ligands, because it involves emission from the ligand's $2s$ level to the metal's $1s$ core hole, and allows one to distinguish O from N and C ligands. This *ligand identification* is of interest as it gives information complimentary to that obtained from EXAFS, which can typically not discriminate between coordination by ligands from the same row of the periodic table, see text. Examples are the variation in the number of O ligands to Mn in the so-called Kok cycle in Photosystem II (Yano and Yachandra, 2008), and the identification of central atom bound to Fe in the Fe, Mo cofactor of nitrogenase as C (Lancaster et al., 2011). A typical theoretical approach calculation of the multiplet 'ligand field multiplets' to interpret Kα and Kβ main lines, and molecular orbital theory for the Kβ satellites (Glatzel and Bergmann, 2005); both are outside the scope of this chapter.

BOX 6.2—cont'd

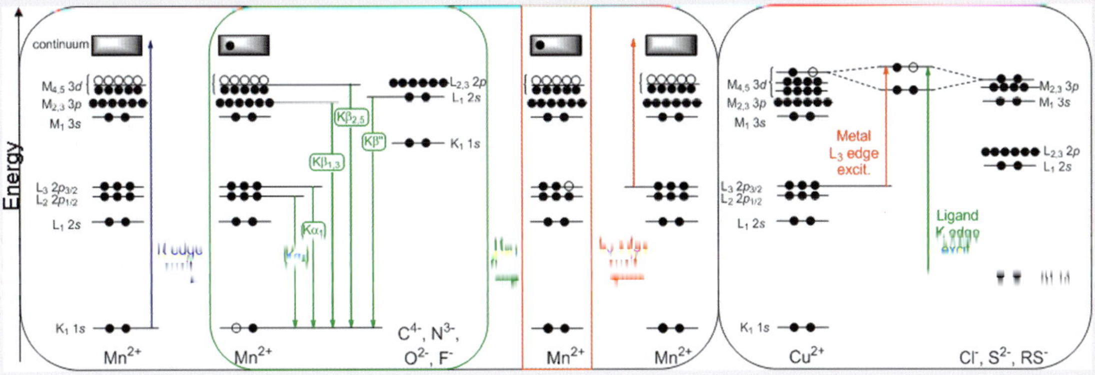

FIGURE 6C Overview of ligand edge, L edge, and X-ray emission transitions; the *y*-axis only gives relative energies and is not to scale. Left panel, effect of 'low-Z' ligands (C, N, O, F) on the X-ray emission of a Mn^{2+} complex. K edge excitation (blue) leads to a 1*s* core hole intermediate state (central green oval), which can emit X-ray fluorescence at various wavelengths. The final state obtained with $K\alpha_1$ fluorescence (red box) is identical to that obtained by direct L_3 edge excitation (red). Right panel, illustration of probing mixed 3*p* orbitals of Cl or S ligands with a transition metal's 3*d* orbital, in this case the singly occupied $3d_{x^2-y^2}$ orbital of Cu^{2+} (other Cu 3*d* orbitals grouped together for clarity), by either K edge or L_3 edge XAS.

X-RAY ABSORPTION SPECTROSCOPY: X-RAY-INDUCED ELECTRON DIFFRACTION

The EXAFS extracted as $\chi(E)$ from the normalised X-ray absorption spectrum (Box 6.3, Figure 6D) is usually presented as $\chi(k)$ with the energy axis converted from energy (E, eV) to wave vector (k, Å^{-1}), using

$$k = \sqrt{\frac{2m_e(E - E_0)}{\hbar^2}} \tag{6.2}$$

in which m_e is the mass of the electron, E_0 is the threshold energy (7120 eV for the Fe example in Box 6.3), and $\hbar$ is $h/2\pi$. This choice of x-axis has the advantage of showing the EXAFS oscillations as sinusoids. To offset the damping of the oscillations, the fine structure is also k^3-weighted, so that the oscillations now have more or less constant amplitude over the k range shown.

In Figure 6.4, various situations above the edge, where the X-ray energy is more than enough to liberate the electron to the continuum, and the excess energy goes into the kinetic energy of the photoelectron wave, are illustrated for an Fe ion surrounded by four nitrogen ligands (as for Fe in haemin, the example in Box 6.3). The wavelength of the photoelectron wave depends on the kinetic energy, i.e. on the difference between the energy supplied by the X-rays and that required to liberate the electron. On the top in Figure 6.4, we have a situation where an integer number of wavelengths exactly fits the Fe—N distance; this means that the electron wave going out from the absorber atom and that backscattered by the backscatterer atom are exactly in phase, as indicated by the lines depicting the wave fronts. As a result, the electron waves interfere constructively at the absorber atom, resulting in a high-electron density and hence a high-absorption coefficient for X-rays. At the bottom in Figure 6.4, we have a situation with higher X-ray energy, with a higher kinetic energy and a larger electron wavelength as a result. Now the wavelength does not fit the Fe—N distance so well, and the interference of the outgoing and backscattered photoelectron waves at the central atom is destructive, resulting in a relatively low electron density,

BOX 6.3 Data Reduction

The initial data reduction, including checks of the individual detector contributions to the individual scans and the alignment of the scans by energy calibration, involves some computation which can in principle be automated. In order for the EXAFS to be interpreted by simulation, it has to be extracted as χ by the procedure shown for the K edge of iron in aqueous haemin chloride (containing the protoporphyrin IX ligand, the Fe-coordinating cofactor present in haem proteins and enzymes) at around 7120 eV in Figure 6D. The experiment has been set up in such a way that a sufficiently long energy range in the pre-edge region is recorded so that it can be extrapolated to give the background absorption of the other elements in the absence of the central absorber atom Fe, μ' (blue line in Figure 6D, top left). Subtraction of this background μ' from the experimental μ gives the background-corrected spectrum μ_A. The next step is to construct the hypothetical X-ray absorption spectrum of Fe in the absence of surrounding atoms (atomic absorption, XAS of Fe as if it were a monoatomic noble gas), μ_0. This is done by instructing the computer programme for background subtraction to find a polynomial through the fine structure (red line in Figure 6D), which is more or less an extreme smoothing. The fine structure χ is then calculated relative to this polynomial as $\mu-\mu_0$, using the value of 7120 eV (Fe K edge) for the threshold energy (E_0),and normalised relative to the edge step ($\mu_0-\mu'$). This yields the fine structure $\chi(E)$ as oscillations around the zero level represented by the green line in the bottom left panel of Figure 6D, which is equivalent to the red line in the top left panel.

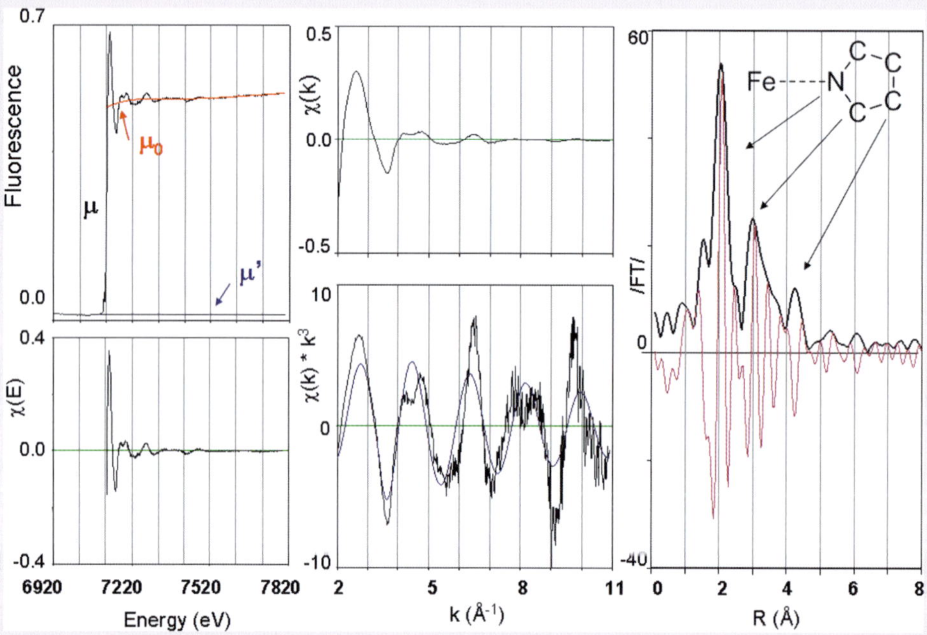

FIGURE 6D Data reduction for aqueous haemin chloride. Left panel, top: Iron K edge X-ray absorption spectrum (μ_f which is for thick samples independent of its thickness x) with μ' (blue), extrapolated pre-edge, representing the theoretical spectrum of the sample without the Fe absorber; μ_0 (red), polynomial resulting from a smoothened fine structure, representing the theoretical X-ray absorption of a Fe atom without neighbouring atoms. Left panel, bottom: extracted fine structure (χ), with the green zero line corresponding to μ_0 (red) in the top panel. Middle panel, extracted fine structure χ in k space, without (top) and with k^3-weighting (bottom); the blue line corresponds to a simulation representing 4 N ligands at 2 Å from Fe. Right panel, phase shift-corrected Fourier transform of the k^3-weighted χ, inset: structure of coordinated pyrrole moiety; black and purple lines represent the modulus and the imaginary parts, respectively.

(Continued)

BOX 6.3—cont'd

As shown in the middle panel of Figure 6D, the EXAFS is usually presented as $\chi(k)$ with the energy axis converted from energy (E, eV) to wave vector (k, Å^{-1}), using Eqn (2) in the text, which has the advantage of showing the EXAFS oscillations as sinusoids (Figure 6D, middle, top). To offset the damping of the oscillations, the fine structure is also k^3-weighted, so that the oscillations now have more or less constant amplitude over the k range shown (Figure 6D, middle, bottom). With this weighting the experimental noise in the high-energy range is also amplified, and it is advisable to set up the experiment in such a way that some extra time is spent on recording this part of the spectrum, and to choose the energy distance between the points such that they are equidistant in k space after the conversion.

In Figure 6D, right, we have not only plotted the real part or modulus of the phase shift-corrected Fourier transform of the EXAFS, as is customary in most literature, but also the imaginary part, which contains information on the phase relation between EXAFS and FT. Good agreement between experiment and simulation in both the real and imaginary parts of the Fourier transform inevitably implies good agreement with the EXAFS. When we compare the positions of the peaks in this radial distribution function to the known structure of haemin (see inset) we note that no contributions of atoms in the structure beyond 4.3 Å from the central Fe atom are observed. Indeed it is exceptional to detect a contribution beyond the first shell of atoms at such a distance as observed for the specific case of the pyrrole unit here; the factors that determine whether a long-range contribution will be detected or not are the electron mean free path, the presence of a rigid chemical system and the Debye–Waller factors, which will be discussed in more detail in Box 6.5.

and a low absorption coefficient for the X-rays. It can be seen that when the X-ray energy is increased, going from the situation at the left in Figure 6.4 to that at the right, the X-ray absorption coefficient goes from a minimum to a maximum. Indeed when the energy is scanned over a longer range, we will go through various situations where the electron wavelength varies such that the X-ray absorption coefficient goes through minima and maxima. This results in the EXAFS which can now be seen to depend on the distance of the absorber atom to the closest backscatterers. This observed interference between the electron waves is in fact a very convincing demonstration of the proposed (de Broglie) wave character of electrons, in addition to the diffraction patterns that can be observed in an electron microscope. Although X-rays are used in X-ray absorption spectroscopy, and it is clearly a spectroscopy that yields structural information that is comparable to or complementary with X-ray crystallography, it does not involve X-ray diffraction, but X-ray-induced electron diffraction.

The distance between absorber and backscatterer determines the frequency of the EXAFS oscillation; if the neighbour atom is close to the absorber (trace element), the frequency is low, and if the neighbour atom is remote, the frequency is high. The mathematical tool of choice to analyse oscillations is the so-called Fourier transformation; applied to the EXAFS with its energy axis in the wave vector dimension (reciprocal length unit), it gives a radial distribution function with maxima at the distances from the central atoms where shells of atoms occur (Sayers et al., 1971). For a system with two shells of atoms at close and remote distances we observe a fine

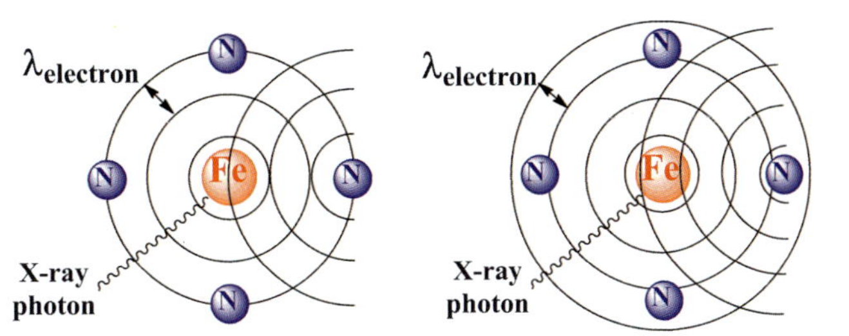

FIGURE 6.4 Examples of interference patterns of electron waves between absorber and backscatterer atoms, leading to (left) constructive interference and maximum electron density at the absorber, and a maximum in the X-ray absorption coefficient μ and the EXAFS χ, or (right) destructive interference and minimum electron density at the absorber, and a minimum in μ and χ.

structure that is an interference pattern of oscillations with low and high frequency, respectively. These appear in the Fourier transform at their respective distances, provided that the phase shift, which will be discussed next, is taken into account.

PHASE SHIFTS AND EFFECT OF ATOM TYPE

What we have not mentioned so far is that the photoelectron wave undergoes a *phase shift* while travelling through the atomic potentials of absorber and backscatterer; there is a phase shift when the wave leaves the absorber atom, another one when it goes back and forth through the potential of the backscatterer atom, and then yet another one when it returns to the absorber atom. Such phase shifts occur as well in refraction, when photons enter a medium of different optical density. In EXAFS, this causes the non-phase shift-corrected R' (also indicated as $R + \phi$) found in the radial distribution function immediately after Fourier transformation to be an underestimation of the true (phase shift-corrected) R by 0.2–0.5 Å depending on the type of backscatterer. It follows that for the EXAFS to be interpreted in terms of absorber–backscatterer distances, knowledge of the phase shifts of the absorber–backscatterer pair is required, either from model compounds or from calculations, as discussed in Box 6.4.

BOX 6.4 Phase Shift Calculations

As mentioned before, knowledge of the phase shifts of the absorber–backscatterer pair is required for the EXAFS to be interpreted in terms of absorber–backscatterer distances. The same is true for the backscatterer amplitudes with respect to number of atoms (coordination number, occupancy), and the variation of the backscatterer amplitude with k to identify the atom type. In early studies, EXAFS was simulated using phase shifts and backscattering amplitudes that were extracted from model compounds, which were shown to be transferable to unknown systems. Nowadays, theoretical calculations using the programmes FEFF (Rehr and Albers, 1990) and EXCURVE (Gurman et al., 1984 & 1986) are accurate and accessible enough for practical use, but of course it is still good practice to validate the results on model compounds of known structure. Such calculations of phase shifts and backscattering amplitudes, usually collectively known as 'phase shift calculations', require knowledge of the potential at every place in the system of absorber and backscatterers; this is approximated by considering the individual atoms and the surrounding cloud of electrons as potential wells in an area of constant interstitial potential, in the so-called *muffin-tin* approximation (Figure 6E). The advantage of a constant interstitial potential is that in simulations the distance between atoms can be varied in order to find one which optimally reproduces the observed EXAFS, with no need to repeat the phase shift calculation for every distance. At the current level of theory, it is not necessary any more to empirically adjust or refine the *amplitude reduction factor* S_j nor the *electron mean free path* λ_j (see Eqn. 4.4).

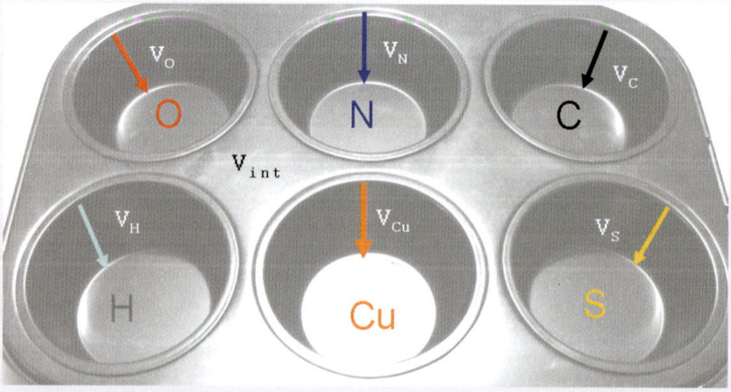

FIGURE 6E 'Muffin tin' illustrating the type of potential approximation used for phase shift calculations; V_{Cu} etc., individual atom potentials; V_{int}, interstitial potential.

The type of backscatterer has an effect on its phase shift, because this is directly affected by the shape of its potential. In addition, there is an effect on the intensity of the backscattered photoelectron wave. This intensity depends on the photoelectron energy and thus the X-ray energy. The envelope (variation of the backscattering amplitude, independent of the oscillatory variation due to the distance to the absorber) describing this dependency varies with the type of backscatterer, as is illustrated in Figure 6.5, left panel. First of all, it is obvious that the contribution of H to the backscattering is very weak compared to that of other backscatterers for most of the *k* range, so it is usually neglected in EXAFS simulations. The similarity of the envelope for elements that are in the same row of the periodic table, such as the examples of C, N, O, and F, or S and Cl, in Figure 6.5, makes it difficult to discriminate between them; the light biological backscatters of the 1st row in the periodic table, C, N, and O, are therefore often collectively referred to as 'low-Z' ligands. The reason is that the backscattering amplitudes and the phase shifts are comparable, due to the similarity in the sizes of the nuclei and the surrounding electron clouds, respectively. It is found that the backscattering amplitude envelope is similar for C, N, O, and F (a decay with increasing *k* for most of the *k* range), and that the differences in backscattering amplitude and phase shift are too subtle to be of diagnostic value in routine cases.

The envelopes in Figure 6.5 (left) also indicate that it should be possible to discriminate between atoms from different rows of the periodic table, for example for the halogens (F, Cl, Br, and I) which are elements from the same column (see Figure 6.1). Figure 6.5 (right) shows the EXAFS (top) and phase shift-corrected Fourier transform (bottom) of these elements when placed at 2.0 Å (not realistic for all elements) from a Zn ion. The EXAFS panel confirms that the maximum in the backscattering amplitude envelope at low (4–5 $Å^{-1}$) *k* for F shifts to higher values for Cl (6), Br (10) and I (11–12 $Å^{-1}$), with an additional maximum at low *k* for I (5–6 $Å^{-1}$). Interestingly, when the phase-relationship between the EXAFS and the Fourier transform is inspected,

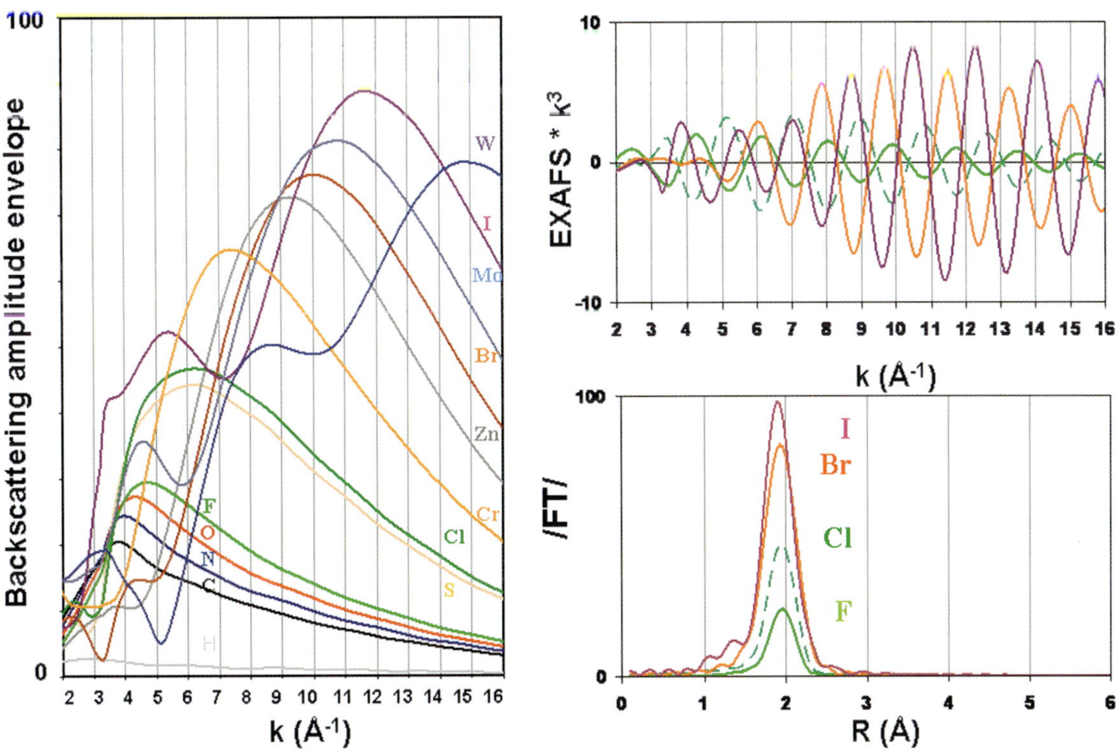

FIGURE 6.5 Effect of atom type; left, backscattering amplitudes of selected atoms; right, EXAFS and phase shift-corrected FT of elements from the same (2nd) row and column (halogens) of the periodic table at 2.0 Å from Zn.

it turns out that the EXAFS of F is approximately π out of phase (has opposite phase) with that of Cl, which in turn has opposite phase to that of Br, which is again out of phase with I. As a result, the contribution of F is approximately in phase with that of Br, and that of Cl with that of I. As will be discussed below, the phase-relationship offers another possibility to identify contributions of elements from different columns, in addition to the inspection of the backscattering amplitude envelope, as will be discussed below for a biological example.

Backscatterers with very high *Z* have multiple maxima in their backscattering envelope. This can be seen when going from Br to I in Figure 6.5, left; the larger backscatterer, I, has actually a weaker backscattering amplitude in the *k* range 7–9 $Å^{-1}$. When the group 6 transition metals are compared, it can be observed that Cr has a larger backscatterer amplitude than Mo and W in the *k* ranges 5–8 and 4–10 $Å^{-1}$, respectively. The backscatterer with the largest Z included in Figure 6.5, W (Z = 74) has the strongest backscattering amplitude only at high *k*; therefore, substitution of Mo by W as the other metal X in mixed $[FeXS_2]$ and $[Fe_3XS_4]$ clusters paradoxically leads to a loss of backscattering power in the Fe EXAFS when investigated over a short *k* range (Antonio et al., 1985).

Upon going from nitrogen to an element from the next row, sulfur, the amplitude of EXAFS and Fourier transform becomes bigger (as expected due to the larger nucleus and number of electrons), but just as in the example of fluorine and chlorine discussed above, the EXAFS is also π out of phase when placed at the same distance; in addition the maximum in the backscattering amplitude envelope shifts towards higher *k*. Figure 6.6A illustrates that the combination of two simulated contributions of equal size but opposite phase, e.g. 2N (blue) and 1S (orange) at the same distance (2.0 Å) from a Zn absorber, can interfere destructively and lead to a very weak total EXAFS signal (black). In reality this rarely happens, not only because of the subtle shift in maximum in the backscattering amplitude envelope, but also due to the difference in ionic radius, N and S are never at exactly the same distance from, e.g. Zn. This is illustrated in Figure 6.6B for the simulated example of 2N at 2.0 and 1S at 2.3 Å from Zn, which predicts that the amplitude envelope of the total EXAFS is not as weak as in the case of the atoms at the same distance, and will go through minima and maxima over the experimental *k* range. This is confirmed in Figure 6.6C for a biological system, the coordination of Zn^{2+} ion in a protein (HIV-2 integrase (Feiters et al., 2003)) by 2N(imidazole) ligands at 2.0 and 2S at 2.3 Å.

The experimental *k* range (2–16 $Å^{-1}$) in Figure 6.6C includes some relatively noisy data at high *k* (also because of the k^3-weighting) but is long enough to resolve the N and S contributions to the major shell. Compared

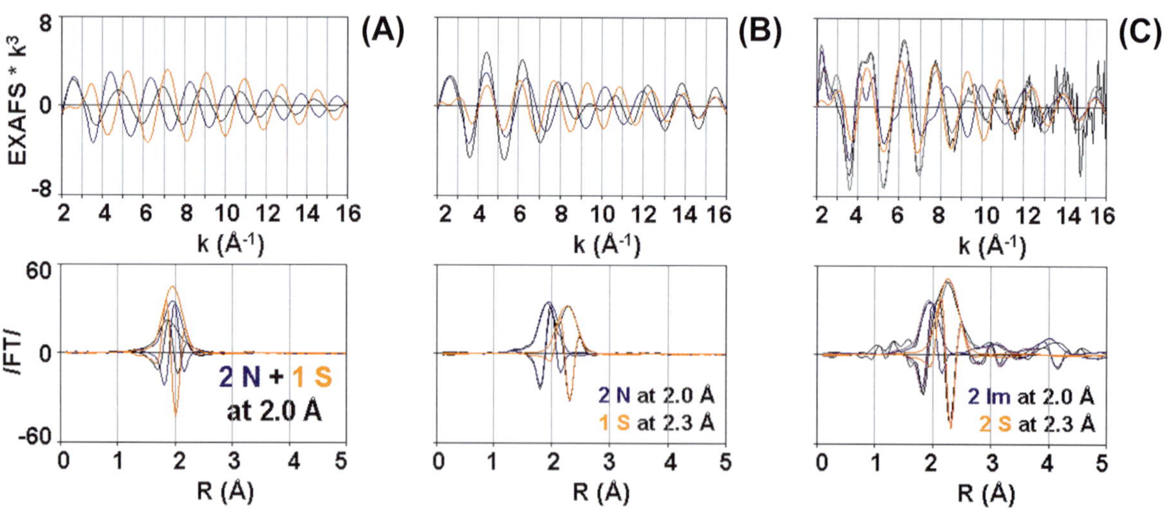

FIGURE 6.6 Dependence of phase shift on atom type. Top panels, k^3-weighted EXAFS taken at the Zn K edge; bottom panels, phase shift-corrected Fourier transform of (A) sum (black) of 2N atoms (blue) and 1S atom (orange) at 2.0 Å, (B) sum (black) of 2N atoms (blue) at 2.0 Å and 1S atom (orange) at 2.3 Å, (C) Zn K edge experimental EXAFS of HIV-2 integrase (Feiters et al., 2003) (black) with theory (pink) calculated for two imidazole ligands at 2.0 Å (blue) and 2 sulfur ligands at 2.3 Å (orange).

to other techniques where Fourier transformation is used, such as pulse nuclear magnetic resonance (NMR) and Cyclotron Mass Spectrometry, the range over which the Fourier transform is taken is relatively small in EXAFS, because the experimental k range is usually limited to a few (<10) oscillations. This means that the resolution, ΔR, of EXAFS is rather limited, according to its dependence on the k range:

$$\Delta R = \pi/2\Delta k(\text{Å}) \tag{6.3}$$

which means that for a k range, Δk, of 14 Å^{-1}, the distance between two shells that can be resolved ΔR is equal to $\pi/2 * 14 = 0.11$ Å. So the paradoxical situation exists that the power of EXAFS to measure distances with an accuracy of ± 0.02 Å is limited by the fact that in order to be resolved in EXAFS, two shells have to be >0.1−0.2 Å, i.e. an order of magnitude more, apart.

PLANE-WAVE AND MUFFIN-TIN APPROXIMATION

As discussed before, the EXAFS can be conceived as the sum of oscillations that are resolved in the Fourier transform and represent a number of shells (j) of backscatterer atoms of a certain type at certain distances from the absorber. This is summarised in the expression for the EXAFS in the so-called plane-wave approximation which is given below as Eqn (6.4), which gives us the opportunity to discuss a number of parameters and their effects on EXAFS simulations and the accuracy of the results obtained. In this approximation, the curvature of the electron wave is neglected; its derivation from the accurate description of the process given by *Fermi's Golden Rule* can be found elsewhere (Burge, 1993). It can be seen that the formula for the plane-wave approximation for EXAFS:

$$\chi(k) = {}_{\mathrm{j}}\sum \underbrace{N_j \cdot S_{\mathrm{i}}(k) \cdot F_{\mathrm{j}}(k)}_{\text{amplitude}} \cdot \underbrace{e^{-2k^2\sigma_{\mathrm{j}}^2} \cdot e^{-2r_{\mathrm{j}}/\lambda}}_{\text{damping}} \cdot \underbrace{\sin(2kr_{\mathrm{j}} + \phi_{\mathrm{j}}(k))/kr_{\mathrm{j}}^2}_{\text{oscillatory}} \tag{6.4}$$

contains an oscillatory part (the sine term) and an amplitude part, which in turn contains pure amplitude parts and damping factors. In the amplitude part, F_{j} is the *backscattering amplitude* of each of the N_{j} (*coordination number, occupancy*) backscattering atoms of type i in shell j. Each shell also has a damping part characterised by the *Debye−Waller factor* $2\sigma_{\mathrm{j}}^2$ (discussed in Box 6.5) and an oscillatory part determined by the distance r_j and the total phase shift ϕ_{j} (2 times that of the absorber + once that of the backscatterer). Because of the definition of k, the choice of the *threshold energy* (Δ)E_0 has its largest effect in the oscillatory part; this is a refinable parameter in the simulations as ΔE_0. There remains in the amplitude part the *amplitude reduction factor* S_{i}, which corrects for X-ray absorption processes not contributing to the EXAFS, such as multiple excitation effects, and in the additional damping factor the *electron mean free path* λ (see Box 6.4).

MULTIPLE SCATTERING IN BIOLOGICAL SYSTEMS

It is of interest to look a little more in detail at the relation between the EXAFS and its Fourier transform. When the FT of a number of imidazole complexes are compared the pattern of imidazole coordination is always the same, independent of metal ions. It is worth noting that the FT patterns for $Zn(im)_4$ and $Zn(im)_6$ are comparable (although shifted by approximately 0.2 Å, because the imidazoles are forced to be further away from the metal ions due to the stronger steric hindrance in 6-coordinated complexes), whereas the appearance of the EXAFS is different (Figure 6.7A); the so-called camel back feature at 4−5 Å^{-1} that is characteristic of 4-coordinate metal imidazole complexes (typical metal−imidazole N distance 2 Å) is absent in the EXAFS of the 6-coordinate complex. This underlines the diagnostic value of the Fourier transform. Figure 6.7A shows the experimental spectrum of $Zn(imidazole)_4$ diperchlorate together with a simulation based on the crystal structure. We have not only plotted the real part or modulus of the Fourier transform, as is customary in the literature, but also the imaginary part, which contains information on the phase relation between EXAFS and FT. There is a good

BOX 6.5 The Debye–Waller Factor

As mentioned when introducing the formula for the resolution (Eqn (6.3)), the power of EXAFS to measure distances with an accuracy of ±0.02 Å is limited by the fact that in order to be resolved in EXAFS, two shells have to be >0.1–0.2 Å, i.e. an order of magnitude more, apart. In cases where shells are not resolved, the distance found in the simulation for the combination of unresolved shells will be an average distance. The increased disorder in a shell that is composed of non-resolved contributions (static disorder) is noted as a more rapid decline of the EXAFS amplitude at higher energy, and a broadening of the peak in the Fourier transform; in the refined simulation, this is reflected in a larger value for the Debye–Waller factor, which was introduced in the formula of the planar wave approximation (Eqn (6.4)) as $2\sigma_j^2$.

In crystallography a displacement factor is used to describe deviations of an atom's position from its lattice point. The Debye–Waller factor or $2\sigma_j^2$ used in EXAFS simulations is related to this crystallographic parameter, but it always relates to at least a *pair* of atoms, i.e. an absorber and backscatterer, or an absorber and more backscatterers. The Debye–Waller factor describes effects of *static and thermal disorder* on the EXAFS spectrum. A high value for the Debye–Waller factor can be caused by a variance in the ligand distances (*static disorder*), as in the example of the unresolved shells discussed above. It can also be caused by disorder due to *thermal effects*, i.e. oscillations in the absorber–backscatterer distance. Whether the origin of the disorder is static or thermal can be probed by temperature variation; upon lowering the temperature, the value that $2\sigma_j^2$ refines to in the simulation should go down in the case of thermal disorder, because the oscillations that cause this disorder become weaker, whereas for static disorder it stays the same (Scherk et al., 2001).

It should be noted that a high σ_j for a single absorber–backscatterer pair can be due to uncorrelated motion because of a weak chemical bond. As mentioned above, a high value for $2\sigma_j^2$ is reflected in a rapid decrease (damping) of the EXAFS signal in k space, and a broadening/decrease in amplitude for the peak in the FT. The bond between an absorber and backscatterer can be so weak that its contribution to the EXAFS is virtually wiped out because of the high value of $2\sigma_j^2$. In fact the observation of shells beyond the first shell of ligand donor atoms, leading to the characteristic patterns for the haemin in Figure 6D and for the imidazole ligand to Zn in the integrase in Figure 6.6, is exceptional, and reserved for rigid systems like ligands with strong bonds (CO, CN) or rings (porphyrin, imidazole) only.

The reasons why EXAFS dies out after one or a few shells can be summarised as follows:

- The electron mean free path λ is limited
- Disorder of a static or thermal origin
- Destructive interference of contributions of opposite phase
- A weak bond between absorber and backscatterer, or none at all

As outcome of the EXAFS simulations, the (bio)chemist is interested in the type of atom (this determines the back-scattering amplitude F_j; it must be chosen, and reconsidered if not adequate for the simulation) and its number N_j and distance r_j (which are set at a reasonable starting value, and then iteratively refined in simulation). Unfortunately, refinement of (and correlation with!) physical parameters which are chemically less interesting, such as the Debye–Waller factor and ΔE_0, cannot be avoided. Please note that the both coordination number N and the exponential Debye–Waller factor are amplitude factors. They are correlated in the analysis and the degree of correlation depends on the length of the energy range, because only the exponential function including the Debye–Waller factor depends on the wave vector.

agreement between experiment and simulation in both the real and imaginary parts of the Fourier transform, and this inevitably means that there is also good agreement with the EXAFS.

Like the aqueous haemin chloride of Figure 6D, the metal imidazole complexes in Figure 6.7 are examples of systems with rigid (heteroaromatic) ligand systems which typically allow a characteristic pattern of shells to be observed. In such systems, besides the sum of single scattering pathways of the photoelectron wave

$$\text{A(bsorber)-B(ackscatterer)-A(bsorber)} \qquad \text{and} \qquad \text{A(bsorber)-R(emote backscatterer)-A(bsorber)}$$

such as represented in Figure 6.4, and represented by the equation for the plane-wave approximation given above, multiple scattering pathways of the kind

$$\text{A(bsorber)-B(ackscatterer)-R(emotebackscatterer)-A(bsorber)}$$

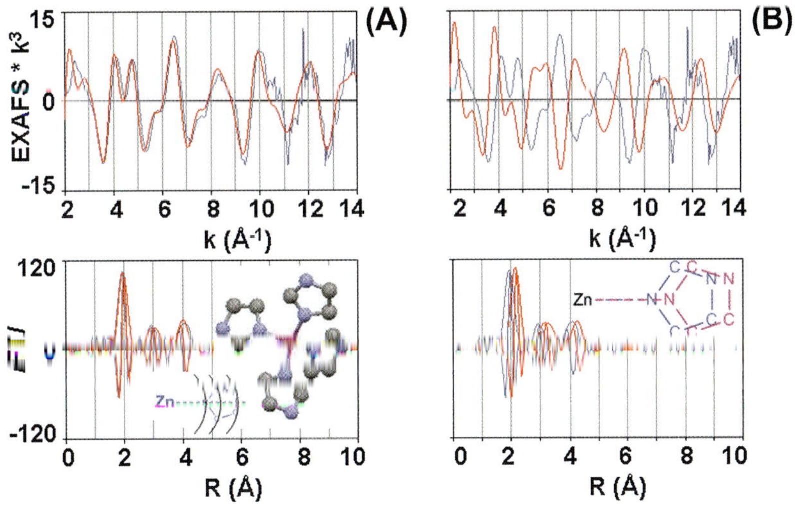

FIGURE 6.7 k^3-weighted Zn K EXAFS (top) and phase shift-corrected Fourier transform (bottom) of Zn imidazole complexes. (A) Zn(imidazole)$_4$ diperchlorate (blue grey) with simulation (red) based on the crystal structure, (B) comparison of Zn(imidazole)$_4$ diperchlorate experimental with imidazoles at 2.0 Å (blue grey) with simulation (violet) of 6 imidazoles at 2.2 Å.

may exist. For many systems, these are only important at low k, i.e. in the XANES region (0−50 eV above edge); this is why this part of the XAS is much more affected by the two- and three-dimensional order in the molecular structure than the EXAFS, which basically depends on the radial distribution function.

However, multiple scattering pathways are important for the whole k range of the EXAFS of complexes with rigid ligand systems where the angle A−B−R−A approaches 180° (>140°). Examples are coordinating cyanide, isocyanide, or carbon monoxide ligands (Korbas et al., 2006), or coordinating rigid heteroatomic ligands, such as pyridine, imidazole (Binsted et al., 1992), pyrrole, and porphyrin, as shown in Table 6.1. The absorber atom can also be at the centre of a multiple scattering unit itself, e.g. in a coordination geometry with perfect octahedral symmetry,

TABLE 6.1 Geometries of Biologically Relevant Ligands with their Multiple Scattering Pathways

Example	Single scattering	Multiple scattering
CO (X=O) and CN (X=N) ligands to Fe in hydrogenase CODH I (Korbas et al., 2006)	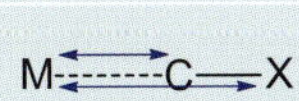	
Pyrrole (Y=C) part of porphyrin; imidazole (Y=N) (Binsted et al., 1992)	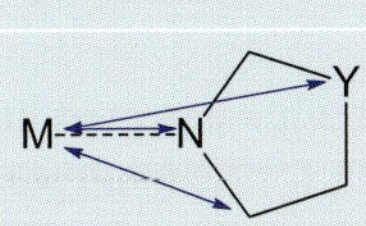	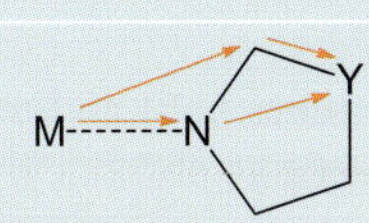
Tungstate or molybdate bound to protein ligands (Hollenstein et al., 2009), square planar Ni in dithionite-reduced CO-dehydrogenase II (Ha et al., 2007)	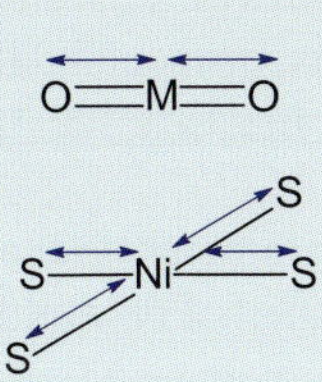	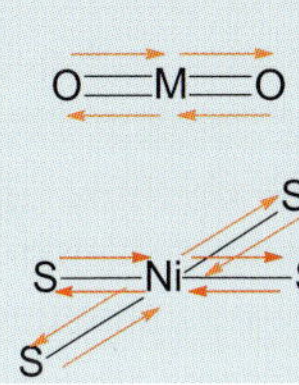

such as the tungstate bound by two monodentate carboxylate ligands in the bacterial tungstate-binding protein WtpA (Hollenstein et al., 2009), or a square planar geometry, such as Ni in dithionite-reduced CO-dehydrogenase II (Ha et al., 2007). Synthetic coordination complexes, which are model complexes in the structural and not necessarily in the functional sense, in which one or more aspects of the (expected) ligand environment in the protein, such as type and/or geometry of the ligands, are mimicked, are important as reference compounds in order to extract phase shift information, or to test the validity of *ab initio* approaches with single and/or multiple scattering.

The important difference between the single scattering pathway A—R and the multiple scattering pathways A—B—R is the presence of the extra atom B in the pathway, whose electrons give an extra contribution to the phase shift. Attempts to simulate the multiple scattering system A—B—R as the sum of two single scattering systems A—B and A—R ignore the effect of the position of the atom B between A and R and therefore lead to anomalous amplitude and phase effects, and unrealistic results for distances and occupancies (Strange et al., 1987). The imaginary part of the Fourier transform of the Zn imidazole model compound in Figure 6.7A shows that the amplitude of the shell at approx. 3 Å is anomalously low compared to that of the major shell at 2 Å (considering that it represents 2C atoms per imidazole ring), but its phase is similar, whereas the shell at approx. 4 Å has an anomalously large amplitude, and a different phase-relationship with the EXAFS. In order to calculate the multiple scattering, geometric two-dimensional information of the imidazole unit has to be used in the simulation, in particular the Zn—N—C and Zn—N—N angles, which are derived from the crystal structures. Only if the shell at approximately. 4 Å is reproduced with its correct phase amplitude is the 'camel back' feature at 4—5 $Å^{-1}$ in the EXAFS, which we know to be characteristic of imidazole coordination at 2.0 Å, correctly reproduced.

STRATEGY FOR THE INTERPRETATION OF EXAFS

Having established the most important aspects of EXAFS simulations, it is time to discuss the strategy. After elementary data reduction, the k^3-weighted EXAFS will be simulated. A numerical indication of the quality of the simulations is the so-called fit index (FI), which is a measure of the difference between experimental and simulated spectrum over the whole data range. Based on the chemical information already available on the system, atom types will be chosen for a calculation of the phase shift and backscattering amplitude. It is possible but not necessary to isolate the shells that contribute to the EXAFS and are resolved in the Fourier transform by a process called Fourier filtering, i.e. Fourier transformation, selection of the R' range of the shell, and back transformation. For every shell, a reasonable choice of atom type and of the four most important parameters: ΔE_0, threshold energy; R, distance absorber—scatterer; N, occupancy; and a ($=2\sigma^2$), Debye—Waller factor, is made and the EXAFS is calculated to see if there is a reasonable agreement with the experimental EXAFS and FT. If such an agreement cannot be obtained, even by adjusting the parameters, it is time to reconsider the atom type.

Once a full starting model is available, it is time to instruct the simulation programme on the computer to start looking for the best possible fit by a process called iterative refinement. It means that the computer programme will start moving the parameters in small steps and calculate the spectrum and, most importantly, compare it to the experiment by the FI to see if this has improved. If the FI has decreased, the computer programme for EXAFS simulations will continue to move the parameter in the same direction; if not, it will try the other direction. This loop continues until no further decrease in the FI is detected. It is expected that the computer programme will then have reached an absolute minimum in the multidimensional parameter/FI space. It is wise to check whether the computer programme reaches the same minimum from a different set of starting parameters. The minimum should correspond to a simulation with good agreement to the experimental data. As in any simulation of experimental data, the resulting parameters should be critically evaluated. In case a satisfactory answer cannot be obtained at this stage, it is time to consider the atom types again.

In single scattering simulations, the parameters that are allowed to float freely (that are refined) are ΔE_0 for the complete simulation, and R for every shell. For multiple scattering simulations, the parameters are ΔE_0 for the complete simulation, and R as well as a, angle M—A—B for every shell. For unknown systems, in addition the occupancy of each shell might be refined. Generally, the number of parameters that is refined should be kept as low as possible; the refinement of too many parameters might lead to an overinterpretation of the data. In EXAFS data analysis the number of parameters that can be refined should always be less than the number of independent data points N(ind), which depends on the k and R range that are fitted according to:

$$N(\text{ind}) = (2\cdot\Delta k\cdot\Delta R)/\pi + 2 \tag{6.5}$$

In the case of heteroaromatic ligands, such as the porphyrin (pyrrole) and imidazole examples from Table 6.1, this means that one should take as much advantage as one possibly can from knowledge about the geometry of the multiple scattering unit (the aromatic ring) that is defined for the multiple scattering calculations. The atom-to-atom distances within the ring are not expected to change much with coordination to metals or with different orientations with respect to the metal-donor atom (typically N) vector, and it is more important to put emphasis on an independent measurement of metal donor atom distance and metal-donor atom-other atom angles. In *constrained refinement*, the distances within the unit are fixed, and the parameters refined are ΔE_0 for the complete simulation, one occupancy and angle for any unit, and one R and one σ for every shell.

This is too rigid for most simulations, and *restrained refinement* (imported from protein crystallography) is often preferred. In this approach, idealised values (restraints) for the distances in the unit are given, the metal-unit atom distances are allowed to vary freely in the refinement, but if this variation results in deviations from the restraints, a penalty (which can be weighed) is added to the FI, to discourage the computer from looking further in the direction where the unit is distorted (Binsted et al., 1992). The parameters refined are ΔE_0 for the complete simulation, one occupancy for every unit, and one R, one angle, and one for every shell. Independent information is obtained for the number of units, their distance to the metal, and (if angles are refined) the orientation with respect to the metal-donor atom vector. In all simulations it is necessary to make reasonable choices for the values of the Debye–Waller factor before refinement. In multiple scattering units this means that it increases with the distance of the shell to the metal ion. To apply the same value of the Debye–Waller factor to shells at similar distances is a useful way to limit the number of parameters in all refinements.

VALIDATION AND AUTOMATION OF EXAFS DATA ANALYSIS

The structural model resulting from EXAFS data analysis or other methods such as protein crystallography should always be compared to prior knowledge on the system under study, its chemistry and established knowledge. Ignoring the scientific scepticism might lead to the publication of crystal structures traced backwards, unreasonable metal–ligand distances caused by low occupancy of the metal binding site or in case of EXAFS to wrong metal binding motifs due to problems with sample quality or wrong metal–ligand distances caused by photoreduction.

The challenge in data analysis is to avoid such pitfalls. Therefore, chemical knowledge and criteria indicating potential problems are applied. In automation, such criteria become part of the routines applied searching for the structural model representing the data best. One approach in automation is based on modelling of Debye–Waller factors or the introduction of boundary conditions, e.g. by the bond valence sum method (see below), and to use these values as boundary conditions in the refinement. In contrast, the second approach favours a shot-gun strategy: One selects all potential ligands and their occupancy range; on this basis potential starting models for the EXAFS refinement are designed; all models are compared to the EXAFS data by refining distances, Debye–Waller factors and ΔE_0, but no occupancies; the parameters are analysed with the help of criteria, that one can as well use as quality indicators:

(i) Do the obtained distances fall into the interval of reported metal–ligand distances? Note that these values depend on the metal oxidation state (and the ligand's chemistry). Reference data for frequently used metal ions are critically summarised by several authors (Harding, 2004) or can be extracted from small molecule databases (e.g. Cambridge structural database).

(ii) Bond valence sum analysis (BVSA). This formalism which is described in detail elsewhere (Thorp, 1998) takes advantage of established empirical correlations between a) the oxidation state of a metal, b) its metal–ligand

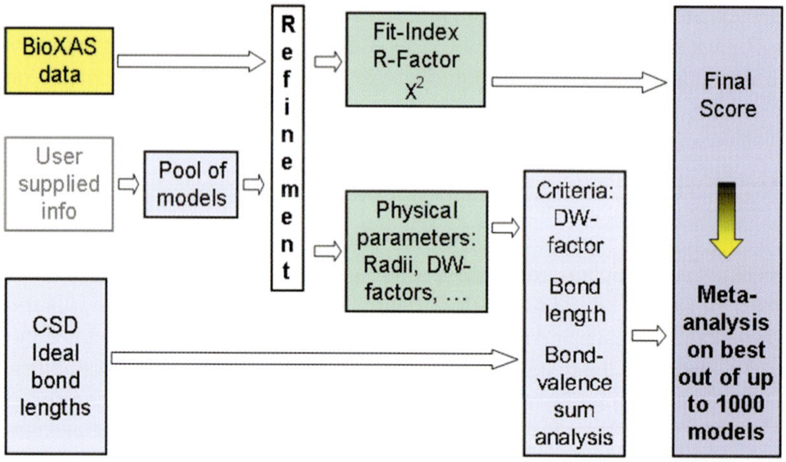

FIGURE 6.8 Typical criteria helpful to computer programs and novice EXAFS groups comprise bond valence sum analysis, the comparison of oxidation state and coordination number dependent metal–ligand distances taken from Cambridge structural data base, Debye–Waller criterion, and ΔE_0 shift. On a computer cluster easily up to 1000 structural models can be compared to the data. A meta-analysis can extract common features of the good models (Wellenreuther et al., 2010)

distances, and c) its coordination number. Its application to results of EXAFS studies is particularly appropriate, because it can be a way to get a better indication for the coordination number (which is relatively inaccurate from EXAFS) based on the distance information (which is relatively accurate from EXAFS).

(iii) Debye–Waller factor criterion. Disorder in biological systems is typically larger than in small molecules, which is reflected by the Debye–Waller factors resulting from EXAFS refinement of metalloproteins. But too large Debye–Waller factors artificially decrease the contribution by the corresponding shell, whereas too small Debye–Waller factors artificially enlarge it. For automation this criterion is based on experience reflecting published values for similar systems.

(iv) The shift of the energy threshold (ΔE_0) should be similar for similar samples – typically less than 1 eV. Note that samples differing in spin or oxidation state are not considered being similar in this context.

In automated refinement (see Figure 6.8 for a schematic overview) for each criterion a parameter is defined, which can vary in value between 1 – entirely fulfilled – and 0 – complete failure. These criteria are weighed and multiplied; thus failure in one criterion results in the rejection of the structural model and only models with a good FI that also fulfil the criteria will be ranked high. Those models ranked high will have many details in common. This information is extracted by a meta-analysis, resulting e.g. in a well defined number of low-Z ligands without claiming to differentiate their similar backscattering potentials. (Wellenreuther et al., 2010)

XANES SIMULATIONS WITH THREE-DIMENSIONAL MODELS

As explained in the Introduction, most theoretical approaches for XANES (the low energy range of XAS) are based on orbital theory, whereas those for EXAFS (the high-energy range) are based on electron diffraction. This raises the question up to what energy the orbital theory should be applied, and from which energy the electron diffraction approach is valid, and this is not easy to answer. In the preceding sections we have already discussed EXAFS as a sum of oscillations due to electron diffraction phenomena involving shells of backscatterers in a one-dimensional radial distribution function, and the single scattering theory that describes the electron diffraction phenomena at the higher energies (high k). We have also seen that for low energy (low k) the multiple scattering pathways, involving scattering of the electron wave from one atom to another, before returning to the absorber atom, are more important.

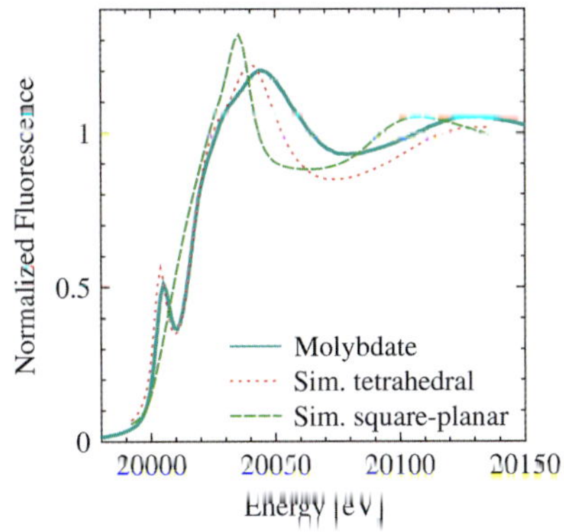

FIGURE 6.9 Comparison of the experimental XANES of aqueous molybdate (turquoise, cf. Figure 6.5) with simulations with four O in a tetrahedral (red dots) and square planar (green dashed) arrangement.

In order to describe the low energy spectra accurately with the electron diffraction theory, two- and even three-dimensional information will have to be taken into account; this is equivalent to (but more complicated than) discussing the spectra in terms of orbital theory. As an example of the limitations and opportunities, simulations for the Mo K edge XANES of molybdate are shown in Figure 6.9; an analysis of the corresponding EXAFS just gives the distance of the Mo to the four oxo ligands, without any two- or three-dimensional information.

The well-known tetrahedral alignment of the four oxo ligands yields the dotted red curve, whereas a square planar alignment of these atoms results in the dashed green curve. Although the red curve does not perfectly resemble the main features of the measured XANES it is not a bad fit to the data. But in this *ab initio* simulation no parameter has been optimised and thus the fact that the green curve shows no similarity to the molybdate XANES allows such a co-ordination to be ruled out. In this manner, XANES serves as a finger print in both the qualitative and quantitative analysis, and XANES simulations and their refinement are of increasing importance. In any case, one should keep in mind that the number of independent data points is more limited than for EXAFS and thus the result might depend on the assumptions used in the refinement; compared to EXAFS one tries in fact to derive more (two- and three-dimensional) information from a shorter range of data, without the guidance provided by the Fourier transformation for the EXAFS. An advantage of the shorter data range is that one does not have to take the dampening due to the Debye–Waller factor (see Box 6.5) into account.

METAL–METAL DISTANCES IN METAL CLUSTERS

In EXAFS contributions at higher distances frequently originate from multiple scattering or backstattering from a metal ion. In our introductory example on oxidised CO-dehydrogenase from *Oligotropha carboxidovorans*, metal–metal contributions are present in both the Cu edge EXAFS as well as in the Mo edge EXAFS. In Figure 6.10, the individual contributions to the EXAFS are shown. For the Mo edge EXAFS (top panel) the spectrum is dominated by the two oxo ligands at 1.74 Å. The second nearest shell corresponds to a single sulfur ligand at about 2.28 Å, which, based on comparison to other Mo enzymes is identified by its bond length as the ligand bridging both metal ions. The two subsequent sulfur ions at 2.50 Å belong to the pterin cofactor, and the last identifiable contribution is refined as Cu backscattering at 3.70 Å. The Cu EXAFS (bottom panel) is dominated by the sulfur contributions at 2.18 Å. Again, metal–metal backscattering can be identified at 3.70 Å, this time with Cu as the absorber and Mo as the backscatterer. The analysis of the individual traces is a good example of the general features discussed above: i) the shorter the bond length, the lower the frequency of the oscillations; ii) the heavier the backscatterer, the more the maximum of the EXAFS intensity shifts to a higher value of the wave vector k; iii) the closer a ligand the more intense the resulting EXAFS contribution will be; iv) the more backscatterer atoms in a shell, e.g. 2 vs. 1 S, the stronger the contribution to the EXAFS, and v) metal–metal contributions show a high frequency in the EXAFS and require

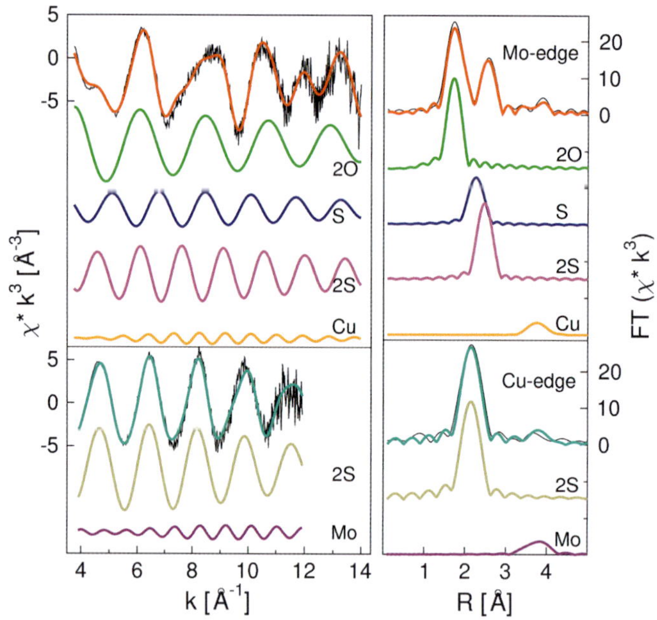

FIGURE 6.10 EXAFS (left) and phase-corrected Fourier transform (right) of experimental (black) and simulated (colours) Mo (top) and Cu (bottom) data of oxidised CO-dehydrogenase from *O. carboxidovorans*. The complete simulation is shown along with the experimental data; other traces represent the contributions of individual components to the model.

a reasonably long energy range for unambiguous identification. They are sufficiently specific to rule out the presence of Se (in case of the Mo EXAFS); the backscattering amplitudes of Cu and Se resemble those given for Zn and Br, respectively, in Figure 6.5, left, and there are also diagnostic differences in the phase-relationship between EXAFS and Fourier transform such as discussed for the halogens in Figure 6.5, right. The long energy range also helps to distinguish metal backscatterers from contributions of a unit with multiple scattering in the same region, such as those observed at just above 4 Å in the Fourier transform for porphyrins (Figure 6D) and imidazole (Figure 6.7); multiple scattering contributions of systems with low-Z backscatterer atoms are very intense in the lower *k* range. In this example, both spectra have been refined simultaneously. Thus Debye–Waller factor as well as the distance of the metal ions was fitted together increasing robustness and consistency of the resulting structural model. In case of homodinuclear metal sites a similar strategy can be applied, but here it is even more important to determine independently the metal content in the sample under study, because this will be reflected by the occupancy of individual binding sites (Svetlitchnyi et al., 2004).

NON-METAL TRACE ELEMENTS: HALOGENS

Although most XAS studies have focussed on the chemical environment of transition metal cations, increasing attention has been paid in recent years to metalloids (such as selenium and arsenic), halogens, and oxyanions. We discuss here features common to the studies of cations and anions, as well as differences. Transition metal cations interact in a dynamic way with a number of electron-donating ligands in order to achieve a favourable coordination number (or secondary valence); halogens are subject to solvation and H-bonding in their anionic form, and can form single covalent bonds (primary valence) in systems where they can still be H-bond acceptors or themselves be a part of a halogen bond. As illustrated in Figure 6.11, where the effect of 6-membered aromatic rings is compared, the Fourier transform of the EXAFS of a coordination complex of Cu with pyridine (Feiters et al., 1999) resembles that of a compound where I or Br is covalently attached to a phenyl ring. As for the 5-membered ring systems highlighted above, the EXAFS can only be simulated satisfactorily with a multiple

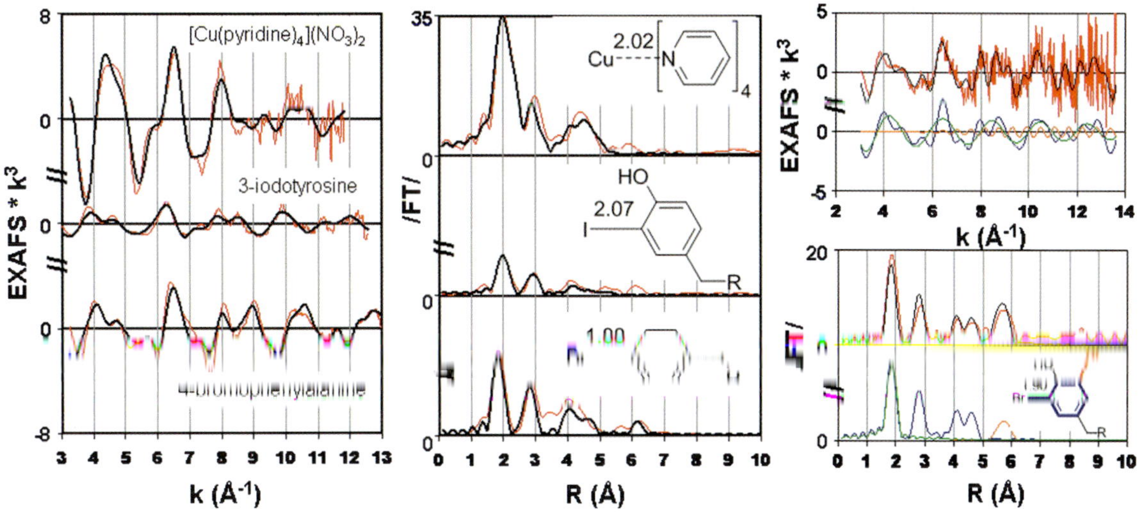

FIGURE 6.11 Left and middle panels: Experimental (thin red lines) and simulated (solid lines) EXAFS (left panel) and corresponding FT (middle panel) of Cu K edge of $[Cu(pyridine)_4](NO_3)_2$ (top), I K edge of 3-iodotyrosine (middle), and Br K edge of 4-bromophenylalanine (bottom). Insets, structures on which the simulations are based, including distances (in Å) of the respective absorbers to the nearest atom. For the Cu spectrum, two oxygen atoms of the weakly coordinating nitrate anion were also included in the simulation at 2.5 Å (Feiters et al., 1999); for the Br spectrum, the *R* group was not included, whereas for the I spectrum none of the other ring substituents was included (Feiters et al., 2005). Right panel: Br K edge EXAFS (top panel) and Fourier transform (bottom panel) of native bromoperoxidase from *Ascophyllum nodosum* (Feiters et al., 2005). Top traces: experimental (thin red line) and complete simulation (solid line); bottom traces: contributions of aromatic ring component (blue), nearest C (green), and bromine (orange line). Inset: dibromotyrosine structure highlighting the aromatic ring; substituents other than Br were not included in the simulation.

scattering approach in both cases. The most important difference between metal and halogens is the decrease in amplitude of the EXAFS because the metal ion has four identical ligands, whereas each halogen has only a single covalent bond. There are also subtle differences in the relation between EXAFS and FT due to the subtle but significant differences in the distances between the absorber and the first backscatterer. Interestingly, the accurate determination of the halogen−carbon distance by EXAFS allows one to discriminate between halogens bound to sp^2- and sp^3- hybridised carbons (such as occur in aromatic/olefinic and aliphatic halocarbons, respectively), because the carbon−halogen bond becomes shorter with increasing *s*-character of the bonding orbital on C (Feiters et al., 2005). Brown algae such as *Laminaria digitata* (oarweed) accumulate iodine to concentrations of 10^6 times that of surrounding seawater, and XAS is ideal as a non-invasive technique to study this system. Iodine in *Laminaria* gave a weak EXAFS even when compared to NaI in water, and was shown to represent iodide ions with their solvation shell displaced by H-bonding to biomolecules at a comparable distance (3.5−3.6 Å) (Küpper et al., 2008). On this basis a physiological role for accumulated iodide as an inorganic oxidant is proposed.

The enzymes responsible for incorporation of the accumulated halide ions into biomolecules are the so-called haloperoxidases. Figure 6.11 illustrates one of the ways by which the bromoperoxidase of *Ascophyllum nodosum* (knotted wrack), can be involved in the halogen chemistry of the algae. The Br K edge EXAFS of the native enzyme (without added Br) reveals a spectrum that is characteristic of a 1,3-dibrominated aromatic ring, such as present in 3,5-dibromotyrosine, with contributions of the ring at 1.90 (nearest C atom) and a remote bromine at 5.70 Å (Feiters et al., 2005). The spectrum is almost entirely accounted for by the ring contribution, except for the Br atom which becomes stronger at high *k*; in addition, the calculated contribution of the Br is much stronger when it is part of the ring, because of the forward scattering and focussing effects of the ring carbons, than when it is on its own. The identification of this post-translational amino acid modification implied an important addition to the protein crystal structure, in which the electron density had been modelled with tyrosine singly substituted with I.

SUMMARY: STRENGTHS AND LIMITATIONS

X-ray absorption spectra arise because of the excitation of electrons from the inner shells of atoms with X-rays of the right energy, and can be recorded with monochromatised synchrotron radiation. The interesting part of an XAS can be divided in two regions, the XANES and the extended x-ray absorption fine structure (EXAFS). The information to be obtained from X-ray absorption spectrum is summarised in Table 6.2. In the typical BioXAS

TABLE 6.2 Information to be Obtained from Certain Features of the XANES and EXAFS Regions of XAS as well as XES

Spectral feature	Information	Accuracy and correlations	Other XAS	Other techniques
XANES				
Edge position	Oxidation state	Relate to model compounds; be aware of correlation with average *R*	Δ*R*: EXAFS	UV-visible spectra, EPR
Pre-edge features	Ligand geometry	Relate to well-characterised model compounds	*N* from EXAFS	UV-visible spectra, EPR, crystallography
	Covalency of metal–ligand bond	–	XANES of other metal- and ligand edges	EPR hyperfine structure
XES				
Kβ′ line	Metal ion spin state	–	–	EPR, magnetic susceptibility
Kβ″ line	Ligand identity	–	EXAFS ($Z \pm 1$ accuracy)	Crystallography
EXAFS				
Amplitude	Coordination number *N*	20% (±1), correlation with Debye–Waller factor	Ligand geometry from XANES	Crystallography
Decay of amplitude with *k*	σ*: static or thermal disorder, distinguish by *T* variation	Correlated with/spoils accuracy of *N*		
Periodicity	Distance *R* of scatterers	±0.02 Å if shell resolved, correlation with threshold energy ΔE_0	Edge shift from XANES	Crystallography
Phase	Backscatterer atom type: C, N, O ('low-*Z*') vs. S	Different to distinguish atom types adjacent in periodic table	Unambiguous from XES Kβ″ line	Crystallography

*Debye–Waller factor

experiment, the fine structure is measured by collecting the X-ray fluorescence, which is dominated by the Kα lines, in a solid-state detector. Detailed investigations of the Kβ main and satellite lines by X-ray emission spectroscopy (XES) give information on the metal ion's spin state and the type of ligand(s), respectively.

The position of the edge in the XANES goes to higher energy with higher oxidation state and with decreasing average metal-to-ligand distance. The fine structure can be interpreted in terms of transitions to empty orbitals of the metal and the transition probabilities depend on the ligand geometry. When transitions to molecular orbitals of the complex are involved, the covalency of the metal–ligand bond also has an influence; this can also be probed by metal L edge or ligand K edge XANES. An interpretation of the XANES in terms of a three-dimensional model extrapolated from the electron diffraction theory for the EXAFS is also possible. As the XANES energy range is short and the fine structure is relatively strong compared to that in the EXAFS region, this is the region of choice to explore in time (reaction kinetics) and space (element chemical state imaging) resolved studies.

The fine structure in the EXAFS is caused by X-ray-induced electron diffraction phenomena. Backscattering amplitudes and phase shifts for absorber and backscatterer atoms can be derived from model compounds, or calculated using the muffin-tin approximation (and validated on model compounds).

The strength of EXAFS is the accurate determination of the distance R. Limitations in this aspect are the poor resolution (ΔR approx. 0.15 Å depending on the k range) and the unreliable observation of weakly bonded atoms. The best data collection strategy is to collect as long a range of data as possible (to improve the resolution) at as low a temperature as possible (to reduce thermal vibrations and disorder).

Another strength of EXAFS is the identification of atom type of the backscatterer. A limitation is that atom types that are close in Z are difficult to distinguish, e.g. 'low-Z atoms', C, N, or O. We have seen (Box 6.2) that this is a particular strength of the newly emerging X-ray emission techniques.

The determination of coordination numbers is not a particular strength of EXAFS, in view of the limitation that it is strongly correlated with Debye–Waller factor. This parameter needs to be incorporated in the EXAFS theory, and its value iteratively refined in the simulation, in order to account for disorder of static and/or thermal origin. It is possible to use other features of the XAS spectrum to get a more accurate idea of the coordination number. One is to interpret the geometrical information from XANES: tetrahedral geometry means 4-, octahedral 6-coordination. The other is to use the accurate distance information to take advantage of established correlations between ligand distance and coordination number by the so-called BVSA.

The Debye–Waller factor and other amplitude effects limit the accuracy of the determination of coordination numbers by EXAFS to approx. ±20 %. It should be kept in mind, however, that coordination numbers may be inferred from XANES based on ligand symmetry arguments, as discussed in the XANES section, or from the aforementioned *bond valence sum analysis* formalism.

EXAFS gives an indication of the presence and orientation (with respect to the metal-N bond) of heteroatomic ligands (imidazoles, porphyrins). A limitation is that in addition to the parameters normally refined in single scattering simulations, the angles will also have to be refined. In such cases the procedure of restrained refinement can be used. The strength and limitations of the two regions of the XAS spectrum (EXAFS and XANES) and their relations with each other and other techniques are summarised in Table 6.2.

CONCLUSIONS: RELATIONS WITH OTHER TECHNIQUES

Metals in proteins would appear to be minor components, but are usually at the active site which makes a study of their close environment by X-ray absorption spectroscopy particularly relevant. Even when a structure of the whole protein is already known from protein X-ray crystallography (PX) or NMR, it is usually worthwhile to study it by XAS for a number of reasons. i) Typically the error in the determination of metal–ligand distances is much larger for macromolecular (protein) crystallography than it is for small molecules crystallography (estimate ±0.1 vs. <0.01 Å), whereas for EXAFS it is the same (±0.02 Å); this implies that the accuracy of the geometry of the metal environments in crystallographic studies can be improved with input from EXAFS. By using XAS results in the refinement of the crystal structure it is possible to resolve the metal environment at higher resolution, and the judicious combination of PX and XAS can even lead to structures with subatomic resolution (Strange et al., 2005). ii) With XAS it is much easier to study the effect of other reactants/circumstances on the metal environment which might be difficult to achieve in the crystalline state, or result in magnetism of the metal site interfering with the structure elucidation by NMR. Thus the important advantage of XAS over other structural and spectroscopic techniques for biological as well as other applications is that the environment of any trace element can be studied irrespective of the physical state (crystalline, solid, liquid, frozen liquid) of the sample, or the chemical state (oxidation state, spin state, magnetism) state of the element.

A number of other techniques whose results overlap or are complimentary with those from XAS studies, such as UV-vis spectroscopy, Electron Paramagnetic Resonance (EPR), and magnetic susceptibility measurements are given in the appropriate places in Table 6.2. An important feature of XAS analysis that was inspired by crystallographic examples is the use of restraints (Binsted et al., 1992) for distances of ligands of well-known geometry, such as imidazoles or porphyrins, in order to reduce the number of independent parameters to be refined in an EXAFS simulation. High-resolution protein crystal structures have been surveyed to identify typical metal–ligand distances in proteins (Harding, 2004). EXAFS is also the method of choice to experimentally validate the geometrical results of computational studies on metalloproteins.

REFERENCES

Antonio, M. R., Teo, B. K., & Averill, B. A. (1985). *J. Am. Chem. Soc., 107*, 3583–3590.

Ascone, I., Meyer-Klaucke, W., & Murphy, L. (2003). *J. Synchrotron. Radiat., 10*, 16–22.

Binsted, N., Strange, R. W., & Hasnain, S. S. (1992). *Biochemistry, 31*, 12117–12125.

Burge, R. E. (1993). In A. G. Michette, & C. J. Buckley (Eds.), *The interaction of X-rays, in X-Ray Science and Technology.* Bristol: Institute of Physics.

Colpas, G. J., Maroney, M. J., Bagyinka, C., Kumar, M., Willis, W. S., Suib, S. L., Baidya, N., & Mascharak, P. K. (1991). *Inorg. Chem., 30*, 920–928.

Cramer, S. P., Ralston, C. Y., Wang, H. X., & Bryant, C. (1997). *J. El. Spectr. Rel. Phen., 86*, 175–183.

Dey, A., Jenney, F. E., Jr., Adams, M. W. W., Babini, E., Takahashi, Y., Fukuyama, K., Hodgson, K. O., Hedman, B., & Solomon, E. I. (2007). *Science, 318*, 1464–1468.

Duhme, A.-K., Meyer-Klaucke, W., White, D. J., Delarbre, L., Mitchenall, L. A., & Pau, R. N. (1999). *J. Biol. Inorg. Chem., 4*, 588–592.

Feiters, M. C., Klein Gebbink, R. J. M., Solé, V. A., Nolting, H.-F., Karlin, K. D., & Nolte, R. J. M. (1999). *Inorg. Chem., 38*, 6171–6180.

Feiters, M. C., Eijkelenboom, A. P. A. M., Nolting, H.-F., Krebs, B., van den Ent, F. M. I., Plasterk, R. H. A., Kaptein, R., & Boelens, R. (2003). *J. Synchrotron. Radiat., 10*, 86–95.

Feiters, M. C., Leblanc, C., Küpper, F. C., Meyer-Klaucke, W., Michel, G., & Potin, P. (2005). *J. Am. Chem. Soc., 127*, 15340–15341.

Feiters, M. C., Küpper, F. C., & Meyer-Klaucke, W. (2005). *J. Synchrotron. Radiat., 12*, 85–93.

Glatzel, P., & Bergmann, U. (2005). *Coord. Chem. Rev., 249*, 65–95.

Gnida, M., Ferner, R., Gremer, L., Meyer, O., & Meyer-Klaucke, W. (2003). *Biochemistry, 42*, 222–230.

Gurman, S. J., Binsted, N., & Ross, I. (1984). *J. Phys. C: Solid State Phys., 17*, 143–151.

Gurman, S. J., Binsted, N., & Ross, I. (1986). *J. Phys. C: Solid State Phys., 19*, 1845–1861.

Ha, S.-W., Korbas, M., Klepsch, M., Meyer-Klaucke, W., Meyer, O., & Svetlitchnyi, V. (2007). *J. Biol. Chem., 282*, 10639–10646.

Harding, M. M. (2004). *Acta Cryst. D Biol., 60*, 849–859.

Hollenstein, K., Comellas-Bigler, M., Bevers, L. E., Feiters, M. C., Meyer-Klaucke, W., Hagedoorn, P.-L., & Locher, K. P. (2009). *J. Biol. Inorg. Chem., 14*, 663–672.

Kau, L.-S., Spira-Solomon, D. J., Penner-Hahn, J. E., Hodgson, K. O., & Solomon, E. I. (1987). *J. Am. Chem. Soc., 109*, 6433–6442.

Korbas, M., Vogt, S., Meyer-Klaucke, W., Bill, E., Lyon, E. J., Thauer, R. K., & Shima, S. (2006). *J. Biol. Chem., 281*, 30804–30813.

Küpper, F. C., Carpenter, L. J., McFiggans, G. B., Palmer, C. J., Waite, T., Boneberg, E.-M., Woitsch, S., Weiller, M., Abela, R., Grolimund, D., Potin, P., Butler, A., Luther, G. W., III, Kroneck, P. M. H., Meyer-Klaucke, W., & Feiters, M. C. (2008). *Proc. Natl. Acad. Sci. USA, 105*, 6954–6958.

Lancaster, K. M., Roemelt, M., Ettenhuber, P., Hu, Y., Ribbe, M. W., Neese, F., Bergmann, U., & DeBeer, S. (2011). *Science, 334*, 974–977.

Rehr, J. J., & Albers, R. C. (1990). *Phys. Rev. B, 41*, 8139–8149.

Sayers, D. E., Stern, E. A., & Lytle, F. W. (1971). *Phys. Rev. Lett., 27*, 1204–1207.

Scherk, C. G., Ostermann, A., Achterhold, K., Iakovleva, O., Nazikkol, C., Krebs, B., Knapp, E. W., Meyer-Klaucke, W., & Parak, F. G. (2001). *Eur. Biophys. J., 30*, 393–406.

Shadle, S. E., Penner-Hahn, J. E., Schugar, H. J., Hedman, B., Hodgson, K. O., & Solomon, E. I. (1993). *J. Am. Chem. Soc., 115*, 767–776.

Strange, R. W., Blackburn, N. J., Knowles, P. F., & Hasnain, S. S. (1987). *J. Am. Chem. Soc., 109*, 7157–7162.

Strange, R. W., Ellis, M., & Hasnain, S. S. (2005). *Coord. Chem. Rev., 249*, 197–208.

Svetlitchnyi, V., Dobbek, H., Meyer-Klaucke, W., Meins, T., Thiele, B., Romer, P., Huber, R., & Meyer, O. (2004). *Proc. Natl. Acad. Sci. USA, 101*(2), 446–451.

Thorp, H. H. (1998). *Inorg. Chem., 37*, 5690–5692.

Wellenreuther, G., Parthasarathy, V., & Meyer-Klaucke, W. (2010). *J. Synchrotron. Radiat., 17*, 25–35.

Westre, T. E., Kennepohl, P., DeWitt, J. G., Hedman, B., Hodgson, K. O., & Solomon, E. I. (1997). *J. Am. Chem. Soc., 119*, 6297–6314.

Yano, J., & Yachandra, V. K. (2008). *Inorg. Chem., 47*, 1711–1726.

EXERCISES

1) What are the differences in origin and application of K and L edges in X-ray absorption spectroscopy?
2) How is the wavelength selected in the X-ray absorption experiment? Which fundamental law determines the energy of the photons on the sample?
3) a) Estimate the errors for signals collected in an X-ray absorption experiment in transmission mode. Typically, a small signal is measured on a large background. Background: 100,000 counts per second; Signal: 1000 counts per second.
   b) For how many seconds per data point the data collection should be set up in order to obtain a signal which at the edge will be 10 times of the background's error.

   Hint: For large numbers use Poisson statistics (square root of the number estimates the error margin).
4) An unknown 'blue copper' protein is suspected to have Cu–N and Cu–S bond lengths of 1.95 and 2.10 Å, respectively.
   a) What kind of energy range, in wave vector (k, $Å^{-1}$) and energy (E, eV), is required for the EXAFS to be able to resolve these shells?
   b) Are there problems to be expected if the protein is contaminated with Zn to an amount of 1 % compared with the Cu?
5) The XAS studies of the Mn_4Ca cluster in the oxygen evolving centre of the photosystem II are hampered by the presence of Fe, which has its K absorption edge 573 eV above the Mn K edge. How might a long (>800 eV) range of Mn EXAFS be recorded without the interference of the Fe?
6) Synchrotron radiation from a bending magnet is 'white' – a continuous spectrum ranging from UV-vis to hard X-rays – and needs to be monochromatised for most techniques, including EXAFS. The most used monochromator material is single crystalline Si; the most used reflection is (111).
   a) Using Bragg's law, calculate the angle of incidence needed to generate a monochromatic beam with a wavelength of 0.8 Å. The spacing of the (111) planes in Si is 3.14 Å.
   b) Usually a so-called double crystal monochromator is used (see Figure 6A). Why it is convenient?
   c) The first crystal of a double crystal monochromator requires cooling, the second does not. Why?
7) In Figure 6.12 (next page) the k^3-weighted Fe K edge EXAFS (top panel) and its Fourier transform (bottom panel) of solid haemin chloride (**1**) and frozen aqueous haemin (dotted lines) are given, along with simulations

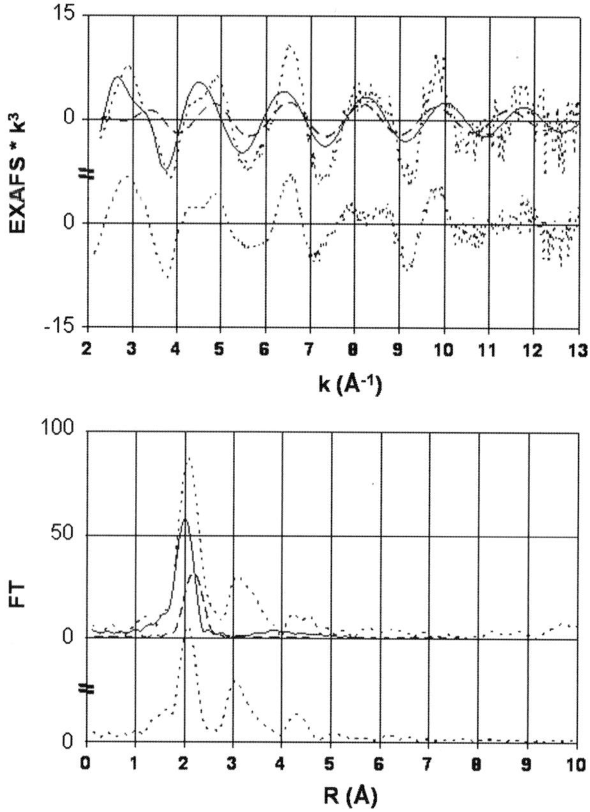

FIGURE 6.12 (Exercise 7). k^3-weighted Fe K EXAFS (top panel) and phase-corrected Fourier transform (bottom panel) of experimental (dotted lines) of solid haemin chloride (**1**, top traces) and aqueous haemin (bottom traces) are given, along with simulations for the nitrogen (4N at 2.06 Å, solid line) and chlorine (1Cl at 2.23 Å, dashed line) contributions to the haemin chloride EXAFS.

for the nitrogen (4N at 2.06 Å, solid line) and chlorine (1Cl at 2.23 Å, dashed line) contributions to the haemin chloride EXAFS.

1

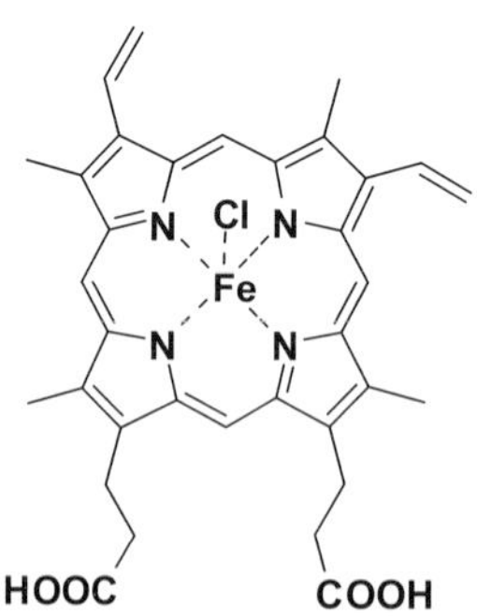

a) What problems are to be expected in the analysis of the close Fe—N and Fe—Cl contributions in the Fourier transform in the range of 2.0–2.3 Å?

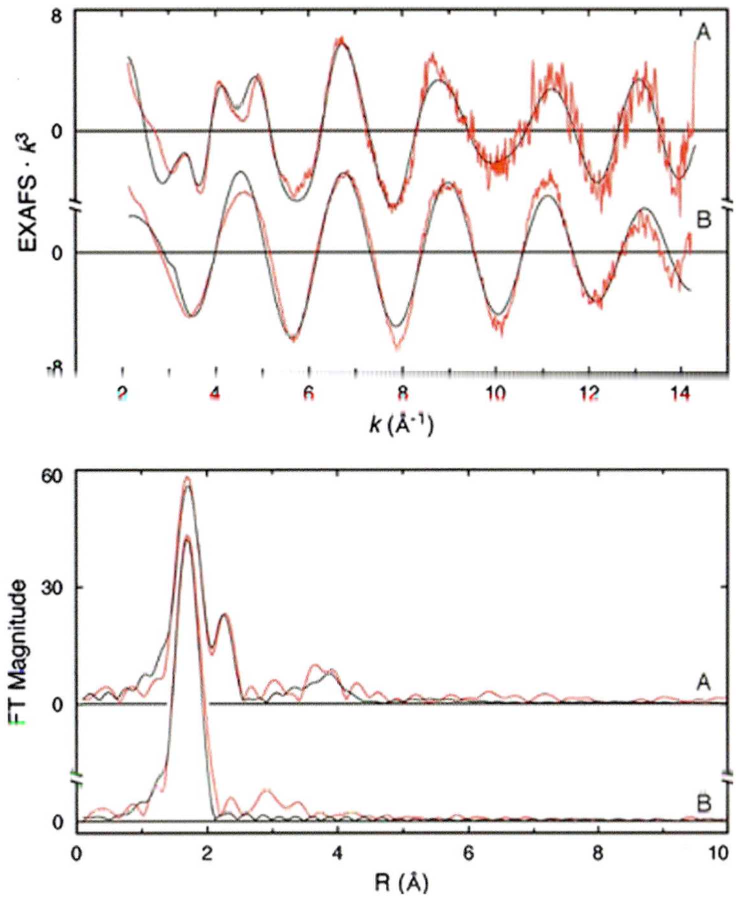

FIGURE 6.13 (Exercise 8). k^3-weighted extended X-ray absorption fine structure (EXAFS) spectra (upper panel) and corresponding phase-corrected (assuming oxygen in main shell) Fourier transforms (FT; lower panel) at the tungsten L_3 edge of tungstate in the binding protein from *A. fulgidus* (traces A) and buffer (traces B). Red line experimental data, black line simulation with the following parameters [distances in angstroms, Debye–Waller-type factors as $2\sigma^2$ in square angstroms in parentheses, fit index (FI) with k^3-weighting]: trace A, threshold energy shift $\Delta E_0 = -14.619$ eV, 3O at 1.786 Å (0.008), 1O at 2.060 Å (0.002), 2O at 2.241 Å (0.003), FI 0.29989×10^{-3}; trace B, $\Delta E_0 = -13.54$ eV, 3.7O at 1.770 Å (0.005), FI 0.14069×10^{-3}.

b) Considering the geometry of the pyrrole units of the porphyrin, would an analysis in which only the single scattering is simulated be appropriate?

c) What happens upon dissolution of haemin chloride in water?

M.C. Feiters, N. de Val, L. Toussaint, M. Cuypers, R.R. Crichton, W. Meyer-Klaucke, Hasylab Annual Report II (2005), 505–506.

8) The binding of tungstate by the bacterial tungstate (WO_4^{2-}) binding protein from *Archaeoglobus fulgidus* has been studied by EXAFS (Figure 6.13). The tungstate oxyanion in solution is a tetrahedron with four extremely short (1.77 Å) W=O bonds (traces B). The protein provides two additional ligands which results in the lengthening of one W=O bond, and the formation of two W–O bonds of 2.24 Å (traces A).

a) What experimental precaution has to be taken in order to be able to resolve the contributions at 1.77 and 2.24 Å in a spectrum like A?

b) Spectrum A shows a high-frequency oscillation at low (3–6 $Å^{-1}$) k. The position of the corresponding shell in the Fourier transform close to 4 Å corresponds to the sums of some of the W=O and W–O distances (1.77 + 1.77 Å; 1.77 + 2.24 Å). What assumptions about the geometry of the unit and the nature of the scattering phenomena within it have to be made in order to explain this part of the spectrum with just the [WO_6] unit?
K. Hollenstein, M. Comellas-Bigler, L.E. Bevers, M.C. Feiters, W. Meyer-Klaucke, P.-L. Hagedoorn and K.P. Locher, J. Biol. Inorg. Chem. 2009, ***14****, 663–672.*

Chapter 7

Other Spectroscopic Methods for Probing Metal Centres in Biological Systems

Robert R. Crichton[a] and Ricardo O. Louro[b]
[a] Batiment Lavoisier, Université Catholique de Louvain, Louvain-la-Neuve, Belgium
[b] Instituto de Tecnologia Química e Biológica, Universidade Nova de Lisboa, Oeiras, Portugal

Chapter Outline

OPTICAL SPECTROSCOPY

Electromagnetic radiation can be considered to behave as two wave motions, one magnetic and one electric, at right angles to one another. Light is one form of electromagnetic radiation, which enables us to see objects in everyday life because they reemit part of the light that falls on them from a light source such as the sun. Of course, we also know from everyday life that interpretation of this reflected or transmitted light can give us a wealth of information about the object with which the light interacts. In this section we will briefly discuss the use of ultraviolet and visible electromagnetic radiation, while infrared (IR) radiation will be discussed in a later section of the chapter.

Electronic spectra involve transitions between different energy levels of electrons in molecules. Electromagnetic radiation causes the electrons to oscillate in response to the incident electrical field. Such electronic transitions result in the characteristic manifestations of electromagnetic radiation with matter that we observe on a daily basis such as colour and vision. Electronic spectroscopy experiments are simple and cheap to perform, yet they can be carried out with great precision, and, in the case of metalloproteins, can frequently give invaluable information.

The energies involved in electronic transitions correspond to the absorption of photons in the visible (400−700 nm) and UV (200−400 nm) regions of the electromagnetic spectrum. The energy of electromagnetic radiation is related to the wavelength (λ) and the frequency (ν) by the relationship:

$$E = hc/\lambda = h\nu$$

where h is the Planck's constant (6.636×10^{-34} J s) and c is the speed of light (3×10^{8} ms^{-1}). Molecular energy levels represent specific states, reflecting the overall structural, electronic and motional properties of the molecule. The lowest energy state is the *ground state*, and absorption of a discrete amount of energy, corresponding to radiation of a defined frequency (ν) can promote the molecule to a state of higher energy, the *excited state*. If two or more states have the same energy, they are termed *degenerate*.

Practical Approaches to Biological Inorganic Chemistry, 1st Edition. http://dx.doi.org/10.1016/B978-0-444-56351-4.00007-5.

In optical spectroscopy the absorbed energy can move an electron from a lower to a higher energy state. Electronic spectra are usually presented as a plot of absorbance vs the wavelength of irradiation (usually expressed in nanometres). Two parameters are used to characterise a particular absorption band — its position at the maximum (λ_{max}) and its intensity. The experimental measure of intensity at a particular wavelength is the **extinction coefficient** (ε), given by the Beer–Lambert law:

$$\log_{10}(I_o/I_t) = \varepsilon cl$$

where I_o is the radiation incident on the compound, I_t is the radiation transmitted by the compound, i.e. ($I_o - I_t$) is the radiation absorbed, c is the concentration of the compound, and l is the path length (the length of the cell through which the radiation actually travels). The variation of ε with wavelength constitutes the absorption spectrum, and the Beer–Lambert law is often expressed in the form:

$$A = \log_{10}(I_o/I_t) = \varepsilon cl$$

where A is called the **absorbance** or **optical density** of the compound (incidentally A is dimensionless).

Both the energy (represented by λ_{max}) and the intensity (represented by ε_λ) give useful information on the types of ligand surrounding a metal ion in a metalloprotein as well as on their geometry (Solomon, 2006). Light absorption is accompanied by rearrangements of electrons over the orbitals of the metal ion and the surrounding ligands. There are three principal sources of electronic spectra in metalloproteins, internal ligand bands ($\pi \rightarrow \pi^*$ transitions) in porphyrins, transitions associated solely with metal orbitals, like d $\rightarrow$ d transitions and charge-transfer bands between metal and ligand. The most commonly observed electronic transitions in metal complexes are outlined in Figure 7.1.

The electronic spectra of haem proteins represent a classic example of the usefulness of UV–visible spectroscopy. An important class of haemoproteins with hexa-coordinate low-spin iron are the cytochromes. First discovered by McMunn in 1884, they were rediscovered in 1925 by David Keilin. Using a hand spectroscope he observed the characteristic absorption (Soret) bands of the three cytochromes *a*, *b* and *c* in respiring yeast cells, which disappeared upon oxygenation, and concluded correctly that they transferred electrons from substrate oxidation to the terminal oxidase, cytochrome *c* oxidase as we know it today. The cytochromes vary in the nature of their haem group — *b*-type haems have protoporphyrin IX, as in haemoglobin, *c*-type cytochromes have

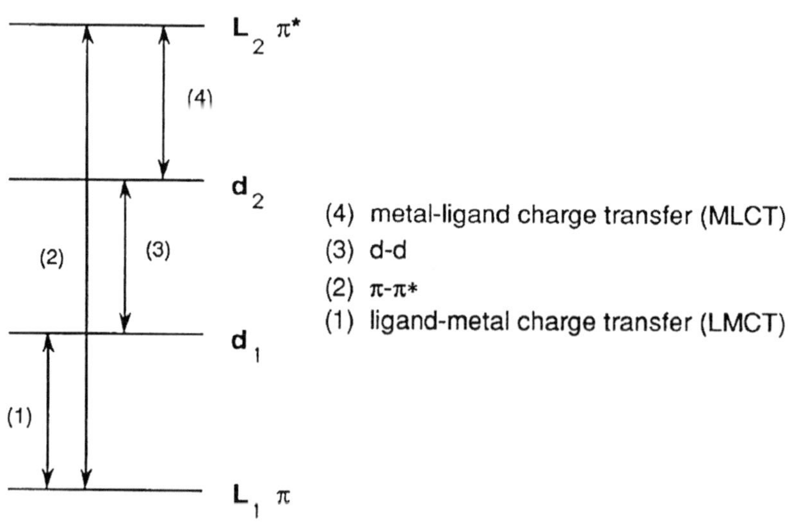

FIGURE 7.1 Electronic transitions commonly observed in metal complexes. *(From Cowan, 1997. Reproduced with permission from John Wiley & Sons, New York.)*

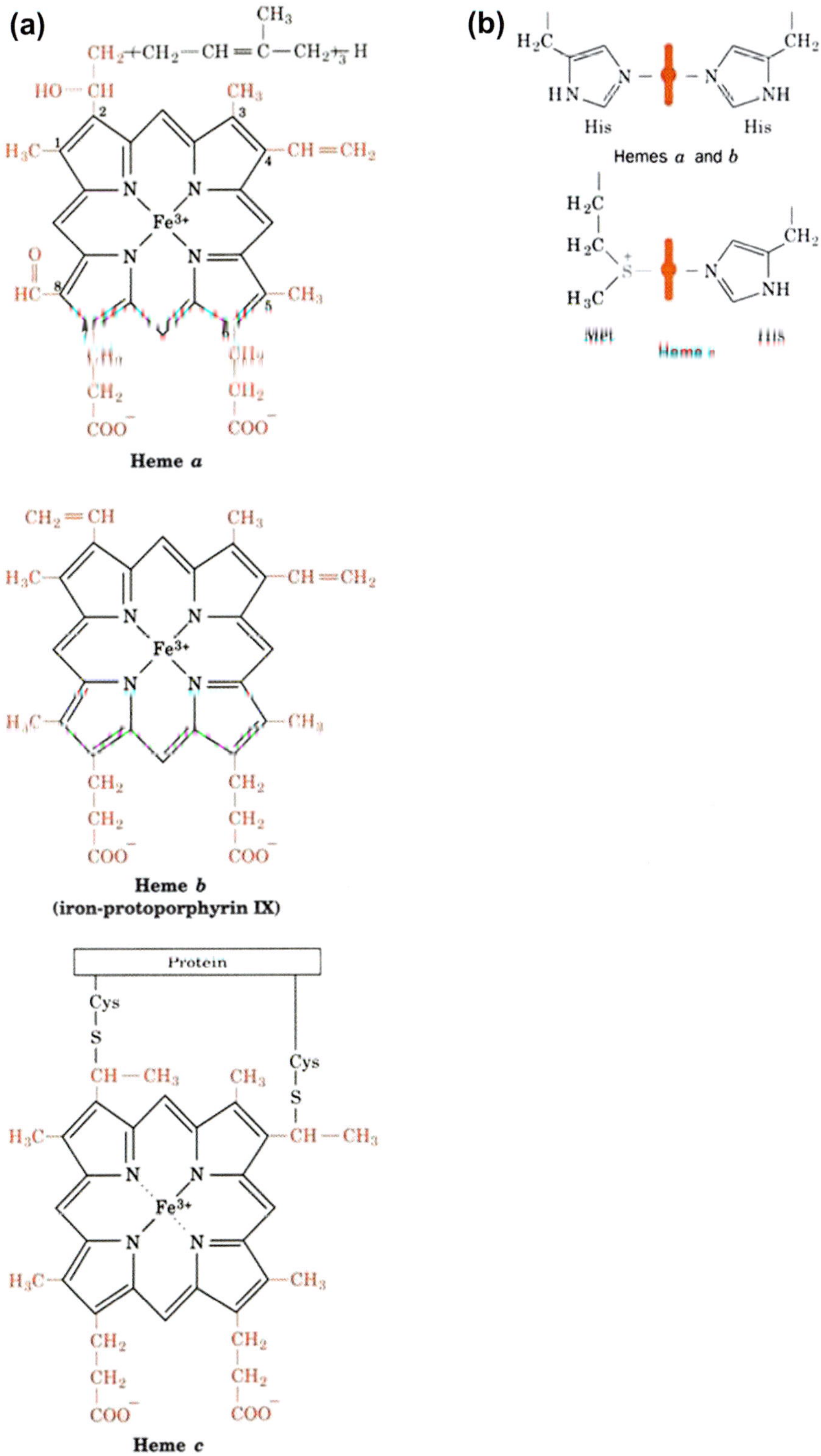

FIGURE 7.2 The chemical structures (a) and axial liganding (b) of the haem groups contained in cyochromes *a*, *b*, and *c*.

protoporphyrin IX in which the vinyl groups form covalent thioether bonds with cysteine residues of the protein, while *a*-type haems contain a long hydrophobic tail of isoprene units attached to the porphyrin, as well as a formyl group in place of a methyl substituent. The axial ligands of haem iron vary with cytochrome type (Figure 7.2). In cytochromes *a* and *b*, both ligands are usually His residues, whereas in cytochrome *c*, one is His and the other is often Met. They have a wide distribution, functioning as electron transporters in mitochondria, chloroplasts, endoplasmic reticulum and bacterial redox chains. The iron in all cytochromes can alternate between an oxidised Fe(III) low-spin state with a single unpaired electron and a formal charge of +1, and a reduced Fe(II) low-spin form with no unpaired electrons and a net charge of 0. Since the iron remains low spin, electron transfer is greatly facilitated.

The characteristic visible absorption spectrum of reduced cytochrome *c* arises from $\pi \rightarrow \pi^*$ transitions and is shown in Figure 7.3, with its characteristic α, β and γ absorption bands (the latter called the *Soret* band, after its discoverer). The absorption maxima for cytochromes *a*, *b*, *c* and c_1, all found in the mitochondrial respiratory chain, are listed. The three separate *a* bands in the visible absorption spectra of beef heart mitochondrial membranes indicate the presence of cytochromes *a*, *b*, and *c*.

The d^{10} configuration of the divalent zinc ion has two important consequences. Its complexes are not subject to ligand-field stabilization, such that Zn^{2+} has no ligand-field constraints on its coordination geometry, and because it has no d−d transitions, it has no absorption spectroscopy. Thus, the study of zinc enzymes has frequently resorted to substitution by more amenable transition elements such as cobalt, as illustrated in Figure 7.4. Two cobalt(II)-substituted *Haemophilus influenzae* β-carbonic anhydrases (HICAs) were produced (Hoffmann et al., 2011), one a deep blue enzyme solution with 99 ± 1% Co (Co-HICA), the other a clear, green solution with 52 ± 5% Co and 48 ± 5% Zn (CoZn-HICA). The Co(II)-substituted HICA (Co-HICA) has comparable catalytic activity to that of wild-type enzyme and its X-ray crystal structure is similar to the zinc enzyme. The absorption spectrum of Co-HICA is consistent with four-coordinate geometry. pH-dependent changes in the absorption spectrum of Co-HICA, including an increase in molar absorptivity and a red shift of a 580 nm peak with decreasing pH, correlate with the pH dependence of k_{cat}/K_m. The addition of bicarbonate ion at pH 8.0 triggers

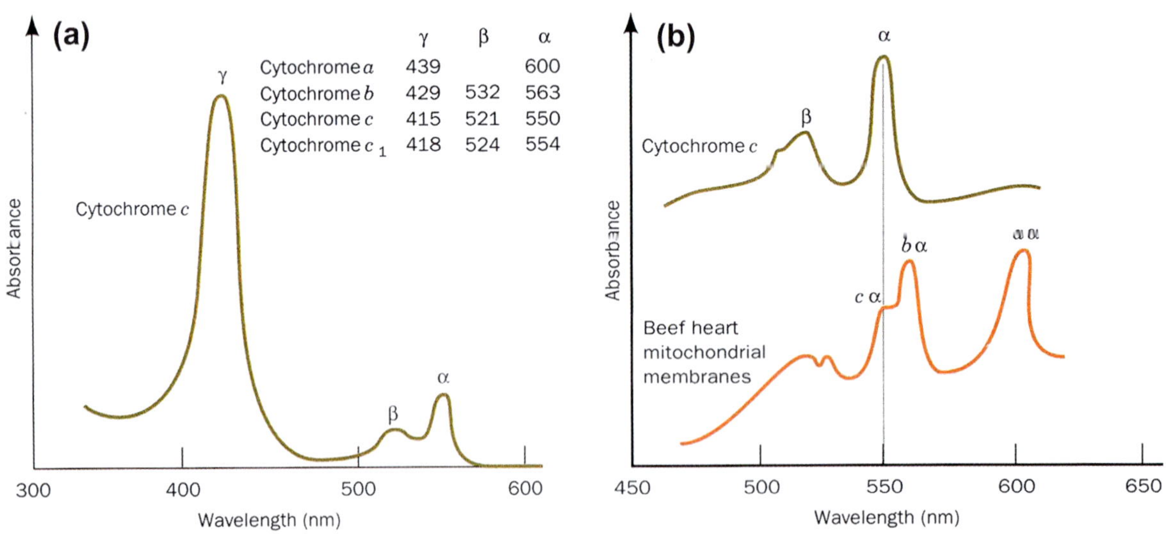

FIGURE 7.3 Visible absorption spectra of cytochromes. (a) Absorption spectrum of reduced cytochrome *c* showing the characteristic α, β and γ absorption bands. The absorption maxima for cytochromes *a*, *b*, *c* and c_1 are listed. The three separate *a* bands in the visible absorption spectra of beef heart mitochondrial membranes indicate the presence of cytochromes *a*, *b*, and *c*. The spectrum of purified cytochrome *c* (*above*) is shown for reference. Adapted from Voet and Voet, 2004.

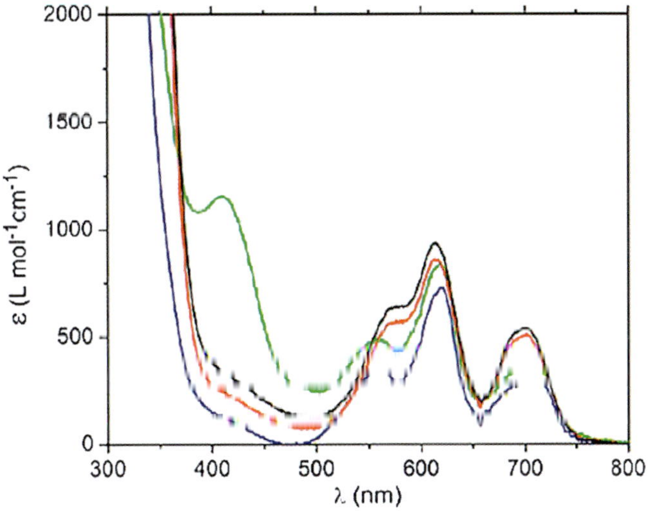

FIGURE 7.4 UV–visible spectral features of Co-HICA and CoZn-HICA. Absorption spectra: CoZn-HICA, pH 8.00 (green); Co-HICA, pH 8.00 (blue); Co-HICA, pH 7.53 (red); Co-HICA, pH 8.00 + 25 mM bicarbonate (black).

spectral changes in the metal coordination sphere that mimic that of lowering pH, supporting its hypothesised role as an allosteric inhibitor of HICA.

Azurin (Az) is a member of a large family of copper electron transfer proteins which contain mononuclear type-1 copper centres, including ceruloplasmin, laccase, plastocyanin, rusticyanin and stellacyanin. They are often referred to as blue copper proteins, on account of their intense blue colour, due to a ligand to metal charge-transfer band near 625 nm ($\varepsilon \approx 5000\ M^{-1}\ cm^{-1}$). All of them have three coordination positions contributed by two His and one Cys, similar to the copper coordination chemistry in azurin (Figure 7.5), yet they span a range of redox potential from less than 200 mV to more than 1000 mV. The copper–cysteine bond is unusually short (~2.1 Å) and highly covalent. It is this bond that generates the intense charge-transfer band characteristic of blue copper centres. Interactions of the axial ligand (Met in the case of azurin) with its blue copper centre are known to be important in tuning spectroscopic

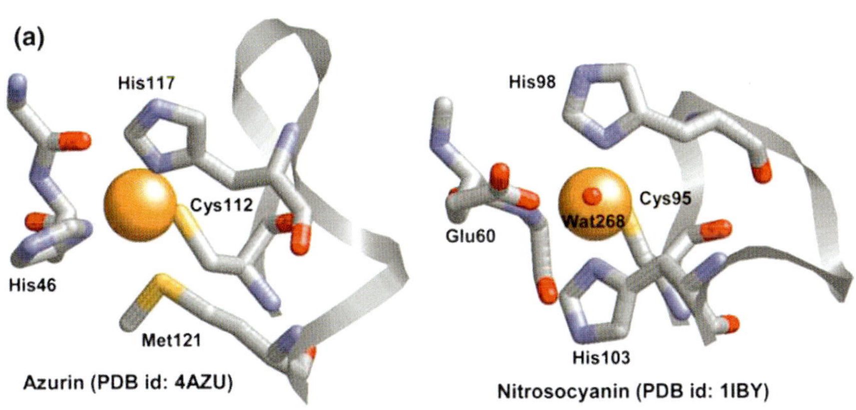

FIGURE 7.5 The active site structures of azurin (*left*, PDB 4AZU) and nitrosocyanin (*right*, PDB 1IBY).

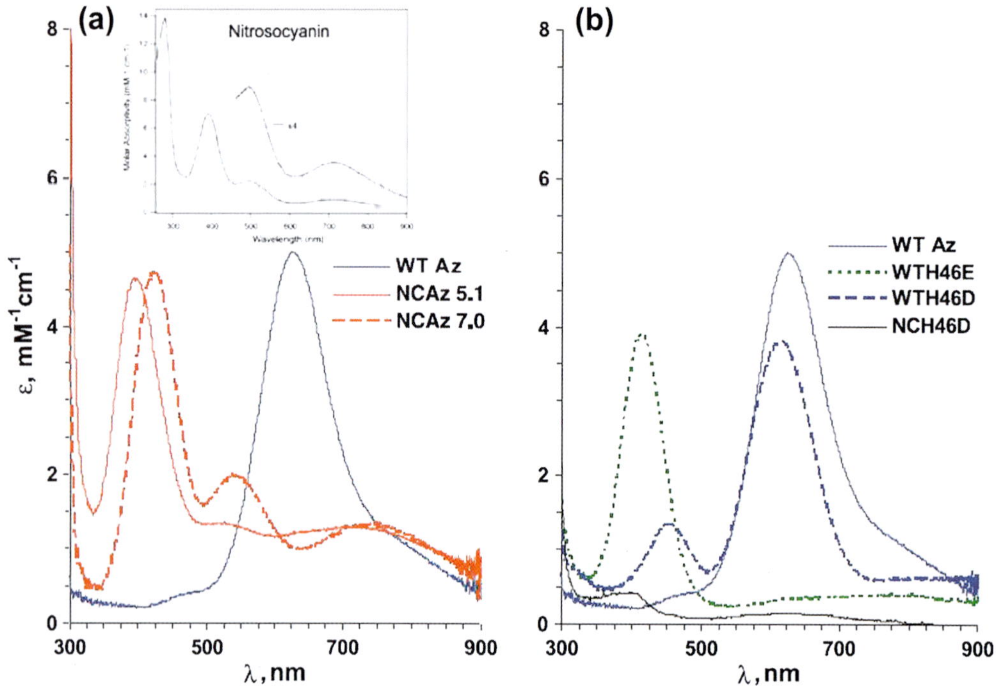

FIGURE 7.6 UV–visible absorption spectra of (a) wild-type (WT) azurin (*WT Az*), NC-azurin pH 5.1 (*NCAz5.1*), and NC-azurin pH 7.0 (*NCAz7.0*). The *inset* shows the spectrum of nitrosocyanin. (b) Spectra of WT azurin (*WT Az*), WT-His46Glu azurin (*WTH46E*), WT-His46Asp azurin (*WTH46D*), and NC-azurin-His46Asp (*NCH46D*). *(From Berry et al., 2010. Reproduced with permission from Springer.)*

and redox properties of cupredoxins. The copper-binding loop which provides three of the four copper ligands (Figure 7.5) in azurin has been replaced by the corresponding loop from the red copper protein nitrosocyanin (Berry et al., 2011). The chimeric protein (Figure 7.6) displayed a red colour, and the UV–visible absorption and electron paramagnetic resonance (EPR) spectra closely resembled those of the loop parent, nitrosocyanin.

MAGNETIC CIRCULAR DICHROISM

Magnetic circular dichroism (MCD) is a powerful probe of both the electronic and the geometric structures of metal centres in metalloproteins. Molecules are said to be optically active if they interact differently with left (**L**)- and right (**R**)-handed circularly polarised light, giving rise to two related but separate phenomena known as **optical rotatory dispersion** and **circular dichroism** (CD). These phenomena are usually observed for electronic transitions, and measurements of optical activity in the wavelength range 200–700 nm, are sensitive probes of molecular conformation. In some cases they can give useful information concerning the secondary structures of biological macromolecules. CD arises from the differential absorption of left- and right-handed polarised light, and can be measured by using a device which can be alternately switched between **L** and **R**.

Optical activity can also be induced by the application of a magnetic field, which perturbs the energy levels of the system, as Michael Faraday first observed in 1845. This is called the Faraday effect, and is the basis of MCD. Although the differential absorption observed is the same quantity which is measured in CD spectroscopy, the origins of CD and MCD are quite different. A CD spectrum is observed for molecules which are inherently chiral, whereas an MCD spectrum arises if the applied magnetic field influences the electronic states of the chromophore in such a way that the **L** and **R** beams are differentially absorbed. Unlike CD, MCD is largely insensitive to molecular conformation, but is sensitive to the total concentration of the MCD-active chromophore. MCD has therefore long been (Cheesman et al., 1991), and still continues to be (McMaster and Oganesyan, 2010) a powerful tool for probing the electronic and geometric structures of metal centres in metalloproteins.

The MCD technique relies on the measurement of the differential absorption of **L** and **R** which is induced in a sample by the application of a strong magnetic field, which is oriented in the same direction as the light propagation (Figure 7.7a). There are three distinct effects a magnetic field can induce on transitions,

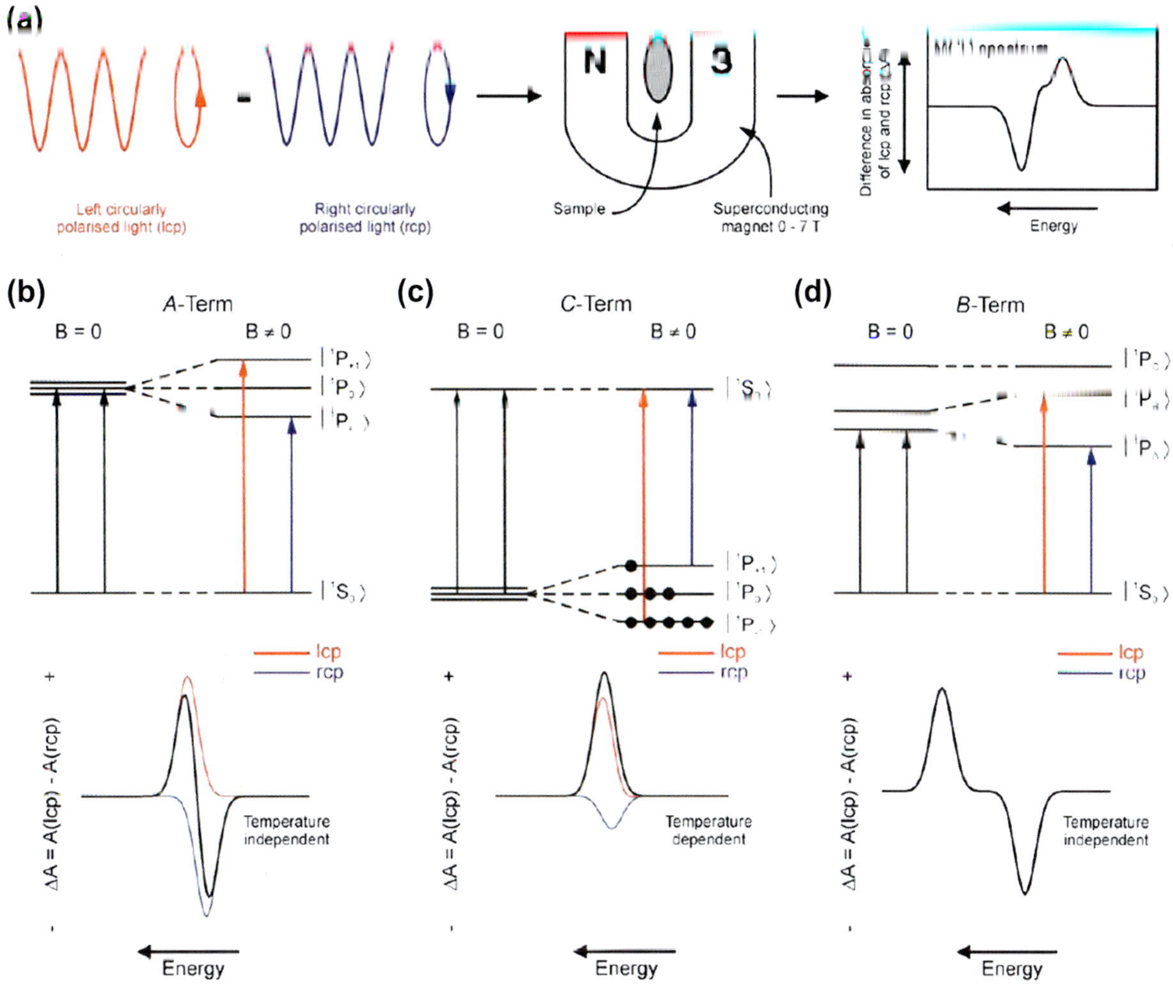

FIGURE 7.7 (a) A schematic of the MCD experiment. (b) Energy level diagrams showing the origin of *A*-, *B*- and *C*-term intensity involving orbitally non-degenerate ($|S^1\rangle$) and orbitally degenerate states ($|P^1\rangle$). For *A*- and *C*-terms, the orbitally degenerate $|P^1\rangle$ state is subject to Zeeman splitting on the application of a magnetic field (*B*). This gives rise to temperature independent derivative-shaped dispersion for *A*-term intensities (b) and to temperature-dependent bands for *C*-term intensities (c). The temperature dependence of the *C*-term intensity is pre-determined by the thermal population of the lowest energy states, shown as black circles in (c). *B*-term intensity (d) arises from the field-induced mixing of ground and/or excited state wavefunctions through the second-order Zeeman effect. *(From McMaster and Oganesyan, 2010. Reproduced with permission from Elsevier.)*

designated as *A*-, *B*- and *C*-terms. (Figure 7.7b). The *A*-term arises in situations where, in the absence of an applied field ($B = 0$), there is a degenerate excited state (Figure 7.7b). In the presence of the magnetic field, the energy levels are split and each of the two resulting absorption bands is optically active. The resulting curve arises because the two states absorb **L** and **R** with equal intensity but opposite sign. The *C*-term arises when the ground state is initially degenerate and is split by the applied field (Figure 7.7c). The differential populations of the ground state are temperature dependent because their levels are relatively close in energy; this implies that the *C* state MCD in spectra is temperature dependent. The temperature dependency is predetermined by the thermal population of the lowest ground state, shown as black circles (Figure 7.7c). The *B*-term intensity (Figure 7.7d) arises from mixing of states, and is not so temperature dependent as the *C*-term.

The low symmetry of metal centres in metalloproteins usually results in orbitally non-degenerate ground states, and therefore no *C*-term intensity. Paramagnetic metal centres however do possess spin degeneracy, and the orbital degeneracy necessary for *C*-term intensity for these low symmetry metal sites is partly 'regained' via spin–orbit coupling, which mixes the orbital and spin states of an electron. Thus, in general, only paramagnetic metal centres exhibit temperature-dependent *C*-term intensity and this *C*-term intensity becomes overwhelmingly dominant at low temperature (ca. 5 K). These features that make MCD spectroscopy such a powerful tool to probe metalloproteins that contain paramagnetic metal centres.

The MCD spectra of metalloprotein sites are usually interpreted in parallel with other available spectroscopic data to extract the full information content of each spectrum (Lehnert et al., 2001). This makes MCD spectroscopy a very powerful tool for probing metal centres in proteins. Thus, the signs of the MCD bands may aid in the assignment of closely overlapping charge-transfer and ligand-field transitions in the absorption spectrum of a metal centre in a protein. Variable-field and temperature measurements, and their effect on *C*-term intensity, can provide considerable insight into the nature of the ground state of a metal centre in a protein, yielding *g* values, zero-field splitting constants and exchange couplings for direct comparison with those obtained by EPR spectroscopic and magnetic susceptibility measurements. In addition, the temperature dependence of the *C*-term intensity may also be exploited to probe, selectively, weakly absorbing paramagnetic metal centres in the presence of strongly absorbing diamagnetic chromophores present in the same protein sample.

General theoretical treatments for the analysis of total MCD intensities and saturation behaviour of spin-allowed transitions for $S \geq 1/2$ paramagnetic transition metal centres in the non-linear regions of magnetic field and temperature have been developed (Organesyan et al., 1999; Neese and Solomon, 1999), and more recent developments including ab initio/DFT computational approaches have been recently summarised (McMaster and Oganesyan, 2010).

Haemoproteins have been extensively studied by MCD, which has been widely used to classify haemoproteins by comparison of their MCD spectra with those of other haemoproteins and of model compounds (Cheesman et al., 1991; Cheek and Dawson, 2000). The MCD technique has the capacity to define haem oxidation, spin state, geometry, oxidation state (ferrous, ferric and ferryl), and axial ligands (number and type) in solution and at ambient temperature of haem centres in proteins when combined with UV–visible spectroscopies.

Unfortunately, the resolving power of MCD has rarely been applied to the intermediate redox states accessible to multihaem proteins. A solution to this problem has been developed, and its application to the cytochrome *bd* complex is illustrated in Figure 7.8 (Marritt et al., 2006). The MCD spectrum of cytochrome *bd* equilibrated anaerobically at 325 mV shows the features reported previously for the aerobic enzyme. The dominant feature is the derivative centred on 414 nm, providing a good indication that only one low-spin ferric haem, i.e. b_{558}, exists in the sample under these conditions. The smaller features resolved between 460 and 600 nm reflect overlapping contributions from both low- and high-spin ferric haems. At longer wavelengths,

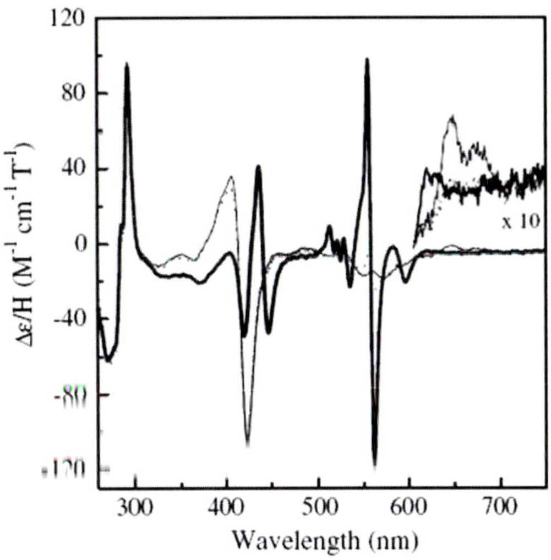

FIGURE 7.8 MCD spectra of *Escherichia coli* cytochrome *bd* equilibrated at defined potentials by MOTTLE. Shown is the as-prepared sample equilibrated at +325 mV (light solid line), after equilibration at +100 mV (heavy solid line), and after equilibration at +348 mV (broken line). Buffer–electrolyte was 20 mM HEPES, 1 mM EDTA, and 50 mM NaCl (pH 7.0). *(From Marritt et al., 2006. Reproduced with permission from Elsevier.)*

peaks at 650 and 680 nm are assigned to the ferrous-oxy and oxo-ferryl forms of haem *d* respectively. After sample equilibration at 100 mV, significant reduction of haem b_{558} is reflected in the emergence of an intense sharp derivative centred on 558 nm together with changes in the Soret region. The small broad derivative centred on 595 nm indicates the presence of high-spin ferrous haem b_{595}. A broad feature (620–630 nm) is resolved and is likely to originate from ferrous haem *d*. After sample reoxidation by equilibration at 348 mV, the MCD spectrum shows little evidence of ferrous haems b_{558}, b_{595}, or *d*. corresponding to the clean conversion of as-prepared oxygenated cytochrome *bd* to its all-ferric state. Other recent developments include time-resolved MCD resulting from in situ chemical manipulation of a metalloprotein sample (Figure 7.9) and

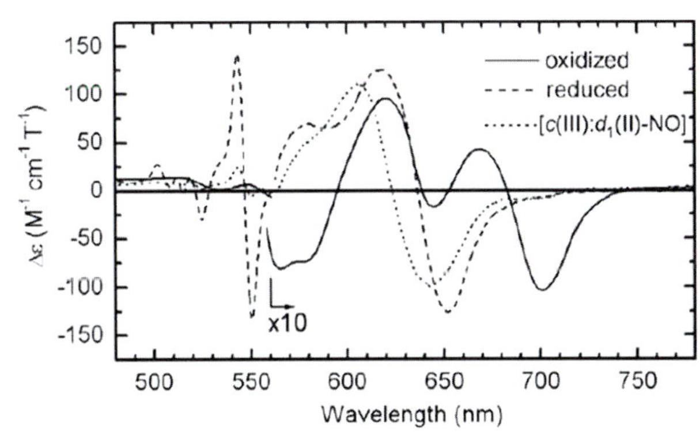

FIGURE 7.9 In situ reduction and partial reoxidation of cytochrome cd_1. Experimental parameters: cd_1 monomer concentration, 30 μM; magnetic field, 8 T; path length, 1 cm. MCD spectra of cd_1 in the oxidised state (——), reduced by sodium dithionite (- - - -), and following partial reoxidation with sodium nitrite (......). *(From Bradley et al., 2011. Reproduced with permission from Elsevier.)*

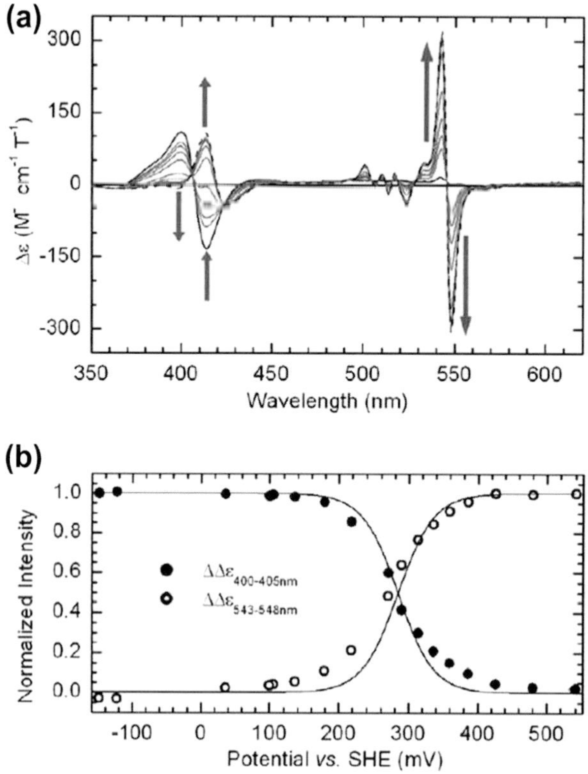

FIGURE 7.10 Potentiometric titrations of cytochrome *c*, in the stirred electrochemical cell, monitored by MCD spectroscopy. (a) MCD spectra for sample equilibrated at a range of potentials between the extremes of +540 mV (——) and −149 mV (- - - -). Arrows indicate the change in intensity on reduction. (b) Plots vs potential of the variation in peak-to-trough intensity of the bisignate MCD feature at 545 nm (○), the intensity of the 400-nm MCD feature measured as $\Delta\Delta\varepsilon_{400-405\mathrm{nm}}$ (●), and the behaviour predicted by the Nernst equation for a single one-electron centre with a midpoint potential of $E_m = +280$ mV (——). *(From Bradley et al., 2011. Reproduced with permission from Elsevier.)*

the parallel development of an electrochemical cell, allowing true potentiometric titrations to be performed (Figure 7.10) within the MCD spectrometer (Bradley et al., 2011).

Another technique which has proved to be a powerful tool, particularly to probe geometric and electronic the structures in mono and dinuclear non-haem iron centres metalloproteins is variable-temperature, variable-field MCD spectroscopy (Solomon et al., 2000, 2009). An example is the di-iron centre of the soluble methane monoxygenase from *Methylococcus capsulatus* (Figure 7.11). The di-iron centre resides in the hydroxylase component (MMOH) of the enzyme and requires binding of the effector protein, MMOB. Binding of MMOB induces a structural change at the Fe2 centre that results in a decreased ligand field splitting from >4900 to ~3700 cm^{-1}, whereas there is negligible change observed for the Fe1 centre. The structural change that provides the best agreement between theory and the spectroscopic data involves an increase in the Fe—Fe—Glu209 angle resulting in a more open structure which permits the binding of O_2 and formation of a bridged peroxo-Fe^{III}—Fe^{III} species (Mitic et al., 2008).

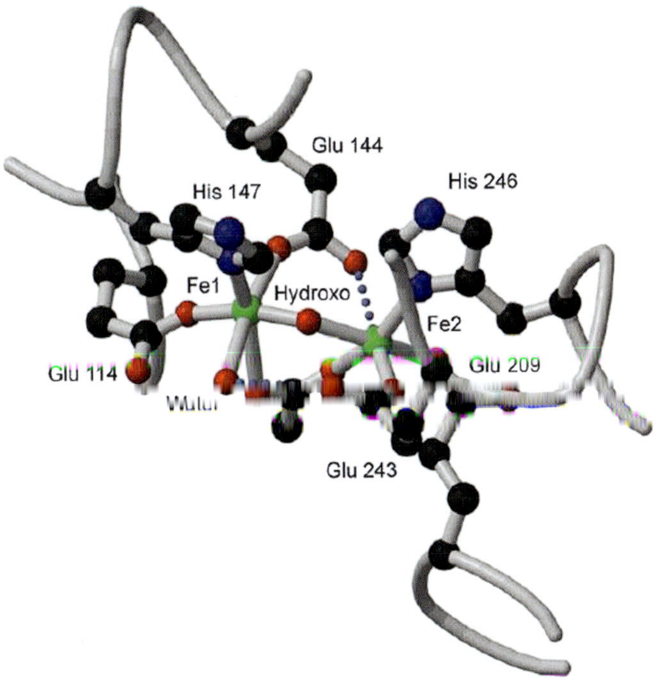

FIGURE 7.11 The dinuclear active site of methane monooxygenase hydroxylase (MMOH) component from *Methylococcus capsulatus* (Bath) (PDB 1MMO).

VIBRATIONAL SPECTROSCOPIES

Bonds between atoms have quantised vibrational energy levels that can be probed by methods that employ electromagnetic radiation of the correct frequency to promote transitions between the different levels. Vibrational spectroscopies explore the molecular structure by analysing the energy of transition between these vibrational levels.

The number of vibrational levels depends on the number of atoms (N) in the molecule and on their spatial organisation. In the case of non-linear molecules there are $3N-6$ vibrational states (and three rotational states), whereas for a linear molecule there are $3N-5$ vibrational states (and two rotational states). These are called the *normal modes* of vibration and can give rise to observable spectroscopic signals depending on selection rules that are specific for each spectroscopic method.

The energy associated with each vibrational mode can be interpreted in a classical way treating atoms and bonds as harmonic oscillators. Therefore it is proportional to the mass of the atoms involved (m_i) and to the spring constant k (Eqn 7.1).

$$E = f\left(\sqrt{\frac{k(m_1+m_2)}{m_1 m_2}}\right) \tag{7.1}$$

The vibrational levels can have different effects on the interaction of matter with electromagnetic radiation depending on the frequency of the radiation used and the phenomenon probed (Figure 7.12).

Three different phenomena are associated with vibrational levels and can be used to probe molecular structure:

- IR absorption of radiation,
- Raman scattering of radiation and
- Fluorescence emission of radiation.

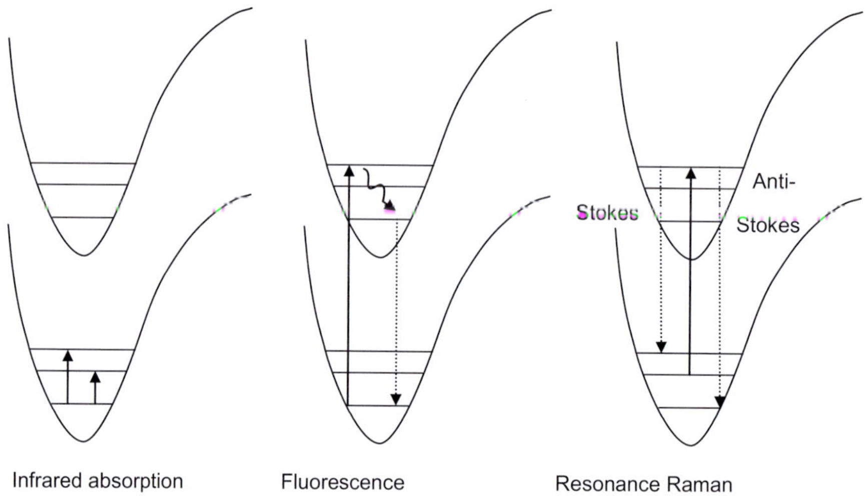

FIGURE 7.12 Diagram representing two electronic energy states with some of the associated vibrational levels, and the phenomena that can be optically observed. Continuous lines are radiation absorption, whereas dashed lines are radiation emission. Wavelike line is the dissipation of vibrational energy in the excited electronic state. The two modes of Raman radiation scattering are depicted.

Figure 7.12 shows a diagram of each of these phenomena. In this section IR spectroscopy and Raman spectroscopy will be described.

IR SPECTROSCOPY

The use of IR radiation to analyse biological samples dates back to the beginning of the twentieth century. When applied to biological macromolecules, IR spectroscopy probes the vibrational properties of the sample under investigation. This is made by exposing the sample to a source of IR radiation and observing the absorption of radiation at specific frequencies. Each absorption band is characterised by its frequency and amplitude. Its energy matches the energy of transition between different vibrational states of chemical bonds within the sample. Absorption of IR radiation increases the frequency of vibration of the bond. Spectroscopic selection rules state that not all vibration modes lead to absorption of IR radiation. Vibrations that are symmetric with respect to the centre of mass are forbidden. Therefore, absorption of IR radiation gives rise to changes in the dipole moment, and the absorption intensity is larger for larger changes in the dipole moment. This means that IR spectroscopy can be used to observe polar functional groups in biological macromolecules.

The introduction of interferometers had a profound impact in IR spectroscopy because it allowed the simultaneous detection of energy absorption across the entire spectral range. The signal is then subjected to a mathematical procedure called Fourier transformation to obtain the absorption bands in the frequency scale, and the method is called Fourier-transformed infrared (FTIR) spectroscopy. This allows for faster acquisition and for multiple acquisitions to be summed up improving the signal to noise of the spectrum for dilute samples. Furthermore, one of the greatest advantages of FTIR is that it can be applied to solids, dehydrated samples, and aqueous or organic solutions (Jackson and Mantsch, 1995).

The IR range of the electromagnetic spectrum is generally described as divided in three regions: near IR from 750 to 2500 nm, mid-IR from 2500 to 25,000 nm, and far IR from 25,000 to 100,000 nm. However, the spectral scale in IR spectroscopy is typically not reported in wavelength as in UV–visible spectroscopy but in reciprocal wavelength in cm^{-1}. The most interesting range for studying biological macromolecules is the mid-IR but the presence of metals in biomolecules may bring some absorption bands into the near IR region. Given the low energy of radiation involved, IR spectroscopy is inherently a non-destructive technique.

IR absorption can provide structural information on biological macromolecules. It is known since the 1950s that there is an empirical correlation between two IR absorption bands typically found between 1500 and 1700 cm^{-1} called amide I and amide II, and the predominant secondary structure elements in a protein (Elliott and Ambrose, 1950). The basis for this correlation appears to arise from the fact that amide I is dominated by C=O stretching and amide II is dominated by N—H bending and therefore both are sensitive to the hydrogen bonding environment that establishes the secondary structural elements. The stronger the hydrogen bonding environment, the lower the amide I band position in reciprocal centimetres. IR spectroscopy is not limited by the size of the macromolecules to be studied or by the rates of interconversion between different forms that may be in dynamic equilibrium, making this method the technique of choice in many applications. When dealing with liquid samples such as water-soluble biological material, the strong absorption of the solvent can be a nuisance. This can be avoided by using deuterated solvents or working with dry material. One particular application using dry films is attenuated total reflectance (ATR), which is performed on material deposited on the surface of prisms that are transparent to IR radiation. This configuration can be slightly modified by covering the reflective surface of the prism with a thin metallic film with a rough surface that allows the enhancement of the IR radiation absorption by the molecules on the surface, giving rise to surface-enhanced infrared absorption spectroscopy. The interaction of the deposited molecules with the rough surface gives rise to an enhancement of the absorption band of the

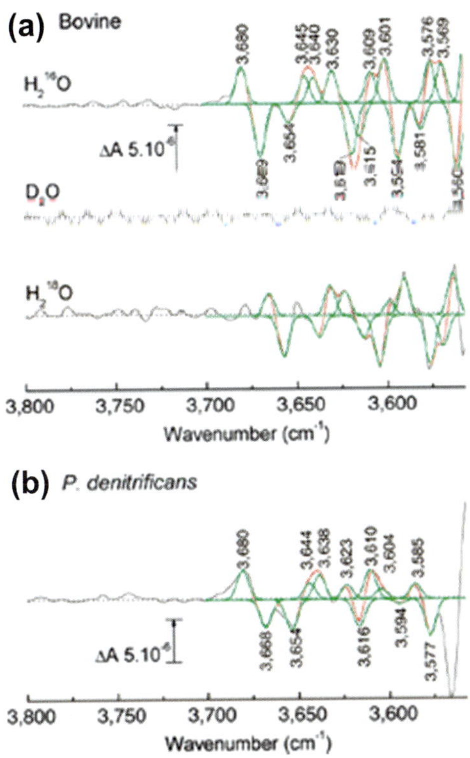

FIGURE 7.13 Comparison of light-induced CO-photolysis transmission FTIR difference spectra of fully reduced CcO-CO from bovine and *Paracoccus denitrificans* CcOs. (a) Bovine CcO: FTIR difference spectra (black traces) were recorded in $H_2^{16}O$ (top trace, average of 4000 transitions), D_2O (middle trace, average of 984) and $H_2^{18}O$ (lower trace, average of 1500) media. Data in $H_2^{16}O$ and $H_2^{18}O$ media were fitted with combinations of Gaussian components all with a fixed full width at half maximum (FWHM) of 6 cm^{-1} (red traces; individual Gaussians shown in green). (b) *P. denitrificans* CcO: FTIR difference spectrum (black trace) in $H_2^{16}O$ at pH 8.5 (average of 1400 transitions). Data were fitted with combinations of Gaussian components all with a fixed FWHM of 6 cm^{-1} (red traces; individual Gaussians shown in green). *(Reproduced with permission from Maréchal and Rich, 2011.)*

molecules that decay rapidly within approximately 10 nm making this set-up ideal for the study of molecular monolayers (Ataka et al., 2010). This experimental set-up also has an added advantage for the study of metalloproteins because the presence of the metallic film provides the means to perform electrochemistry concurrently with the spectroscopic measurement. Such measurements can reveal the occurrence of redox-linked conformational changes in redox proteins due to the sensitivity of the IR absorption bands to the secondary structural elements in the protein.

Such a set up was employed to obtain electrochemically induced ATR FTIR difference spectra of bovine cytochrome *c* oxidase (Maréchal and Rich, 2011). This set-up allowed the photolysis of CO bound to cytochrome *c* oxidase using laser light and recording of difference spectra between the CO-bound form and the photolysed form. Data obtained with cytochrome *c* oxidase solubilised in normal and isotopically labelled water showed that there are weakly bound water molecules in the structure of cytochrome *c* oxidase that respond to CO photolysis from the binuclear centre and to redox changes of the redox centres (Figure 7.13).

These changes are likely relevant for the molecular bases of the electron/proton coupling mechanism that sustain energy transduction by cytochrome *c* oxidase. Comparison with the bacterial cytochrome *c* oxidase from *Paracoccus denitrificans* showed a similar pattern of changes in water signals which suggests a common mechanism for the electron/proton coupling mechanism, using hydrophilic channels that are very similar in both proteins (Figure 7.14).

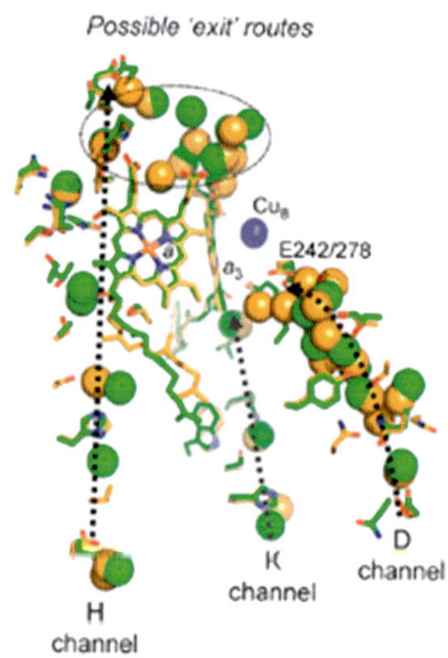

FIGURE 7.14 Comparison of the hydrophilic channels and their associated water molecules in bovine and *Paracoccus denitrificans* CcOs. Crystal structures of subunit I of oxidised CcOs from bovine [PDB ID code 1V54 (26), green] and *P. denitrificans* [PDB ID code 3HB3 (13), orange] were aligned on their a_3 haems. Amino acids and crystallographically resolved water molecules within the D, K, and H channels are displayed as spheres (green, bovine CcO; orange, *P. denitrificans* CcO). *(Reproduced with permission from Maréchal and Rich, 2011.)*

RAMAN SPECTROSCOPY

This method utilises the radiation scattered by the molecules under analysis to explore their structure. Scattering occurs when the size of the particles is smaller or of the same order of magnitude as the wavelength. Lasers are used as radiation source and detection is made perpendicular to the direction of the beam. The interaction between the photons in the laser light and matter can be interpreted as the collision of photons with particles that are vibrating. When the collision is elastic, the energy of the particles and of the photons does not change and the scattered radiation is of the same frequency as the exciting radiation. This is called Rayleigh scattering, and corresponds to the most frequent outcome of the interaction between the electromagnetic radiation and the matter. However, a very small fraction of the molecules in the sample (1 in 10^{10}) absorb or release energy upon interaction with the laser beam. In this case, the scattered radiation does not have the same frequency as the incident radiation. The scattered radiation in this case is called Raman scattering in honour of Chandrasekhara Venkata Raman, the physicist that first observed the phenomenon and subsequently won the physics Nobel Prize in 1930. The photons can have lower energy than the incident radiation and this is called Stokes radiation, or can have higher energy than the incident radiation and is called anti-Stokes radiation. The difference between the incident and the scattered radiation matches the difference of a molecular vibration of the sample. However, not all vibration modes can be detected by Raman spectroscopy. Selection rules state that vibrations that are anti-symmetric with respect to the symmetry centre are forbidden.

Raman spectra plot the intensity of the scattered radiation vs the shift in wave number (measured in cm^{-1}) between the incident radiation and the scattered radiation. Each line corresponds to the energy of a vibrational mode of the molecule. The intensity of the Raman scattering is dependent on a number of factors: It is directly proportional to the sample concentration and to the intensity of the incident light (I_0). This is the reason for using lasers as the light source. It is proportional to the fourth negative power of the wavelength (λ) of the incident radiation. It is also dependent on the intrinsic scattering properties of the sample, that is on its polarisability (α). The polarisability is the capability of inducing a dipole, which does not have to be permanent. It scales with size and therefore larger particles typically produce more scatter. Finally it depends on the angle of the scattered radiation vs the direction of the incident radiation (θ). Equation (7.2), which describes the scattered intensity also includes the dependence of the distance to the detector (r).

$$I_s = \frac{8\pi^4\alpha^2}{\lambda^4 r^2}(1 + \cos^2\theta)I_0 \qquad (7.2)$$

Detecting the very weak Raman scattering in the background of the Rayleigh scattering, and eventual fluorescence of the sample when excitation is made using UV—visible light, can be challenging and two experimental arrangements were devised that enhance Raman scattering associated to specific vibrational modes that are relevant for metal-containing biomolecules.

Resonance Raman

The resonance Raman spectroscopy is a particular application of the general Raman spectroscopy where the incident laser radiation has a frequency that matches the energy of an electronic transition in the sample. It is applied to molecules that absorb light in the visible range such as those containing transition metal centres. However, metal centres that are spectroscopically silent in the visible frequency range such as Zn^{2+} centres cannot be probed this way. The Raman scattering arising from symmetric vibrations of the chromophore is enhanced by 10^2—10^6. This allows for its discrimination from the non-resonant scattering arising from the remainder of the molecule.

In metal-containing biomolecules, resonance Raman spectroscopy observes the vibrational frequencies of the metal-associated chromophore that is in resonance with the incident radiation. The coordination environment and geometry of the metal can be probed and the strength of the bonds between the metal and the ligands can be determined from the vibrational frequencies. Because changes in redox state of the metal affect these parameters redox titrations can be followed spectroscopically using this method. If various chromophores with different absorption maxima exist within a molecule, a different resonance Raman spectrum can be obtained by changing the frequency of the incident laser light to match the respective absorption maxima of each chromophore. Nonetheless, because the spectral lines are relatively sharp, it is possible to discriminate the signals of different chromophores within the same molecule, even if they share the same absorption maximum. This property also makes this technique extremely sensitive to small structural changes, and these can be probed in a vast range of timescales that can be as short as the picosecond range when applying time-resolved methods (Schelvis et al., 1997).

Resonance Raman is one of the most versatile methods for the study of metal-containing biological molecules. It requires a very small volume (in the tens of microlitre range) of pure sample with as high concentration as possible, given the very small proportion of Raman scattering. The measurement can be made using liquid or frozen solutions of the target molecule, using solids or even crystals. Furthermore, because the measurement has to be done in conditions where sample degradation by the laser radiation has to be avoided, the method is inherently non-destructive allowing for the sample to be recovered afterwards.

However, the method cannot be used in a quantitative way because the mechanisms for resonance enhancement of specific vibrational modes are not understood in detail. Furthermore, resonance assignment requires isotopic substitution when dealing with a sample with totally new metal centre, in order to assign the spectral lines to specific vibrations in the structure. A final concern with respect to experimental set-up, intrinsic fluorescence of the target molecule or the presence of fluorescent contaminants will make the detection of the resonance Raman signals difficult or even impossible.

Surface-Enhanced Raman Scattering

This phenomenon was accidentally discovered in the 1970s and provides enhancements of the order of 10^6 but can be as high as 10^{14}, allowing the study of single molecules. It occurs when molecules are in contact with metallic surfaces that are not smooth. This makes it selective for what is happening at the surface up to a distance of tens of nanometres. Surface-enhanced Raman scattering is usually performed on gold, silver or copper roughened surfaces.

The enhancement can be electromagnetic or chemical, with the former the most common. The intensity of Raman scattering is proportional to the square of the intensity of the electromagnetic field acting on the sample. Molecules that are in contact with a rough metallic surface are subject to the electromagnetic field arising from the laser light plus the field associated with the rough features at the surface. Therein lies the origin of the enhancement, which clearly is dependent on the shape and size of the features in the roughened surface. This

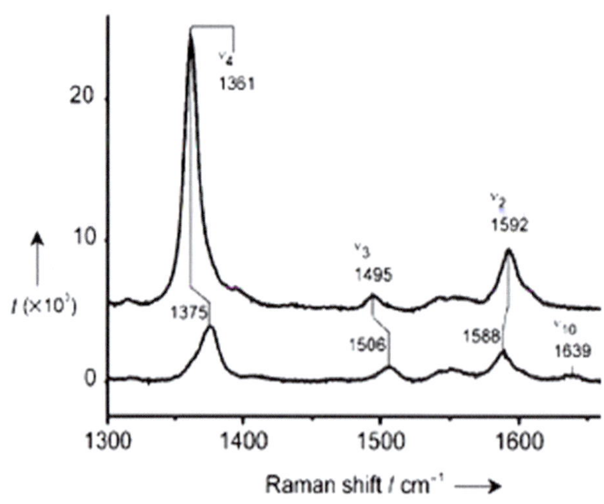

FIGURE 7.15 SERR spectra of the reduced (upper spectrum) and oxidised (lower spectrum) outer membrane cytochromes from *Geobacter sulfurreducens*, obtained at −425 and 0 mV, respectively. The spectra were obtained with excitation at $L = 413$ nm, laser power of 1 mW and acquisition time 90 s. Potentials refer to the Ag/AgCl (KCl 3.0 M) reference electrode (210 mV vs SHE). *(Reproduced with permission from Millo et al., 2011.)*

reveals the importance of the surface preparation to produce features that are of similar shape size and orientation in order to respond to a single excitation wavelength. If the surface is prepared in a way that makes it responsive to a wavelength that matches one of the electronic transitions of the molecule to be analysed, then an additional enhancement is obtained and surface-enhanced resonance Raman scattering (SERRS) is obtained. It should be noted that if aggregation of the molecule to be analysed occurs the enhancement is less efficient and therefore samples should not be too concentrated (Figure 7.15).

One of the most interesting applications of resonance Raman to metal-containing biomolecules is the study of redox proteins and enzymes with simultaneous spectroscopic and electrochemical characterisation. This allows the probing of the changes in the chromophores when they undergo changes in redox state. This combination of techniques was applied to the redox characterisation of multihaem cytochromes present at the surface of cells in a biofilm of *Geobacter sulfurreducens* grown on the surface of the SERRS silver electrode. This organism is capable of resisting the toxicity of nanostructured silver and the results allowed the identification of two distinct redox couples involved in direct electron transfer to the electrode (Millo et al., 2011).

REFERENCES

Ataka, K., Kottke, T., & Heberle, J. (2010). Thinner, smaller, faster: IR techniques to probe the functionality of biological and biomimetic systems. *Angew. Chem. Int. Ed. Engl., 49*, 5416–5424.

Berry, S. M., Bladholm, E. L., Mostad, E. J., & Schenewerk, A. R. (2011). Incorporation of the red copper nitrosocyanin binding loop into blue copper azurin. *J. Biol. Inorg. Chem., 16*, 473–480.

Bradley, J. M., Butt, J. N., & Cheesman, M. R. (2011). Electrochemical titrations and reaction time courses monitored in situ by magnetic circular dichroism spectroscopy. *Anal. Biochem., 419*, 110–116.

Cheek, J., & Dawson, J. H. (2000). In K. Kadish, K. Smith, & R. Guilard (Eds.), *Handbook of Porphyrins and Related Macrocycles* (pp. 339–369). Academic Press.

Cheesman, M. R., Grenwood, C., & Thomson, A. J. (1991). Magnetic circular dichroism of hemoproteins. *Adv. Inorg. Chem., 36*, 201–255.

Elliott, A., & Ambrose, E. J. (1950). Structure of synthetic polypeptides. *Nature, 165*, 921–922.

Hoffmann, K. M., Samardzic, D., Heever, K., & Rowlett, R. S. (2011). Co(II)-substituted *Haemophilus influenzae* β-carbonic anhydrase: spectral evidence for allosteric regulation by pH and bicarbonate ion. *Arch. Biochem. Biophys., 511*, 80–87.

Jackson, M., & Mantsch, H. A. (1995). The use and misuse of FTIR spectroscopy in the determination of protein structure. *Crit. Rev. Biochem. Mol. Biol., 30*, 95–120.

Lehnert, N., George, S. D., & Solomon, E. I. (2001). Recent advances in bioinorganic spectroscopy. *Curr. Opin. Chem. Biol., 5*, 176–187.

McMaster, J., & Oganesyan, V. S. (2010). Magnetic circular dichroism spectroscopy as a probe of the structures of the metal sites in metalloproteins. *Curr. Opin. Struct. Biol., 20*, 615–622.

Maréchal, A., & Rich, P. R. (2011). Water molecule reorganization in cytochrome c oxidase revealed by FTIR spectroscopy. *Proc. Natl. Acad. Sci. U.S.A., 108*, 8634–8638.

Marritt, S. J., van Wonderen, J. H., Cheesman, M. R., & Butt, J. N. (2006). Magnetic circular dichroism of hemoproteins with in situ control of electrochemical potential: "MOTTLE". *Anal. Biochem., 359*, 79–83.

Millo, D., Harnisch, F., Patil, S. A., Ly, H. K., Schroder, U., & Hildebrandt, P. (2011). In situ spectroelectrochemical investigation of electrocatalytic microbial biofilms by surface-enhanced resonance Raman spectroscopy. *Angew. Chem. Int. Ed. Engl., 50*, 2625–2627.

Mitic, N., Schwartz, J. K., Brazeau, B. J., Lipscomb, J. D., & Solomon, E. I. (2008). CD and MCD studies of the effects of component B variant binding on the biferrous active site of methane monooxygenase. *Biochemistry, 47*, 8386–8397.

Neese, F., & Solomon, E. I. (1999). MCD C-Term Signs, Saturation Behavior, and Determination of Band Polarizations in Randomly Oriented Systems with Spin S >/= (1)/(2). Applications to S = (1)/(2) and S = (5)/(2). *Inorg. Chem., 38*, 1847–1865.

Schelvis, J. P. M., Deinum, T., Varotsis, C. A., Ferguson-Miller, S., & Babcock, G. T. (1997). Low-power picosecond resonance Raman evidence for histidine ligation to heme a_3 after photodissociation of CO from cytochrome *c* oxidase. *J. Am. Chem. Soc., 119*, 8409–8416.

Solomon, E. I., Brunold, T. C., Davis, M. I., Kemsley, J. N., Lee, S. K., Lehnert, N., Neese, F., Skulan, A. J., Yang, Y. S., & Zhou, J. (2000). Geometric and electronic structure/function correlations in non-heme iron enzymes. *Chem. Rev., 100*, 235–349.

Solomon, E. I. (2006). Spectroscopic methods in bioinorganic chemistry: blue to green to red copper sites. *Inorg. Chem., 45*, 8012–8025.

Solomon, E. I., Wong, S. D., Liu, L. V., Decker, A., & Chow, M. S. (2009). Peroxo and oxo intermediates in mononuclear nonheme iron enzymes and related active sites. *Curr. Opin. Chem. Biol., 13*, 99–113.

Voet, D., & Voet, J. G. (2004). *Biochemistry* (3rd ed.). Hoboken, NJ: John Wiley and Sons. p. 1591.

Chapter 8

An Introduction to Electrochemical Methods for the Functional Analysis of Metalloproteins

Christophe Léger
Laboratoire de Bioénergétique et Ingénierie des Protéines, CNRS and Université Aix-Marseille, Marseille, France

Chapter Outline

INTRODUCTION

Direct electron transfer (ET) to proteins (without the need for mediators) was first reported in the 1970s, opening the way for detailed studies of biological reactions, and electrochemical investigations of large redox enzymes are now common.

Determining reduction potentials is only one application of the method; in studies of redox proteins or small molecules, electrochemical techniques are used for learning not only about the thermodynamics but also the kinetics of chemical reactions that immediately precede or follow ET (e.g. protonation or substrate binding). Using direct electrochemistry, the turnover rate of enzymes can also be measured with very high temporal resolution and potential control. This greatly broadens the possibilities of enzyme kinetics. Direct electrochemistry has indeed been used to study all aspects of catalysis: interfacial and intramolecular ET, substrate diffusion along substrate channels, active site chemistry, mechanism of reaction with inhibitors, redox-driven (in)activation processes, etc.

The proteins or enzymes that can be studied using direct electrochemistry have at least one surface-exposed redox centre, which is the entry point for electrons from the electrode. The chance of success is greater when the protein of interest is small and hydrophilic or, if it is an enzyme, when it has a large turnover rate. The amount of protein required depends on which method is used, but it can be as small as a few picomoles for a series of experiments carried out with the same 'film' of adsorbed proteins. The electrochemical equipment is particularly inexpensive (compared with many biophysical techniques) and available in most chemistry labs.

Practical Approaches to Biological Inorganic Chemistry, 1st Edition. http://dx.doi.org/10.1016/B978-0-444-56351-4.00008-7.

An extensive description of most of the electrochemical techniques will be found in refs. Bard and Faulkner (2004) and Compton and Banks (2011). The physical aspects of electrochemistry, including hydrodynamics, are discussed elsewhere (Newman and Thomas-Alyea, 2004; Levich, 1962). Savéant provided a comprehensive discussion of voltammetric wave shapes under various conditions, with emphasis on the case where the electrocatalyst diffuses in solution (Savéant, 2006). References Cornish-Bowden (2004) and Laidler and Bunting (1973) are insightful textbooks on enzyme kinetics. There are many comprehensive reviews on the use of direct elecrochemistry to probe the mechanism of redox proteins and enzymes (Armstrong et al., 1997; Elliott et al., 2002a; Léger et al., 2003; Murgida and Hildebrandt, 2005; Armstrong, 2005; Vincent and Armstrong, 2005; Hirst, 2006; Bernhardt, 2006; Vincent et al., 2007; Léger and Bertrand, 2008; Jeuken, 2009; Armstrong et al., 2009; Gates et al., 2011; Lojou, 2011). References Vincent et al. (2007) and Armstrong et al. (2009) focus on hydrogenases. Armstrong and co-workers discuss, in ref. Cracknell et al. (2008), the applications of redox enzymes in electrochemical devices. Reference Jeuken (2009) summarises recent advances in the electrochemistry of *membrane-bound* redox enzymes. Lojou reviews in ref. Lojou (2011) the strategies for connecting hydrogenases to electrodes.

BASICS

Redox Thermodynamics: The Nernst Equation

Consider a reaction mixture containing the oxidised and reduced forms of two different species referred to by subscripts 1 and 2:

$$\mathrm{Ox_1} + \mathrm{Red_2} \rightarrow \mathrm{Red_1} + \mathrm{Ox_2} \tag{8.1}$$

The free energy of the reaction ($\Delta_r G$, in units of J/mol) is given by

$$\Delta_r G = \Delta_r G^0 + RT \ln \frac{[\mathrm{Red_1}][\mathrm{Ox_2}]}{[\mathrm{Ox_1}][\mathrm{Red_2}]} \tag{8.2}$$

where R is the gas constant, T the absolute temperature and $\Delta_r G^0$ is the (tabulated) standard free energy of the reaction ('standard conditions' means that the activity of all constituents is unity, and the pressure equals 1 bar).

If equilibrium between the different species is reached, $\Delta_r G = 0$, and the ratio of concentrations (the reaction quotient) is linked to $\Delta_r G^0$ by the relation:

$$K_{\mathrm{eq}} = \frac{[\mathrm{Red_1}]_{\mathrm{eq}}[\mathrm{Ox_2}]_{\mathrm{eq}}}{[\mathrm{Ox_1}]_{\mathrm{eq}}[\mathrm{Red_2}]_{\mathrm{eq}}} = \exp\left(-\frac{\Delta_r G^0}{RT}\right) \tag{8.3}$$

Initially upon mixing $\mathrm{Ox_1}$ and $\mathrm{Red_2}$, the concentrations of various species in the solution do not satisfy Eqn (8.3), and $\Delta_r G$ is nonzero: the system is not at equilibrium. Thermodynamics predicts that reaction (8.1) will spontaneously proceed in the direction given by $\mathrm{d}G = \Delta_r G\,\mathrm{d}\xi < 0$, where $\mathrm{d}\xi$ is the change in the extent of reaction, until the reaction quotient equals K_{eq}.

During the reaction, $\mathrm{Red_2}$ is oxidised and gives electrons to $\mathrm{Ox_1}$. The overall reaction can be written as the sum of two 'half-reactions':

$$\mathrm{Ox_1} + ne^- \rightarrow \mathrm{Red_1} \tag{8.4a}$$

$$\mathrm{Red_2} \rightarrow \mathrm{Ox_2} + ne^- \tag{8.4b}$$

It is possible to measure the flux of electrons from $\mathrm{Red_2}$ to $\mathrm{Ox_1}$ and therefore the *rate* $\mathrm{d}\xi/\mathrm{d}t$ of the overall reaction, by placing the species in a a cell made of two compartments connected by a salt bridge. ($\mathrm{Ox_1}$ in one compartment and $\mathrm{Red_2}$ in the other). Place on each side an electrode with which the species can interact. When the two electrodes are connected together, a current flows, while the system evolves towards equilibrium. This current results from a potential difference between the two electrodes, $V = E_2 - E_1$, the value of which can be predicted applying the *Nernst equation* to each of the two electrodes.

$$\boxed{E = E^0 + \frac{RT}{nF}\ln\frac{[\mathrm{Ox}]}{[\mathrm{Red}]}} \tag{8.5}$$

E^0 is the standard reduction potential of the redox couple Ox/Red. $F = 96500\mathrm{C}$ is the Faraday constant. The electrode potential, given by Eqn (8.5), cannot be measured; only the difference between the potentials of two electrodes can.[1] The potential difference between the electrodes in compartments 1 and 2 is:

$$V = E_2 - E_1 = E_2^0 - E_1^0 + \frac{RT}{nF}\ln\frac{[\mathrm{Ox}_2][\mathrm{Red}_1]}{[\mathrm{Red}_2][\mathrm{Ox}_1]} \tag{8.6}$$

The electrons are going to flow from the cell whose electrode potential is the lowest to the other until V and $\Delta_r G$ are zero and the concentrations satisfy

$$K_{\mathrm{eq}} = \frac{[\mathrm{Red}_1]_{\mathrm{eq}}[\mathrm{Ox}_2]_{\mathrm{eq}}}{[\mathrm{Ox}_1]_{\mathrm{eq}}[\mathrm{Red}_2]_{\mathrm{eq}}} = \exp\left(-\frac{nF}{RT}\left(E_2^0 - E_1^0\right)\right) \tag{8.7}$$

This is equivalent to Eqn (8.3), since reduction potentials and free energies are linked by

$$\boxed{\Delta_r G^0 = -nF\left(E_1^0 - E_2^0\right)} \tag{8.8}$$

The Nernst equation can therefore be used to determine the direction in which a redox reaction will proceed spontaneously. Reaction (8.1) will proceed forward significantly (K_{eq} will be large) only if $E_1^0 > E_2^0$. If E_1^0 is 'high', Ox_1 is called a strong oxidant and Red_1 is a weak reductant. But it may not be enough to compare standard reduction potentials, since the sign of $\ln\frac{[\mathrm{Ox}_2][\mathrm{Red}_1]}{[\mathrm{Ox}_1][\mathrm{Red}_2]}$ in Eqn (8.6) can change the sign of V (and $\Delta_r G$) and thus the direction of the current flow. Also remember that thermodynamics predicts the direction but not the rate of the reaction: a 'spontaneous' reaction might not happen because its rate is very small, in which case the equilibrium cannot be reached.

It makes more sense, in our opinion, to call E^0 a standard *reduction* potential than a standard *redox* potential: the sign of $\Delta_r G^0$ relates to a reaction proceeding in a certain direction, and the sign of a standard reduction potential (as defined in Eqn (8.8)) is what it is because the associated half-reaction is a reduction.

The reader should make sure he/she distinguishes between an *electrode* potential (the difference between two electrode potentials is measured using a voltmeter, it has an unit of Volt, and relates to the free energy of the reaction, in units of J/mol) and a standard *reduction* potential (a thermodynamic property which is related by Eqn (8.8) to a standard free energy of reaction). This should not be confused with the *electrochemical potential*, in units of J/mol, which is the equivalent of the partial free energy (or *chemical* potential) of a charged species with an additional term that accounts for the effect of the electric potential ($\mu_i = \mu_i^0 + z_i F \Phi_S$).

Reference Potential and Reference Electrodes

If one is interested in studying only one half-reaction, it is convenient to make sure that the potential difference between the two electrodes reflects the potential of the electrode one is interested in. This can be done by using, in the other compartment, an electrode designed to have a constant potential: this is called a '*reference electrode*'.

1. Note that the usage is to call E the 'electrode potential', but strictly speaking, it is the difference between the electrical potential of the metal electrode and the electrical potential of the solution adjacent to the metal, $\Phi_M - \Phi_S$ (Compton and Banks, 2011). The term Φ_S cancels in an expression of the potential difference between two electrodes, as in Eqn (8.6).

The *Standard Hydrogen Electrode* (SHE) is one of these. It consists of a platinum electrode immersed in a pH = 0 electrolyte under 1 bar of H_2.

A real SHE is rarely practical. Instead, one uses convenient reference electrodes such as:

$Hg/Hg_2Cl_2/KCl$ (Standard Calomel Electrode, SCE, E(SCE) = 241 mV vs. SHE.) This is the most commonly used reference electrode.

$Hg/Hg_2SO_4/K_2SO_4$, used when chloride ions must be avoided. E(SCE) = 615 mV vs. SHE.

Ag/AgCl (E = 200 mV vs. SHE). Silver chloride electrodes are easy to make (East and del Valle, 2000).

Since only potential differences can be measured, reduction potentials can only be reported against a certain reference electrode. The International Union of Pure and Applied Chemistry requires that this primary reference electrode be the SHE. All reduction potentials tabulated in the literature are (or should be) quoted versus the SHE. Therefore, by convention, the potential of the SHE is zero.

The Biological Redox Scale

Because standard conditions (pH 0!) are not suitable for biological reactions, reduction potentials are usually stated for pH 7 and termed $E^{0'}$ or $E_{m,7}$.

Figure 8.1 gives an idea of the range of reduction potentials spanned by biologically important redox cofactors and redox couples. It is important to realise that biological processes occur in water and that the oxidation and reduction of water set the limits of the narrow range of relevant potentials. The reaction of a biological molecule under extreme (very oxidising or very reducing) conditions may be interesting from the point of view of chemistry but physiologically irrelevant.

Influence of Coupled Reactions (e.g. Protonation or Ligand Binding) on Reduction Potentials

Redox reactions can be coupled to other chemical equilibria such as ligand binding (e.g. protons, substrate, inhibitor, etc.) or conformational changes.

A very common (and physiologically important) coupled reaction is protonation, as represented in the square scheme in Figure 8.2. K_{Ox} and K_{Red} are the acidity constants for Ox and Red. Utilising the principle of thermodynamic cycles (the sum of $\Delta_r G^0$ values round the square is zero), these acidity constants can be linked to the reduction potentials of the protonated and unprotonated redox couples. Note that potentials alone cannot be summed; they must be scaled by n.

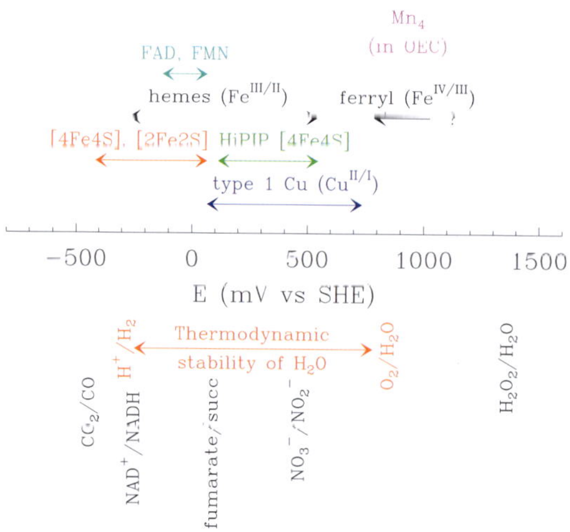

FIGURE 8.1 The biological redox scale at pH 7. 'OEC' stands for the 'oxygen evolving centre' of Photosystem II.

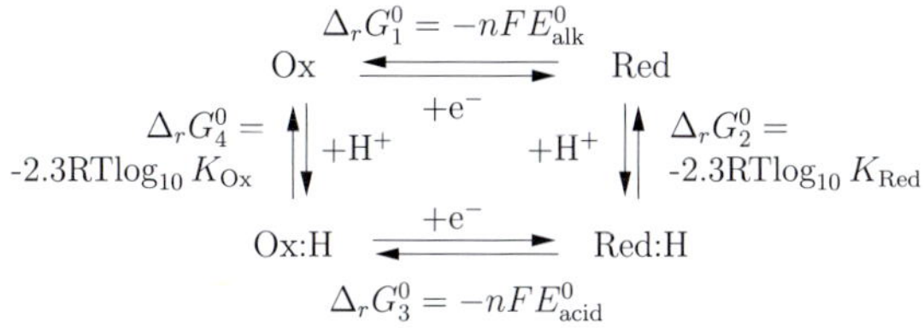

FIGURE 8.2 Square scheme for a protonation reaction coupled to a redox process.

For an n-electron, one-proton process, the whole pH-dependence of the reduction potential is given by

$$E^{0'}([H^+]) = E^0_{alk} + \frac{2.3RT}{nF}\log_{10}\left(\frac{1+\frac{[H^+]}{K_{Red}}}{1+\frac{[H^+]}{K_{Ox}}}\right) \tag{8.9}$$

This equation is demonstrated in the appendix A2. Usually, Red is a better base than Ox, so it has a higher pK_a, i.e. pK_{Ox} < pK_{Red}.

For pH < pK_{Ox}, both Ox and Red are protonated.

$$\text{Ox : H} + ne^- \rightarrow \text{Red : H} \tag{8.10}$$

The reduction potential is pH independent, and equals E^0_{acid} (Figure 8.2).

For pH between pK_{Ox} and pK_{Red}, Ox is not protonated but Red is.

$$\text{Ox} + ne^- + H^+ \rightarrow \text{Red : H} \tag{8.11}$$

The reduction potential is pH-dependent: it decreases by $2.3RT/nF$ Volts per pH unit (−59/n mV/pH at 25 °C). This is simply understood as follows: the reduction potential increases as the concentration of protons increases because protonation makes reduction *easier* by consuming the reduced species.

For pH > pK_{Red}, neither Ox nor Red are protonated, the redox process is

$$\text{Ox} + ne^- \rightarrow \text{Red} \tag{8.12}$$

The reduction potential is pH independent and equals E^0_{alk} (Figure 8.2).

Redox-linked protonations are conveniently represented by a Pourbaix diagram, a plot of $E^{0'}$ as a function of pH, as schematised in Figure 8.3A.

In general, for a redox process involving n electrons and m protons, the maximal pH-dependence is

$$-\frac{2.3RT}{F}\frac{m}{n}\ \text{V/pH unit} \tag{8.13}$$

As an example, Figure 8.3B (Jeuken et al., 2002) shows the dependence on pH of the reduction potential of the type-I copper site of amicyanin, determined using protein film voltammetry (PFV). The data can be analysed in terms of the ET being coupled to the protonation of a single group with pK_{Red} = 6.3 and p$K_{Ox} \leq 3.2$. Protonation occurs on one of the two histidine ligands of the copper ion.

It is essential to understand that these considerations apply for any ligand that binds a redox centre: provided that the dissociation constants from the reduced and oxidised forms are different, the reduction potential depends on the concentration of ligand, and the dissociation constants can be measured by determining how the reduction potential depends on the concentration of ligand. The latter may be the proton, the substrate/product or inhibitor of an enzyme, or even the apoprotein.

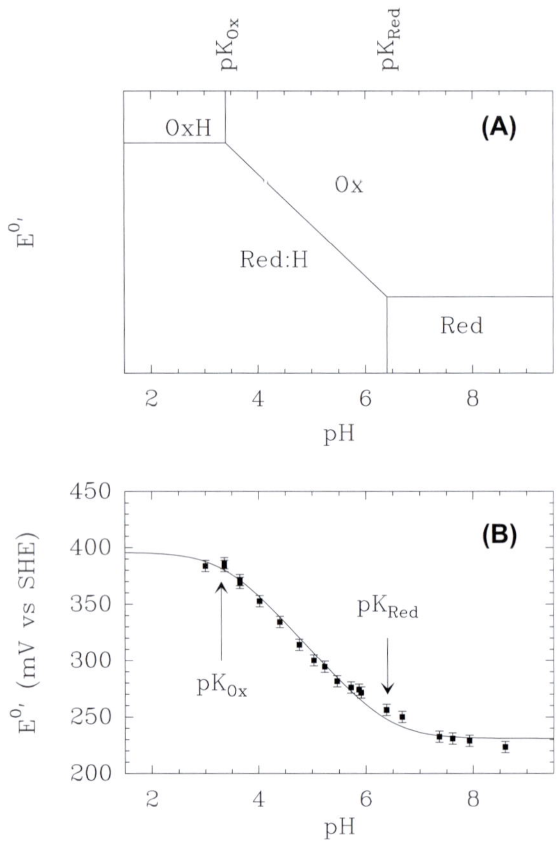

FIGURE 8.3 (A) Schematic Pourbaix diagram for the $1e^-$:$1H^+$ reaction of amicyanin, a type-I blue copper protein. (B) Reduction potential vs. pH for amicyanin from *Paracoccus versatus* at 2 °C. The line is a best fit to Eqn (8.9). The values of pK_{Red} and pK_{Ox} can easily be measured from the fit to Eqn (8.9), with $n = 1$. From ref. Jeuken et al. (2002).

Figure 8.4 illustrates the succinate-concentration-dependence of the reduction potential of the active site flavin in *Escherichia* fumarate reductase, determined using experiments such as those in Figure 8.22A. In this case, the reduction potential decreases as the concentration of succinate is raised (from right to left in this figure) because binding of succinate to the oxidised form of the enzyme is tighter than to the reduced form ($pK_{Ox} > pK_{Red}$).

References Zu et al. (2001); Butt et al. (1993), (1997) and Fawcett et al. (1998) provide examples of the effect of protonation or binding of metal or exogenous thiolate on the reduction potential of FeS clusters.

ET Kinetics

Electrons are transferred between the redox centres of redox proteins and enzymes involved in respiration and photosynthesis, and along chains of redox centres within certain enzymes. We first define the parameters that determine the *rates* of these *homogeneous* ET events before discussing interfacial ET.

Let us consider a certain ET step, from a donor D to an acceptor A.

$$D_{red} + A_{ox} \underset{k_{A \to D}}{\overset{k_{D \to A}}{\rightleftharpoons}} D_{ox} + A_{red} \tag{8.14}$$

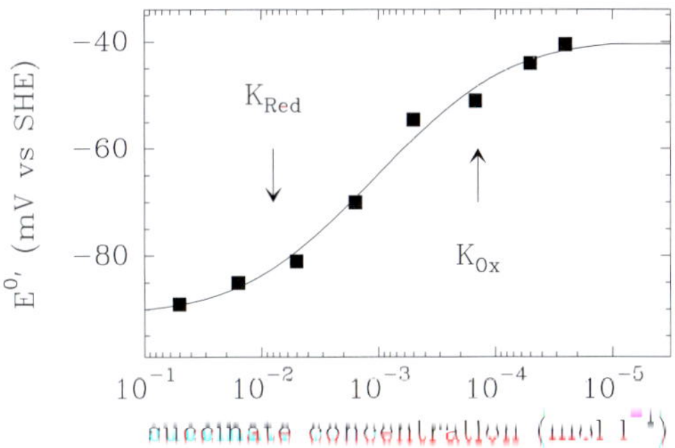

FIGURE 8.4 Pourbaix-like diagram for the change in the reduction potential of the flavin cofactor of *E. coli* fumarate reductase as a function of succinate concentration (increasing from right to left) at 20 °C, pH 7. Line is best fit to Eqn (8.9), with $n = 2$, but with [succinate] instead of $[H^+]$. From ref. Léger et al. (2001).

Thermodynamics predicts the ratio $k_{D\rightarrow A}/k_{A\rightarrow D}$ (Eqn (8.7)) but not the values of the two rate constants. The rate of ET from D to A depends on the reduction potentials of D and A, but also on other parameters. This was established in the 1960s by Rudolph Marcus, who developed a model based on a molecular description of ET between small molecules in solution and received the 1992 Nobel prize in chemistry for his theory. He showed that this process requires the formation of a transient complex in which the kinetics of the ET step can be described by an equation of the form

$$k_{D\rightarrow A} = C\exp\left(-\frac{(\Delta_r G^0 + \lambda)^2}{4\lambda RT}\right) \tag{8.15}$$

- $\Delta_r G^0$ is the standard free energy of the reaction, which is related to the standard reduction potentials of the donor D and the acceptor A according to $\Delta_r G^0 = F(E_D^0 - E_A^0)$ (cf. Eqn (8.8)).
- The parameter λ, called 'reorganisation energy,' is all the greater that large molecular rearrangements accompany the transfer (both the geometries of the molecules that are oxidised or reduced, and the polarisation of the surrounding solvent are considered). Biological ETs usually take place in the 'normal region' ($|\Delta_r G^0| < \lambda$), where the rate constant is increased when ΔE^0 increases and λ decreases, but the 'inverted region' plays a very important role in photosynthetic ET.
- The expression of the preexponential factor *C* depends on the strength of the electronic coupling between the acceptor and the donor. If it is strong enough ('adiabatic' transfer), *C* simply equates kT/h, as given by the classical transition state theory. When it is weak (this is so for long-distance, 'nonadiabatic' ET), *C* depends on the overlap of the molecular wave functions of D and A and therefore on the nature of the redox centres, on their distance and on the intervening medium. An exponential decrease of *C* with distance is observed (Gray and Winkler, 2005). In the literature, nonadiabatic transfers are often referred to as electron tunnelling processes.

The relation between thermodynamics and kinetics is understood by calculating an equilibrium constant from the ratio $k_{D\rightarrow A}/k_{A\rightarrow D}$ (then compare with Eqn (8.3)):

$$\frac{k_{D\rightarrow A}}{k_{A\rightarrow D}} = \frac{C\exp\left(-\frac{(\Delta_r G^0 + \lambda)^2}{4\lambda RT}\right)}{C\exp\left(-\frac{(-\Delta_r G^0 + \lambda)^2}{4\lambda RT}\right)} = \exp^{-\Delta_r G^0/RT} \tag{8.16}$$

Now, consider the interfacial ET between a molecule and a *metallic electrode* (the case of semiconducting electrodes is treated, e.g. in ref. Bard and Faulkner (2004)).

$$\text{Red} \underset{k_{red}}{\overset{k_{ox}}{\rightleftharpoons}} \text{Ox} + e^-(\text{electrode}) \tag{8.17}$$

The equation that gives the rate of ET is a complex function of the reduction potential of the molecule $E^{0'}$, the electrode potential E and the reorganisation energy λ (Savéant, 2006; Léger and Bertrand, 2008) but a simplified rate equation, known as the 'Butler Volmer equation,' predicts that the rates of oxidation and reduction are independent on λ and exponentially increase and decrease (respectively) as the electrode potential increases:

$$k_{ox} = k_0 \exp^{\frac{F}{2RT}(E-E^{0'})} \tag{8.18a}$$

$$k_{red} = k_0 \exp^{-\frac{F}{2RT}(E-E^{0'})} \tag{8.18b}$$

The preexponential factor k_0 depends on the coupling between the electrode and the redox molecule. The greater k_0 the faster the electron exchange between the electrode and the redox molecule; of course, this parameter has no physiological relevance. k_0 is often called the rate of electron transfer, but it is important to understand that rates of ET depend on electrode potential, and the rate of ET equates k_0 only when there is no driving force, at $E = E^0$ (in which case $k_{ox} = k_{red} = k_0$).

The relation between thermodynamics and kinetics is understood by calculating the ratio k_{ox}/k_{red} (then compare with Eqn (8.5)):

$$\frac{k_{ox}}{k_{red}} = \frac{k_0 \exp^{\frac{F}{2RT}(E-E^{0'})}}{k_0 \exp^{-\frac{F}{2RT}(E-E^{0'})}} = \exp^{\frac{F}{RT}(E-E^{0'})} \tag{8.19}$$

The more sophisticated Marcus–Hush model of interfacial ET takes into account the distribution of energy states in the metal; it gives Eqn (8.18) justification when the overpotential is small (at $E \approx E^0$), but in contrast, it predicts that the rate asymptotically approaches a plateau value k_{max} when the overpotential is greater than λ. An inverted region does not occur for an electrode reaction at a metal electrode. For further elaboration, see sections 1.4.2 in ref. Savéant (2006) or 3.6.2 in ref. Newman and Thomas-Alyea (2004).

Since, according to the Marcus–Hush model, the rate of interfacial ET depends on E and λ, the dependence of the ET rate on E could be used to determine λ, which is biologically a relevant parameter. However, a complication arises from the fact that interfacial ET may be limited ('gated') by a process that is distinct from ET. In that case, the plateau of the apparent ET rate may be incorrectly interpreted as the ET rate is reaching the maximal value predicted by Marcus–Hush theory, and the reorganisation energy may be greatly underestimated (for a discussion, see ref. Jeuken et al. (2002) and refs therein).

Kinetics of Proton-Coupled ET: Stepwise versus Concerted Mechanisms

When a redox process is coupled to proton transfer (judging from the pH-dependence of the corresponding reduction potential), one can discern between *stepwise* mechanism (whereby ET either precedes or follows proton transfer) and *concerted* mechanism (whereby both transfers occur simultaneously). The mechanisms can sometimes be discriminated on the basis of kinetic isotope effects or from the dependence of rates on T, buffer, pH and/or ΔG (or E). Electrochemical investigations can be highly informative in this respect (Costentin et al., 2010; Anxolabéhère-Mallart et al., 2011). We will discuss in section 'Fast-Scan Voltammetry to Determine the Rates of Coupled Reactions' simple electrochemical experiments that establish the stepwise mechanism of proton transfer/electron transfer to a buried FeS cluster. Various examples of biologically relevant coupled-proton−ETs are discussed in ref. Costentin et al. (2010).

ELECTROCHEMISTRY UNDER EQUILIBRIUM CONDITIONS: POTENTIOMETRIC TITRATIONS

The reduction potential of a redox couple is determined by recording the ratio [Ox]/[Red], after allowing the system to equilibrate with the electrode at different potentials.

The potential is usually varied by repeatedly adding small amounts of an oxidant or a reductant. Unlike small molecules, protein redox centres do not generally react rapidly with the measuring electrode and equilibrium is not established quickly. To overcome this problem, small redox agents called mediators are added to the solution to transport electrons between the active site and the electrode. For best results, these should have reduction potentials close to that of the active site being studied, mixtures of mediators are often employed to cover a wide range. A short list of mediators and their reduction potentials is given in Table 8.1. See also ref. Wardman (1989).

The ratios [Ox]/[Red] are typically determined by examining the optical or EPR spectra of the sample. Very often, the concentration of Ox or Red is measured as a function of the electrode potential, and the data are fitted with

$$[\mathrm{Red}] \propto \frac{1}{1+\exp\left(\frac{nF}{RT}\left(E-E^{0'}\right)\right)} \tag{8.20a}$$

$$[\mathrm{Ox}] \propto \frac{\exp\left(\frac{nF}{RT}\left(E-E^{0'}\right)\right)}{1+\exp\left(\frac{nF}{RT}\left(E-E^{0'}\right)\right)} \tag{8.20b}$$

Figure 8.5 is the result of a potentiometric titration of the 'blue' (type-I) copper site in an azurin mutant, followed by UV−Vis spectroscopy. The oxidised copper site absorbs at ≈600 nm due to a Cys-S to Cu ligand to metal-charge-transfer, and the intensity of this band is therefore proportional to the concentration of oxidised copper site.

Performing such experiment under fully anaerobic conditions require a glove-box … or skills. Dutton described the glassware that can be used on the bench for the anaerobic potentiometric preparation of samples to be examined by EPR or UV−Vis spectroscopy (Dutton, 1971, 1978).

TABLE 8.1 Standard Reduction Potentials at pH 7, 25 °C, of Common Mediators

Mediator	$E^{0'}$ (mV vs SHE)
Benzyl viologen	−360
Lapachol	−170
Methylene blue	10
Ferricyanide ($K_3Fe(CN)_6$)	360
Hexachloroiridate (Na_2IrCl_6)	870

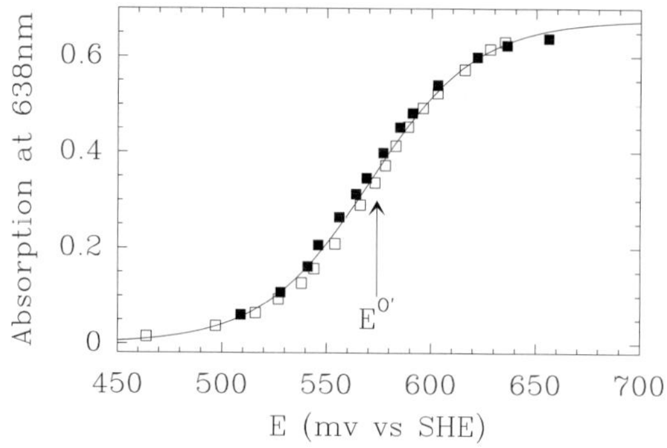

FIGURE 8.5 Potentiometric titration of *Pseudomonas aeruginosa* His117Gly azurin, a type-I blue copper protein (0.1 mM protein, 20 mM MES, pH 6, 33 mM Na_2SO_4, 1 M NaCl, 20 °C.) The different symbols correspond to step-wise reduction and step-wise reoxidation. Plain line is best fit to Eqn (8.20b) with $n = 1$. The adsorption at 638 nm is characteristic of the oxidised (blue) form of the copper centre. Mediators: $K_3Fe(CN)_6$, Na_2IrCl_6, 1,2-ferrocene dicarboxyl acid. From ref. Jeuken et al. (2000).

DYNAMIC ELECTROCHEMISTRY

Distinction between Equilibrium and Dynamic Electrochemistry

In potentiometry experiments, the measurement of the electrode potential is carried out under equilibrium conditions: the stepwise addition of titrant makes the concentrations change, but when equilibrium is reached, the rates of oxidation and reduction exactly cancel each other and there is no net transformation. The reduction or oxidation is detected by spectroscopy, and therefore, this approach requires that the redox centre has a distinct spectroscopic signature.

A completely different approach consists in forcing the electrode potential to take a value that is different from the equilibrium potential. In that case, the system may evolve towards equilibrium by taking electrons from (of giving electrons to) the electrode. This reduction (resp. oxidation) is detected as a current (in units of Amperes, that is Coulomb/second) that measures the amount of electric charge passing the electrode per unit time. Therefore, the current is proportional to the *rate* of reduction or oxidation.

Electrodes for Electron Transfer to/from Proteins

To be successful, electrodes must exchange electrons quickly with the proteins and preserve their native properties. These electrodes may resemble natural environments or reaction partners for the protein.

ET can be achieved if the redox centre is exposed at the protein surface (or not too deeply buried). In the case of enzymes whose active site is buried in the protein, a favourable situation occurs when it is 'wired' to the surface by a chain of redox cofactors; having one of these centres exchanging electrons with the electrode is enough for achieving an electric connection of the active site. Direct ET to enzymes was first reported in the late 1970s (Tarasevich et al., 1979; Yaropolov et al., 1984) and is now very common, but not all redox enzymes can be electrically connected to electrodes.

Protein/electrode interactions may be tailored to be weak or strong. Weak interactions might ideally give rise to diffusion-controlled electrochemistry (see 'Diffusion-Controlled Voltammetry' section), whereas with strong interactions, the experiment may address just a small sample ('film') of protein molecules adsorbed on the electrode (see 'Voltammetry of Adsorbed Proteins: PFV' and 'Catalytic Voltammetry and Chronoamperometry with Adsorbed Redox Enzymes' sections).

Electrode surfaces for which protein electrochemistry is commonly observed are listed below:

- Metal (Au, Pt, Ag) surfaces onto which a monolayer of adsorbate is self-assembled ('Self-Assembled Monolayers' or SAMs (Love et al., 2005)). The adsorbate is a bifunctional molecule of the type X-(CH_2)$_n$-Y, where X is a substituant that anchors the molecule on the metal electrode surface (e.g. a thiol) and Y is a functional group that interacts with the protein (typically carboxyl for cytochromes *c*, or amino for acidic proteins such as plastocyanin or ferredoxins.)
- Pyrolytic graphite edge (Blanford and Armstrong, 2006) or basal plane electrodes provide hydrophilic or hydrophobic interactions, respectively. The former is sometimes used with co-adsorbates (aminocyclitols, polymyxin, polylysine), which probably form cross-linkages between the protein and the electrode surface.
- Graphite or carbon nanotubes can be functionalised by reducing a diazonium salt (Allongue et al., 1997) to expose aromatic functionalities (Blanford et al., 2008) or amino groups (Rudiger et al., 2005; Alonso-Lomillo et al., 2007) that interact favourably with hydrophobic or carboxylate-rich patches on the protein surface. Carbodiimide coupling can then be used to form an amide bond between amino groups on the electrode surface and protein carboxylates (or vice versa) (Baffert et al., 2012).
- Various electrode materials covered which either a film of nonbiological surfactant (e.g. DDAB) or layers of polyions have been used to incorporate large membrane-bound proteins, but catalytic activity was generally greatly impaired. Suitable electrodes for integral membrane enzymes are described in refs Jeuken (2009), Gutiérrez-Sánchez et al. (2011).

When the protein that is studied is an enzyme, proof that it is not denatured on the electrode is that it can still catalyse the tranformation of its physiological substrate *at a reasonable rate and in a range of electrode potential that is consistent with what we know about the catalytic cycle and the redox properties of the cofactors*. Checking the effects of known specific inhibitors can also be useful. Upon starting a new project, and before embarking on detailed electrochemical studies, it is essential to make sure that the catalytic properties of the adsorbed enzyme bear some resemblance to those determined in biochemical experiments. In this chapter, we illustrate this by also discussing the results of conventional experiments, which confirm certain unexpected results obtained in electrochemical investigations of redox enzymes.

Electrochemical Equipment

The experiment is carried out using an electrochemical analyser in conjunction with the cell. The cell consists of three electrodes. The reference electrode is often contained in a side arm linked to the main compartment by a capillary tip, called a Luggin (after the glassblower who invented it). The tip is positioned close to the working electrode. To avoid passing current through the reference electrode (this would change its potential and also damage it), a third electrode, called auxiliary or counter electrode, is used. The working electrode can also be rotated (see 'Diffusion-Controlled Voltammetry at Rotating Electrodes' and 'Catalytic Voltammetry and Chronoamperometry with Adsorbed Redox Enzymes' sections) to control mass transport of solution species.

The analyser measures the current registered in response to the potential that is applied. In general, the potential of the working electrode (versus the reference electrode) is modulated (e.g. in a linear sweep) and the current flowing between the working electrode and the counter electrode is recorded. Since the electrode potential is swept forward and back, the technique is called 'cyclic voltammetry'. The scan rate ν (in units of V/s) is a very important parameter that determines the time scale of the experiment and, therefore, the time constant of the processes that can be resolved (Savéant, 2006). A voltammogram is a plot of current against electrode potential. Alternatively, in an experiment called 'chronoamperometry', the electrode potential is held at a fixed value (or sometimes stepwise changed), and the current is recorded as a function of time.

Vocab and Conventions

A cathodic process is a reduction; the cathode is the electrode on which the reduction occurs. An anodic process is an oxidation, occurring at the anode.

In Europe, a cathodic current is counted as negative and an oxidation results in a positive current. Americans and software manufacturers often use the opposite convention for the sign of the current (see e.g. Figure 8.7).

The Capacitive Current

The measured current is usually the sum of a Faradaic current (which reveals the redox transformations of molecules that come sufficiently close to the electrode) and a capacitive current i_c, which does not involve the passage of electrons across the electrode–solution interface. The capacitive current (or 'charging current') arises as a consequence of the variation of the electrode potential, and it is proportional to the electrode surface A:

$$i_c = C\frac{\mathrm{d}E}{\mathrm{d}t} = Cv \tag{8.21}$$

where C, the capacitance of the electrode/electrolyte interface, is proportional to A.

In a voltammetric experiment, the magnitude of the capacitive current is therefore proportional to the scan rate v. It is positive if E is increasing, and negative if the electrode potential is swept down.

There is no capacitive contribution to the current if the potential is constant, but potential steps (in chronoamperometry experiments) result in current transients that are approximately exponential ($i_c < 0$ if the potential is stepped down, $i_c > 0$ if the potential is stepped up). This current usually decays in less than a few seconds.

The capacitive current must be subtracted from the total current to obtain the Faradaic contribution. It can sometimes be determined from a control experiment where there is no Faradaic current, or extrapolated from the part of the signal where there is no Faradaic contribution (see e.g. Figure 8.10) (Fourmond et al., 2009b).

DIFFUSION-CONTROLLED VOLTAMMETRY

Diffusion-Controlled Voltammetry at Stationary Electrodes

We consider a solution containing only the reduced form of a soluble electroactive species, and the potential is swept linearly in time, as shown in Figure 8.6A, starting from a low potential. The current response as a function of time is plotted in Figure 8.6B, and the cyclic voltammogram (current against potential) in Figure 8.6C.

While the potential is lower than $E^{0'}$, no oxidation occurs and no current is measured [see (a) in Figure 8.6]. This is because the rate of oxidation $k_{\mathrm{ox}} = k_0 \exp^{\frac{F}{2RT}(E-E^{0'})}$ is much lower than the rate of reduction $k_{\mathrm{red}} = k_0 \exp^{-\frac{F}{2RT}(E-E^{0'})}$ (see 'ET Kinetics' section). When the electrode potential approaches $E^{0'}$, Red starts being oxidised into Ox, giving electrons to the electrode. This is measured as a (positive) current which increases as E and the rate of oxidation increase (b) However, the electrode oxidises only species adjacent to it, and the interface is soon depleted. The current reaches a maximum before it starts to decrease (c) It tends to zero like $t^{-1/2}$ (noting that E changes in proportion to time t). This decrease in current reveals that the size of the diffusion layer (i.e. the zone of the solution adjacent to the electrode, where the concentration of species differ from that in the bulk) increases by diffusion like the square root of time and the concentration gradient that drives the diffusion of Red from the bulk of the solution to the interface and therefore decreases like $t^{-1/2}$.

While a positive current is being measured, Ox produced by the reaction accumulates near the electrode and diffuses slowly towards the bulk.

After the scan is reversed, (d), the current is still positive and decreasing: Red species are still being oxidised since the electrode potential is above the reduction potential. Near the reduction potential, the Ox species which

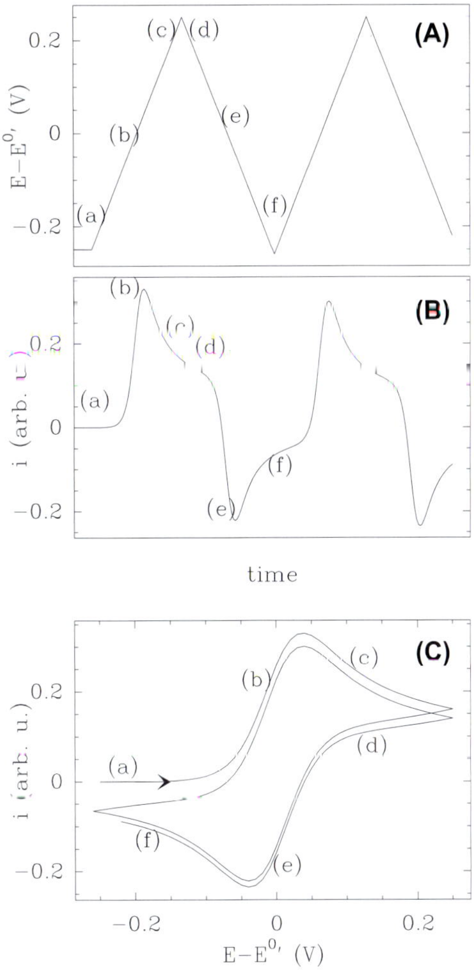

FIGURE 8.6 Cyclic voltammetry for a *redox species free to diffuse in solution*, to and from a stationary electrode. In a typical voltammetric experiment, the electrode potential is swept linearly in time (A), and the current recorded as a function of time (B) A convenient and usual way of displaying the results is to plot the current against potential (C) The labels (a)–(f) are referred to in the text. 'arb. u.' means 'arbitrary units'.

have accumulated are now being reduced and a negative current is observed (e) until the concentration of Ox near the interface drops down (f), and so does the magnitude of the current.

This results in a peak-like response in both directions. The modelling of voltammograms is complex (chapter 6 in Bard and Faulkner (2004)), because the current response depends on the free diffusion of the species in solution, and the differential equations coupling the reaction and diffusion are, in general, difficult to solve.

For an n-electron reaction, the separation between cathodic (reduction) and anodic (oxidation) peaks is given by

$$\Delta E_{\mathrm{p}} = E_{\mathrm{p,a}} - E_{\mathrm{p,c}} = 2.218\frac{RT}{nF} = \frac{57}{n}\,\mathrm{mV\ at\ 298\ K} \tag{8.22}$$

The peak current is proportional to the bulk concentration of species C and to the square root of the potential scan rate v:

$$i_{\mathrm{p}} = 2.69\,10^5 n^{3/2} A D^{1/2} \times C\sqrt{v} \tag{8.23}$$

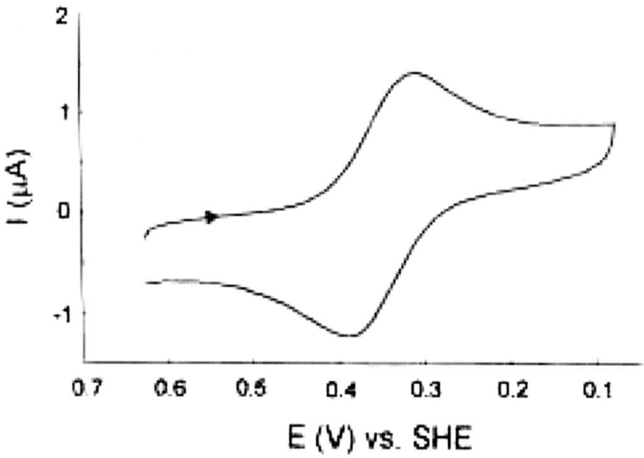

FIGURE 8.7 Data for diffusion-limited voltammetry of a redox protein, the cytochrome c_2 from *Rps. palustris*. $C = 0.2$ mM, cell volume 0.5 mL. From ref. Battistuzzi et al. (1997). Mind the axis! The voltammogram is plotted with the American convention: the high electrode potentials are on the left, and the current is positive for a reduction. (C) American Chemical Society 1997. Reprinted from ref. Battistuzzi et al. (1997) with permission.

This is the Randles–Sevcik equation. A is the electrode surface, D is a diffusion coefficient. A linear plot of i_p against $\sqrt{v}$ is the criterion used to identify when the redox species are diffusing from the bulk to the electrode.

A system that conforms to the above criteria is said to be *reversible* and *diffusion controlled*, and the reduction potential is obtained from the average of the cathodic and anodic peak potentials.

$$E^{0'} \approx \frac{E_{p,a} + E_{p,c}}{2} \tag{8.24}$$

Figure 8.7 shows a voltammogram for the reversible oxidation and reduction of a cytochrome. Note that the concentration of the protein sample must be high. For example, the experiment in Figure 8.7 used 100 nmol of cytochrome.

Deviation from this ideal behaviour might arise when interfacial ET is slow (in which case, the ideal peaks broaden and tend to separate) or when one of the redox species is irreversibly transformed on the voltammetric time scale (in which case, the signal might become asymmetrical) (see chapter 2 in ref. Savéant (2006)) for the effect of follow-up reactions on the voltammetry of diffusive species and see below for an example with an adsorbed protein).

Diffusion-Controlled Voltammetry at Rotating Electrodes

The peak shape of the diffusion-limited voltammogram at a macroelectrode is due to the depletion of electroactive species near the electrode surface as they are consumed by the redox reaction.

There are many electrochemical techniques in which the solution moves with respect to the electrode. In the most popular configuration, the electrode (called a 'rotating disc electrode') is rotated along its axis in the solution. This introduces a *convective* movement of the solution which increases the efficiency of the transport of species from the bulk towards the electrode. Because the depletion layer can no longer spread in the solution, the current reaches a limiting value i_{lim} at high driving force, Figure 8.8.

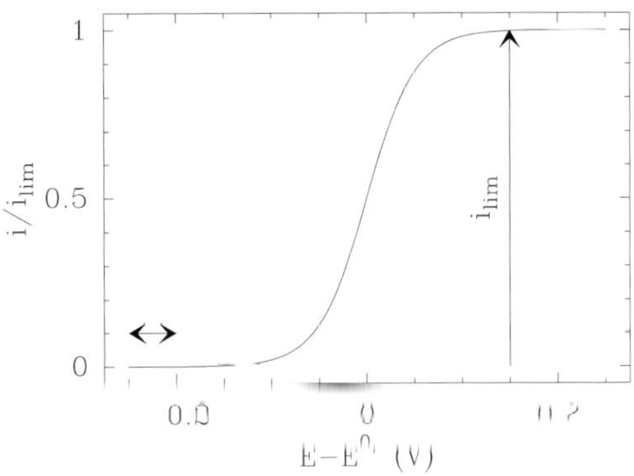

FIGURE 8.8 Sigmoidal wave obtained in a cyclic voltammetry of an *electroactive species in solution* at a rotating disc electrode.

The Levich equation predicts that the limiting current is proportional to the concentration of electroactive species C and to the square root of the electrode rotation rate ω:

$$i_{\text{lim}} = 0.620nFA\nu_s^{-1/6}D^{2/3} \times C\sqrt{\omega} \tag{8.25}$$

In this equation, ν_s is the kinematic viscosity of the solution. (The kinematic viscosity is the ratio of the viscosity over the density. For example, the viscosity of pure water at 20 °C is 10^3 μPa·s, and its density ≈ 1 g/cm^3; this gives a kinematic viscosity $\nu_s \approx 10^{-2}$ cm^2/s.) A plot of i_{lim}^{-1} against $\omega^{-1/2}$ is called a Koutecky–Levich plot. The reduction potential $E^{0'}$ is simply given by the half-wave potential $E_{1/2}$, the potential at which the current reaches half its limiting value.

$$\boxed{E^{0'} = E_{1/2}} \tag{8.26}$$

The scan rate ν and direction do not enter the measurement if ν is small. When the current depends on electrode potential but is independent of time, the voltammogram is said to be at *steady state*.

This configuration is not practical for measuring the reduction potential of redox proteins, because rotating the electrode in the solution requires that the volume of the electrochemical cell be large, but it is important to understand the difference between this experiment and that shown in Figure 8.22A, since they give similar electrochemical responses for completely different reasons.

The above considerations apply only to 'macro'electrodes (i.e. when the diameter of the electrode is larger than the typical size of the diffusion layer). With a 'micro'electrode, whose typical size is of the order of a few micrometres, the voltammogram may have a sigmoidal shape even in the absence of convection (Section 5.2.2 in Bard and Faulkner (2004)). Reference Kudera et al. (2000) shows cyclic voltammograms of amicyanin from *P. denitrificans* at a 3-μm gold micro electrode. Microelectrodes have also been used in the context of cell biology; their sizes make them suitable to detect electroactive species released by a single cell. See e.g. ref. Amatore et al. (2008) for a recent review.

VOLTAMMETRY OF ADSORBED PROTEINS: PROTEIN FILM VOLTAMMETRY (PFV)

When the protein is immobilised on the electrode surface, diffusion is eliminated and much greater thermodynamic and kinetic resolutions can be obtained with extremely small sample quantities. This approach was developed by F. Armstrong (now in Oxford) in the 1970s with small redox proteins and, since the beginning of the 1980s, with large redox enzymes. Over the last years, this technique was used to study all sorts of aspects of the mechanism of redox proteins and enzymes (Léger and Bertrand, 2008): proton transfer (Chen et al., 2000), inter and intramolecular ET (Dementin et al., 2006; Léger et al., 2006), diffusion along substrate channels

(Leroux et al., 2008; Liebgott et al., 2010), etc. The rest of this article will focus on the principle and applications of PFV.

'Noncatalytic' Voltammetry at Slow Scan Rates to Measure Reduction Potentials

Figure 8.9 shows the ideal shape of a cyclic voltammogram when the redox species is adsorbed onto an electrode (Bard and Faulkner, 2004).

Starting again at an electrode potential lower than E^0, the redox centres are fully reduced [see (A) in Figure 8.9]. Sweeping the potential towards high values, the protein starts being oxidised when the electrode potential approaches the reduction potential (B); a positive current is then measured, which drops down to zero at high potential when all the adsorbed protein molecules have been oxidised (C). On the reverse scan, a reductive

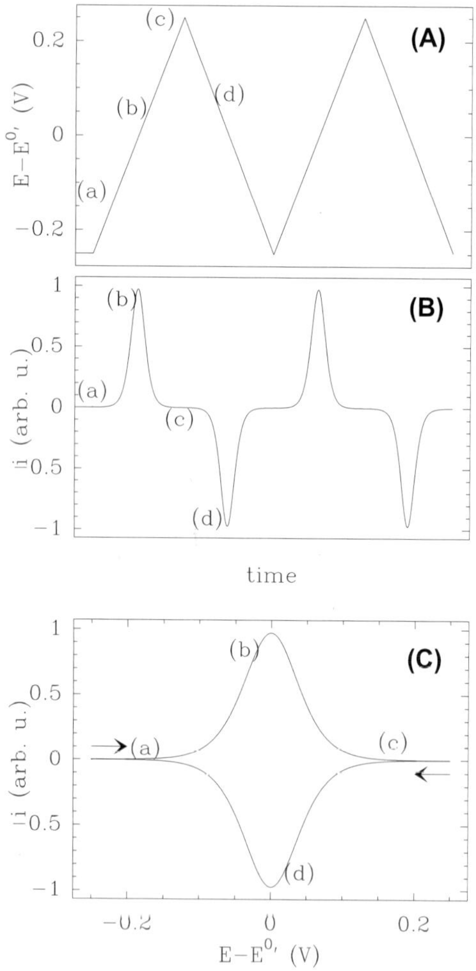

FIGURE 8.9 Cyclic voltammetry for a *redox species adsorbed on an electrode surface*. Rotation of the electrode should make no difference. In a typical voltammetric experiment, the electrode potential is swept linearly in time (A), and the current recorded as a function of time (B) Panel C shows the cyclic voltammogram.

(negative) current is observed when the electrode potential matches the reduction potential of the protein (d) until the entire sample has been reduced and the current vanishes.

For an ideal, reversible system, the signal consists of symmetrical oxidation and reduction peaks centred at the reduction potential $E^{0'}$.

$$\boxed{E^{0'} = E_{\mathrm{p,c}} = E_{\mathrm{p,a}}} \tag{8.27}$$

The area under the peak ($\mathscr{A}$, in units of VA) gives the charge passed for that redox couple (nF electrons per mole of adsorbed centres, ΓA, A is the electrode surface). It should be the same for the oxidative and for the reductive peaks.

$$\mathscr{A} = nFA\Gamma\nu \tag{8.28}$$

The peak current is proportional to the scan rate ν, to the surface concentration of electroactive species Γ and to the square of n. Therefore, the electroactive coverage must be high enough for a current to be observed. Typically, a coverage higher than 0.5 pmol/cm^2 will suffice.

$$i_p = \frac{n^2F^2A\Gamma\nu}{4RT} \tag{8.29}$$

A linear plot of i_{p} against ν proves that the redox species are adsorbed onto the electrode.

The peak width at half-height, δ, is

$$\boxed{\delta \approx 3.53\frac{RT}{nF}} \tag{8.30}$$

(91/n mV at 25 °C). Note that the expected dependence on temperature is not observed in experiments (McEvoy and Armstrong, 1999; Chobot et al., 2007).

Cooperative two-electron transfers (n=2) gives signals with up to four times the height and half the width of one ETs, they are, therefore, more easily distinguished.

Figure 8.10 shows a noncatalytic voltammogram that makes it possible to determine the redox potential of the copper site of azurin. In experiments, there is a capacitive contribution resulting from ‘electrode charging’ (see ‘The Capacitive Current’ section). The dashed line shows the interpolated capacitive current, which has to be subtracted to obtain the Faradaic current alone.

Figure 8.11 shows the voltammetry of the 7Fe ferredoxin of *Sulfolobus acidocaldarius*, which contains one [3Fe4S] cluster, with redox transitions $[3Fe4S]^{+/0}$ and $[3Fe4S]^{0/2-}$, and one [4Fe4S] cluster with a 2+/+ redox

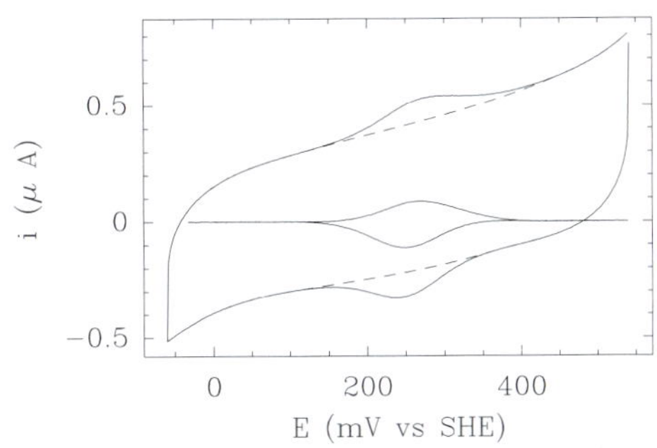

FIGURE 8.10 Cyclic voltammogram for *Pseudomonas aeruginosa* azurin adsorbed at a pyrolytic graphite electrode. The dashed line is the baseline and the inset shows the baseline subtracted current (the Faradaic current). 0 °C, pH 8.5. ν = 20 mV/s. $\Gamma A \approx 5.5$ pmol.

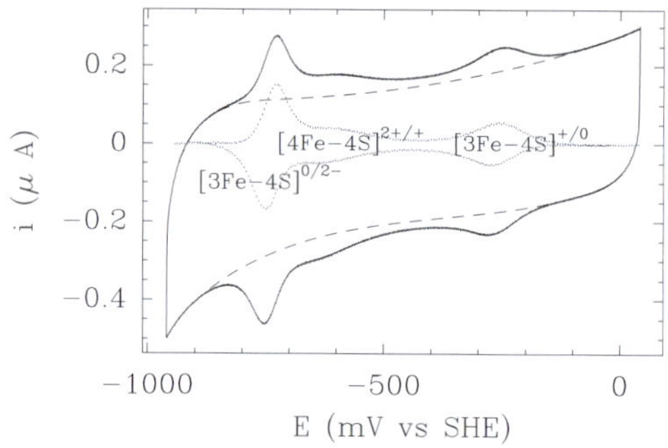

FIGURE 8.11 Cyclic voltammogram for *S. acidocaldarius* 7Fe ferredoxin adsorbed at a pyrolytic graphite electrode. The dashed line is the charging current. 0 °C, pH 8.5, $v = 20$ mV/s. $\Gamma A \approx 3.5$ pmol. (McEvoy and Armstrong, 1999).

transition. When different centres are present in a protein, the different redox transitions appear as multiple peaks, the areas of which reveal the stoichiometry of the redox processes. For example, the area under the low-potential $[3Fe4S]^{0/2-}$ peak is twice as much as that of the high-potential $[3Fe4S]^{+/0}$ peak. This figure illustrates also the fact that two-electron redox processes give prominent signals (cf. Equations (8.29) and (8.30)).

The fumarate reductase from *E. coli* contains 3 FeS clusters and one flavin cofactor. The two-electron signal associated with the flavin is easily distinguished from the three one-electron peaks due to the FeS clusters (Figure 8.12). The data clearly show that the reduction potential of the FAD is strongly pH-dependent, as expected for a reduction process coupled to protonation (cf. see 'Influence of Coupled Reactions on Reduction Potentials' section). The fact that the peak shifts about −60 mV between pH 7 and 9 shows that flavin reduction is a two-electron−one-proton reaction (cf. Eqn (8.13)).

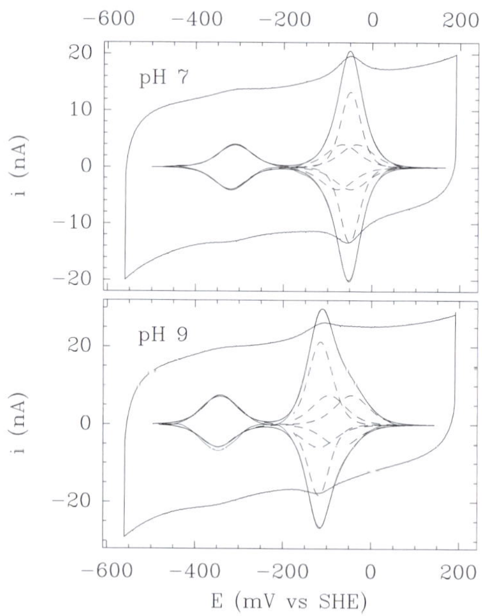

FIGURE 8.12 Cyclic voltammograms (raw data out of scale, base-line subtracted and deconvoluted signals) for *E. coli* fumarate reductase (FrdAB) adsorbed at a pyrolytic graphite edge electrode. This enzyme contains 3 FeS clusters and a flavin cofactor. 20 °C, pH 7 (top panel) & 9 (bottom panel), $v = 10$ mV/s. At pH 7, FADox/FADred −50 mV vs SHE, $[2Fe2S]^{2+/+}$ − 40 mV, $[4Fe4S]^{2+/+}$ −305 mV, $[3Fe4S]^{+/0}$ −65 mV, $\Gamma A \approx 0.4$ pmol. Note the strong pH-dependence of the FAD signal. (C) American Chemical Society 2001. Reprinted from Léger et al. (2001), with permission.

Voltammetry is now becoming a routine technique to measure reduction potentials and offers several advantages over potentiometric titrations described above:

- *No need for a spectroscopic 'handle'*. There is no requirement for a distinctive and unambiguous change in some spectroscopic parameter. The counterpart of this advantage is obvious: voltammetry provides no structural information.
- *Sample economy*. A few picomoles of protein/enzyme are adsorbed onto the electrode (although making a film sometimes requires a larger amount than that, typically ten to one hundred times as much, this is still much less than required for bulk titration).
- *Quick measurements*. Recording a cyclic voltammogram usually takes a few seconds to a few minutes.
- *Instantaneous dialysis*. The electrode can be transferred in a solution of different composition/pH and the measurement repeated with the same sample. The protein might even survive long enough in a hostile environment for measurements to be performed before the electrode is taken back into a more gentle solution (see e.g. ref. Zu et al. (2001)) for the measurement of the reduction potential of a [2Fe2S] cluster in the pH range from 2 to 14).
- *Easier analysis & modelling of data*. The baseline (the charging current) is easily removed, and interpretation of noncatalytic data does not require solving any transport (diffusion/convection) equation.
- *In situ measurements*. Reduction potentials can be measured as a function of temperature (McEvoy and Armstrong, 1999; Hagedoorn et al., 1998; Park et al., 1991; Brereton et al., 1998) or even pressure (Gilles de Pelichy and Smith, 1999).

Fast-Scan Voltammetry to Determine the Rates of Coupled Reactions

The measurement of a reduction potential (an equilibrium property) theoretically requires that the system is at equilibrium. This appears to be in contradiction to measuring a current. Indeed, the flow of electrons, when the protein is oxidised or reduced, results from the fact that the system is driven out of equilibrium when the electrode potential is changed around the (equilibrium) reduction potential. In practice, this does not matter too much if the scan rate (and therefore the current) is slow enough that the system is *nearly* at equilibrium (Laviron, 1979). When the scan rate is raised, however, departure from equilibrium can be observed, and a great deal of information about the *kinetics* of redox processes can be gained by looking at the scan rate dependence of the voltammograms.

Theoretical voltammograms for an uncoupled one-electron redox process are plotted in Figure 8.13A. At slow scan rate, the oxidation peak occurs at $E^{0'}$. If the scan rate is high, because it takes time for the ET between the electrode and the redox centre to occur, the maximal current is observed *after* the redox potential has been reached, i.e. at higher electrode potential. The same reasoning applied to the reductive process predicts that $E_{p,c}$ is lower than $E^{0'}$.

When measurements are performed over a large range of scan rates, the results can be displayed by plotting $E_{p,a}$ and $E_{p,c}$ as a function of scan rate, on a log scale, as shown in Figure 8.13B (Laviron, 1979). This has been called a trumpet plot. The more efficient the ET between the electrode and the active site (the greater k_0 in Eqn (8.18)), the higher the scan rate at which the oxidative and reductive peaks start to separate (peak separation occurs when the scan rate is greater than about k_0RT/F (Laviron, 1979)). For adsorbed redox proteins, reported values of k_0 vary greatly, from a few inverse seconds to 15,000 s^{-1}, in which case the peaks remain visible at scan rates as high as 3000 V/s (Baymann et al., 2003; Hirst et al., 1998a). Needless to say, the situation of fast ET is more desirable if the focus is on studying biologically relevant processes rather than interfacial electrochemistry.

Fast-scan voltammetry also gives information about the rates of the reactions that are coupled to ET. If these coupled reactions (e.g. (de)protonation) are fast on the voltammetric time scale, their effect is to shift the reduction potential of the redox couple (see 'Influence of Coupled Reactions on Reduction Potentials' section) and also to decrease the apparent rate of ET, the parameter noted k_0 in Eqn (8.18) (the latter effect is discussed in a series of

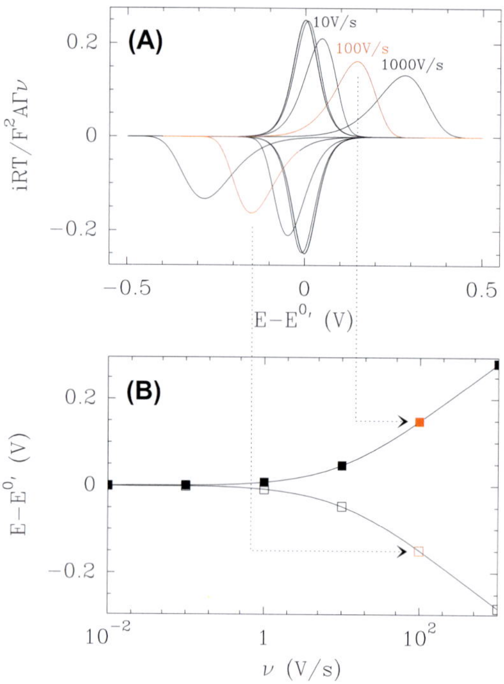

FIGURE 8.13 Effect of scan rate on the voltammetry of a redox species undergoing a one-electron, no-proton redox process, with $k_0 = 0.1\ s^{-1}$. A: calculated cyclic voltammograms at different scan rates (from 1000 V/s to 10 mV/s). The difference between $E_{p,a}$ and $E_{p,c}$ increases as the scan rate is raised. B: 'Trumpet plot': $E_{p,a}$ (filled squares) and $E_{p,c}$ (empty squares) as a function of the log of the scan rate. The slower ET rate is, the peak will separate at a slow scan rate and the trumpet plot will shift to low scan rates.

papers written by Etienne Laviron in the 1980s (Laviron, 1980; Meunier-Prest and Laviron, 1992), see also Anxolabéhère-Mallart et al. (2011) and section 2.1.2.1 in ref. Léger and Bertrand (2008)). In that case, the voltammogram remains reversible at very low scan rate and symmetrical but with greater peak separation when the scan rate is increased (Figure 8.13).

Recording voltammograms at increasing scan rates can be used for determining the value of k_0 but also to determine the rates of chemical processes that are coupled to ET. From a biological point of view, proton transfer is certainly the most important reaction coupled to ET, because the synthesis of ATP in most organism is coupled to long range proton transfers across biological membranes (Nicholls and Ferguson, 2002), but the kinetics and mechanism of proton transfer are difficult to study using conventional techniques. It is remarkable that electrochemistry proved very useful in this context (Chen et al., 2000; Hirst et al., 1998b), as described below.

Figure 8.15 illustrates the voltammetric study of the $[3Fe4S]^{+/0}$ one-electron one-proton reaction (Figure 8.14), for a mutant of *Azotobacter vinelandii* ferredoxin I. High-resolution crystal structures reveal that

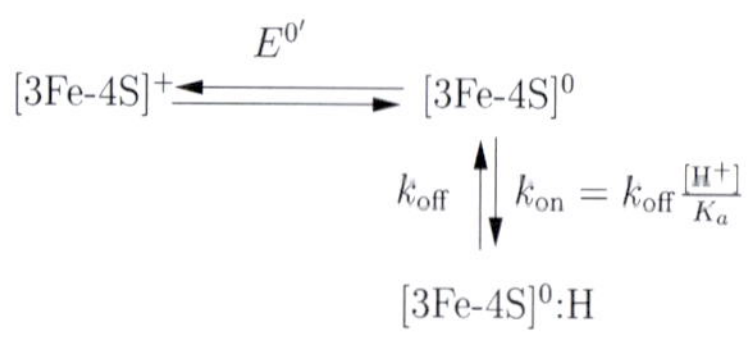

FIGURE 8.14 L-shape scheme used to interpret the fast-scan voltammetry of a [3Fe4S] cluster. Electrochemists call this an 'EC' mechanism (Compton and Banks, 2011; Savéant, 2006).

the [3Fe4S] is buried with no access to water molecules and that a carboxylate group from an aspartate (D15) is located close to the cluster on the protein surface. It was suggested that a movement of the position 15-side chain may transfer a proton from the solvent to the cluster. The experiments depicted in Figure 8.15 illustrate the use of fast-scan voltammetry to determine the kinetics of protonation of the [3Fe4S] cluster in a mutant where D15 is replaced with a glutamate. They were performed at pH 5.4, greater than pK_{Ox} and 1.3 pH unit lower than the $pK_{Red} = 6.7$ of the [3Fe4S] cluster, the reduced form of which is therefore protonated at equilibrium. The scans were started from the high potential limit, and only the first scan is considered.

- At slow scan rates (panels A and B in Figure 8.15), oxidation and reduction peaks for the $[3Fe4S]^{+/0}$ appear at the same electrode potential (≈ -350 mV). Under these 'close-to-equilibrium' conditions, the reduction is followed by protonation, and oxidation proceeds along the reverse route (Figure 8.14).
- When the scan rate is increased, oxidation and reduction peaks start to separate (Figure 8.15C).

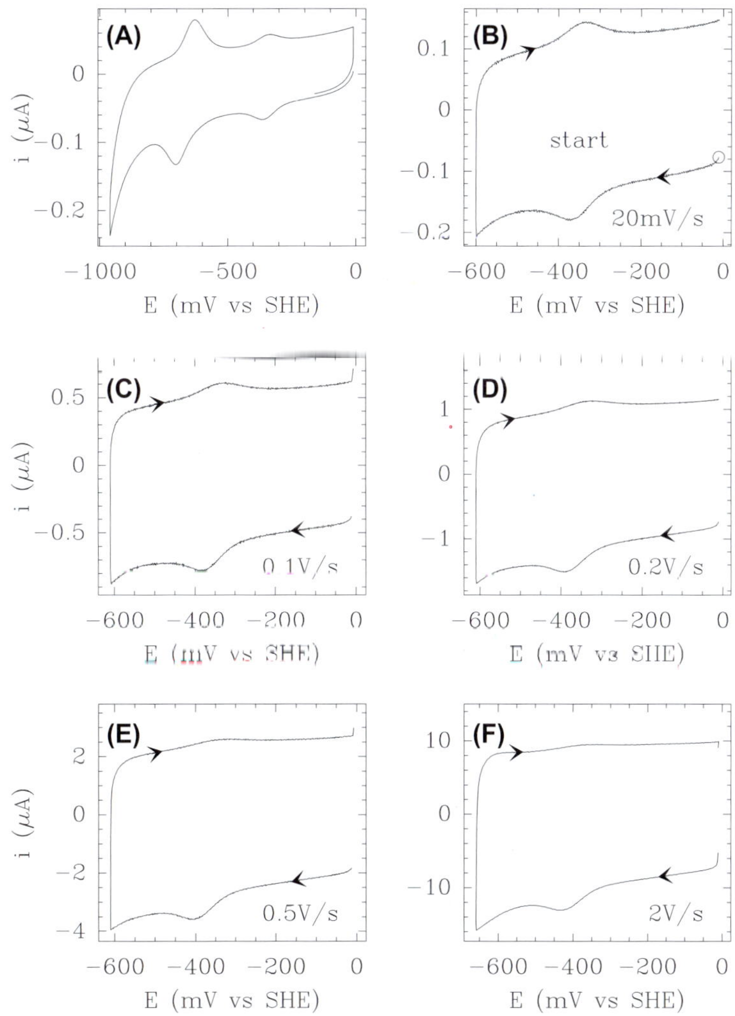

FIGURE 8.15 Effect of scan rate on the voltammetry of a redox species undergoing a one-electron–one-proton redox process. The data are for the $[3Fe4S]^{+/0}$ redox couple of a slow proton-transfer mutant of *A. vinelandii* ferredoxin I (D15E) at low pH (Chen et al., 2000; Hirst et al., 1998b).

- At scan rates between 1 and 10 V/s, the reductive peak is still clearly visible, but the oxidation peak vanishes (Figure 8.15D–F), because the cluster is trapped in the protonated form: it is quickly protonated upon reduction, but the rate of deprotonation, k_{off} in Figure 8.14, is too small for the cluster to be deprotonated during the fast oxidative scan. The deprotonation of the cluster 'gates' its reoxidation.
- At very high scan rates, 20V/s and above (not shown), both peaks are observed, but the average reduction potential is lower than at low scan rate and matches the alkaline limit: this is because the scan is reversed before the reduced cluster is protonated, and the ET is therefore not coupled to protonation.

These very simple experiments determine the rates of (de)protonation. Electrochemistry can achieve this because the time scale of potential modulation can be changed over orders of magnitude (1 min at 10 mV/s to 1 ms at 1000 V/s) to match that of the chemical events. In conjunction with the site-directed mutagenesis, crystallography, and molecular dynamics simulations, these experiments made it possible to obtain very original information about the molecular mechanism of protonation (Chen et al., 2000; Hirst et al., 1998b).

Several examples of voltammetric studies of coupled reactions, involving cytochromes and FeS clusters, have been reported [see e.g. (Jeuken et al., 2002, 2000; Armstrong et al., 2001)].

CATALYTIC VOLTAMMETRY AND CHRONOAMPEROMETRY WITH ADSORBED REDOX ENZYMES

Principle and General Comments

In the absence of substrate and at sufficiently high coverage, a redox enzyme immobilised onto an electrode gives peak-like signals, resulting from the reversible transformation of its redox centres (Figure 8.12). Upon adding substrate, the nonturnover peaks are transformed to sizeable 'catalytic waves' (Limoges and Savéant, 2004): reaction with substrate transforms the active site, which is regenerated by electron exchange with the electrode in a succession of catalytic cycles. The magnitude of the current is proportional to electroactive coverage and to turnover rate, and so, the relationship between driving force (potential) and catalytic activity is traced in a single voltammetric experiment. Note that catalysis may be observed even if coverage is too low to observe noncatalytic signals (as is unfortunately often the case).

We shall discuss qualitatively a series of voltammograms and chronoamperograms obtained with three different enzymes and selected because they illustrate the variety of experiments that can be carried out and the variety of information that can be obtained. We shall emphasise what makes each result particularly relevant in the context of mechanistic studies. These three enzymes are

- The soluble fraction of *E. coli* fumarate reductase (FrdAB), a flavoenzyme that has a electron transfer chain consisting of three FeS clusters (Iverson et al., 1999). Its physiological function is to reduce fumarate to succinate, but it can also oxidise succinate both in vitro and in vivo, when it replaces succinate dehydrogenase (complex II).
- Various hydrogenases (from *Allochromatium vinosum*, *D. fructosovorans* and *A. aeolicus*) in which the NiFe active site is connected to the protein surface by a chain of three FeS clusters and a gas channel (Leroux et al., 2008; Fontecilla-Camps et al., 2007). These enzymes reversibly catalyse the oxidation of H_2.
- The periplasmic nitrate reductase from *Rhodobacter sphaeroides* (NapAB), which irreversibly reduces nitrate into nitrite. This enzyme houses a buried molybdenum active site, one 4Fe4S cluster and two surface-exposed haems (Arnoux et al., 2003).

The noncatalytic voltammetry of *E. coli* fumarate reductase has been discussed above (Figure 8.12). Figure 8.16 shows again the noncatalytic voltammetry of this enzyme in panel A and the catalytic signal that is observed when the solution contains both fumarate and succinate and the electrode is rotated at a high rate to avoid mass-transport control (panel B). The flavin cofactor is oxidised at high electrode potential, giving two electrons to the electrode. The oxidised enzyme can bind succinate, and the oxidation of succinate results in the reduction of

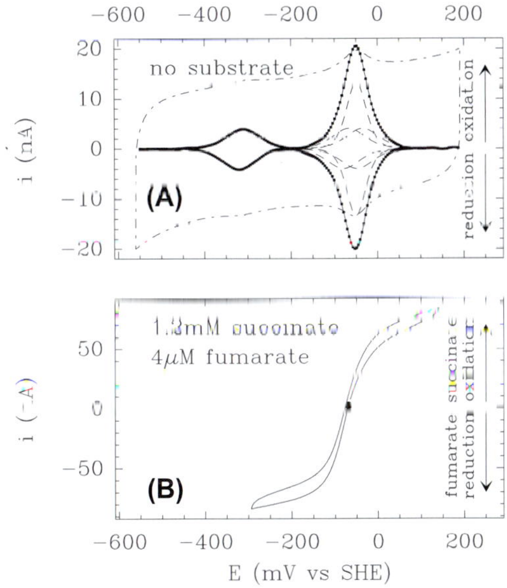

FIGURE 8.16 (A) Noncatalytic voltammogram obtained for *E. coli* fumarate reductase (FrdAB) adsorbed at a pyrolytic graphite edge electrode in the absence of substrate. The raw voltammogram (outer dash–dot line) is not to scale. Inset: background corrected current (small dots) and deconvoluted data (dashed lines). $v = 10$ mV/s, 20 °C, pH 7. See also Figure 8.12. (B) Catalytic wave showing reversible succinate oxidation and fumarate reduction by adsorbed FrdAB in a solution containing succinate and fumarate, with the electrode rotating at a high rate. $v = 1$ mV/s, 20 °C, pH 7, $\omega = 3000$ rpm. (C) American Chemical Society 2001. Reprinted from ref. Léger et al. (2001), with permission.

the flavin. The reduced FAD can be reoxidised, giving electrons to the electrode and so on. This results in a steady-state flux of electrons from succinate in solution to the electrode, via the adsorbed enzyme, which is measured as a steady-state positive current. At low electrode potential, a reductive (negative) current is observed which is proportional to the rate of fumarate reduction.

The filled circle in Figure 8.16B indicates the potential of zero net current (or 'open-circuit potential' (OCP) at −51 mV. This corresponds to the reduction potential of the fumarate/succinate couple, which can be calculated for any concentration ratio using the Nernst equation (Eqn (8.5)) and the published value $E^{0'} \approx +20$ mV at 25 °C, pH 7 ($E^{0'} + RT/2F \ln([F]/[S]) = -46$ mV). As a general way of things, provided the solution contains both the oxidised and reduced substrates, and that the adsorbed enzyme is able to catalyse both the oxidation and reduction reactions at significant rates, the value of the OCP is given by the Nernst equation and does not tell us anything about the enzyme. It characterises only the substrate/product redox couple. In contrast, the value of the current on either side of the OCP and its dependence on electrode potential reveal the enzyme's intrinsic properties (Abou Hamdan et al., 2012).

Catalytic electrochemistry offers several advantages with respect to conventional solution assays of the enzyme's activity.

- The temporal resolution of the activity is very high (the current can easily be sampled every 0.1 s or faster), which is useful if the activity of the enzyme is evolving. For reasons which will become clear below, this makes it possible to study very easily the reaction of certain enzymes with their gaseous substrates or inhibitors (CO, H_2, etc).
- In contrast to certain solution assays, anaerobicity is not required (at least on condition that the electrode potential is high enough that oxygen is not reduced on the electrode), which makes it possible to study the reaction of the enzyme with oxygen; this has proved very useful recently in the field of hydrogenase research (Vincent et al., 2007; Liebgott et al., 2010).

- Last, and most importantly, the electrode potential is a very useful control parameter, which influences the redox state of the enzyme and determines the driving force for the catalytic reaction and for certain redox-driven (in)activation processes. The dependence of activity on electrode potential can be very complex, and in some cases, efforts are still being made to make sense of the data in relation to the catalytic mechanism.

The magnitude and shape of the catalytic signal depend on a number of factors because catalysis involves many consecutive reactions or processes: transport by convection and diffusion of the substrate in solution towards the adsorbed enzyme, substrate diffusion *within* the enzyme, its binding and transformation at the active site, product release and diffusion away from the enzyme, regeneration of the redox state of the active site upon intramolecular ET (assuming there is a redox chain in the enzyme) and interfacial ET between the electrode and a redox centre that is exposed at the protein surface.

The overall turnover rate depends on the slowest of these steps, noting that changing the electrode potential changes the rates of the redox processes. This means that if one is interested in studying the catalytic mechanism, one should try to reach a situation where interfacial ET and mass transport in solution do not limit the current. A good kinetic model will not include all steps (this would lead to indetermination) but only those which influence the turnover rate. Often the examination of the catalytic signal gives very useful information about which steps matter, as discussed below. Although kinetic models exist and are adapted to many different situations, they will not be described below, and we shall refer the reader only to the primary literature. However, it is important to acknowledge that enzyme kinetics and electrochemistry are quantitative sciences, and that by looking at the data too superficially or qualitatively, one may miss important information or misinterpret the data.

Mass-Transport Controlled Catalytic Voltammetry

Figure 8.17A illustrates the voltammetry for hydrogen oxidation by *A. vinosum* NiFe hydrogenase adsorbed at a rotating disc electrode. In this experiment, the positive current at high potential is proportional to the rate of catalytic H_2 oxidation and the negative current results from proton reduction (H_2 evolution). The current tends to a limiting value at high potential, which increases dramatically as the electrode rotation rate ω is raised (Pershad et al., 1999). This is because, during turnover, the concentration of hydrogen near the electrode decreases, and the enzyme is able to consume H_2 faster than it is brought to the electrode by the convective motion of the solution. The greater the rotation rate, the more efficient the transport of hydrogen from the bulk solution towards the enzyme, and the greater the current (see 'Diffusion-Controlled Voltammetry at Rotating Electrodes' section). At infinite rotation rate, the catalytic current is finite: mass transport is no longer rate limiting, and the extrapolated current reveals the intrinsic efficiency of the enzyme. The Koutecky–Levich plot in Figure 8.17B appears to follow

$$\frac{1}{i_{\lim}} \approx \frac{1}{nFA\Gamma \times (\text{turnover rate})} + \frac{\text{constant}}{\sqrt{\omega}} \tag{8.31}$$

The equation above emphasises departure from mass-transport control at high ω (compare to Eqn (8.25)), but it is not rigorous (see Heering et al. (1997) and section 2.3.1 in ref. Léger and Bertrand (2008)).

This limitation by mass transport is all the more influential that the enzyme has high activity, that the electrode coverage is high, and that the concentration of substrate is small compared with the Michaelis constant (indeed, under saturating conditions, a small decrease of interfacial substrate concentration should have no effect on turnover rate).

The magnitude of the current is proportional to $A\Gamma$ times the turnover rate, and can give an estimate (sometimes only a lower estimate) of the turnover rate. Interestingly a study of hydrogenase showed that the turnover number of the enzyme is significantly higher than that observed in solution assays, using oxidising dyes (Pershad et al., 1999); in the latter case, it becomes evident that turnover in solution assays is limited by ET to the soluble electron partner and that the electrode is a much faster electron acceptor than the soluble dye.

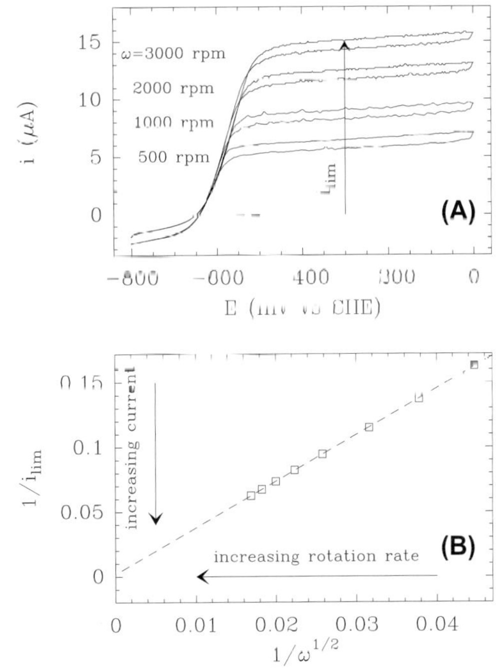

FIGURE 8.17 (A) Influence of electrode rotation rate (ω, given in units of revolution per minute) on the catalytic current measured for hydrogen oxidation by *A. vinosum* NiFe hydrogenase adsorbed at a rotating graphite electrode. (B) Koutecky–Levich plot (Eqn (8.31)) showing small but nonzero intercept at infinite rotation rate, allowing one to estimate k_{cat}. 0.1 bar H_2, $\nu = 100$ mV/s, $T = 45$ °C, pH = 6.5. Adapted from ref. Pershad et al. (1999).

Chronoamperometry to Measure Michaelis and Inhibition Constants

In conventional biochemistry experiments, the rate of turnover is measured as a function of substrate concentration to determine the Michaelis–Menten parameters (k_{cat}, the maximal turnover rate, and K_m, the Michaelis constant):

$$\boxed{\text{turnover rate} = \frac{k_{cat}}{1 + \dfrac{K_m}{[S]}}} \tag{8.32}$$

The same parameters can be determined from chronoamperometric experiments looking at the substrate-concentration dependence of the current recorded at a fixed potential

$$i = nFA\Gamma \frac{k_{cat}}{1 + \dfrac{K_m}{[S]}} \tag{8.33}$$

In practice, this kind of measurement is far from easy. (i) The limiting current is proportional to $A\Gamma$, the total amount of enzyme adsorbed, which can be determined (with a very relative accuracy) only when the electrode coverage is high enough for noncatalytic signals to be measured in the absence of substrate (Eqn (8.28)). (ii) The measurement of K_m can be performed without knowing the exact electroactive coverage. This requires, however, that the adsorbed film is stable enough as a function of time for the coverage to be constant when currents are measured with the same film in solutions of different substrate concentrations. Reference Fourmond et al. (2009b) describes methods for correcting the effect of film desorption. (iii) Last, Eqn (8.33) does not take into account mass transport of substrate in solution; this is correct only if there is no depletion of substrate near the electrode. This should be checked for by looking at the rotation rate dependence of the current, or using the rotation rate dependence to extrapolate the current at infinite rotation rate.

Figure 8.18 shows the result of this simple chronoamperometric experiment, in a case where it is particularly informative: nitrate reduction by the molybdoenzyme *R. sphaeroides* periplasmic nitrate reductase. The nitrate reduction rates (negative current) shown in the middle row were measured at two different electrode potentials. The top panels show the concentration of nitrate against time (each step corresponds to the injection in the electrochemical cell of a small amount of a concentrated stock solution of nitrate). The bottom panels show the steady-state current at the end of each step plotted against nitrate concentration. At low potential (left column), the change in current simply follows Michaelis–Menten kinetics, whereas under less reductive conditions (right

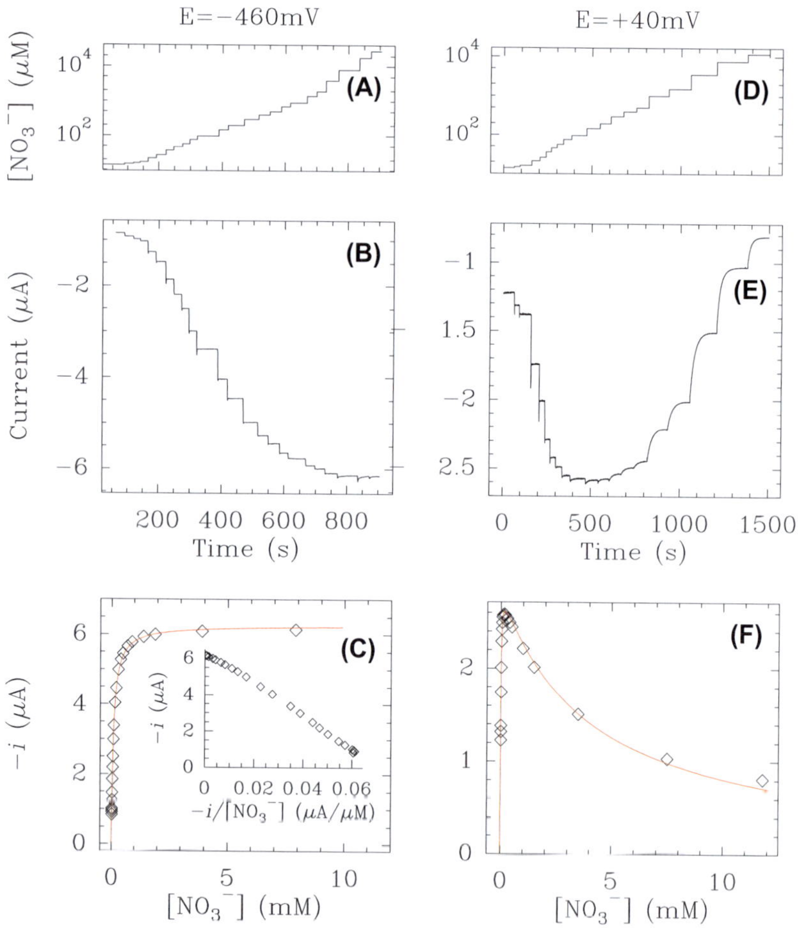

FIGURE 8.18 Dependence of the rate of nitrate reduction on nitrate concentration, with nitrate reductase adsorbed at a rotating disc electrode spun at a high rate. The left- and right-hand sides correspond to a redox poise at −460 and +40 mV vs SHE, respectively. Conditions: pH 6; 25 °C; 3 krpm. Panels A and D show the evolution of nitrate concentration against time, when the concentration is stepwise increased by adding aliquots of a stock solution of potassium nitrate (note the logarithmic Y scale). Panels B and E show the resulting change in catalytic current. Panels C and F show the catalytic current reached at the end of each step as a function of nitrate concentration. The fit of the data in C to the Michaelis–Menten returns the value of $K_m \approx 85$ μM. The inset shows a Eadie–Hofstee plot (Cornish-Bowden, 2004). The red line in panel F is the best fit to an equation accounting for substrate inhibition, with $K_m = 10$ μM and $K_i \approx 4$ mM. (C) American Chemical Society 2008. Reprinted from Fourmond et al. (2010a), with permission.

column), high concentrations of nitrate inhibit the enzyme. This is particularly relevant because experiments aimed at trapping catalytic intermediates before they are characterised in spectroscopy are often carried out with very high concentrations of substrate and in the case of nitrate reductase, moderately reducing conditions are needed to detect a Molybdenum(V) intermediate by EPR (the fully reduced Mo(IV) state is EPR-silent). Figure 8.18 demonstrates that these conditions favour the formation of an inactive enzyme rather than a catalytic intermediate. Solution assays with two electron donors having different reduction potentials fully supported these electrochemical results (Fourmond et al., 2010a).

The high temporal resolution of the activity measurement is also useful for probing the reaction of redox enzymes with gaseous substrates and inhibitors. Many redox enzymes use, consume, or are inhibited by small molecules like O_2, CO, N_2, H_2, NO, etc. The fact that these molecules tend to escape the solution and equilibrate with the gas phase above it makes certain experiments difficult. For example, to establish the competitive character of the inhibition of hydrogenase by CO by carrying out normal (solution) assays, one must measure the H_2-oxidation turnover rate as a function of the concentrations of H_2 and CO dissolved in solution, but setting their concentrations precisely and maintaining them constant may be tricky and time-consuming. This problem is easily solved using electrochemistry, because it is possible to change the concentrations of dissolved gas in a controlled manner and to simultaneously monitor the change in activity, as illustrated in Figure 8.19 (Léger et al., 2004).

Figure 8.19A shows, in black, the change in concentration of dissolved H_2 against time that is obtained when, after having maintained an atmosphere of 1 bar of H_2 until $t = 0$, a small tube is suddenly plunged into the buffer and used to bubble argon in the electrochemical cell. This flushes hydrogen away from the cell solution, and its concentration decreases exponentially with time, with a time constant that we denote by τ_{H_2} (Léger et al., 2004). The red curve in panel B shows the sigmoidal change in current for hydrogen oxidation by *D. fructosovorans* NiFe hydrogenase in the same experiment. The equation used to fit this data (black dashed line) is simply obtained by inserting into the Michaelis–Menten equation a time-dependent substrate concentration: $[H_2](t) = [H_2]_0 \exp(-t/\tau_{H_2})$.

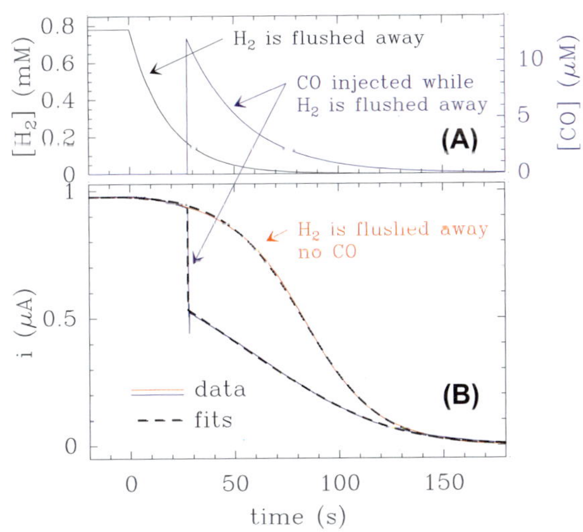

FIGURE 8.19 Hydrogen oxidation by WT *D. fructosovorans* NiFe hydrogenase adsorbed at a rotating disc electrode: measurement of the Michaelis constant relative to H_2 and of the inhibition constant relative to CO. Panel A shows the change in hydrogen (black) and CO (blue) concentrations against time. The red line in panel B shows the change in current against time in the experiment where H_2 is flushed away at $t > 0$. The blue line is the result of a similar experiment, but a solution saturated with CO was injected at $t \approx 30$ s while the concentration of H_2 was decreasing. The fits to Eqns (8.31) and (8.32) are shown as dashed lines. (C) American Chemical Society 2001. Reprinted from ref. (Léger and Bertrand, 2008), with permission.

$$i(t) = \frac{i_{\max}}{1 + \frac{K_{\mathrm{m}}}{[\mathrm{H_2}]}\exp(t/\tau_{\mathrm{H_2}})} \tag{8.34}$$

In the second experiment, shown as a blue curve in Figure 8.19B, an aliquot of CO-saturated solution is injected at $t \approx 30$ s while the hydrogen concentration is decreasing (blue line in panel A). In this experiment, the concentrations of both CO and H_2 decrease with time, but the equation for the transient current is again simply obtained by using the rate equation that considers competitive inhibition by CO, in which we insert exponential decays of both H_2 and CO:

$$i(t) = \frac{i_{\max}}{1 + \frac{K_{\mathrm{m}}}{[\mathrm{H_2}](t)}\left(1 + \frac{[\mathrm{CO}](t)}{K_i}\right)} \tag{8.35}$$

The fit is shown as a dashed curve, and simultaneously determines K_{m} for H_2 and K_i relative to CO, from an experiment that lasts no longer than a few of minutes! However, note that the sort of experiments shown in Figure 8.19 cannot be very accurate; indeed, the concentrations varying exponentially with time, only the log (that is, the order of magnitude) of the values of K_{m} and K_i can be determined.

Chronoamperometry to Resolve Rapid Changes in Activity

Nitrate reductase is reversibly inactivated at high concentration of nitrate and moderate potential (Figures 8.18 and 8.22F); in addition to that, it also irreversibly activates the first time it is reduced. This is clear from the chronoamperogram in Figure 8.20, which shows the response of a fresh film of periplasmic nitrate reductase immersed into a solution of nitrate when the potential is stepped as indicated in the upper panel. When the potential is poised at $E = -160$ mV (at $t = 40$ s), the current is essentially constant. On the step to -460 mV, at $t = 80$ s, the activity first instantly decreases (the current becomes less negative) and then slowly increases before it stabilises. This slow change in current demonstrates that the enzyme activates at low potential. The magnitude of the activation phase accounts for 20% of the current reached after activation. The instant decrease in activity when the potential is first stepped from -160 to -460 mV is not surprising considering the steady state profile in Figure 8.22D (see below). When this sequence of potential steps is repeated with the same film of enzyme (from $t = 480$ s), the

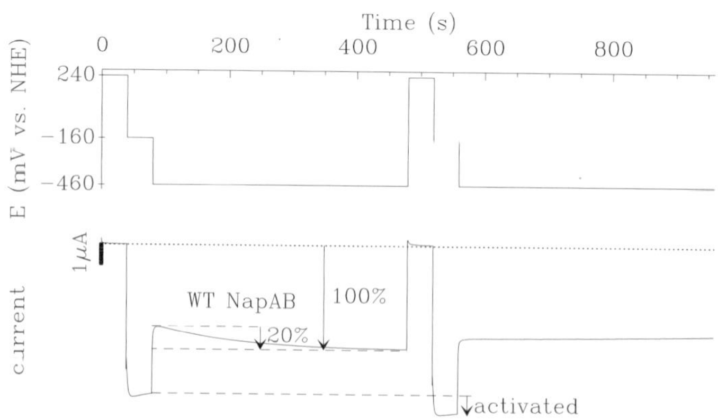

FIGURE 8.20 Chronoamperometric experiments demonstrating the irreversible reductive activation of NapAB. The top panel shows the sequence of potential steps which was applied to a film of as-prepared WT NapAB. The horizontal dotted line is for $i = 0$. $\omega = 5$ krpm, pH 6, 25 °C (Fourmond et al., 2008).

activity detected at $E = -160$ mV is greater than at the same potential in the first experiment, as indicated by the downward arrow, and no further activation occurs on the second step to −460 mV, at $t \approx 560$ s. Therefore, the activation proceeds only once, on the first step to −460 mV, and it is not reversed by taking back the enzyme to oxidising conditions (+240 mV): it is irreversible.

This irreversible reductive activation cannot be detected in solution assays of the enzyme, because assaying the enzyme requires that it is reduced, and this reduction activates the sample. Therefore, the enzyme molecules that are already active and those that require reductive activation are indistinguishable in solution assays. In contrast, the temporal resolution of the activity measurement in PFV is high enough that the activity can be measured while the enzyme is activating (Figure 8.18). This is reminiscent of the 'redox-cycling' experiment with *Rhodobacter capsulatus* DMSO reductase: the enzyme has greater activity for DMS oxidation after it has been reduced (Bray et al., 2000). Furthermore, the reductive activation of nitrate reductase is very relevant because its amplitude correlates with the concentration of a certain Mo(V) EPR signal (the so-called 'high-g' signal) present in the sample before activation, demonstrating that, unlike all expectations, this signature arises from an inactive state which activates the first time the enzyme is reduced (Fourmond et al., 2008).

On a different note, chronoampetrometry experiments with hydrogenase make it easy to characterise quantitatively the kinetics of inhibition by CO (and also O_2). The experiment in Figure 8.21 consists in monitoring the H_2-oxidation current after the concentration of CO first suddenly increases when an aliquot of solution saturated with inhibitor is injected in the electrochemical cell and then slowly returns to zero as the buffer is flushed by a stream of H_2 (ref. Léger et al. (2004) and section 2.4.2 in ref. Léger & Bertrand (2008)). The concentration of H_2 is nearly constant, and the inhibitor concentration follows an exponential decay (panel A). The concentration of inhibitor need not be independently measured because its change against time is defined by the amount of inhibitor that is injected and by the time constant of the decay, which is determined by fitting the change in current (Léger et al., 2004). If binding and/or release of the inhibitor is

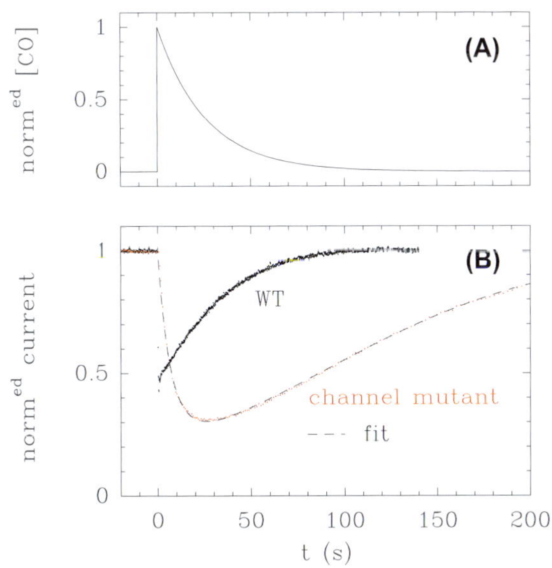

FIGURE 8.21 Inhibition by CO of H_2 oxidation by the native form of *D. fructosovorans* NiFe hydrogenase and a variant where a double mutation narrows the gas channel (Leroux et al., 2008; Liebgott et al., 2010). Panel A: CO concentration against time. Panel B: the change in current against time for the WT enzyme (black) and the mutant (red). The modelling of these data directly gives the rates of diffusion along the channel (Almeida et al., 2007; Leroux et al., 2008; Liebgott et al., 2010). The aliquot of solution saturated with CO was injected at $t = 0$. 1 bar H_2, $E = -160$ mV, pH 7, 40 °C, 3 krpm.

slow, the change in activity is delayed from the time of injection, and the *rates* of inhibition can be measured (Leroux et al., 2008).

In the case of the WT enzyme from *D. fructosovorans* (at 40 °C), the decrease in activity after CO injection is too fast to be resolved (black squares in Figure 8.21B). Also, the recovery of activity follows exactly the decrease in CO concentration as the latter is flushed away from the cell. In contrast, the red signal was obtained with a mutant of the enzyme, where amino acids whose side chains point inside the substrate channel connecting the active site to the solvent have been substituted (Leroux et al., 2008). Clearly, the binding and release of CO are much slower, suggesting that CO diffuses slowly within the mutant, to and from the active site. The data can be fit to the model in ref. Almeida et al. (2007) to measure the rate constants corresponding to diffusion in the channel in either direction. Independent experiments, based on solution assays of the isotope-exchange reaction, confirmed the effects of the mutations on the rates of intramolecular diffusion and proved that the difference between the two signals in Figure 8.21 is not an artefact resulting from the presence of the electrode (Leroux et al., 2008; Abou Hamdan, et al., 2012).

Similar electrochemical experiments could be carried out to determine how the mutations of amino acids whose side chains point into the channel of hydrogenase affect the rates of binding and release of CO, O_2 and H_2, and PFV is now a unique tool for studying how the structure of gas channels affects intramolecular diffusion rates (Leroux et al., 2008; Liebgott et al., 2010; Lautier et al., 2011).

Determining the Reduction Potentials of an Active Site Bound to Substrate

Chronoamperometry experiments are usually easy to analyse, whereas understanding the profile of activity against potential sometimes demands effort. Figure 8.22 shows various catalytic voltammograms obtained with FrdAB, NapAB and hydrogenases adsorbed at rotating disk electrodes. We shall now discuss their *shape* (as opposed to merely their magnitude) and explain what can be learned about the enzyme from each signal.

We examine first succinate oxidation by *E. coli* fumarate reductase, FrdAB (Figure 8.22A). Like all other data in this figure, this signal was obtained with the electrode rotating at a very high rate so that mass transport was not influential. The background (capacitive) current is shown as a dotted line. At low electrode potential, the enzyme and its active site are reduced and unable to oxidise succinate – there is no activity and no Faradaic current. At very high potential the current tends to a limit (a plateau), *which is independent of electrode rotation rate*, and represents the maximal turnover rate, when the enzyme is oxidised at a very high rate and remains fully oxidised in the steady state. In between the two, the activity increases when the electrode potential becomes high enough that the active site flavin becomes oxidised. The position of the inflection point of the main catalytic wave is a phenomenological parameter often called the 'catalytic potential'. By using a very simple kinetic model that uses the Nernst equation to relate the electrode potential to the redox state of the flavin, the *position* and precise *shape* of the voltammogram can be interpreted to determine the two reduction potentials of the active site flavin, under turnover conditions and in the presence of substrate. In contrast, in equilibrium titrations, the reduction potential of the active site flavin can only be measured in the absence of substrate, otherwise the enzyme would turnover and equilibrium could not be reached. In ref. Léger et al. (2001), voltammograms such as that in Panel A were used for determining the dependence of the reduction potential of the active site flavin on succinate concentration (shown in Figure 8.4) and pH, from which the affinity for succinate and pKa of the flavin in its different redox states could be determined using equations such as Eqn (8.9). These thermodynamic properties of the active site are relevant to the catalytic cycle, regarding the protonation and binding states of the FAD intermediates.

The Effect of Slow Intramolecular ET

The catalytic potential matches the reduction potential of the active site only if the ET between the electrode and the active site (both interfacial and intramolecular ET) is fast. If the enzyme has a chain of redox cofactors that relay electrons internally, and if one of the ET transfers along this chain is slow, then the catalytic potential may be shifted from the reduction potential of the active site and approach the reduction potential of a relay

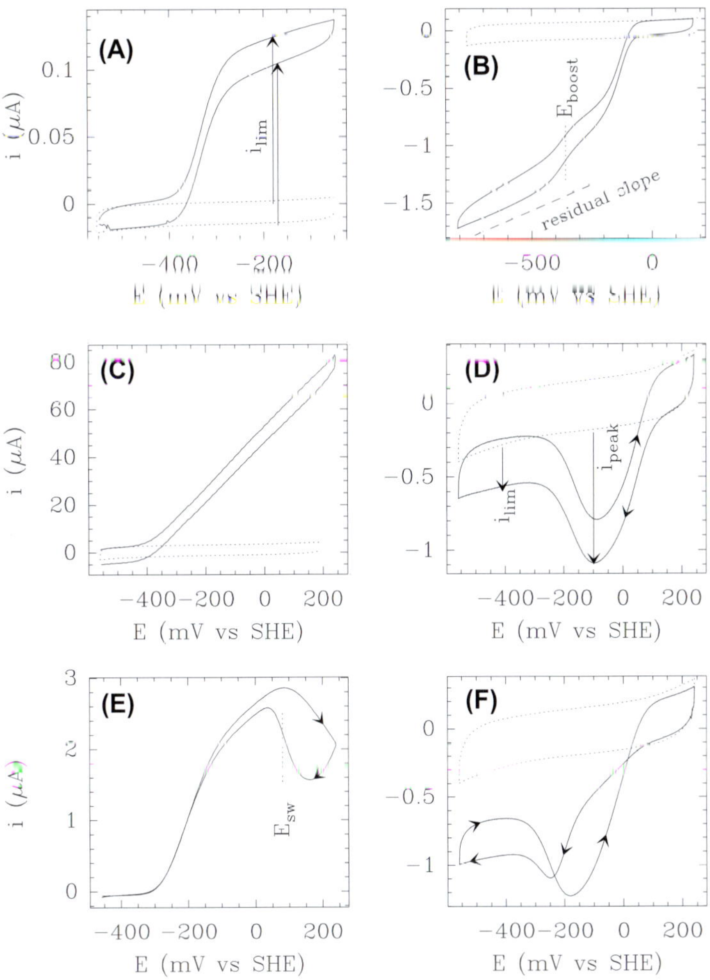

FIGURE 8.22 A small collection of cyclic voltammograms, obtained with various enzymes adsorbed at a rotating electrode spun at a high rate (so that substrate transport towards the electrode is fast). In most panels, the voltammogram plotted with a dotted line is a blank, recorded with no adsorbed enzyme: the contribution of the enzyme is obtained by subtracting this capacitive current. A: succinate oxidation by *E. coli* fumarate reductase $v = 1$ mV/s, $\omega = 3$ krpm, 20 °C, 16 mM succinate, pH 7.5 (Léger et al., 2001) B: fumarate reduction by *E. coli* fumarate reductase. (Hudson et al., 2005) C: H_2-oxidation by *A. vinosum* NiFe hydrogenase, at high temperature (60 °C) and fast-scan rate ($v = 1$ 1V/s), $\omega = 2.5$ krpm, 1 bar H_2, pH 7 (Léger et al., 2002). D: nitrate reduction by *R. sphaeroides* periplasmic nitrate reductase, at very low nitrate concentration (10 μM) $v = 20$ mV/s, 25 °C, pH 6 (Bertrand et al., 2007). E: H_2-oxidation by *A. aeolicus* NiFe hydrogenase, at slow scan rate ($v = 0.3$ mV/s) (Fourmond et al., 2010a,b,c); compare with the signal in panel C. F: nitrate reduction by *R. sphaeroides* periplasmic nitrate reductase, at very high nitrate concentration (24 mM) $v = 20$ mV/s, 25 °C, pH 6 (Fourmond et al., 2010a); compare with the signal in panel D.

(Elliott et al., 2002b). This is indeed the prediction of models that take into account intramolecular ET along the chain (Léger et al., 2006). How the catalytic potential compares with the reduction potentials of the relays may therefore inform on the rate of the intramolecular ET.

For reasons that are not entirely clear yet, the shape of the catalytic wave may also have a particular feature at a potential that corresponds to one of the redox relays. For example, the fumarate reductase discussed above has a series of three FeS clusters that connect the active site flavin to the solvent. The medial relay is a low potential [4Fe4S] cluster. The catalytic signal for fumarate reduction in Figure 8.22B shows a first wave at the potential of the flavin and a 'boost' of activity at the potential of the [4Fe4S] cluster (Hudson et al., 2005).

Slow Interfacial Electron Transfer

In Figure 8.22A and B, it also appears that the current does not reach a plateau at low potential, but instead it keeps increasing linearly. In some cases, this linear increase in current hides the underlying sigmoidal wave: this is so, for example in Figure 8.22C, which shows H_2 oxidation by a NiFe hydrogenase (Léger et al., 2002). A residual slope at high driving force is observed when (1) the rate of the interfacial ET is not very fast compared with the enzyme's turnover *and* (2) not all enzymes are adsorbed in exactly the same orientation, which results in a distribution of interfacial ET rates.

The catalytic current corresponding to the assembly of enzyme molecules can be derived by averaging over the distribution of interfacial ET rate constants, and a linear change in current as a function of electrode potential at high driving force can indeed be predicted (Léger et al., 2002). This results from the contribution of enzyme molecules having low k_0 values that contribute only at high driving force. This effect, which blurs the signal and hides its interesting features, is all the more pronounced that the dispersion of orientation is wide and that interfacial ET (k_0) is slow compared to the intrinsic turnover rate of the enzyme.

Slow Substrate Binding

Many redox enzymes exhibit complex activity profiles as a function of the electrode potential (Elliott et al., 2002a).

As an example, Figure 8.22D shows a catalytic voltammogram corresponding to the reduction of nitrate by *R. sphaeroides* nitrate reductase (NapAB) at very low nitrate concentration. The activity of the enzyme drops down (the current becomes less negative) when the electrode potential is decreased below −400 mV, and it is recovered on the reverse scan.

One may first wonder whether this could be an artefact, resulting, for example, from the reorientation of the enzyme on the electrode at low potential. However, the decrease in activity under very reducing conditions is also observed in solution assays carried out with reduced methyl viologen (MV): under conditions where the concentration of reduced MV greatly decreases during the assay, so that the driving force for the reduction of nitrate decreases as a function of time, the activity measured in solution *increases* before MV is completely exhausted (Fourmond et al., 2010c, see also ref. Sucheta et al. (1992) for similar experiment with complex II). This mirrors the observation in Figure 8.22D: starting from the low potential limit, the activity increases when the electrode potential is increased. This acceleration of turnover upon consumption of reduced MV was also observed with the periplasmic nitrate reductase from *P. pantotrophus*, which exhibits the same voltammetry as in Figure 8.22D (Gates et al., 2008) and *R. capsulatus* periplasmic DMSO reductase (Adams et al., 1999), which belong to the same family of molybdoenzymes. The membrane bound DMSO reductase from *E. coli* was the first molybdoenzyme for which a signature like that in as Figure 8.22D was observed (Heffron et al., 2001). Therefore, this peculiar relation between driving force and activity appears to be a common property of all reductases from the DMSO reductase family.

Although the concept of pH optima for enzyme activity is well established (Cornish Bowden, 2004; Laidler and Bunting, 1973), the possibility that activity of redox enzymes might be optimised within a narrow range of potential has been explored only recently. While pH optima give mechanistic information on protonation equilibria during catalysis, potential optima provide information on the roles played by different oxidation states of redox-active sites. Certain redox transitions may regulate electron flow to or from the active site, or substrate binding or atom transfer may occur only when the active site is in a certain oxidation state. These relationships are difficult to observe by conventional techniques but can be revealed by PFV, due to its ability to measure subtle changes in catalytic activity as the electrode potential is varied.

Considering the signal in Figure 8.22D, we now wonder what makes the activity drop under very reducing conditions. It has been proposed that the 'switch off' results from the reduction of the [4Fe4S] cluster that relays electrons towards the active site (Anderson et al., 2001). However, there appears to be no correlation between the reduction potential of the [4Fe4S] cluster and the position of the low potential feature in the voltammogram

(Fourmond et al., 2010c). It has also been proposed that the complex shape may result from the fact that catalysis can follow distinct routes, as substrate binding either precedes or follows the reduction of the active site, and the relative rates of the two reactions determine which track is used (Heffron et al., 2001). Indeed, changing the electrode potential changes the rate of reduction of the active site, which may make ET faster or slower than substrate binding. When substrate binding to any redox state of the active site is taken into account in kinetic models, all sorts of complex wave shapes can indeed be predicted (see Figure 8.3 in ref. Bertrand et al. (2007)), and the signal in Figure 8.22D can indeed be accurately predicted (Bertrand et al., 2007; Frangioni et al., 2004).

Slow, Redox-Driven (in)Activation

So far, we have considered only steady state voltammograms, where the current depends on electrode potential but not on time, and the catalytic signal is therefore independent of scan rate and sweep direction. For example, the switch 'on and off', observed in Figure 8.22D, is observed at the same potential irrespective of whether the potential is scanned towards positive of negative potential, suggesting that whichever reaction triggers this modulation of activity is fast on the time scale of the voltammetric experiment (not necessarily so on the time scale of turnover).

Sometimes, the enzyme undergoes slow activation or inactivation processes as the potential is varied, and this results in a strong hysteresis: the magnitude of the current depends on time and sample history, and the voltammogram depends on scan rate and direction (but this is not caused by the diffusion of the substrate, contrary to the situation discussed in section 'Diffusion-Controlled Voltammetry').

For example, NiFe hydrogenases reversibly deactivate under oxidising conditions due to the overoxidation of the active site. The enzyme has to be reduced (e.g. incubated with hydrogen or reduced MV) for the activity to be recovered. Figure 8.22E is a catalytic voltammogram showing H_2 oxidation with *A. aeolicus* NiFe hydrogenase. It is recorded essentially under the same conditions as in Figure 8.22C but for a slow scan rate (recording a whole catalytic voltammogram at this scan rate takes 1 h). The activity of the enzyme, measured as a positive current, decreases above −200 mV, and the activity is recovered as the potential is taken down. However, because the inactivation and reactivation are slow on the time scale of the voltammetry, the (in)activation reactions lag behind the change in electrode potential, resulting in a characteristic hysteresis. This inactivation was not observed in the electrochemical experiments depicted in Figure 8.22C because the scan rate was too fast: the enzyme was taken to high potential and back before the inactivation reaction could proceed (Limoges and Savéant, 2004). One can also design experiments in which the electrode potential is held and the slow change in current resulting from the (in)activation measured, to study the kinetics and mechanism of (in)activation (Fourmond et al., 2010b; Jones et al., 2003).

The shape of the signal in Figure 8.22E is now fully understood, and a very important conclusion from the theoretical study is that the potential where the activity is recovered on the scan towards low potential is *not* the reduction potential of the inactive state (Fourmond et al., 2010b, Abou Hamdan et al., 2012) (a clear simple observation suggesting that it is not a thermodynamic quantity is indeed that its value is greatly dependent on a scan rate over the entire range of accessible scan rates). This illustrates a pitfall of the qualitative interpretation of the voltammograms: an inflection point in a voltammogram cannot always be equated to a reduction potential.

The catalytic voltammogram in Figure 8.22F may be the most complex reported to date. It shows how the nitrate reductase signal in Figure 8.22D is deformed by a reversible (in)activation process that decreases the activity at high potential and increases it under reducing conditions. The resulting strong hysteresis at high potential is detected only at a high concentration of nitrate (1 M in panel E, compared with 10 μM in panel D). This is another manifestation of the inhibition of the enzyme by its substrate that we discussed in relation to Figure 8.18, right column (Fourmond et al., 2010a).

SOFTWARES

An open-source program called SOAS (Fourmond et al., 2009a) has been developed, first in the group of F. Armstrong in Oxford and then in our lab in Marseilles, with the aim of analysing one-dimensional signals. It offers a large set of commands for handling voltammetric and chronoamperometric data. It is free of charge and can be compiled on computers running any flavour of Unix or Linux (including Mac OS X). Binary installers for Mac OS X and Debian/Unbuntu distributions of Linux are available. See installation instructions on our web page at http://bip.cnrs-mrs.fr/bip06/software.html.

A new version of the program, QSOAS, is currently being developed in our group. It will eventually become available for all OS (including Windows). Check our web pages for updates.

PFV QUIZ

In Figure 8.23, we have plotted two voltammograms recorded with a film of *E. coli* fumarate reductase in contact with a solution containing only succinate at a concentration $[S] = 50$ mM (approximately 200 times the Michaelis constant). The plain line corresponds to a stationary (nonrotating) electrode, whereas the dashed voltammogram was recorded with the electrode rotating at 3000 rpm. (i) Why is the current at high potential rotation-rate independent in this case? (ii) Why are the shapes of the voltammograms different?

ACKNOWLEDGEMENTS

I am very grateful to F. Armstrong (Oxford University, UK) for introducing me to this research field. Some of the data shown herein were obtained in his lab. I also thank him and A. K. Jones (formerly in Oxford, now at Arizona State University, USA) for proofreading an early version of this document. I thank Anne K. Jones, Harsh Pershad, Raoul Camba, James Mc Evoy and Lars J. C. Jeuken for kindly providing some of the data shown herein.

Our work in Marseilles is funded by the CNRS (mainly), Aix-Marseille Université, the Agence Nationale de la Recherche, the Region Provence-Alpes-Cote d'Azur and the city of Marseilles.

APPENDIX

A.1 Notations and Abbreviations

$\mathscr{A}$: area under a pic (in units of VA). A: electrode surface. C: concentration of species, or capacitance (Eqn (8.21)), or coupling prefactor (Eqn (8.15)). D: diffusion coefficient. δ: peak width at half-heigth. η: overpotential. E: electrode potential. E^0: reduction potential. E_p: peak potential. ET: electron transfer. F: Faraday constant. Γ: electroactive coverage. i: current. i_c: capacitive current. K_{m}: Michaelis constant. K_i: inhibition constant. k: rate constant. λ: reorganisation energy. m number of protons. MV: methyl viologen. n number of electrons. ν: scan rate. ν_s: kinematic viscosity (Eqn (8.25)). OCP: open circuit potential. ω: electrode rotation rate. R: gas constant. RDE: rotating disc electrode. T: temperature (in K). t: time. V: potential difference. ξ: extent of reaction.

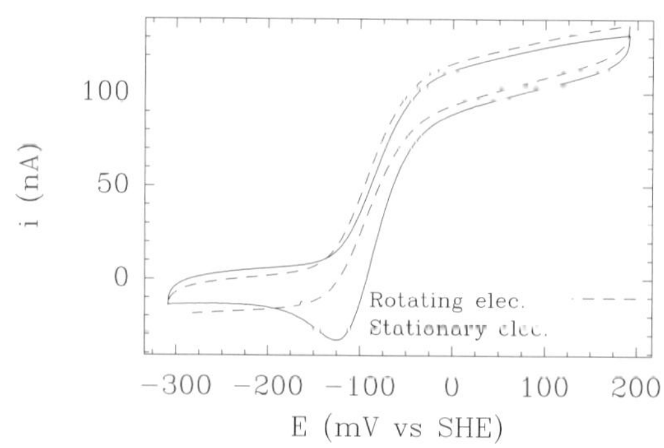

FIGURE 8.23 Effect of electrode rotation on the catalytic voltammetry of *E. coli* fumarate reductase (FrdAB) in a solution initially free of fumarate. $[S] = 50 \text{ mM} \times K_{\mathrm{m}}^{\mathrm{succinate}}$.

A.2 Derivation of Eqn (8.9)

We write the Nernst equation first for the alkaline couple Ox/Red, and then for both forms (protonated and deprotonated) of the redox couple:

$$E = E^0_{\text{alk}} + \frac{RT}{nF}\ln\left(\frac{[\text{Ox}]}{[\text{Red}]}\right) \tag{8.36a}$$

$$E = E^{0'}\left([\text{H}^+]\right) + \frac{RT}{nF}\ln\left(\frac{[\text{Ox}]+[\text{OxH}]}{[\text{Red}]+[\text{RedH}]}\right) \tag{8.36b}$$

We rewrite Eqn (8.36b) as follows:

$$E = E^{0'}\left([\text{H}^+]\right) + \frac{RT}{nF}\ln\frac{[\text{Ox}]\left(1+\frac{[\text{H}^+]}{K_{\text{Ox}}}\right)}{[\text{Red}]\left(1+\frac{[\text{H}^+]}{K_{\text{Red}}}\right)} \tag{8.37a}$$

$$E = E^{0'}\left([\text{H}^+]\right) + \frac{RT}{nF}\ln\frac{[\text{Ox}]}{[\text{Red}]} + \frac{RT}{nF}\ln\frac{1+\frac{[\text{H}^+]}{K_{\text{Ox}}}}{1+\frac{[\text{H}^+]}{K_{\text{Red}}}} \tag{8.37b}$$

Equating Eqn (8.37b) and Eqn (8.36a) gives Eqn (8.9):

$$\boxed{E^{0'}\left([\text{H}^+]\right) = E^0_{\text{alk}} + \frac{2.3RT}{nF}\log_{10}\left(\frac{1+\frac{[\text{H}^+]}{K_{\text{Red}}}}{1+\frac{[\text{H}^+]}{K_{\text{Ox}}}}\right)} \tag{8.38}$$

Check that $E^{0'}([\text{H}^+])$ tends to E^0_{alk} when $[\text{H}^+]$ is small. Using

$$E = E^0_{\text{acid}} + \frac{RT}{nF}\ln\frac{[\text{OxH}]}{[\text{RedH}]} \tag{8.39}$$

instead of Eqn (8.36a) gives

$$\boxed{E^{0'}\left([\text{H}^+]\right) = E^0_{\text{acid}} + \frac{2.3RT}{nF}\log_{10}\left(\frac{1+\frac{K_{\text{Red}}}{[\text{H}^+]}}{1+\frac{K_{\text{Ox}}}{[\text{H}^+]}}\right)} \tag{8.40}$$

Check that $E^{0'}([\text{H}^+])$ tends to E^0_{acid} when $[\text{H}^+]$ is large.
The relation between E^0_{alk} and E^0_{acid} is simply obtained by equating 36a and 39:

$$E^0_{\text{acid}} = E^0_{\text{alk}} + \frac{RT}{nF}\ln\frac{K_{\text{Ox}}}{K_{\text{Red}}} \tag{8.41}$$

Check that with $\text{p}K_{\text{Ox}} < \text{p}K_{\text{Red}}$, $E^0_{\text{acid}} > E^0_{\text{alk}}$.

REFERENCES

Adams, B., Smith, A. T., Bailey, S., McEwan, A. G., & Bray, R. C. (1999). *Biochemistry, 38*, 8501–8511.

Abou Hamdan, A., et al. (2012). *J. Am. Chem. Soc, 134*, 8368–8371.

Allongue, P., Delamar, M., Desbat, B., Fagebaume, O., Hitmi, R., Pinson, J., & Savéant, J.-M. (1997). *J. Am. Chem. Soc., 119*, 201–207.

Almeida, M. G., Guigliarelli, B., Bertrand, P., Moura, J. J. G., Moura, I., & Léger, C. (2007). *FEBS Lett., 581*, 284–288.

Alonso-Lomillo, M. A., Rudiger, O., Maroto-Valiente, A., Velez, M., Rodriguez-Ramos, I., Munoz, F. J., Fernandez, V. M., & de Lacey, A. L. (2007). *Nano Lett., 7*, 1603–1608.

Amatore, C., Arbault, S., Guille, M., & Lemaitre, F. (2008). *Chem. Rev., 108*, 2585–2621.

Anderson, L. J., Richardson, D. J., & Butt, J. N. (2001). *Biochemistry, 40*, 11294–11307.

Anxolabéhère-Mallart, E., Costentin, C., Policar, C., Robert, M., Savéant, J.-M. M., & Teillout, A.-L. L. (2011). *Faraday Discuss., 148*, 83–95.

Armstrong, F. A. (2005). *Curr. Opin. Chem. Biol., 9*, 110–117.

Armstrong, F. A., Heering, H. A., & Hirst, J. (1997). *Chem. Soc. Rev., 26*, 169–179.

Armstrong, F. A., Camba, R., Heering, H. A., Hirst, J., Jeuken, L. J. C., Jones, A. K., Léger, C., & McEvoy, J. P. (2001). *Faraday Discuss., 116*, 191–204.

Armstrong, F. A., Belsey, N. A., Cracknell, J. A., Goldet, G., Parkin, A., Reisner, E., Vincent, K. A., & Wait, A. F. (2009). *Chem. Soc. Rev., 38*, 36–51.

Arnoux, P., Sabaty, M., Alric, J., Frangioni, B., Guigliarelli, B., Adriano, J. M., & Pignol, D. (2003). *Nat. Struct. Mol. Biol., 10*, 928–934.

Baffert, C., et al. (2012). *Anal. Chem, 84*, 7999–8005.

Bard, A. J., & Faulkner, L. R. (2004). *Electrochemical Methods. Fundamental and Applications* (3rd ed.). John Wiley & Sons, Inc.

Battistuzzi, G., Borsari, M., Sola, M., & Francia, F. (1997). *Biochemistry, 36*, 16247.

Baymann, F., Barlow, N. L., Aubert, C., Schoepp-Cothenet, B., Leroy, G., & Armstrong, F. A. (2003). *FEBS Lett., 539*, 91–94.

Bernhardt, P. V. (2006). *Aust. J. Chem., 59*, 233–256.

Bertrand, P., Frangioni, B., Dementin, S., Sabaty, M., Arnoux, P., Guigliarelli, B., Pignol, D., & Léger, C. (2007). *J. Phys. Chem. B, 111*, 10300–10311.

Blanford, C., & Armstrong, F. (2006). *J. Solid State Electrochem., 10*, 826–832.

Blanford, C. F., Foster, C. E., Heath, R. S., & Armstrong, F. A. (2008). *Faraday Discuss., 140*, 319–335.

Bray, R. C., Adams, B., Smith, A. T., Bennett, B., & Bailey, S. (2000). *Biochemistry, 39*, 11258–11269.

Brereton, P. S., Verhagen, M. F. J. M., Zhou, Z. H., & Adams, M. W. W. (1998). *Biochemistry, 37*, 7351–7362.

Butt, J. N., Sucheta, A., Armstrong, F. A., Breton, J., Thomson, A. J., & Hatchikian, E. C. (1993). *J. Am. Chem. Soc., 115*, 1413–1421.

Butt, J. N., Fawcett, S. E. J., Breton, J., Thomson, A. J., & Armstrong, F. A. (1997). *J. Am. Chem. Soc., 119*, 9729–9737.

Chen, K., Hirst, J., Camba, R., Bonagura, C. A., Stout, C. D., Burgess, B. K., & Armstrong, F. A. (2000). *Nature, 405*, 814–817.

Chobot, S. E., Hernandez, H. H., Drennan, C. L., & Elliott, S. J. (2007). *Angew. Chem. Int. Ed. Engl., 46*, 4145–4147.

Compton, R. G., & Banks, C. E. (2011). *Understanding Voltammetry*. Imperial College Press.

Cornish-Bowden, A. (2004). *Fundamental of Enzyme Kinetics*. Portland Press.

Costentin, C., Robert, M., & Savéant, J.-M. (2010). *Acc. Chem. Res., 43*, 1019–1029.

Cracknell, J. A., Vincent, K. A., & Armstrong, F. A. (2008). *Chem. Rev., 108*, 2439–2461.

Dementin, S., Belle, V., Bertrand, P., Guigliarelli, B., Adryanczyk-Perrier, G., Delacey, A., Fernandez, V. M., Rousset, M., & Léger, C. (2006). *J. Am. Chem. Soc., 128*, 5209–5218.

Dutton, P. (1971). *Biochim. Biophys. Acta Bioenerg., 226*, 63–80.

Dutton, P. L. (1978). In*Redox Potentiometry: Determination of Midpoint Potentials of Oxidation-reduction Components of Biological Electron-transfer Systems*, vol 54. Elsevier.

East, G. A., & East and del Valle, M. A. (2000). *J. Chem. Educ* 77–97.

Elliott, S. J., Léger, C., Pershad, H. R., Hirst, J., Heffron, K., Blasco, F., Rothery, R., Weiner, J., & Armstrong, F. A. (2002a). *Biochim. Biophys. Acta, 1555*, 54–59.

Elliott, S. J., McElhaney, A. E., Feng, C., Enemark, J. H., & Armstrong, F. A. (2002b). *J. Am. Chem. Soc., 124*, 11612–11613.

Fawcett, S. E. J., Davis, D., Breton, J. L., Thomson, A. J., & Armstrong, F. A. (1998). *Biochem. J., 335*, 357–368.

Fontecilla-Camps, J. C., Volbeda, A., Cavazza, C., & Nicolet, Y. (2007). *Chem. Rev., 107*, 4273–4303.

Fourmond, V., Burlat, B., Dementin, S., Arnoux, P., Sabaty, M., Boiry, S., Guigliarelli, B., Bertrand, P., Pignol, D., & Léger, C. (2008). *J. Phys. Chem. B, 112*, 15478–15486.

Fourmond, V., Hoke, K., Heering, H. A., Baffert, C., Leroux, F., Bertrand, P., & Léger, C. (2009a). *Bioelectrochemistry, 76*, 141–147.

Fourmond, V., Lautier, T., Baffert, C., Leroux, F., Liebgott, P.-P., Dementin, S., Rousset, M., Arnoux, P., Pignol, D., Meynial-Salles, I., Soucaille, P., Bertrand, P., & Léger, C. (2009b). *Anal. Chem., 81*, 2962–2968.

Fourmond, V., Sabaty, M., Arnoux, P., Bertrand, P., Pignol, D., & Léger, C. (2010a). *J. Phys. Chem. B, 114*, 3341–3347.

Fourmond, V., Infossi, P., Giudici-Orticoni, M.-T., T., Bertrand, P., & Léger, C. (2010b). *J. Am. Chem. Soc., 132*, 4848–4857.

Fourmond, V., Burlat, B., Dementin, S., Sabaty, M., Arnoux, P., Etienne, E., Guigliarelli, B., Bertrand, P., Pignol, D., & Léger, C. (2010c). *Biochemistry, 49*, 2424–2432.

Frangioni, B., Arnoux, P., Sabaty, M., Pignol, D., Bertrand, P., Guigliarelli, B., & Léger, C. (2004). *J. Am. Chem. Soc., 126*, 1328–1329.

Gates, A. J., Richardson, D. J., & Butt, J. N. (2008). *Biochem. J., 409*, 159–168.

Gates, A. J., Kemp, G. L., To, C. Y. Y., Mann, J., Marritt, S. J., Mayes, A. G., Richardson, D. J., & Butt, J. N. (2011). *Phys. Chem. Chem. Phys., 13*, 7720–7731.

Gilles de Pelichy, L. D., & Smith, E. T. (1999). *Biochemistry, 38*, 7874.

Gray, H. B., & Winkler, J. R. (2005). *Proc. Natl. Acad. Sci. USA, 102*, 3534–3539.

Gutiérrez-Sánchez, C., Olea, D., Marques, M., Fernández, V. M., Pereira, I. A., Vélez, M., & De Lacey, A. L. (2011). *Langmuir, 27*, 6449–6457.

Hagedoorn, P. L., Driessen, M. C. P. F., vanden Bosch, M., Landa, I., & Hagen, W. R. (1998). *FEBS Lett., 440*, 311–314.

Heering, H. A., Weiner, J. H., & Armstrong, F. A. (1997). *J. Am. Chem. Soc., 119*, 11628–11638.

Heffron, K., Léger, C., Rothery, R. A., Weiner, J. H., & Armstrong, F. A. (2001). *Biochemistry, 40*, 3117–3126.

Hirst, J. (2006). *Biochim. Biophys. Acta, 1757*, 225–239.

Hirst, J., Jameson, G. N. L., Allen, J. W. A., & Armstrong, F. A. (1998a). *J. Am. Chem. Soc., 120*, 11994–11999.

Hirst, J., Duff, J. L. C., Jameson, G. N. L., Kemper, M. A., Burgess, B. K., & Armstrong, F. A. (1998b). *J. Am. Chem. Soc., 120*, 7085–7094.

Hudson, J. M., Heffron, K., Kotlyar, V., Sher, Y., Maklashina, E., Cecchini, G., & Armstrong, F. A. (2005). *J. Am. Chem. Soc., 127*, 6977–6989.

Iverson, T. M., Luna-Chavez, C., Cecchini, G., & Rees, D. C. (1999). *Science, 284*, 1961–1966.

Jeuken, L. J. (2009). *Nat. Prod. Rep., 26*, 1234–1240.

Jeuken, L. J. C., van Vliet, P., Verbeet, M. P., Camba, R., McEvoy, J. P., Armstrong, F. A., & Canters, G. W. (2000). *J. Am. Chem. Soc., 122*, 12186–12194.

Jeuken, L. C., Camba, R., Armstrong, F. A., & Canters, G. W. (2002a). *J. Biol. Inorg. Chem., 7*, 94–100.

Jeuken, L. J. C., McEvoy, J. P., & Armstrong, F. A. (2002b). *J. Phys. Chem. B, 106*, 2304–2313.

Jones, A. K., Lamle, S. E., Pershad, H. R., Vincent, K. A., Albracht, S. P. J., & Armstrong, F. A. (2003). *J. Am. Chem. Soc., 125*, 8505–8514.

Kudera, M., Aitken, A., Jiang, L., Kaneko, S., Hill, H. A. O., Dobson, P. J., Leigh, P. A., & McIntire, W. S. (2000). *J. Electroanal. Chem., 495*, 36–41.

Laidler, K. J., & Bunting, P. S. (1973). *The Chemical Kinetics of Enzyme Action*. Oxford: Clarendon Press.

Lautier, T., Ezanno, P., Baffert, C., Fourmond, V., Cournac, L., Fontecilla-Camps, J. C., Soucaille, P., Bertrand, P., Meynial-Salles, I., & Léger, C. (2011). *Faraday Discuss., 148*, 385–407.

Laviron, E. (1979). *J. Electroanal. Chem., 101*, 19–28.

Laviron, E. (1980). *J. Electroanal. Chem., 109*, 57–67.

Léger, C., & Bertrand, P. (2008). *Chem. Rev., 108*, 2379–2438.

Léger, C., Heffron, K., Pershad, H. R., Maklashina, E., Luna-Chavez, C., Cecchini, G., Ackrell, B. A. C., & Armstrong, F. A. (2001). *Biochemistry, 40*, 11234–11245.

Léger, C., Jones, A. K., Albracht, S. P. J., & Armstrong, F. A. (2002). *J. Phys. Chem. B, 106*, 13058–13063.

Léger, C., Elliott, S. J., Hoke, K. R., Jeuken, L. J. C., Jones, A. K., & Armstrong, F. A. (2003). *Biochemistry, 42*, 8653–8662.

Léger, C., Dementin, S., Bertrand, P., Rousset, M., & Guigliarelli, B. (2004). *J. Am. Chem. Soc., 126*, 12162–12172.

Léger, C., Lederer, F., Guigliarelli, B., & Bertrand, P. (2006). *J. Am. Chem. Soc., 128*, 180–187.

Leroux, F., Dementin, S., Burlat, B., Cournac, L., Volbeda, A., Champ, S., Martin, L., Guigliarelli, B., Bertrand, P., Fontecilla-Camps, J., Rousset, M., & Léger, C. (2008). *Proc. Natl. Acad. Sci. USA, 105*, 11188–11193.

Levich, V. G. (1962). *Physical Hydrodynamics*. Prentice-Hall, Inc.

Liebgott, P.-P., et al. (2010). *Nat. Chem. Biol., 6*, 63–70.

Limoges, B., & Savéant, J.-M. (2004). *J. Electroanal. Chem., 562*, 43–52.

Lojou, E. (2011). *Electrochimica Acta 56*. 10385–10397.

Love, J. C., Estroff, L. A., Kriebel, J. K., Nuzzo, R. G., & Whitesides, G. M. (2005). *Chem. Rev., 105*, 1103–1170.

McEvoy, J. P., & Armstrong, F. A. (1999). *Chem. Commun.* 1635–1636.

Meunier-Prest, R., & Laviron, E. (1992). *J. Electroanal. Chem., 328*, 33–46.

Murgida, D. H., & Hildebrandt, P. (2005). *Phys. Chem. Chem. Phys., 7*, 3773–3784.
Newman, J., & Thomas-Alyea, K. E. (2004). *Electrochemical Systems*. Wiley Interscience.
Nicholls, D. G., & Ferguson, S. J. (2002). *Bioenergetic 3*. Academic Press.
Park, J. B., Fan, C. L., Hoffman, B. M., & Adams, M. W. W. (1991). *J. Biol. Chem., 266*, 19351–19356.
Pershad, H. R., Duff, J. L. C., Heering, H. A., Duin, E. C., Albracht, S. P. J., & Armstrong, F. A. (1999). *Biochemistry, 38*, 8992–8999.
Rudiger, O., Abad, J. M., Hatchikian, E. C., Fernandez, V. M., & de Lacey, A. L. (2005). *J. Am. Chem. Soc., 127*, 16008–16009.
Savéant, J. M. (2006). *Elements of Molecular and Biomolecular Electrochemistry*. John Willey & Sons, Inc.
Sucheta, A., Ackrell, B. A. C., Cochran, B., & Armstrong, F. A. (1992). *Nature, 356*, 361–362.
Tarasevich, M. R., Yaropolov, A. I., Bogdanovskaya, V. A., & Varfolomeev, S. D. (1979). *J. Electroanal. Chem. Interf. Electrochem., 104*, 393–403.
Vincent, K. A., & Armstrong, F. A. (2005). *Inorg. Chem., 44*, 798–809.
Vincent, K. A., Parkin, A., & Armstrong, F. A. (2007). *Chem. Rev., 107*, 4366–4413.
Wardman, P. (1989). *J. Phys. Chem. Ref. Data, 18*, 1637–1755.
Yaropolov, A. I., Karyakin, A. A., Varfolomeev, S. D., & Berezin, I. V. (1984). *Bioelectrochem. Bioenerg., 12*, 267–277.
Zu, Y., Fee, J., & Hirst, J. (2001). *J. Am. Chem. Soc., 123*, 9906–9907.

Chapter 9

X-ray Crystallography

Jose A. Brito and Margarida Archer
Membrane Protein Crystallography Laboratory, Instituto de Tecnologia Química e Biológica, Universidade Nova de Lisboa, Av. República, EAN 2780-157, Oeiras, Portugal

Chapter Outline

QUESTIONS

- Describe briefly the main steps involved in the three-dimensional structure determination of macromolecules by X-ray crystallography.
- Comment on the major advantages and limitations of macromolecular crystallography.
- Enumerate the main factors that contributed to the development of crystallography.
- Understand the rationale of the crystallisation process and the parameters that may influence it.
- Distinguish between co-crystallisation and soaking techniques, and when one is preferable over the other.
- Compare the different methods used to solve the phase problem in X-ray crystallography.
- Specify the reasons why membrane proteins are more difficult to manipulate than soluble proteins and why crystals generally diffract poorly.

INTRODUCTION

X-ray crystallography is one of the most commonly used techniques to determine the three-dimensional structure of biological macromolecules, such as proteins, nucleic acids, or viral particles. Structure determination by X-ray

Practical Approaches to Biological Inorganic Chemistry, 1st Edition. http://dx.doi.org/10.1016/B978-0-444-56351-4.00008-7.

crystallography begins with growing a single crystal of the macromolecule whose structure is to be determined. An X-ray beam is then passed through the crystal. The X-rays interact with the electron clouds of the atoms in the crystal and their regular and repeating atomic arrangement gives rise to a complex pattern of diffracted beams which are recorded by a detector as spots (the diffraction pattern). Encoded in this pattern is information about the positions of all the atoms in the crystal. The nature of the diffraction phenomenon is such that the information available from the diffraction pattern is not enough to recover this information. Additional experiments and a considerable amount of mathematical computation are needed to obtain a map of electron density, displayed as contour maps. Ideally, the peaks in the electron density map correspond to the atomic positions in the molecule. This map is interpreted by building an atomic model of the molecule into it, and this model is then refined against the experimental data until a final model of good quality is obtained.

Crystallography can give reliable answers to many structure-related questions, from global folds to atomic details of bonding. The three-dimensional structure provides detailed information on the atom positions, specific atomic interactions (intra- and inter-molecular hydrogen bonds, salt bridges, hydrophobic pockets, etc.), as well as hints about the flexibility or mobility of the molecule. It can also give insights into the active site centres and reaction mechanism of enzymes, conformational changes occurring upon ligand binding, effects of point mutations in the protein fold and their repercussion on its function. The knowledge of accurate molecular structures is an important prerequisite for structure-aided drug design and for structure-based reaction mechanisms.

Structural studies in the crystal (by X-ray crystallography) and in solution (by NMR) for the same macromolecule, as well as studies of the same macromolecule crystallised in different conditions, have revealed very similar models. Although conformational changes may occur in some flexible regions of the protein due to crystal packing, X-ray crystallography has proved to be an excellent technique to characterise the three-dimensional structure of macromolecules. Moreover, many proteins were shown to be still active while in the crystal form, allowing further experiments to be performed on the crystalline state including time-resolved crystallography, UV-visible and Raman spectroscopy.

An analogy can be made between X-ray diffraction analysis of crystal structures and a very powerful microscope used to visualise the shape of small objects. In optical microscopy, a beam of visible light strikes the object and is scattered in various directions. A lens then collects the scattered rays and reassembles them to form an image. In turn, the use of electromagnetic radiation to visualise objects requires a wavelength comparable to the features one desires to resolve. Since atoms are separated by distances in the order of the angstrom (Å $= 10^{-10}$ m), X-rays are in the right wavelength range for the resolution of atomic features. X-ray diffraction may be compared with optical microscopy, except that there is no lens to focus the X-rays scattered by the electrons within the crystal, so no X-ray microscope can be built that allows direct visualisation of the protein atoms. Instead, the crystallographer measures the intensity of the scattered X-rays from the crystal and combines them with additional phasing information to compute an electron density map from which a model can be built. Each diffracted beam is a wave, characterised by an amplitude and a phase angle. However, only the amplitude of each wave can be derived from the experimentally measured diffraction intensities – the phase information is lost and must be recovered. This is the so-called 'phase problem' in X-ray crystallography. Since the signal from a single molecule would be too weak to be detected, a crystal is required because it comprises a very large number of molecules ($>10^{15}$) in the same orientation acting as an amplifier (Hickman and Davies, 2001). Also, the scattering of X-rays from a single molecule is a continuous function. However, when one or more molecules (objects) are arranged in a three-dimensional periodic arrangement (i.e. in a crystal), the scattering of X-rays can only occur in specific directions determined by the crystal lattice, whereas the intensity of these scattered beams is determined by the molecules in the crystal.

A brief journey through the history of crystallography with some relevant discoveries and related achievements will be highlighted herein, although many others could also be added. Over the past century, several scientists made important contributions to the crystallography and/or structural biology fields and were awarded Nobel Prizes, mostly in Physics and Chemistry. First, the discovery of radiation called X-rays ('X' meaning 'unknown') carried out by Wilhelm Röntgen in 1895 earned him the Nobel Prize in Physics in 1901, the year the Nobel Prizes were first attributed. Max von Laue (Nobel Prize in 1914) then discovered that X-rays are diffracted by crystals, which opened doors that were tightly closed until then and a new age began where the world of atoms was no longer out of reach. One year later, William H. Bragg and his son William L. Bragg won the Nobel Prize for their work on the crystal structure determination of NaCl. At that time, scientists irradiated all kinds of materials with X-rays (from glasses to fibrous substances, from polycrystalline metals to beeswax). The first X-ray diffraction image of a hydrated protein was taken in 1934 by J.D. Bernal, a doctoral student of W. H. Bragg. Interestingly, two of Bernal's doctoral students won Nobel Prizes due to their work in crystallography: Max Perutz (Chemistry Nobel Prize in 1962), who together with John Kendrew determined the first protein structures (myoglobin and haemoglobin); and Dorothy Hodgkin (Nobel Prize in 1964), for the determination of several important structures (penicillin, insulin, and vitamin B_{12}). The Nobel Prize in Physiology and Medicine in 1962 was attributed to Francis Crick, James Watson, and Maurice Wilkins for their discoveries concerning the molecular structure of nucleic acids (DNA double-helix) and its significance for information transfer in living organisms.

In the last 20 years there has been a considerable increase in the number of structures determined by macromolecular crystallography. As of November 2012, there were around 86,000 structures in the Protein Data Bank (PDB), however only ca. 1.5% correspond to membrane protein structures. The first X-ray structure of a membrane protein was a significant breakthrough in macromolecular crystallography. The structure of the bacterial photosynthetic reaction centre granted Johan Deisenhofer, Robert Huber, and Hartmut Michel the Nobel Prize in 1988. Their work pioneered the use of detergent molecules for the solubilisation and crystallisation of membrane proteins. Until then, attempts to extract proteins from membranes were unsuccessful due to the fact that proteins precipitated/denaturated in aqueous solutions because of the hydrophobic nature of their transmembrane domains. The knowledge obtained by Deisenhofer, Huber and Michel was the basis for all subsequent work in the field, including the determination of the three-dimensional structures of ATP synthase (Nobel Prize to John Walker in 1997), potassium channels (Nobel Prize to Roderick MacKinnon in 2003) and G protein-coupled receptors, GPCRs (Nobel Prize to Robert Letkowitz and Brian Kobika, 2012).

There is almost no limit to the size and complexity of the structures currently characterised by X-ray diffraction, provided suitable crystals are available. These achievements have resulted from: developments in molecular biology and protein engineering, which made it possible for a wide variety of proteins to be produced in sufficient amounts for structural studies; advances in computer technology, both in computational power and crystallographic software; access to increasingly more powerful X-ray sources such as synchrotrons, which provide very intense and tunable X-ray beams; and crystal cryo-cooling techniques that minimise crystal decay due to radiation damage during data collection, and in many cases allow a full data set to be recorded from one single macromolecule crystal.

X-ray structures of several protein–protein and protein–nucleic acid complexes were also characterised which, a few decades ago, would be unthinkable. The work of Roger Kornberg on the structure determination of RNA polymerase (Nobel Prize in 2006), and the studies on the structure and function of the ribosome (pioneered by Ada Yonath in the late 1970s which, together with Venkatraman Ramakrishnan and Thomas Steitz, won the Nobel Prize in 2009), are good examples. The ribosome, a molecular machine that translates the genetic information encoded in the mRNA into proteins, has a very large (>2.5 MDa) and complex macromolecular structure composed of both RNA and proteins molecules (with approximate proportion of 2/3 and 1/3, respectively).

A detailed description of the methods used to solve the three-dimensional structure of biological macromolecules by X-ray crystallography is beyond the scope of this chapter; more exhaustive information can be found in some of the referenced books (Blundell and Johnson, 1976; Drenth, 1999; Rupp, 2010) and reviews (Hickman and Davies, 2001). We aim to provide graduate students or researchers working with metalloproteins with an overview on what needs to be done in order to determine the three-dimensional-structure of proteins, and what information can be extracted from the crystallographic models. We should also point out that throughout this chapter some generalisations are made (where exceptions may exist) and simplifications that may lead to some omissions due to space limitation.

PROTEIN CRYSTALLISATION

Protein Production and Sample Preparation

Procedures on protein purification will not be included in this section since they are protein-specific and beyond the scope of this chapter. It is noteworthy that in the case of proteins with affinity tags attached to one of their termini (or even both), protocols include a selective purification step by affinity chromatography. Short peptide tags such as a stretch of histidine residues or a STREP-tag with eight amino acids (WSHPQFEK) are commonly used. Other tags are also available which are composed of proteins, such as glutathione-S-transferase, maltose-binding protein or green fluorescent protein. These tag proteins are co-expressed with the target protein and used for purification with specific affinity chromatography resins. If tags do not interfere with the folding and bioactivity of the target protein, and suitable crystals are obtained, there is no need to remove them. If not, an additional step of tag removal has to be performed provided a protease cleavage site is present between the tag and the target protein. However, the protein of interest may also suffer proteolysis at internal sites, and sample heterogeneity may occur as a result of incomplete tag cleavage and removal.

Protein Quality Assessment

Regardless of the protein source (e.g. bacterial, yeast, or mammalian) and production methods (native or recombinant protein), the quality of the protein sample needs to be assessed prior to crystallisation experiments. The protein should have a high degree of purity, stability, and homogeneity.

Denaturating gel electrophoresis such as sodium dodecyl sulfate polyacrylamide gel electrophoresis (SDS-PAGE) is usually a first indicator of the protein sample purity. It is important to examine the protein over a range of concentrations: more concentrated samples to visualise contaminating proteins present in low amounts, and more dilute to confirm the existence of only one band concerning the target protein, otherwise two close bands resulting from proteolysis may appear as a single band.

Heterogeneity can arise from different sources: post-translational modifications, deamidation, oxidation of amino acid side-chains, multiple conformations, or oligomeric forms (due to disulfide bridges between 'free' cysteines). Addition of dithiothreitol (DTT) or 2-mercaptoethanol prevents oxidation of the cysteine residues. Multiple conformations of the protein in solution may result from flexible domains. Size-exclusion chromatography is frequently used during protein purification or as a last 'polishing' step, where a sharp and symmetrical peak is a good indication of homogeneity. Furthermore, any aggregated material, if present in the sample, will be eluted in the column void volume. If a calibration curve is performed, the molecular weight of the protein can be estimated as well as its oligomeric form in solution. Dynamic light scattering is another method that can yield information about the sample monodispersity.

Protein samples can be further analysed by native gel electrophoresis, isoelectric focussing, N-terminal sequencing, and mass spectrometry. A stable, monodisperse and pure protein sample is the best starting point for crystallisation trials.

Protein Concentration

Another important parameter to consider is the protein quantity needed in order to proceed to crystallisation trials. As a rule-of-thumb, 10 mg of protein should be available to carry out the initial screens and to perform optimisation around those first hits in case crystalline material appears. Nowadays, with the advent of nanolitre crystallisation robots the amount of protein sample required to conduct screenings may be 10–20 times lower compared to manual screenings. This is particularly important for native or recombinant proteins with low production yields and for membrane proteins. Usually, the protein is at a concentration of 10–20 mg/mL, although proteins are known to have crystallised from samples as low as 2–3 mg/mL and as high as 275 mg/mL. Protein sample manipulations are usually performed at 4 °C to prevent sample degradation and denaturation and to retard bacterial growth. Several methods exist to exchange the sample buffer and to concentrate the protein solution. One of the most commonly used is filtration through a membrane using ultraconcentrators under gas pressure, or smaller concentrators that operate by centrifugation. The membrane cutoff (e.g. 10, 30, 100 kDa) selected depends on the molecular weight (MW) of the protein. It should be noted that some buffer components such as glycerol and detergents tend to concentrate above the membranes. Moreover, some proteins may stick to, or aggregate on the membrane. In these cases, buffer exchange through dialysis or size exclusion chromatography might be an option. Sometimes it is also not possible to concentrate the protein above a certain threshold, so a search for the best buffer composition should be carried out.

In this case, a fluorescence-based thermal stability assay can be used to quickly screen buffer and ligand conditions and investigate factors affecting protein stability by monitoring thermal melting curves. The fluorescence signal as a function of temperature can be followed using a real-time PCR device and a fluorophore (e.g. commercially available dyes or intrinsic probes such as flavin adenine dinucleotide, FAD) bound to the protein. An increase in the melting temperature (T_m) values will identify stabilising conditions for the protein sample. The basic assumption is that a more stable protein is more amenable to crystallise than a less stable one.

As crystallisation trials involve changing the protein environment in a controlled manner, the protein buffer composition should be kept to a minimum, e.g. 10–20 mM of a buffer adjusted to a specific pH. Eventually addition of DTT, 2-mercaptoethanol, ethylenediaminetetraacetic acid (EDTA), salts, glycerol or other agents (such as detergents for membrane proteins) may be required to further stabilise the protein. If the protein tolerates it, it is convenient to freeze small aliquots at −80 °C or in liquid nitrogen followed by storage at −80 °C, so that the samples can be thawed as needed. A cryo-protecting agent, like glycerol, may need to be included in the protein buffer prior to freezing. Some of the analytical methods described above can be used to confirm whether the protein is still in 'good shape' after thawing; and if possible check its activity. Some cases have been reported in the literature, where protein crystals were obtained only if the crystallisation drops were set up immediately after purification.

Crystallisation Techniques and Initial Screens

The first step in the three-dimensional structure determination of a protein involves the preparation of crystals of suitable quality for diffraction data measurements from a highly purified protein sample. Crystallisation is the first and most unpredictable step in protein crystallography, a consequence of the thermodynamic and kinetic variables involved in the formation of a unique crystalline three-dimensional packing. For crystallisation to occur, protein molecules must separate from solution and self-assemble into a periodic crystal lattice. Precipitants are added to the protein solution to reduce its solubility until supersaturation is achieved. In the supersaturated, themodynamically metastable state, nucleation can occur and crystals form.

Although a vast amount of knowledge has been accumulated over the last few decades, it is still not possible to know *a priori* in which conditions a specific protein may crystallise, so very often many different conditions have to be explored before crystals can be grown. Usually a systematic search is performed by varying in conjugated manner parameters like the protein concentration, precipitating agents, temperature, pH, buffers, ionic strength, etc. A wide variety of commercial kits are available to set up the initial crystallisation screens, although homemade screens may also be used (see Table 9.1 – sparse-matrix screen first compiled by Jancarik and Kim (1991)). Many laboratories are now equipped with crystallisation robots which can dispense nanolitre volume drops into 96-well plates. The miniaturisation of crystallisation experiments not only reduces the demand on the amount of purified protein required for screening but also enables extensive exploration of the chemical 'space' to identify optimal conditions for crystallisation.

Crystallisation methods commonly used include vapour diffusion (hanging or sitting drops), batch, dialysis and capillary diffusion. The most frequent one is the **vapour diffusion** technique (Figure 9.1a).

TABLE 9.1 Sparse-matrix Screen

No.	Precipitant	Buffer	Salt
(1)	30% MPD	0.1 M acetate	0.2 M Ca chloride
(2)	0.4 M Na, K tartrate		
(3)	0.4 M NH_4 phosphate		
(4)	2.0 M NH_4 sulfate	0.1 M tris	
(5)	40% MPD	0.1 M hepes	0.2 M Na citrate
(6)	30% PEG 4000	0.1 M tris	0.2 M Mg chloride
(7)	1.4 M Na acetate	0.1 M cacodylate	
(8)	30% 2-propanol	0.1 M cacodylate	0.2 M Na citrate
(9)	30% PEG 4000	0.1 M citrate	0.2 M NH_4 acetate
(10)	30% PEG 4000	0.1 M acetate	0.2 M NH_4 acetate
(11)	1.0 M NH_4 phosphate	0.1 M citrate	
(12)	30% 2-propanol	0.1 M hepes	0.2 M Mg chloride
(13)	30% PEG 400	0.1 M tris	0.2 M Na citrate
(14)	28% PEG 400	0.1 M hepes	0.2 M Ca chloride
(15)	30% PEG 8000	0.1 M cacodylate	0.2 M NH_4 sulfate
(16)	1.5 M Li sulfate	0.1 M hepes	
(17)	30% PEG 4000	0.1 M tris	0.2 M Li sulfate
(18)	20% PEG 8000	0.1 M cacodylate	0.2 M Mg acetate
(19)	30% 2-propanol	0.1 M tris	0.2 M NH_4 acetate
(20)	25% PEG 4000	0.1 M acetate	0.2 M NH_4 sulfate
(21)	30% MPD	0.1 M cacodylate	0.2 M Mg acetate
(22)	30% PEG 4000	0.1 M tris	0.2 M Na acetate
(23)	30% PEG 400	0.1 M hepes	0.2 M Mg chloride
(24)	20% 2-propanol	0.1 M acetate	0.2 M Ca chloride
(25)	1.0 M Na acetate	0.1 M Imidazole	
(26)	30% MPD	0.1 M citrate	0.2 M NH_4 acetate
(27)	20% 2-propanol	0.1 M hepes	0.2 M Na citrate
(28)	30% PEG 8000	0.1 M cacodylate	0.2 M Na acetate
(29)	0.8 M Na,K tartrate	0.1 M hepes	
(30)	30% PEG 8000		0.2 M NH_4 sulfate

TABLE 9.1 Sparse-matrix Screen—Cont'd

No.	Precipitant	Buffer	Salt
(31)	30% PEG 4000		0.2 M NH_4 sulfate
(32)	2.0 M NH_4 sulfate		
(33)	4.0 M Na formate		
(34)	2.0 M Na formate	0.1 M acetate	
(35)	1.6 M Na,K phosphate	0.1 M hepes	
(36)	8% PEG 8000	0.1 M tris	
(37)	8% PEG 4000	0.1 M acetate	
(38)	1.4 M Na citrate	0.1 M hepes	
(39)	2% PEG 400, 2.0 M NH_4 sulfate	0.1 M hepes	
(40)	20% 2-propanol, 20% PEG 4000	0.1 M citrate	
(41)	10% 2-propanol, 20% PEG 4000	0.1 M hepes	
(42)	20% PEG 8000		0.05 M K phosphate
(43)	30% PEG 1500		
(44)	0.2 M Mg formate		
(45)	18% PEG 8000	0.1 M cacodylate	[illegible]
(46)	18% PEG 8000	0.1 M cacodylate	0.2 M Ca acetate
(47)	2.0 M NH_4 sulfate	0.1 M acetate	
(48)	2.0 M NH_4 sulfate	0.1 M tris	
(49)	2% PEG 8000		0.2 M Li sulfate
(50)	15% PEG 8000		0.2 M Li sulfate

PEG - polyethylene glycols, MPD - 2-methyl-2,4-pentanediol
Buffers pH: acetate - 4.5, citrate - 5.5, cacodylate - 6.5, hepes - 7.5, tris - 8.5
Adapted from Jancarik and Kim (1991).

A basic experiment setup is illustrated in Figure 9.1a. Here, a small droplet of protein solution (typically 1−5 μL) is mixed with a similar volume of the crystallising solution and is placed on a siliconised glass cover (in a hanging drop). The reservoir or well, with a much larger volume (usually 500 μL) of precipitating solution, has a greased rim and is sealed with the flipped-over cover slide. It should be noted that the precipitant in the protein-precipitant mixture in the hanging drop was diluted and so is less concentrated than the reservoir solution. The difference in the concentration of the precipitating agent between the drop and the reservoir drives the system towards equilibrium by diffusion through the vapour phase. Water will evaporate from the drop into the reservoir until the concentration of the precipitant in the drop equals the one in the well. Thus, the concentrations of both protein and precipitant slowly increase in the drop. The solution in the drop should become supersaturated, i.e. the protein solubility limit is exceeded, so eventual nucleation can occur and crystals grow (Figure 9.2).

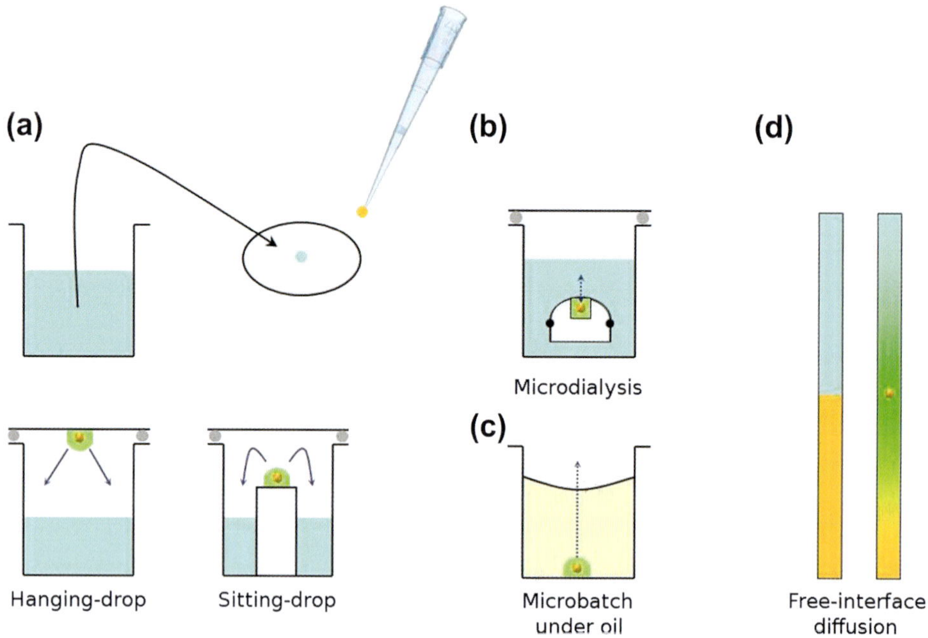

FIGURE 9.1 Schematic representation of crystallisation techniques: vapour diffusion by hanging and sitting drop (a), microdialysis (b), microbatch (c) and free-interface diffusion (d) Solutions are depicted in blue for reservoir, pale yellow for oil, yellow for protein, and green for mixture of protein and reservoir; crystals are represented by an yellow cubic shape.

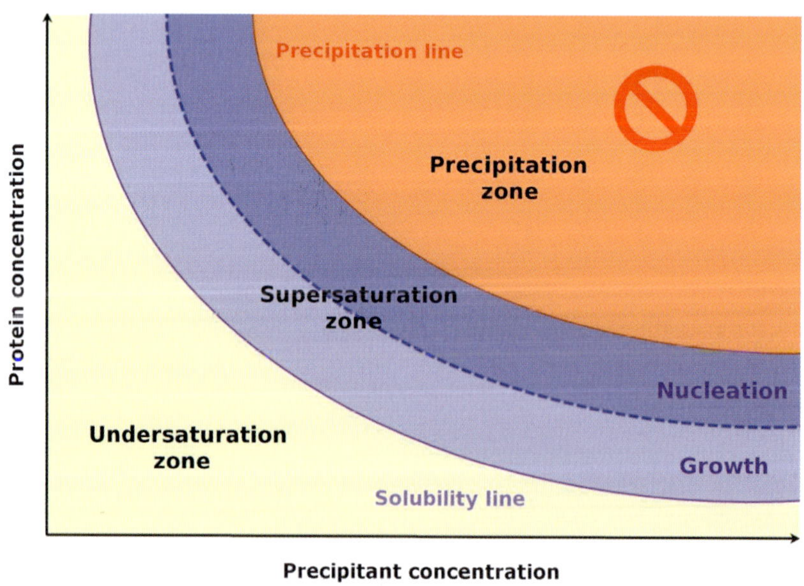

FIGURE 9.2 Solubility phase diagram. In a crystallisation experiment, the protein sample should be in a supersaturated zone (metastable zone) where nucleation may occur and crystals grow.

Dialysis methods (Figure 9.1b) permit the variation of many parameters that can influence the crystallisation process. The protein is separated from a large volume of solvent by a semi-permeable membrane that allows the passage of water and small molecules but blocks protein macromolecules (and also high molecular weight polyethylene glycols – PEGs). Equilibration kinetics depends on the membrane molecular weight exclusion size, ratio of precipitant concentration inside and outside the chamber, temperature and geometry of the dialysis cell. Microdialysis buttons exist in various sizes and forms.

Another crystallisation method employed is the **batch** technique (Figure 9.1c), where precipitant and protein solutions are simply mixed and allow kinetics to take their course. Supersaturation is achieved directly, rather than by diffusion. If very small volumes are used, this technique is called microbatch and the droplets are covered with oil, such as paraffin, to prevent evaporation. Sometimes, silicone oils, which are more water permeable than paraffin, are used to cover the microbatch wells, allowing for a partial exchange of solvent vapours.

In the **free-interface diffusion**, the protein and precipitant solutions are brought into contact in a narrow vessel (e.g. capillaries) without premixing and the components are allowed to equilibrate by diffusion only. Due to the small size of the capillary, the precipitant usually diffuses directly into the protein solution.

Analysis of Crystallisation Trials

Regardless of the chosen methods, the crystallisation setups have to be monitored regularly, and depending on those observations, a decision has to be made on how to proceed next. Vapour diffusion is the most commonly used technique for the initial screenings. Visualisation of the drops under a stereo microscope should be done just after their set up, repeated every other day during the first weeks, and then at longer time-intervals. If most of the drops remain clear, then the precipitant and/or protein concentrations should be increased (undersaturation zone, Figure 9.2). On the contrary, if most drops show heavy precipitation, then the concentration of precipitant and/or protein should be decreased (supersaturation zone, Figure 9.2). It is also important to observe whether the protein tends to denature under a specific pH range or in the presence of certain precipitants (e.g. alcohols, polyethylene glycols, salts). If no positive hits are obtained (no crystalline material observed), the search field should be enlarged: other screenings can be tried, the ratio of protein to precipitant modified (e.g. 1:1, 2:1, or 1:2) or the temperature varied (usually room temperature ~20–25 and 4 °C, and if possible try intermediate and higher temperatures, e.g. 15 and 30 °C). Moreover, the addition of various compounds (e.g. additive screenings), formation of complexes with ligands, cofactors or substrate analogues in the case of enzymes (may decrease protein flexibility and/or increase its stability) can also influence the crystallisation process. At this stage, prior knowledge of the protein and its biochemical/functional characterisation can be very informative, providing additional clues or strategy options for further screening experiments. Since the number of possible combinations is unlimited, crystallisation can be an *ad eternum* process and a tricky question is to decide whether to keep going or give up.

As a final resort, other strategies such as limited proteolysis can still be explored, which can be useful for isolating protein fragments that can fold autonomously and thus behave as protein domains which are often more prone to crystallise in the absence of flexible loops. Site-specific mutagenesis, protein engineering and protein modification (e.g. lysine methylation) can also be useful.

Salt or Protein Crystals?

If crystalline material is observed, it is important to confirm that the crystal is made of protein and not salt (or detergent in the case of membrane proteins). One simple method is to stain the drop with a dye such as methylene blue or IZIT© (Hampton Research). These small molecule dyes diffuse into the solvent channels of protein crystals, colouring the crystals blue with an intensity over and above the background drop colour (Fig. 9.3e). Salt crystals do not possess these large solvent channels and generally stay colourless. Another test is to crush the suspect crystal: protein crystals are usually very sensitive and fragile and easily break with a needle or a stiff fibre, in contrast with salt crystals that are much harder and resistant, although there can be exceptions. Intrinsic ultraviolet fluorescence is another method that may be used to distinguish between salt and protein crystals, or to visualise protein crystals hidden under amorphous precipitates. The simplest setup requires a UV light source coupled to a microscope, or a more sophisticated system consisting of a UV-fluorescence microscope with a CCD (charge-coupled device) camera, although a low protein content in aromatic residues, particularly tryptophan residues, might be a limitation. Moreover, several crystals can also be collected out of the drop with a nylon loop, thoroughly washed and dissolved so that a sample can be run on SDS-PAGE to see whether a band corresponding to the molecular weight of the expected protein is observed. However, the ultimate test will be the X-ray diffraction pattern of the crystals, where protein crystals should have many 'spots' very close to one another, in contrast to only a few spots far away from one another for salt crystals; this observation relates to their different unit cell dimensions (see Figure 9.14).

Protein samples that are coloured due to the presence of chromophore(s), such as metalloproteins or flavin-containing proteins are easily recognised as protein crystals because of their colour (e.g. red/brown for iron, blue for copper and yellow for flavin; a few examples are shown in Figure 9.3). Interestingly, some crystals can turn colourless upon protein reduction or loss of the 'coloured' cofactor or metal.

Crystal Optimisation and Seeding

The positive hits from the initial screenings reveal conditions in which protein crystals can be obtained. Afterwards, a fine-tuning of the experimental parameters is necessary to improve the crystal quality. Usually, narrower grids are tried around the initial conditions, including small variations in the precipitant/protein concentrations, pH, temperature, drop ratio, accompanied by trial of additives and/or ligands. The typical size of a protein crystal is in the order of 0.1–0.2 mm, although even smaller crystals can now be measured at synchrotron facilities using microfocus beamlines (i.e. with a very narrow X-ray beam).

For crystal optimisation, micro- and macro-seeding techniques are often used. **Nucleation** is a key factor to crystallisation and only occurs at supersaturated levels, which often leads to poor results, as too many small crystals in the drop. **Seeding** permits to uncouple nucleation and crystal growth. Seeds from the initial or previous trials are placed in low supersaturated crystallisation solutions (so that there is no spontaneous nucleation) to allow for their controlled and slow growth. In **microseeding**, a stock solution is made containing a few crushed crystal fragments or microcrystals and a series of dilutions are prepared from the initial stock solution. A small aliquot from each solution is then added to new crystallisation drops of reduced supersaturation (less precipitant and/or less protein) and the best dilution will yield a few larger crystals in the droplet. A popular variation of this technique is the **streakseeding**, where a thin whisker or fibre is streaked across a seed crystal and swiped through the new drop(s); sometimes several consecutive passes through various drops are necessary until only a small number of crystals are obtained. In contrast, **macroseeding** involves the transfer of a single well-formed crystal into a new crystallisation drop (of similar condition where it was originally grown) after washing it several times. The idea is to remove imperfections or other micronuclei from the seed crystal surface and have more 'fresh' protein available for further growth.

Co-crystallisation and Soaking

Once native crystals are obtained, it may be of interest to obtain the structure of the protein in complex with other molecules such as ligands, cofactors, non-hydrolysable substrate analogues, drug lead compounds, DNA/RNA oligomers, peptides or even with other proteins. Obtaining crystals of protein–ligand complexes can be achieved by either **soaking** ligands into pre-grown protein crystals or **co-crystallisation**, where the protein and its binding partner are incubated prior to crystallisation setups (Figure 9.4).

Protein crystals contain large solvent channels between the protein molecules which are filled with solvent (crystal solvent content is often around 50%, but can vary from 20 to 80%). The soaked ligand molecules are usually small enough to diffuse through these solvent channels (around 20–100 Å wide) and bind to the protein. In

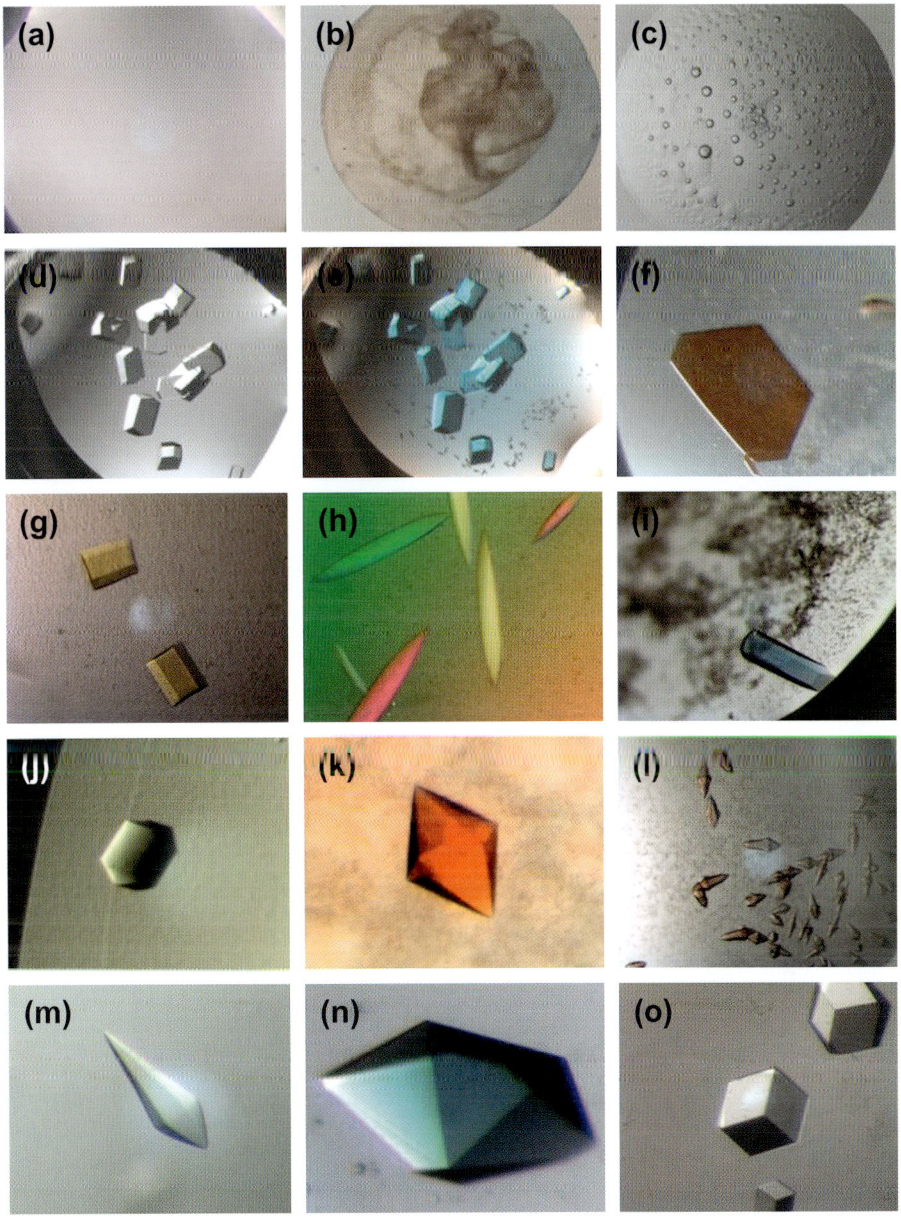

FIGURE 9.3 Images of crystallisation drops: (a) a clear drop (undersaturation zone); (b) drop with amorphous precipitate (precipitation zone); (c) drop with phase separation (metastable zone); and crystals of (d) lysozyme; (e) lysozyme after IZIT staining; (f) Rieske ferredoxin (with [2Fe-2S] centre); (g) sulfide:quinone oxidoreductase (Brito et al., 2006) (FAD cofactor); (h) papain from *Carica papaya* visualised under polarised light (crystals were originally colourless); (i) laccase CotA (Bento et al., 2005) (Cu centres); (j) dissimilatory sulfite reductase (Oliveira et al., 2008) (sirohaems and [4Fe-4S] clusters); (k) aldehyde oxido:reductase (Romao et al., 1995) (molibdopterin and [2Fe-2S] clusters); (l) CbiK chelatase (Romao et al., 2011)(haem *b*); (m) RNaseII (McVey et al., 2006); (n) neelaredoxin (Pinho et al., 2010) (Fe centre); (o) Holliday-junction resolvase mutant.

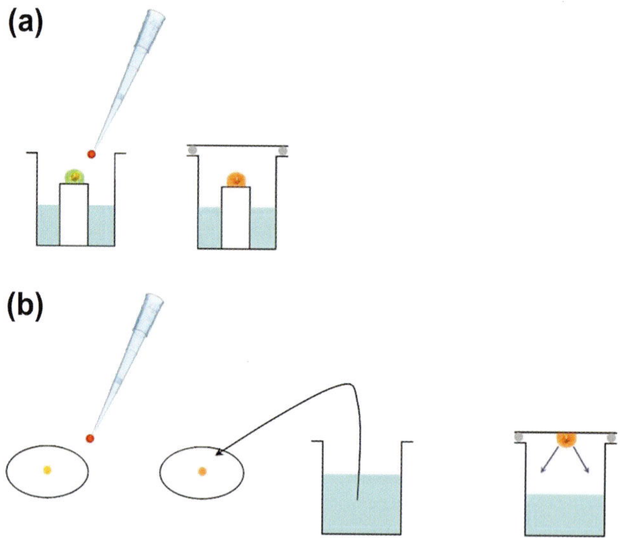

FIGURE 9.4 Schematic representation of (a) soaking, where the ligand solution is added into a drop containing pre-grown protein crystals and (b) co-crystallisation techniques, where the protein and ligand solutions are incubated prior to crystallisation experiments. Solutions are coloured in blue for reservoir, yellow for protein, red for ligand and orange for protein-ligand complex. The yellow cubic-shaped crystal becomes orange upon ligand binding.

soaking experiments, a solution containing the ligand is added directly to a drop with crystal(s) (Figure 9.4a). However, some ligands can induce protein conformational changes upon binding, leading to crystal destruction. Changes in the mother buffer (crystallisation solution) can also lead to crystal cracking or a decrease of their X-ray diffraction power. Stabilisation of the protein crystals by cross-linking with glutaraldehyde can reduce these problems. The soaking time and ligand concentration usually require optimisation. In **co-crystallisation** experiments (Figure 9.4b), the ligand is added to the protein to form a complex that is subsequently used for crystallisation trials. This is often the method of choice when the compounds are quite insoluble or the protein aggregates easily. It should be stressed that the crystallisation conditions of the complexed protein may be different from those of the protein by itself.

Membrane Proteins

It is estimated that about 30% of the human genome encodes for membrane proteins, and over 60% of all current drug targets are membrane receptors. Membrane proteins are key players in many cellular functions, such as energy production, transport, signal transduction, and intercellular communication. Despite the importance of membrane proteins, data analysis of the PDB reveals that very little structural information is available on membrane proteins, representing less than 2% of all known protein structures. The reason behind this huge deficit is due to the generally low abundance of membrane proteins and difficulty in obtaining crystals suitable for X-ray diffraction. Additional complications arise if membrane proteins suffer post-translational modifications, e.g. glycosylation of G-protein coupled receptors (GPCRs).

Integral membrane proteins span across the hydrophobic lipid bilayer (as transmembrane α-helices or forming β-barrels) and are insoluble in aqueous media. During overexpression, only a small portion of the expressed membrane proteins can be transported through the cell and properly incorporated into the host membrane before the cells are broken down. Extraction from the cell membranes and solubilisation usually requires the use of detergents (water-soluble amphipatic molecules, commonly with a hydrophobic tail and polar head). Detergent molecules will associate with the membrane protein forming a micelle complex with their hydrophobic parts facing inwards and the polar groups in contact with the aqueous solution (Figure 9.5). The detergent concentration has to be kept above its critical micelle concentration (CMC), otherwise the micelle complex will be disrupted with the subsequent precipitation of the protein. However, if the detergent concentration is too high above the CMC or if salts are used as precipitants with ionic detergents, a detergent-rich phase tends to form, into which the protein can get partitioned and denaturated. PEGs are therefore the first choice to be used as precipitants for membrane protein crystallisation. In the case of peripheral membrane proteins, which are weakly associated with the membrane, the extraction can be accomplished by the addition of high concentrations of salt, such as NaBr (Figure 9.5).

Detergents tend to inhibit crystallisation and, even if crystals are obtained, their diffraction quality is usually very poor. Among the most important factors contributing to the very weak diffraction of membrane protein crystals are the occlusion of polar amino acid residues by detergent molecules, thus hindering their participation in intermolecular contacts; together with a high solvent content and internal disorder. The primary solubilisation detergent often needs to be exchanged against a milder detergent more suitable for crystallisation, such as sugar-based detergents of the maltoside or glucopyranoside family (e.g. octyl-β-D-glucopyranoside, dodecyl-β-D-maltoside) or zwitterionic phosphocholine detergents. The size, structure and type (ionic versus neutral) of detergents, along with their CMC, are critical parameters for the crystallisation of membrane proteins. Addition of other detergents, lipid molecules or small amphiphiles can help to improve crystallisation. Detergent kits are commercially available that can be tried as additives. Although the solubilised membrane protein can be regarded

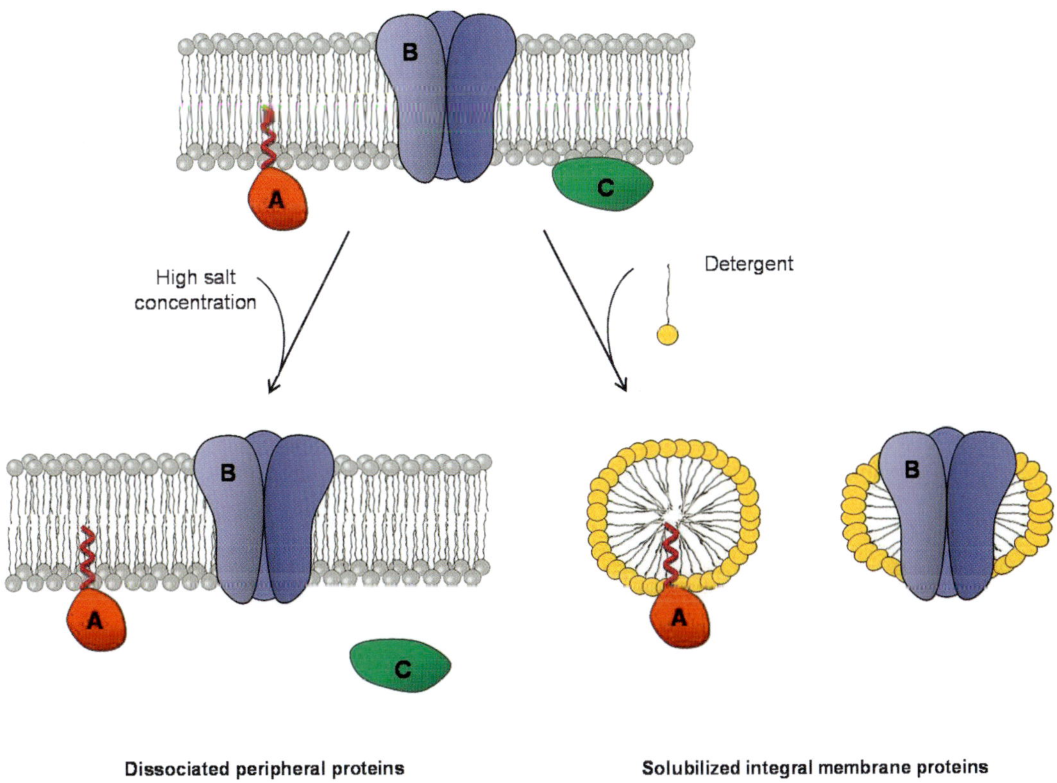

FIGURE 9.5 Solubilisation of membrane proteins. Peripheral membrane proteins (C) can usually be extracted from the membrane with high salt concentrations (e.g. 3 M NaBr), and are soluble without detergents, while integral (B) and associated (A) membrane proteins require detergents for their extraction and solubilisation forming protein–detergent micelles.

as a 'soluble protein', the presence of detergent and the delicate equilibrium of the detergent-membrane protein micelle add another level of complexity to the process.

Another technique that has been developed for membrane proteins is crystallisation in lipid cubic phases. The lipid mono-olein forms a complex phase system with water. One of the phases is a bilayered cubic phase containing 50–80% lipid and interconnected solvent channels. This setup tries to mimic the native lipid bilayer environment (Caffrey, 2009).

Harvesting and Mounting of Crystals

The ultimate test of crystal quality is achieved when X-ray diffraction images are obtained so that its diffraction pattern and resolution limit can be analysed. In order to do this, crystals need to be harvested from the crystallisation drop and mounted on the diffractometer. Protein crystals need to remain surrounded by their mother liquor during crystal mounting and data collection, otherwise they will dry out, lose long-range three-dimensional order and stop diffracting. Crystals can be mounted in quartz capillaries and data collected at room temperature (Figure 9.6a). However, crystals are most frequently pooled from the crystallisation drop using a tiny loop (made of, e.g. nylon fibre, Figure 9.6b) attached to a steel pin which is in turn mounted on a steel base, and flash-cooled in either liquid nitrogen (boiling point of 77 K or −196 °C), liquid propane (~150 K, more rarely used) or under a cryogenic nitrogen gas stream (around 100 K). The crystal-containing pin is then mounted on a diffractometer equipped with a cryostat to keep the low temperature (ca. 100 K) during data acquisition. Alternatively, the flash-cooled crystal can be inserted into a plastic vial filled with liquid N_2 and stored in a cooled Dewar for later measurements.

The main benefit of low-temperature diffraction measurements is the reduction of X-radiation damage to the crystals. As a consequence, complete data sets can be recorded from a single crystal. The dynamic disorder of the crystals may also be decreased (reflected in the thermal motion parameters, see below). However, to prevent the detrimental formation of ice within the crystals and/or mother liquor, a cryo-protectant is required. Therefore, crystals are either grown in solutions already containing a cryo-protectant, such as glycerol, PEG, sucrose or salts in the appropriate concentration, or are briefly swept through a buffer, where such a compound was added prior to freezing. It is important that this addition is done by replacing water in the buffer, so that the concentration of all the other substances remains unchanged.

DATA COLLECTION

Once crystallisation is successful, the next step is to expose a crystal to an X-ray beam and measure the diffracted signal. A **crystal** is a periodic, finite assembly of **unit cells**, the smallest repeating unit that can generate the crystal with only translation operations (Figure 9.7). Each unit cell contains the same number of identically arranged molecules. The **asymmetric unit** is the smallest fraction of the unit cell that can be rotated and translated using

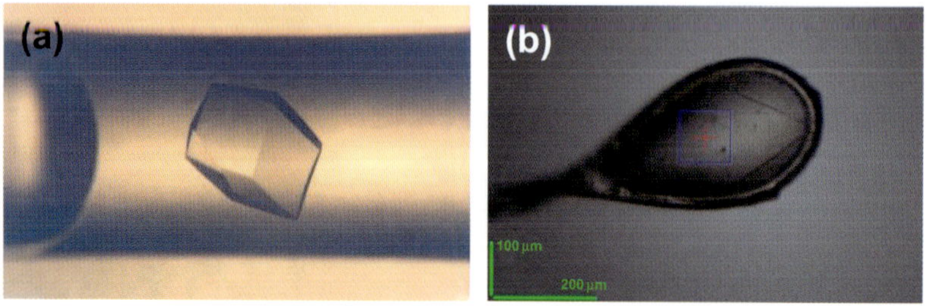

FIGURE 9.6 Crystal mounted in a sealed quartz capillary (a) with reservoir (usually for room temperature data collection), and in a loop (b) surround by a cryo-protectant solution (for cryogenic temperature data collection).

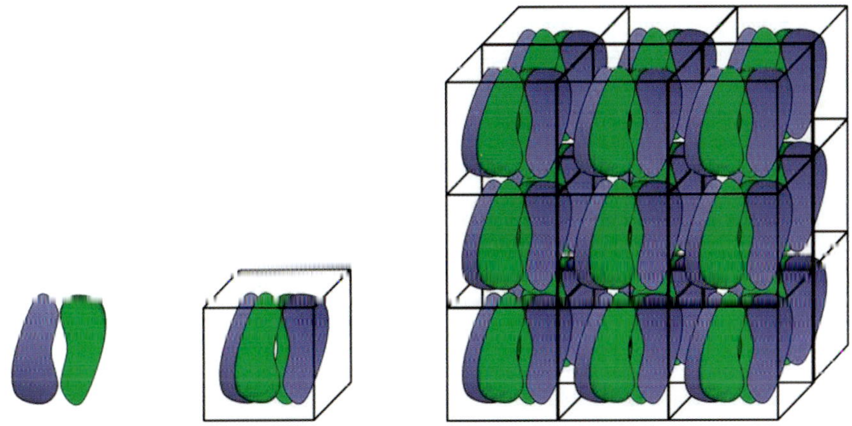

FIGURE 9.7 The asymmetric unit is the smallest portion of a crystal structure to which symmetry operations can be applied in order to generate the complete unit cell (i.e. the crystal repeating unit). In this case, the asymmetric unit also displays Non-Crystallographic Symmetry (NCS), which relates the two molecules by a two-fold NCS operator (180° rotation). Two asymmetric units form the unit cell, which is then translationally repeated in three directions to make a three-dimensional crystal.

only the symmetry operators allowed by the crystallographic symmetry to generate one unit cell. It may contain the whole macromolecule or one (or more) subunit(s) of a multimeric protein. If two or more identical molecules are present in the asymmetric unit, they are usually related by non-crystallographic symmetry (NCS) (Figure 9.7).

The crystal acts as a three-dimensional grating that diffracts the incident beam of X rays only in certain directions. These directions depend on the orientation of the crystal and its unit cell dimensions. The diffraction pattern is a record of the directions and intensities of all the diffracted X-ray beams by a crystal (see Figure 9.14). Scattered waves in phase add up, augmenting the signal to a measurable level in some directions, and cancel one another out in all other directions. That is why the diffraction pattern from a crystal consists of a discrete number of diffracted beams. These beams can be recorded on a detector as diffraction spots. The crystal is rotated relative to the X-ray beam so a complete diffraction pattern can be recorded. A single diffraction data set may comprise tens to hundreds of thousands of reflections, and sometimes more than one data set is needed.

The diffraction data is the experimental 'raw material', from which the protein structure is derived. A good quality data set is characterised by: a low background noise, data extending to high values of the reflection angle θ with no overlaps, a wide range of intensities between the weakest and strongest reflections (which should not be overloaded), a good agreement between the intensities of symmetry-related reflections. The resolution d depends on the angular extent to which the intensity data have been measured (θ is the diffraction angle) and on the wavelength λ of the radiation, according to Bragg's law:

$$\lambda = 2d \sin\theta$$

Resolution (Å) in a protein structure is usually defined as the value of $d_{\min}$ corresponding to the $\theta_{\max}$ of the data set. This parameter corresponds to the level of detail which can be seen in the resulting electron density map, and ultimately determines the quality of the structural model that can be derived from the map.

Usually, in-house X-ray equipment comprises a rotating anode X-ray generator with a copper target which emits a peak of CuKα radiation at 1.5418 Å, a two-dimensional X-ray detector, either an imaging plate or a charge coupled device (CCD), and a cryo-cooling system to keep crystals at temperatures close to that of liquid nitrogen. Imaging-plate detectors are exposed to the scattered X-rays, readout with a laser scanning device and erased with

UV light, so they can be re-exposed. Their disadvantage is the long (2–5 minutes) time required for image readout and plate erasure. More recently, CCD detectors have emerged, offering the ability to record diffraction data at an unprecedented rate, speeding up data collection by two orders of magnitude. Their much faster readout time is also combined with higher sensitivity and a better dynamic range (ratio between the strongest and the weakest measurable intensities). Moreover, synchrotron beamlines are now being equipped with PILATUS detectors, consisting of two-dimensional hybrid pixel array detectors, which operate in single-photon counting mode. These detectors feature several advantages compared to current state-of-the-art CCD, namely no readout noise, superior signal-to-noise ratio, shorter readout time and higher dynamic range. The short readout and fast framing times allows for the collection of diffraction data in continuous mode without opening and closing the shutter for each frame.

The access to synchrotron facilities was and still is very important in the development of X-ray crystallography. Synchrotron radiation sources provide X-ray beams with high intensity, low divergence and a wide range of energy. The high intensity of X-rays increases the rate of data acquisition and allows data to be recorded faster and from smaller crystals than when using a home source. The low divergence reduces the overlap between adjacent reflections enabling data collection from crystals with larger unit cells (e.g. virus particles, large assemblies or complexes). A beam with tunable energy offers the possibility to select a specific wavelength, which is essential for anomalous dispersion experiments (see 'phase determination'). On the other hand, a white X-radiation can yield a complete Laue diffraction pattern for a protein crystal on a millisecond scale. Laue diffraction is most often used in time-resolved crystallographic experiments.

PHASE DETERMINATION

As described in the previous section, the diffraction pattern is a record of the directions and intensities of all the diffracted X-ray beams by a crystal. Each diffracted beam corresponds to the vector sum of the individual contributions from all the atoms in the unit cell, and is characterised by an amplitude $|F_{hkl}|$ (related to the number of electrons in each contributing atom) and a phase α_{hkl} (derived from the atom positions within the unit cell). Just as the atomic positions in a crystal structure can be defined using a three-dimensional coordinate system (x, y, z), each individual diffracted beam (or diffraction spot) can be defined via three integer values (h, k, l), termed Miller (or Bragg) indices.

$\mathbf{F}_{hkl}$, is called the structure factor (depends on the structure of the scattering atoms) and is represented by:

$$F_{hkl} = \sum_h \sum_k \sum_l |F_{hkl}| \exp(i\alpha_{hkl})$$

The structure factor amplitudes $|F_{hkl}|$ are proportional to the square root of the intensity of the corresponding diffracted beams and are experimentally measured, whereas the phase angles are not. An estimate of these 'lost' phases α_{hkl} has to be obtained by 'indirect methods', in order to calculate an electron density map of the crystal structure. Once both the amplitudes and reasonable estimates for the phases of all the diffracted X-ray beams are known, it is possible to calculate the electron density (ρ) at any point xyz within the unit cell, represented by the following equation:

$$\rho(xyz) = \frac{1}{V} \sum_h \sum_k \sum_l |F_{hkl}| \exp(i\alpha_{hkl}) \exp[-2\pi i(hx + ky + lz)]$$

where, V is the volume of the unit cell (the repeating unit of the crystal), and α_{hkl} is the phase associated with the structure factor amplitude $|F_{hkl}|$.

It is important to remember that each $\mathbf{F}_{hkl}$ results from the diffraction of all atoms within the crystal lattice. Therefore, the final three-dimensional structure will be a time- and space-averaged picture of the entire volume that was irradiated by the X-rays.

The so-called 'phase problem' is the last bottleneck that lies between the crystallographer and the three-dimensional structure of a macromolecule. There are three major methods for obtaining phase information: molecular replacement, anomalous scattering and isomorphic replacement. Direct methods may also be a possibility when atomic resolution diffraction data is available.

Molecular Replacement

When the three-dimensional structure of a macromolecule homologous to the one of interest is available, phase determination by **molecular replacement** is frequently the method of choice. In this method, the known structure (search model) is used to estimate the phases for the target unknown structure. First, a rotation function is computed to orient the search model in the new unit cell, followed by a translation function to determine the position of the correctly oriented model in the target cell. Thus, the three-dimensional structure of the known macromolecule is used as starting model to provide phase angles for the observed structure factor amplitudes from the unknown structure. This method is applicable in the case of mutated proteins, proteins complexed with ligands that crystallised in different crystal forms or structurally related proteins. However, since the structural identity is not *a priori* known, the sequence identity is used as a guide. A rule-of-thumb is that above 50% of sequence identity, a molecular replacement solution should be straightforward, since these two proteins should be structurally homologous. If a family of homologous structures is available, an 'average structure' can be used as search model. However, it should be noted that sequence identity is not necessarily an indication of a sufficient structural homology to allow the successful application of this method, and vice-versa. A main drawback of this method is the model bias, which results from using an atomic model to calculate crystallographic phases. The resulting electron density map will tend to have features in the model even if they are not actually present in the structure, so refinement procedures are used to reduce model bias. Many structures are now solved by the molecular replacement method, alone or in combination with other methods (see Case Study below).

Isomorphous Replacement

The **isomorphous replacement** method was first applied by Max Perutz and John Kendrew back in the 1960s. It requires the attachment of a heavy atom (i.e with high atomic number, e.g. Hg, Pt, Au, Pb, Ag) to the protein in the crystal. Usually, the protein crystals are soaked in a solution containing salts of the heavy atom at low concentrations or alternatively, the protein can be co-crystallised with the heavy atom compound. The heavy atom addition/substitution should neither affect the protein structure nor the crystal unit cell dimensions (isomorphism), and so soaking is usually preferred. These heavy-atom derivatised crystals should show measurable changes in the intensities of some reflections in relation to the 'native' crystals, which can be used to deduce the positions of the heavy-atoms. In this method, the interference effects on the intensities of the diffracted beams caused by the addition of heavy atoms to the protein provide the estimates of the protein phase angles. A good single isomorphous derivative may be sufficient for structure determination (single isomorphous replacement, SIR), but usually multiple derivatives are necessary to derive phase information (multiple isomorphous replacement, MIR).

Anomalous Scattering

Phase determination by **anomalous scattering** is based on the fact that heavy atoms at wavelengths near their absorption edges scatter X-rays anomalously with a change in amplitude and phase. The resulting changes in the diffracted intensities can be used to determine the heavy atom sub-structure, from which phase estimates for all structure factor amplitudes can be computed. This method overcomes problems of non-isomorphism (present in SIR/MIR phasing), because all data are collected from the same crystal but requires an accurate measurement of the diffracted intensities; however, crystal decay due to radiation damage is sometimes a problem. Wayne Hendrickson and his colleagues developed a procedure making use of engineered proteins containing selenomethionine instead of methionine residues to provide a suitable anomalous scatterer (Hendrickson and Ogata, 1997). As a general rule-of-thumb, for a successful multiple wavelength anomalous dispersion (MAD) experiment on selenomethionine-containing proteins, at least one selenomethionine per 100 amino acids is required. However, under favourable conditions, it may be possible to succeed with a lower ratio.

This method is the obvious choice for **metalloproteins**, which contain one or more endogenous metals (e.g. Fe, Cu, Mo, Zn, Ni). However, the presence of the metal(s) in the protein is not *per se* a guarantee of success towards phasing using this inherent anomalous scatterer(s). This will depend on the size of the protein, type and number of metals (atomic number and anomalous scattering properties), diffraction quality of the protein crystals and accuracy of X-ray data measurements. Eventually, it may be necessary to search for additional heavy-atom derivatives, where a combination with any of the aforementioned methods may contribute to solve the phase problem.

A few crystallographic basic concepts need to be introduced at this stage to explain the rationale behind anomalous scattering.

The atomic scattering factor has three components: a normal scattering term that is dependent on the Bragg angle and two terms that are not dependent on the scattering angle, but on wavelength:

$$f(\theta, \lambda) = f_0(\theta) + f'(\lambda) + if''(\lambda)$$

The latter two terms represent the anomalous scattering that occurs near the absorption edge when the X-ray energy is sufficient to promote an electron from an inner shell. The dispersive term [$f'(\lambda)$], reduces the normal scattering factor whereas the anomalous term [$if''(\lambda)$], is 90° advanced in phase. This leads to a breakdown in Friedel's law (stating that intensities of the h, k, l and $-h$, $-k$, $-l$ reflections are equal) giving rise to anomalous differences that can be used to locate the anomalous scatterers. It should be noted that these anomalous scattering effects are significant for atoms such as Fe, Cd, and Se, much smaller for S and P, and virtually non-existent for C, N, O, and H.

A typical MAD experiment collects diffraction data sets at three wavelengths: at the peak of the absorption curve (highest f''), at the point of inflection of the absorption curve (minimum f'), and at a remote wavelength, above the absorption edge, where f' is near zero and f'' may still be relatively large. The values for f' and f'' depend on the scatterer and the theoretical values can be plotted, as illustrated in Figure 9.8, for three elements:

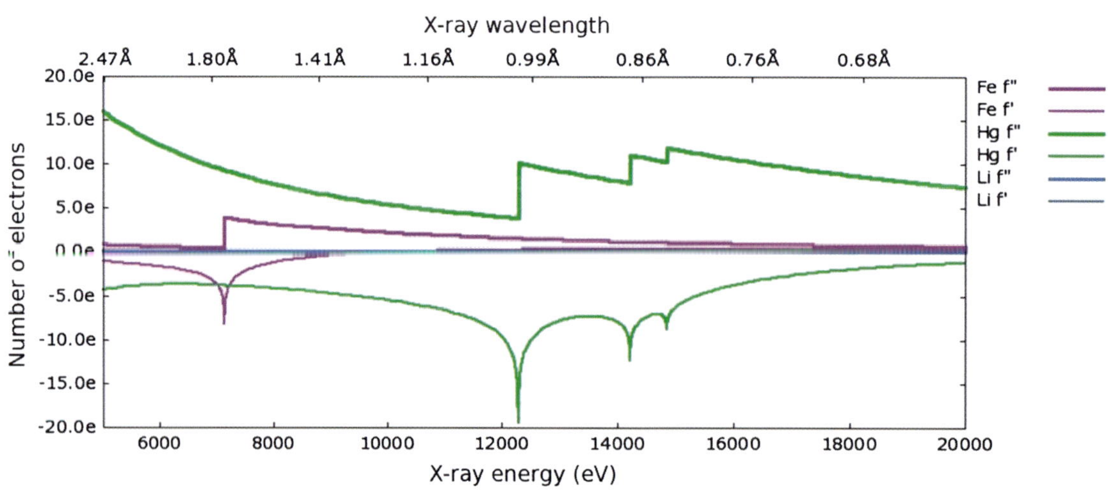

FIGURE 9.8 Plot of theoretical f' and f'' values for Fe, Hg, and Li, in the energy range 5000–20,000 eV, to assess the anomalous scattering properties of the elements *(Adapted from http://skuld.bmsc.washington.edu/scatter/AS_form.html).*

iron, mercury and lithium. However, since they are also dependent on the experimental setup, the actual shape of the absorption curve must be experimentally determined by a fluorescence scan of the crystal at the synchrotron beamline prior to data collection. For example, in the crystals of cytochrome *c* nitrate reductase complex (see Case Study below), the Fe peak and inflection point wavelengths were 1.7393 Å and 1.7408 Å, respectively. As can be observed in Figure 9.8, and in contrast with Fe and Hg, the absorption edge of Li lies outside the represented energy range, which is that most commonly accessible in a tunable beamline at a synchrotron X-ray source. This is also true for many other elements.

In recent years, Single wavelength Anomalous Dispersion (SAD) method is becoming increasingly popular. This method makes use of data collected at just one wavelength, typically at the absorption peak or high energy remote. It minimises problems of radiation damage and non-isomorphism, but requires very accurate measurements. The SAD method relies on density-modification protocols, such as solvent flattening, to break the phase ambiguity problem introduced by using data from only one wavelength, and provide interpretable maps.

Direct Methods

Direct methods are based on the positivity and atomicity of electron density which leads to phase relationships between normalised structure factor amplitudes. If the phases of some reflections are known, or can be assumed as a set of starting values, then the phases of other reflections can be deduced leading to a bootstrapping of phase values for all reflections. This methodology, also called *ab initio* phase determination, has limited applicability since it requires atomic resolution (better than 1.2 Å), but it is used routinely to find the heavy-atom substructures in other methods.

Heavy-Atom Derivatisation

The essential foundations of techniques for heavy-atom crystal derivatisation and its theoretical basis can be found in several books (Blundell and Johnson, 1976; Drenth, 1999) and review articles (Boggon and Shapiro, 2000; Garman and Murray, 2003) and will not be extensively referred here. The International Tables for crystallography – Volume F (Rossmann and Arnold, 2006) also presents valuable information on the subject. A very useful tool prior to any derivatisation experiment is the Heavy-Atom Database (HAD) (Islam et al., 1998), a website that contains information on the preparation and characterisation of heavy-atom derivatives of protein crystals.

Heavy-atom derivatives available for phasing can be divided into seven categories: endogenous metals which can be used directly for SAD/MAD (e.g. haem irons for cytochromes), or substituted by heavier metals with similar valency (e.g. substitution of calcium for strontium which has higher anomalous signal); single metal ions which can be bound electrostatically to the protein; selenium in selenomethionine; metal compounds which require a chemical reaction to take place; multi-metal cluster complexes used for large proteins and multi-protein assemblies (e.g. the $Ta_6Br_{12}^{2+}$ cluster (Knablein et al., 1997)), the noble gases xenon and krypton; and halides and triiodide, including brominated nucleic acids (Garman and Murray, 2003).

Historically, heavy metal salts used to successfully make derivatives are the so-called 'magic seven' compounds (Boggon and Shapiro, 2000), namely K_2**Pt**Cl_4 (platinum potassium chloride), K**Au**$(CN)_2$ (aurous potassium cyanide), K_2**Hg**I_4 (mercuric potassium iodide), **U**$O_2(C_2H_3O_2)_2$ [uranium (VI) oxyacetate], **Hg**Cl_2 (mercuric chloride), $K_3UO_2F_5$ (potassium uranyl fluoride) and *para*-chloromercurybenzoic sulfate. However, the success of the 'magic seven' as derivatives may be consequential to the fact that they have been utilised more often than other compounds.

Careful analysis of the protein primary sequence can give important clues as to which heavy-atoms to try initially: mercurial compounds tend to bind to cysteines (free sulfhydryls and not on disulfide bridges) or histidines, while platinum compounds are more prone to bind to cysteine, histidine, and methionine residues. The usual starting conditions for trying heavy-atom derivatisation are 0.1, 1.0, and 10 mM concentrations of soaking solution for 10 minutes to several days at the highest concentration that the crystal will tolerate. It is generally accepted that concentration of the heavy-atom is a more useful variable than soaking time. Co-crystallisation of the protein with heavy-atoms can also be performed, although it may change the crystal unit cell dimensions (problems of non-isomorphism). Moreover, the composition and pH of the mother liquor can also affect the metal compound used for derivatisation and should be taken into consideration when designing the derivatisation experiment.

MODEL BUILDING AND REFINEMENT

Once phases are known, they can be combined with the structure factor amplitudes derived from the X-ray diffraction data to calculate an electron-density map. Computer graphics are required to build a model of the structure of interest into the electron-density map. This model is then refined, to optimise the agreement between the observed and calculated values of the structure factor amplitudes.

The quality of the electron-density maps depends on the quality of the X-ray diffraction data and phase estimates. Of particular importance is the resolution of the data (Figure 9.9). Maps at 5–6 Å resolution may allow identification of helices and eventually of β-sheets; at 2–3 Å the topology folding can be established, the polypeptide chain be traced and side-chains assigned providing the amino acid sequence is known. At 1.5 Å or better, the individual atoms are almost resolved with well-defined network of water molecules and hydrogen atoms may become visible. In contrast to well-ordered small molecules that often diffract to resolutions better than 0.5 Å, not many macromolecule structure crystals diffract to atomic resolution. This may result from packing defects (not all protein molecules are identically packed within the crystal), weak inter-molecular interactions, protein flexibility, and high solvent content. The diffraction pattern does not abruptly stop at high diffraction angles, but gradually fades away.

In the case of molecular replacement, a preliminary model (based on the similar known structure) is already available, and one has to adjust this model (by moving, adding, or removing atoms or groups of

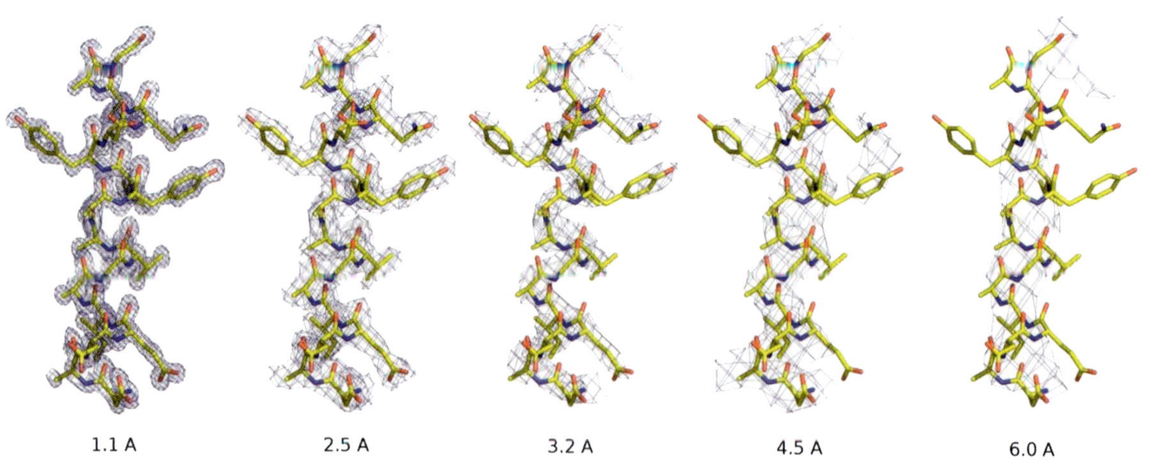

FIGURE 9.9 Electron-density maps displayed for a α-helix at different resolutions. The level of detail fades away with the decrease of the resolution: at 1.1 Å, the electron density perfectly 'wraps' the built model and even the aromatic residues display a doughnut shape with a hole in the middle of the aromatic ring; in contrast at 6 Å only an oblong shaped electron density is visible for the whole α-helix and no side-chains are clearly defined.

atoms, namely side-chains that differ from the search model and the model to be built) into the computed electron-density maps. Model building is facilitated but care must be taken to get rid of model bias arising from the initial phases.

If phases were obtained experimentally (by anomalous scattering or isomorphous replacement methods) the initial map will be 'empty' and the model needs to be built from scratch. The crystallographer can start by inspecting the maps and try to recognise continuous segments with a shape consistent with the geometry of secondary structure elements such as α-helices or β-sheets. It may be possible to build some stretches of poly-alanine chains and to identify residues with distinctive side-chains, such as tryptophan, phenylalanine, histidine, or tyrosine. Selenium (in the case of a Se-Met derivative), metal ions, or heavy atoms make good anchor points in sequence assignment. While some years ago, model building had to be done entirely by hand, at present automated model-building programs are available to help with this task.

Despite our best efforts the model will always contain errors, namely geometry and stereochemistry errors (deviation of bond lengths and angles from expected values, implausible torsion angles and some highly correlated small errors). Therefore, the model needs to be refined against the experimental diffraction data (observed structure factor amplitudes). While earlier methods used an iterative non-linear least-squares procedure, modern refinement programs use more robust maximum-likelihood algorithms to minimise the difference between the observed and calculated structure factor amplitudes. The model parameters being refined are one or more overall scale factors, the atomic positions, B-factors (describing the atomic displacements) and occupancies. The high solvent content of protein crystals also needs to be taken into account. While the ordered solvent molecules can be included in the model being refined, a large fraction is disordered but still contributes to the Bragg scattering, in particular at low resolution. For this reason, parameters describing this contribution are also included in the refinement.

The ratio of observations to variables is usually too small for unrestrained positional refinement due to the typical resolution we obtain for macromolecules. Hence, additional 'observations' are included in the refinement procedure in the form of geometrical restraints for the atomic positional parameters, based on dictionaries of standard geometrical data (e.g. that compiled by Engh and Huber (1991) on geometrical data derived from small molecule structures). If there are several copies of the same molecule in the asymmetric unit, then non-crystallographic symmetry (NCS, see Figure 9.7) restraints among them can also be included in the refinement, making the copies geometrically similar.

As the model improves, so do the phases calculated from it, and so do the electron-density maps. Thus, the overall procedure consists of alternating cycles of restrained refinement with model correction against the improved electron-density maps, until convergence is achieved (no more significant improvements are possible) and a suitable three-dimensional structure of our protein model is ready.

A schematic representation of the various steps involved in structure determination of proteins by X-ray crystallography is presented in Figure 9.10 and herein described.

STRUCTURE ANALYSIS AND MODEL QUALITY

Content of Crystallographic Models

A refined model contains the atomic positions (coordinates *x*, *y*, *z*) of each element present in the asymmetric unit, its occupancy and B-factor. Hydrogen atoms can also be included in idealised calculated positions. At present, most scientific journals require the deposition of the coordinates of the structure(s) to be published in the PDB. It is recommended, although in many journals not yet mandatory, that the experimental data (i.e. the structure factors) are also submitted. Aside from the coordinates, each entry also includes the names of molecules present in the PDB file, primary and secondary structure information, sequence database references, ligand and biological assembly information, details about data collection and structure solution, and bibliographic citations.

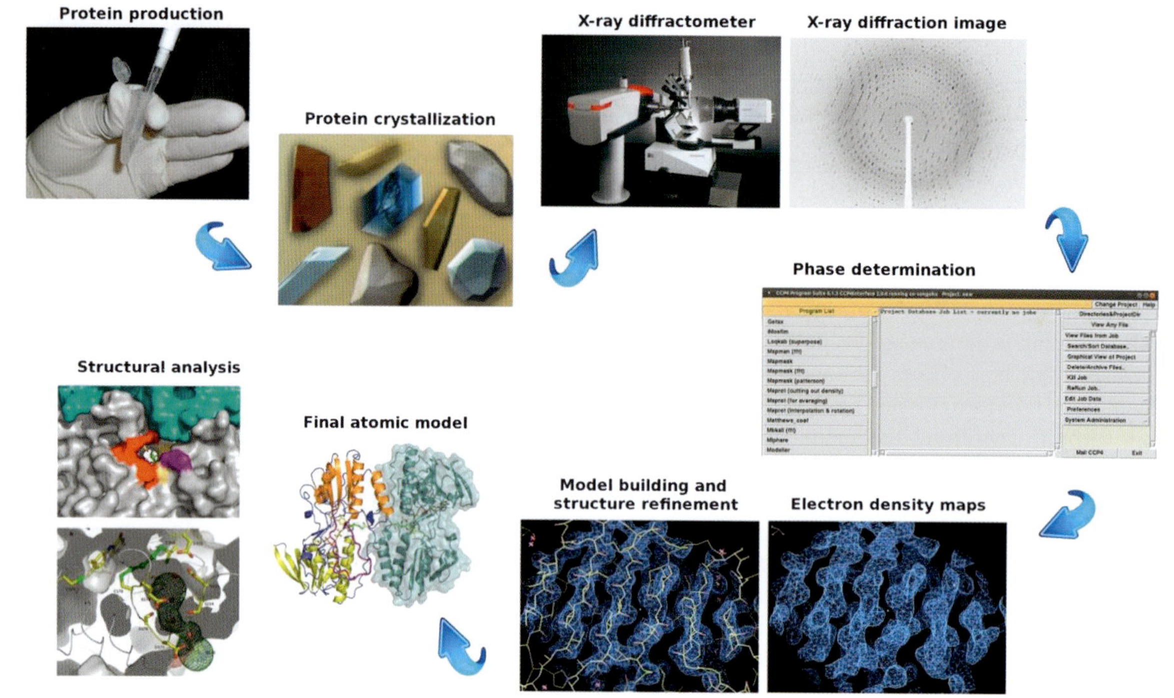

FIGURE 9.10 Schematic diagram of the steps involved in the three-dimensional structure determination of proteins by X-ray crystallography.

A few lines are listed corresponding to the **PDB entry** (2J7A) of cytochrome *c* nitrite reductase complex (see Case Study):

```
HEADER    OXIDOREDUCTASE                          06-OCT-06   2J7A
TITLE     CRYSTAL STRUCTURE OF CYTOCHROME C NITRITE REDUCTASE NRFHA
TITLE    2 COMPLEX FROM DESULFOVIBRIO VULGARIS
...
ATOM      1  N   GLY A  26       6.532  78.390  35.719  1.00 49.72           N
ATOM      2  CA  GLY A  26       5.841  77.273  34.998  1.00 49.81           C
...
HETATM54766 FE   HEM A1001      -6.896  30.107  70.628  1.00 11.26          FE
HETATM54767  CHA HEM A1001      -3.984  29.659  68.883  1.00  6.68           C
HETATM54768  CHB HEM A1001      -7.680  26.759  70.277  1.00  8.96           C
...
HETATM58415 CA    CA Q1006      52.958  -9.260 -72.372  1.00 17.39          CA
...
HETATM58624  O   HOH A2001       0.602  62.423  37.108  1.00 14.08           O
HETATM58625  O   HOH A2002      -3.834  62.200  38.101  1.00 20.04           O
TERM
```

where '6.532, 78.390, 35.719' are the coordinates of the first atom of the model, corresponding to main-chain nitrogen atom (N) of glycine (GLY) 26 from chain A, showing a full occupancy (occ = 1.00) and a B-factor of 49.72 Å^2. ATOM is used for amino acid residues and HETATM for hetero-atoms (e.g. metals, ions, ligands, or solvent molecules). Notice that ATOM 2 represents the C_α of Gly26, whereas HETATM58415 CA corresponds to a calcium ion. The last column lists the atom type, in this case, C (carbon) or CA (calcium).

The **occupancy** parameter measures the fraction of molecules in the crystal in which a certain atom actually occupies the position specified in the model, it can vary between 0 and 1. Some side chains may show alternate conformations, if so the side-chain will be split into parts A and B with their respective occupancies (e.g. 0.4 and 0.6). Some other side chains, such as polar surface residues, or even whole amino acid residues(s), as part of disordered loops, may not be well defined in the electron-density maps. Here, opinions diverge whether crystallographers should try to fit and include them in the model (occupancy of 1) and allow the B factor to refine to high values or set the occupancy to 0. Sometimes, no electron density is visible at all for some parts of the protein, and if so the corresponding part of the protein chain must be omitted from the model, and breaks in the polypeptide chain will be visible while displaying the model on a graphics program.

Another important parameter is the **atomic displacement parameter**, often referred as temperature factor or B-factor. It describes the degree of displacement of an atom within the unit cells of the measured crystal, i.e. how a certain atom oscillates or vibrates around its mean position specified in the model. The lower the B-factor, the more static an atom is; the higher the B-factor, the more mobile an atom is. Very high B-factors (>80 Å^2) are often found in both termini of the model as well as in solvent exposed residues with long side-chains (e.g. arginines, lysines, glutamates). A picture of the crystallographic model coloured according to temperature factors (commonly from low B values in blue to high B values in red) provides insights into which parts of the protein have higher motion/disorder. If atomic resolution is available, B-factors may be refined anisotropically, with six parameters instead of one, and the atomic motion can be displayed as triaxial ellipsoids rather than spheres.

Validation

A concern among the crystallographic community was the development of validation tools to assess the quality of the experimental diffraction data, the refined model, and the agreement between them. All publications involving X ray structure determinations should report the relevant statistical parameters for data collection and processing, and model refinement.

The parameters commonly used to assess the **quality of the diffraction data** include resolution, completeness, multiplicity, signal-to-noise ratio and the merging *R*-factor of the data set.

Resolution (d_{min}, Å) is calculated from the highest Bragg angle θ to which significant diffraction data were measured, and roughly corresponds to the distance range at which two features can be resolved in the corresponding electron density. In practical terms, the resolution determines the level of detail of an electron density maps (see Figure 9.9).

Completeness is defined as the percentage of the measured crystallographic reflections in a data set over the total number of theoretically possible unique reflections, at specified resolution limits. Completeness should be as high as possible, nearly 100%, in all resolution bins of the data set (missing reflections lead to the deterioration of the model parameters).

Redundancy (or **multiplicity**) is calculated as the ratio between the number of measured reflections and the number of unique reflections present at a given resolution range. This calculation gives the average number of independent measurements of each reflection in a crystallographic data set, so the more individual measurements are done, the more precise the average measurement becomes.

Signal-to-noise ratio, ($I/\sigma(I)$), in practical terms, indicates how many times above noise level were the intensities measured. $\langle I/\sigma(I)\rangle$, the mean value of $I/\sigma(I)$, is usually listed for the whole data set and for the highest resolution shell. The diffraction limit of a given crystal is usually set at a resolution where this ratio remains above 1.

$\boldsymbol{R_{merge}}$, also known as $\boldsymbol{R_{sym}}$, measures the agreement between the multiple independent observations of the same reflection. Because this value tends to increase as the data multiplicity increases, two more appropriate statistics were devised, named redundancy-independent R-factor (R_{rim} or R_{meas}) and precision-independent R-factor (R_{pim}) (Diederichs and Karplus, 1997). A good data set should have a low R_{merge}, high values might indicate a sub-optimal data quality or problems with the data processing. In the highest resolution shell, R_{merge} may be as high as 30–40%, although values in the order of 60–70% are still reasonable for high-symmetry space groups.

In addition to assessing the quality of the diffraction data, **model validation** against the experimental data, and examination of stereochemistry during the model building and refinement processes are also of critical importance. This is usually performed using the residual R-factors and by comparison of the model geometry against the standard values. Validation programs currently available include PROCHECK (Laskowski et al., 1993), WHATCHECK (Hooft et al., 1996) and MolProbity (Chen et al., 2010).

***R*-factor** and $\boldsymbol{R_{work}}$ measures the agreement between the experimentally measured structure factor amplitudes, F_{obs}, and those calculated from the model, F_{calc}. The R-factor measures the global relative discrepancy between the 'observed' F_{obs} and calculated F_{calc}.

***R*-free** or free R-factor is computed in the same manner as R-factor, but using a small set of structure factor amplitudes, the so-called 'test set', which are set aside and not used during refinement. They are used only for cross-validation. R-free measures how well the current atomic model compares with that subset of the measured structure factor amplitudes that were not present in the refinement calculations, whereas R measures how well the current model compares with the entire measured data set.

These residual R-factors are used to evaluate the progress of structure refinement, and their final values are important criteria of model quality. The R_{merge} and resolution of the data will influence the value of R_{work}. The higher the resolution, the lower the R-factor is expected to be. A desirable target R-factor for a protein crystal structure refined with data to 2.5 Å is around 20%; structures near atomic resolution should have lower R-factors, around 10–15%. The value of R-free also depends on the data resolution. At lower resolutions, it may be up to 10% higher than R_{work}, whereas at near atomic resolution the difference should be 1–2% at most.

Validation of model geometry compares model properties such as stereochemistry, local chemistry environments, and packing propensity against their expected target values based on prior knowledge. In particular, properties such as bond lengths, bond angles, φ (phi)-ψ (psi) torsion angles, side-chain torsion angles, peptide flips, clashes, C_β deviations, asparagine/histidine/glutamine side-chain flips and local environment profiles can be checked.

As introduced in the refinement section, a restraint is a subsidiary condition imposed on the atomic parameters during crystallographic refinement, based on prior knowledge. For example, in a protein structure, the bond lengths and angles are expected to be within a specified range of values. These values are standard for all structures and are extracted from high-resolution data sets and small-molecule structure databases like the Cambridge Structural Database (CSD). During refinement, the stereochemical deviations from ideal values are minimised along with the deviations between observed and calculated structure factor amplitudes. In order to ensure a well-behaved refinement providing a chemically meaningful structure, different weights are assigned to each term of the minimised function. At low resolution (below e.g. 2.5 Å), a higher weight is given to the stereochemistry, whereas at high resolution (above e.g. 1.5 Å) the crystallographic term becomes more important and the electron density is allowed to be the major driving force in the refinement. Deviations from the standard values indicate how much a model differs from geometrical parameters that are considered typical or represent chemical knowledge. Reasonable root mean square deviation (r.m.s.d.) values are around 0.02 Å for bond lengths and 1.5° for bond angles. Higher values may indicate errors in the model or problems with the refinement protocol.

The **Ramachandran diagram** plots the φ–ψ torsion angles of a polypeptide (Figure 9.11). The rotations of main-chain N–C_α and C_α–C bonds, represented by φ and ψ, respectively, are free to rotate, but many combinations are not possible because of steric repulsion. The Ramachandran plot shows regions where conformations are allowed and others that are sterically disallowed for all amino acids except glycine which is unique in that it lacks a side chain. In fact, glycines are often found in loop regions where the polypeptide chain makes a tight turn. Most, if not all amino acid residues should lie in the favoured regions, although proteins can accommodate a few outliers. Structure publications often include the results of the plot, with an explanation about non-glycine residues that lie in high-energy (not allowed) areas, such as structural constraints that overcome the energetic cost of an unusual backbone conformation, and which may have functional significance.

Once the validation step is accomplished, we can move on to structure–function analysis, i.e. interpreting the final model based on the available experimental data (physicochemical and biochemical information). What can we get out of the three-dimensional structure? Does it correlate with the function? Are there any binding sites? Can we identify interacting partners? Are there any structural homologues? Can we propose a structure-based mechanism?

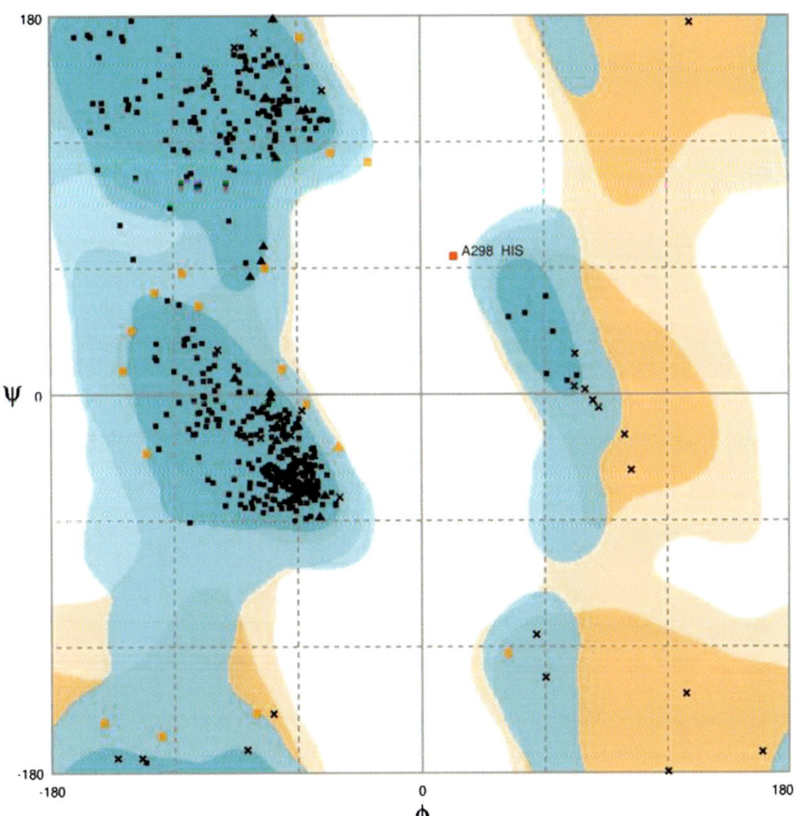

FIGURE 9.11 Ramachadran plot calculated with the program RAMPAGE (Lovell et al., 2003). Different shades of blue refer to favoured (dark blue), and allowed (light blue) regions of the Ramachandran for all residues except glycine. Dark brown and light brown represent favoured and allowed regions of the Ramachandran for glycines, respectively. Squares represent all residues except prolines (triangles), and glycines (crosses).

CASE STUDY

We will briefly describe the procedures involved in the X-ray structure determination of the membrane-bound cytochrome *c* nitrite reductase complex (NrfHA) isolated from *Desulfovibrio vulgaris* Hildenborough. The crystallisation and preliminary crystallographic studies were published in Acta Crystallographica (Rodrigues et al., 2006), the 2.3 Å crystal structure of NrfHA in EMBO Journal (Rodrigues et al., 2006) and the complex structure of NrfHA bound to 2-heptyl-4-hydroxyquinoline-N-oxide (HQNO) inhibitor at 2.8 Å resolution in Journal of Molecular Biology (Rodrigues et al., 2008). We will also herein succinctly describe the structural and functional information derived from these two crystal structures. A more detailed characterisation of NrfH cytochrome *c* quinol dehydrogenase can be found in Handbook of Metalloproteins (Rodrigues et al., 2011).

INTRODUCTION

Cytochrome *c* nitrite reductase complex (NrfHA) was isolated from *D. vulgaris* Hildenborough, a sulfate reducing bacteria. It catalyses the conversion of nitrite to ammonia ($NO_2^- + 6e^- + 8H^+ \rightarrow NH_4^+ + 2H_2O$) via the oxidation of menaquinol (Pereira et al., 2000). The complex is bound to the membrane through the NrfH subunit, a 20 kDa tetraheam cytochrome *c* that belongs to the NapC/NirT family of membrane-anchored quinol dehydrogenases. The soluble catalytic subunit, NrfA, is a pentahaem cytochrome *c* of about 60 kDa that has been shown to form a biological dimer.

PURIFICATION AND CRYSTALLISATION

The NrfHA complex was purified from cell membranes following two protocols involving various chromatographic steps using either the zwitterionic sulfobetaine detergent SB12 [Zwittergent 3-12; 3-(N,N-dimethyldodecylammonio)-propanesulfonate] or the non-ionic detergent n-dodecyl-β-D-maltoside (DDM). The purity of the two protein samples was assessed by UV-visible spectroscopy, SDS-PAGE stained with Coomassie and haem staining. Protein samples were concentrated using an Amicon YM30 Diaflo membrane to about 10 mg/mL in 10 mM Tris-HCl pH 7.5 buffer containing 0.2% Zwittergent 3-12 or 0.03% DDM.

Initial crystallisation experiments were setup using a grid screen with polyethylene glycols as precipitant (PEG 2000 to 8000, concentrations: 5–20% (w/v)), buffers at different pHs and additives, at 4 and 20 °C. The sitting-drop vapour diffusion technique was used by mixing equal amounts (1.5 μL) of protein and reservoir solutions. Crystals were obtained from protein samples with both detergents, but the results with DDM seemed more promising. Reddish-brown crystals appeared within one to three weeks from drops equilibrated against reservoir solutions containing 10% PEG 4000 (w/v) in 0.1 M Hepes pH 7.5 buffer. The presence of both subunits, NrfA and NrfH, was confirmed by SDS-PAGE on the dissolved crystals (Figure 9.12). Previous crystallisation experiments had been reported with the NrfHA complex from *D. desulfuricans* ATCC 27774, where only the soluble NrfA subunit crystallised (Cunha et al., 2003).

The optimisation of the crystallisation conditions of NrfHA complex in order to obtain suitable crystals involved a great experimental effort, due to non-reproducibility of crystal growth and non-reliability of diffraction quality for crystals grown under similar conditions. This phenomenon is particularly well known for membrane protein crystals. Although crystals appeared in several drops containing PEG 4 K at pH around 7.5, most crystals did not diffract or diffracted poorly. Moreover, drops with identical experimental setups yielded different results (e.g. no crystals, a few crystals, many microcrystals). Figure 9.13 illustrates crystals grown in similar conditions, but showing different morphologies. Crystal optimisation involved varying the type of precipitant, concentration and ratio of precipitant and protein, pH, additives and temperature. Different crystallisation methods (vapour diffusion, microbatch, agarose gels) and seeding techniques were also attempted. Many crystallisation trials were carried out and many crystals had to be screened until a crystal with good diffraction quality was found. The growth of good diffracting crystals of the NrfHA complex appeared to be fortuitous; the conditions that favour the formation of well-ordered crystals are still not completely understood.

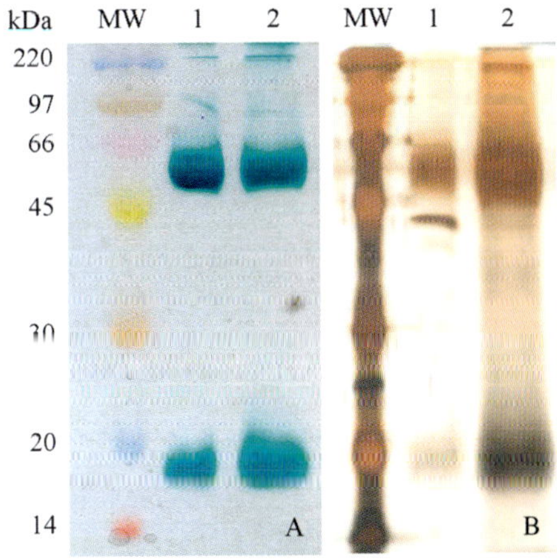

FIGURE 9.12 SDS-PAGE of protein solution and dissolved crystals. Gels stained with haem staining (Panel A) and $AgNO_3$ (Panel B). Gels loaded with protein solution (lanes 1) and dissolved crystals (lanes 2). MW denotes the markers. About 10 needle-like crystals from the same drop were dissolved in the protein buffer solution and loaded onto each gel.

X-RAY DIFFRACTION DATA COLLECTION AND PHASING

NrfHA crystals were also sensitive to the addition of the cryo-protectant solution. Several compounds were tested for an efficient cryo-cooling, including 2,3-butanediol, glycerol, PEG 400, ethylene glycol, MPD and oils (paraffin oil, paratone-N from Hampton Research, and Panjelly™ from Jena Bioscience). In the end, crystals were then cryo protected with a solution containing 20% glycerol, 14% PEG 4 K, 0.1 M Hepes pH 7.5 and flash-cooled in liquid nitrogen prior to X-ray data measurements.

It is worth mentioning that we have tested over 150 crystals until a few crystals were found that diffracted better than 3 Å at a synchrotron source. This implied many crystal pre-screenings using an in-house X-rays generator to evaluate their diffraction quality (the intensity of the X-ray source was too weak to collect data from membrane protein crystals) and several trips to synchrotron facilities. Finally, a crystal was obtained from a sitting-drop vapour-diffusion experiment containing 5 mg/mL protein, 0.015% DDM detergent, 10% PEG 4 K, 0.1 M Hepes pH 7.5 and 30 mM glycyl-glycyl-glycine as an additive, from which data were collected at the Swiss Light Source (Figure 9.13). It is important to stress that the number of NrfHA crystals tried is unusually low compared to most membrane proteins for which the structures have already been determined. This number can easily go up to several thousands of trials needed until one or a few reasonable diffracting crystals are discovered!

X-ray diffraction data were measured at the Swiss Light Source (SLS) synchrotron, on the tunable beamline PX-I (X06AS), which was equipped with a Mar225 mosaic CCD detector. A fluorescence scan around the iron

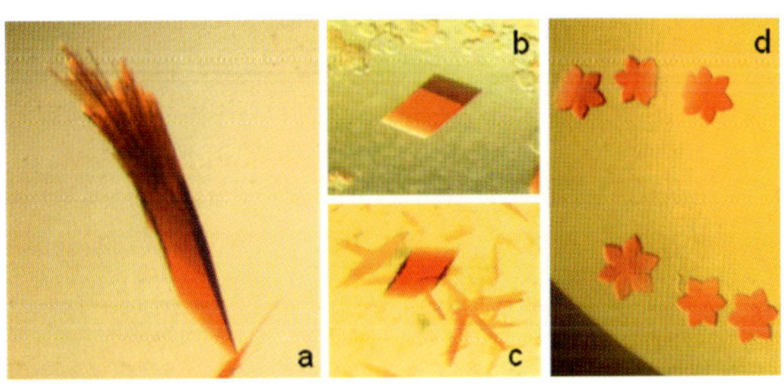

FIGURE 9.13 Crystals of the NrfHA complex crystallised using PEG 4 K as precipitant as long needles with 0.60 × 0.20 × 0.05 mm (a), parallelepipeds to maximum dimensions of about 0.30 × 0.20 × 0.10 mm (b, c) or flower shaped (d).

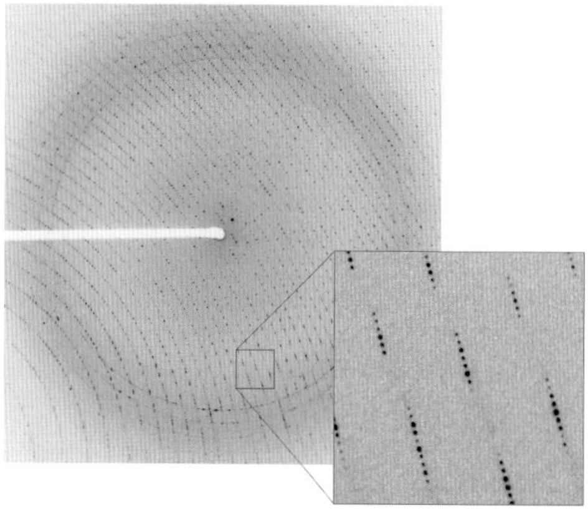

FIGURE 9.14 X-ray diffraction image recorded up to 2.3 Å resolution (0.9000 Å). Inset, is a magnified area of the diffraction pattern at ~4.5–5 Å resolution showing very close reflection spots along the longest cell axis.

K-edge was performed prior to a MAD experiment. X-ray diffraction images were collected from a single crystal just above the iron absorption peak ($\lambda_{pk} = 1.7393$ Å), at the inflection point of the fluorescence curve ($\lambda_{ip} = 1.7408$ Å), and at a high-energy remote wavelength ($\lambda_{rem} = 0.9000$ Å), with an oscillation angle of 0.25° and exposure time of 1 s per frame. Since we had a long needle (~0.6 mm in length), the crystal was translated a few times during data collection so the X-ray beam would irradiate a 'fresher' part of the crystal (not so burnt-out). The crystal diffracted beyond 2.3 Å resolution and belonged to orthorhombic space group $P2_12_12_1$, with unit cell parameters $a = 79.5$, $b = 256.7$ and $c = 578.2$ Å (Figure 9.14). The presence of a very long unit-cell axis of about 580 Å (often observed for viral particles) was a significant obstacle for optimal data collection, since its orientation was not favourable, leading to many spatial overlaps between diffraction spots at higher resolution (Figure 9.14). Also, the remaining beamtime available for these measurements was limited, and did not allow using smaller oscillation angles. Data were processed with MOSFLM and merged with SCALA from the CCP4 program suite (Collaborative Computational Project, 1994). Data collection and processing statistics are listed in Table 9.2.

The three-dimensional structure of *D. vulgaris* NrfHA complex was solved by a combination of molecular replacement and MAD phasing. Molecular replacement was performed with Phaser (McCoy et al., 2005) using the coordinates of the *D. desulfuricans* NrfA dimer (PDB code: 1OAH, 66% of amino acid sequence identity with *D. vulgaris* NrfHA) as search model. Six such NrfA dimers were found in the asymmetric unit of the crystal structure of the *D. vulgaris* NrfHA complex. An electron density map was computed based on the molecular replacement phases with a preliminary model of the NrfA dimer (Figure 9.15). Inspection of the map showed a well-defined electron density corresponding to the soluble catalytic NrfA dimer. In contrast, the electron density for the unknown membrane NrfH subunit was ill defined and not continuous, thus not allowing the tracing of the NrfH polypeptide chain. The α-helix responsible for the membrane attachment could be identified and the stoichiometry of the complex established as 2NrfA:1NrfH, a matter which was under debate at that time. A map was then calculated using the phases from the molecular replacement solution of the NrfA dimer sub-structure and the anomalous coefficients obtained from the peak data set. 24 iron atoms were located from the unknown smaller NrfH subunit. The remaining 60 iron atoms from the NrfA subunit had already been located from the previous map (molecular replacement phased). A total of 84 iron sites were then input to SHARP (de la Fortelle and Bricogne, 1997) for a maximum-likelihood heavy-atom parameter refinement based on the three-wavelength MAD data sets. To minimise bias, the phases from the molecular replacement calculations were not used. The resulting phases were further improved by using a density-modification procedure including solvent content optimisation. These calculations led to an optimised solvent content of 59%. This high solvent content is

TABLE 9.2 Statistics of Data Collection and Processing

Space group	$P2_12_12_1$		
Unit cell dimensions (Å)	$a = 79.4$, $b = 256.8$, $c = 579.2$		
Data processing	**Peak**	**Inflection**	**Remote**
Wavelength (Å)	1.7393	1.7408	0.9000
Resolution range (Å)	64.0–2.60 (2.74–2.60)	55.3–2.60 (2.74–2.60)	54.0–2.30 (2.42–2.30)
Completeness, overall (%)	85.2 (45.2)	85.1 (44.4)	83.9 (67.3)
Completeness, anomalous (%)	62.4 (15.4)	65.7 (17.7)	67.3 (33.1)
No. of observations	972,147 (36,782)	1,023,754 (38,166)	1,279,751 (107,004)
No. of unique reflections	310,637 (23,876)	311,156 (23,563)	443,656 (51,436)
Redundancy, overall	3.1 (1.5)	3.3 (1.6)	2.9 (2.1)
Redundancy, anomalous	1.9 (1.3)	1.9 (1.3)	1.7 (1.5)
R_{merge}* (%)	6.8 (14.7)	7.6 (19.2)	9.8 (31.0)
$<I/\sigma(I)>$	13.7 (4.6)	12.8 (4.0)	8.3 (3.1)

$*R_{merge} = \sum_{hkl}\sum_i \left| I_i(hkl) - \overline{I(hkl)} \right| / \sum_{hkl}\sum_i |I_i(hkl)|$ where $I_i(hkl)$ is the ith measurement and $\overline{I(hkl)}$ is the mean of all measurements of $I(hkl)$.

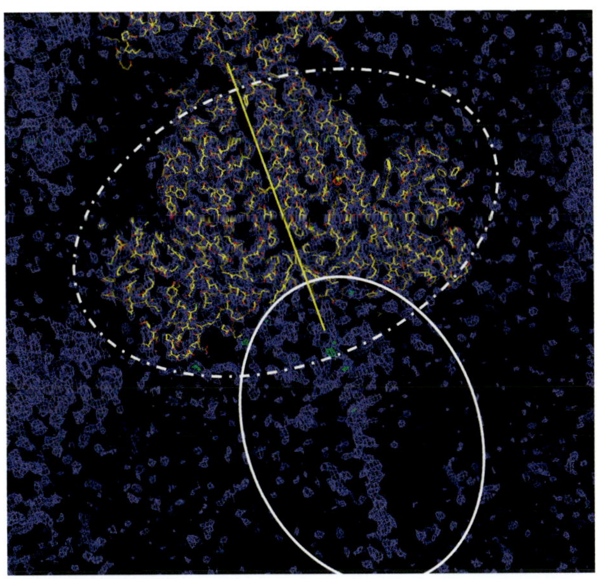

FIGURE 9.15 Electron density map computed after molecular replacement procedure using NrfA dimer from *D. desulfuricans* as search model (two-fold symmetry axis between the two NrfA molecules is displayed by a yellow line). The electron density corresponding to NrfA dimer is surrounded by a white dashed line and NrfH by a continuous white line. Note the 2NrfA:1NrfH stoichiometry.

common for membrane protein crystals, also contributing for their low-diffraction quality. The calculated 2.3 Å electron density map, using only the experimental phases, was of very good quality.

STRUCTURE REFINEMENT

Once the electron density maps were available on a computer graphics workstation, the manual model building of the unknown NrfH and the rebuilding of the NrfA subunit (e.g. correction of side-chains, addition or removal of amino acid residues) commenced using COOT (Emsley and Cowtan, 2004). Model building was carried out for one $NrfHA_2$ assembly, the other being generated by non-crystallographic symmetry operations. An initial restrained refinement was carried out with Refmac5 (Murshudov et al., 1999), after which improved electron density maps were computed, and used to correct and complete the crystallographic model. Water and other solvent molecules were added to the model at a later stage. Further cycles of manual fitting and restrained refinement were performed until convergence was achieved, with R-factor and R_{free} values of 20.1 and 24.0%, respectively, at 2.3 Å resolution (Table 9.3).

The final model comprised of 6,813 amino acid residues, 84 haem groups, 24 calcium ions, belonging to 18 protein chains (six chains were assigned to 'interior' NrfA, named $NrfA_{int}$; six chains to exterior NrfA, named $NrfA_{ext}$, and six chains to NrfH) (Figure 9.16). A calcium-binding site was found near each NrfA haem 1 and

TABLE 9.3 Refinement Statistics

Resolution (Å)	54.0–2.30
No. of reflections	421,309
R_{factor}/R_{free} (%)[§]	20.1/24.0
No. of atoms	
Protein	54,721
Haem	3,612
Calcium ion	24
LMT head	144
Acetate ion	12
Water	2,326
B-factors (Å^2)	
Protein	20.1
Haem	12.5
Solvent, other molecules	24.8
r.m.s deviations	
Bond lengths (Å)	0.012
Bond angles (°)	1.46

[§] $R_{factor} = \sum_{hkl} \left| |F(hkl)_{obs}| - |F(hkl)_{calc}| \right| / \sum_{hkl} |F(hkl)_{obs}|$, *where $F(hkl)_{obs}$ and $F(hkl)_{calc}$ are the observed and calculated structure factors, respectively. R_{free} was calculated for 5% of reflections randomly chosen for cross-validation.*

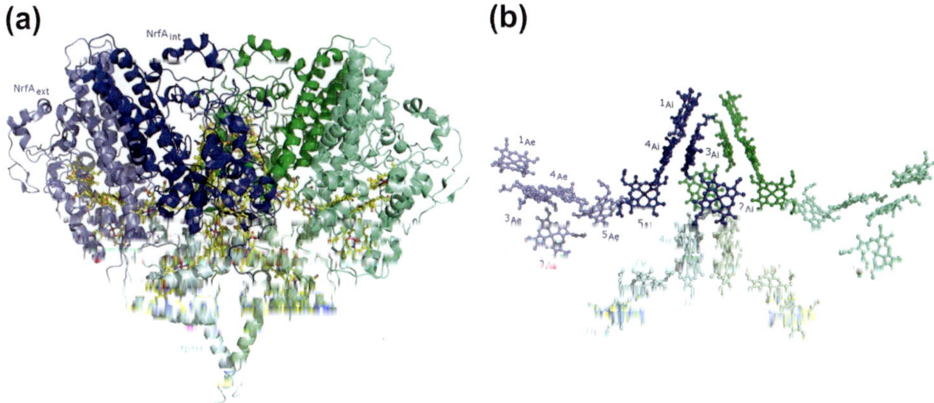

FIGURE 9.16 Cartoon representation of the NrfHA crystal structure (a) One NrfA unit is coloured in dark blue ($NrfA_{int}$) and in lighter blue ($NrfA_{ext}$) and the other in two green shades; NrfH is in pale cyan and pale green; the haems are shown in ball-and-stick with carbon atoms in yellow, oxygen in red, nitrogen in blue and iron in dark red. Haem arrangement (b) in the biological assembly with the same orientation and depicted in the same colour code as the main chains shown in Panel a. All structure figures were prepared with the program PyMol (DeLano, 2002).

a second Ca^{2+} ion was octahedrally coordinated in the vicinity of haems 3 and 4. A blob of electron density, located in a cavity surrounded by the N-terminal tail of $NrfA_{int}$ and an inter-helical loop of NrfH was assigned as the polar head of a dodecylmaltoside detergent molecule. In addition, 2326 water molecules and three acetate ions were also included in the crystallographic model.

The different NrfA molecules start at residues 25 or 26 and end at residues 519–522, depending on the protein chain. A few residues that belong to the β-sheet loop formed by residues 325–331 from $NrfA_{ext}$ monomers, were poorly defined in the electron density maps. Their average thermal motion parameters are around 45 Å^2, which are considerably above the average thermal motion parameters of all NrfA protein residues (20.1 Å^2). The model for the electron-donor subunit NrfH comprises residues from 14 to 158 and has an average overall thermal motion parameter of 27.4 Å^2. The NrfH transmembrane helix is not well ordered, as most of its residues (from 14 to 31) have high-thermal motion parameters (between ~50 and 100 Å^2). A Ramachandran plot analysis with PROCHECK (Laskowski et al., 1993) showed that 89.8% of the total protein residues are within the most favoured regions, 9.6% are in additionally allowed regions, 0.4% in the generously allowed regions and only 0.2% in disallowed ones. The refinement statistics are summarised in Table 9.4. Coordinates of NrfHA structure were deposited with the PDB under the accession code 2J7A.

STRUCTURE ANALYSIS

Complex Architecture

The NrfHA asymmetric unit contains six NrfH and twelve NrfA molecules forming three dimers of $NrfHA_2$ units, where each membrane-anchored NrfH is tightly bound to two NrfA molecules (Figure 9.16a). This provides a rationale for the physiological relevance of the NrfA dimer, which has been reported for all NrfA crystal structures determined so far. We proposed that the striking $\alpha_4\beta_2$ arrangement corresponded to the biological active form of the complex, which was further supported by the determined molecular mass of the complex in solution (ca. 300 kDa) (Rodrigues et al., 2006).

NrfA Structure

The three-dimensional structure of the catalytic NrfA dimer was similar to other NrfA structures which were previously characterised (Cunha et al., 2003; Einsle et al., 2002; Bamford et al., 2002; Holm and Sander,

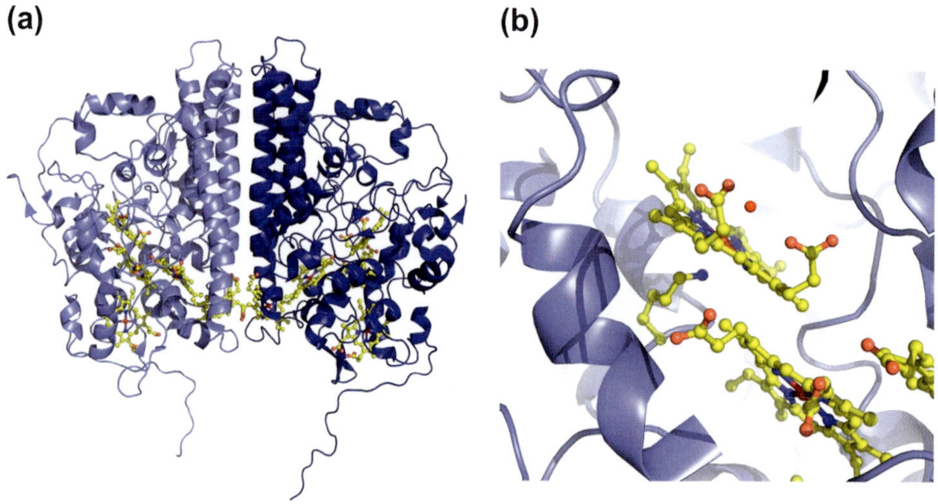

FIGURE 9.17 Cartoon representation of the NrfA dimer with haem cofactors shown in ball-and-stick representation (a) Zoomed view of the catalytic NrfHA haem 1 with Lys331 as proximal ligand and a water molecule in the distal site (b) Colour code as in Figure 9.5a.

1999) (Figure 9.17a). It also shows the characteristic three-helix bundle at the dimer interface, and the catalytic haem 1 has a lysine as the proximal ligand from a CXXCK motif and a water molecule at the distal side (Figure 9.17b). A structure-based mechanism for the six-electron reduction of nitrite to ammonia, without the release of reaction intermediates, has been proposed by Einsle et al. (2002). The four bis-histidinyl-coordinated haems, a calcium-binding site near haem 1 and the putative channels for substrate access and product efflux are also present in the *D. vulgaris* NrfA structure. A second Ca^{2+} is present in the vicinity of haems 3 and 4, as also observed in the *Escherichia coli* (Bamford et al., 2002) and *D. desulfuricans* (Cunha et al., 2003) NrfA proteins.

A relevant feature displayed by the *D. vulgaris* NrfA structure was the elucidation of the conformation adopted by the first amino acids (25–40) of the mature protein, which were not visible in any of the previously available NrfA crystal structures, but are ordered in the *D. vulgaris* NrfHA complex. These amino acid residues form an extended tail that embraces NrfH subunits, allowing several intersubunit interactions and thus contributing to the complex stability. The N-terminal tail of the $NrfA_{int}$ monomer interacts with both NrfH subunits of the $\alpha_4\beta_2$ complex, whereas the tail of the $NrfA_{ext}$ interacts with only one NrfH molecule.

NrfH Structure

NrfH is an α-helical protein (Figure 9.18a) with a novel protein fold, as no significantly similar structure was found by fold recognition analysis using the DALI (Holm and Sander, 1999) server. The transmembrane helix, comprising residues Lys14 to Asp38, is included in our crystallographic model, whereas the cytoplasmic N-terminal region (residues 1–13) is not observed in the electron density maps. The helix from Leu75 to Leu95, which is approximately perpendicular to the transmembrane one, should also be embedded in the membrane, since it is mainly formed by hydrophobic residues and is likely to contribute to the membrane attachment of NrfH. The two NrfH molecules in the complex are oriented so that their transmembrane helices are tilted in the membrane, crossing each other around Gly19. This arrangement brings NrfH haem 1 in close contact with the membrane, in a position that should be optimal for its interaction with menaquinol.

The globular domain of NrfH binds four haems *c*, which are arranged in two pairs (haems 1/2 and 3/4, numbered according to their order of attachment to the protein chain) displaying a typical dihaem parallel stacking motif, whereas haems 2 and 3 are almost perpendicular to each other (Figure 9.18a). This reveals that the haem

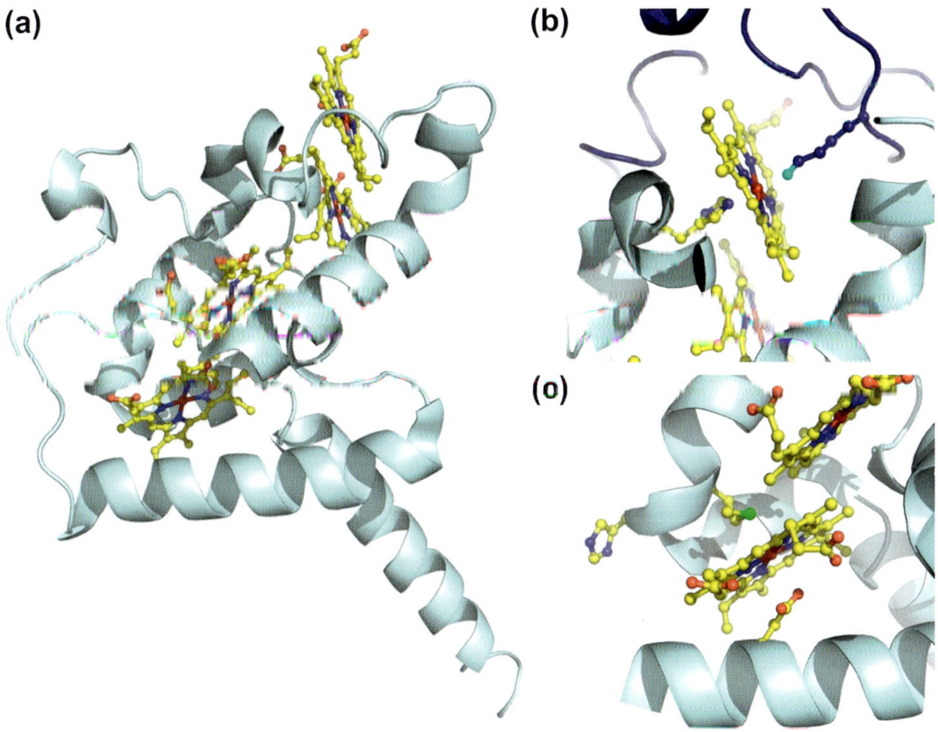

FIGURE 9.18 Cartoon representation of the overall fold of NrfH with haem cofactors shown in ball-and-stick (a) Zoomed view of NrfH haem 4 with proximal site coordinated by His140 and distal site by Lys331 from an internal NrfA subunit (b) (Colour code as in Figure 9.16a, except for Lys331 where carbon atoms are depicted in dark blue and nitrogen in bright cyan). Zoomed view of NrfH haem 1 showing the proximal ligand Met49, Asp89 and His47 from the CXXCHXM motif (c).

arrangement of NrfH is characteristic of a diverse family of multi-haem cytochromes *c*, whose haem groups are arranged in alternated stacking and perpendicular dihaem motifs.

NrfH Haem Coordination

The most remarkable feature of the NrfH structure was its surprising haem coordination. In particular, haem 1 displays an unprecedented ligation, with a methionine residue, Met49 from the CXXCHXM motif, as proximal axial ligand (S_{δ}—Fe distance of about 2.8 Å) rather than the histidine of the CXXCH haem *c* binding motif (Figure 9.18c). The second unexpected feature of NrfH haem 1 was the presence of an aspartate residue (Asp89) occupying the distal ligand position, with its carboxylate group approximately parallel to the haem plane, and with its $O^{\delta 2}$ atom at ~3 Å from the iron (Figure 9.18c). No continuous electron density was observed between the iron and the oxygen atoms, showing that Asp89 was not coordinated to the iron. Interestingly, this aspartate residue is replaced by a histidine in many NrfH proteins or a glutamate among the related NapC proteins. Thus, NrfH haem 1 is a methionine-coordinated high-spin haem, which, to the best of our knowledge, has never been described in biological systems. A pronounced cavity located close to haem 1 was identified in NrfH and proposed to be the menaquinol-binding site.

Another remarkable feature of NrfH is the coordination of haem 4 by a lysine residue (Lys331) from an NrfA protein (Figure 9.18b). This lysine residue, which belongs to the NrfA molecule that is closer to NrfH haems (named internal NrfA, $NrfA_{int}$), is the distal haem ligand, whereas NrfH His140 is the proximal haem ligand. Haem 4, which is the gateway for electrons in transit from NrfH to $NrfA_{int}$, is notably protruding from the NrfH structure towards a groove comprising $NrfA_{int}$ residues. This finding reinforces the structural complementarily between NrfH and NrfA subunits and the stability of the complex, which cannot be dissociated under non-denaturing conditions. Lys331 is conserved in NrfH-interacting NrfA proteins, suggesting that this lysine residue should also be the distal ligand in other NrfH proteins.

NrfH haems 2 and 3 are bis-histidinyl coordinated, with His70 and His120 as proximal ligands, and His145 and His61 as distal ligands, respectively.

Overall Haem Arrangement of the Complex

The formation of a stable complex between NrfH and its physiological partner, NrfA, allows important insights into the electron transfer pathway from menaquinol oxidation to nitrite reduction. The overall haem arrangement of the NrfHA complex is quite surprising, as the NrfH haem groups are non-symmetrically positioned with respect to the two NrfA molecules (Figure 9.16a). Only one of these, the $NrfA_{int}$ has the haems in close proximity with NrfH haems (Figure 9.16b). On the contrary, the haems from the external NrfA ($NrfA_{ext}$), are more distant from the NrfH haems, with minimum edge-to-edge distances of ~17 Å. The proximity of NrfH haem 4 to both haems 2 and 5 of $NrfA_{int}$, with edge-to-edge distances of 12.1 and 8.5 Å, respectively, suggests that both haems can accept electrons through direct electron tunnelling (Page et al., 2003). The electron transfer from $NrfA_{int}$ to the $NrfA_{ext}$ molecule should then occur through both haems 5 of the NrfA dimer, which are in very close contact (Figure 9.16b). Another remarkable aspect regarding haem packing is the close contact and parallel stacking of haems 3 and 4 of the two NrfH molecules in the complex (edge-to-edge distances vary between 4.6 and 10.4 Å), which should allow for fast electron transfer between them. This close proximity and haem arrangement strongly suggest a physiological role for this packing (Figure 9.16b).

Overall, the picture that emerges from the analysis of the 28 haem cluster in the $\alpha_4\beta_2$ complex is that the two NrfH haems 1 are the electron entry points from menaquinol oxidation, from where the electrons can travel very rapidly through the haem network to the four catalytic haems in the complex (NrfA haem 1). The 28 haems may be required for electron storage during the 24-electron reduction of four nitrite molecules occurring at the four catalytic sites.

NfrHA in Complex with Ligands

Aiming at a more detailed characterisation of the putative menaquinol-binding site, many co-crystallisation and soaking experiments were carried out with NrfHA using menaquinone analogues 2,3-dimethyl-1,4-naphthoquinone (DMN) or HQNO (Figure 9.19). The crystallisation conditions were similar to those used for the native protein. For co-crystallisation experiments NrfHA was first incubated, for 30 minutes at 4 °C, with ethanolic solutions of either DMN or HQNO to final concentrations ranging from 0.3 to 0.7 mM (around 10–20× molar excess). In the initial soaking experiments, excess of the quinone analogues was used as the ethanolic stock solutions were directly added to the crystallisation drops. Crystals from different co-crystallisation and soaking trials were tested but the majority presented poor or multiple diffraction patterns. Crystals had a tendency to grow as multiple needles and to be

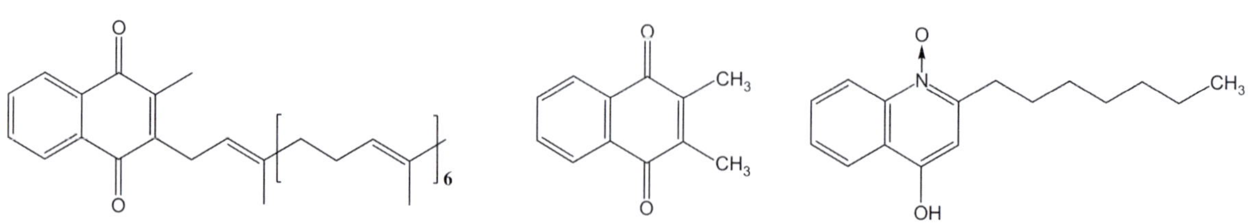

FIGURE 9.19 Structures of menaquinone and analogues 2,3-dimethyl-1,4-naphthoquinone (DMN) and 2-heptyl-4-hydroxyquinoline-*N*-oxide (HQNO) used for co-crystallisation and soaking experiments.

sensitive to soaking experiments, leading to crystal degradation in the drops after ligand addition. Difficulties in obtaining crystals of respiratory membrane enzymes in complex with quinone analogues/inhibitors had already been reported. The resolution of the ligand-bound structures may also be significantly lower than that of the free complexes.

Crystals of NrfHA were grown as previously described. A long needle of ~$1.2 \times 0.08 \times 0.06$ mm^3 appeared in 9% PEG 4 K, 0.1 M Hepes pH 7.5, 100 mM glycyl-glycyl-glycine after several weeks. This crystal was cut into three parts to facilitate crystal manipulation and freezing. Soaking was performed by the gradual addition of small volumes of a cryo protectant solution consisting of 14% PEG 4 K, 0.1 M Hepes pH 7.5, 20% glycerol, 0.7 mM HQNO, 1% ethanol. After about 4 h, the crystal was transferred to a fresh drop of cryo-solution with HQNO and flash-cooled in liquid nitrogen. X-ray diffraction data were collected from this single crystal, which diffracted to 2.8 Å resolution, at the ID14 1 beamline of the European Synchrotron Radiation Facility (ESRF, Grenoble). The NrfHA crystal soaked with HQNO belonged to the orthorhombic $P2_12_12_1$ space group, with unit cell dimensions of $a = 80.1$, $b = 189.1$ and $c = 263.5$ Å, with only one $NrfH_2A_4$ complex in the asymmetric unit, and not isomorphous with the native crystals. Data were processed, scaled and merged with HKL2000 (Otwinowski and Minor, 1997) and relevant statistics are shown in Table 9.4. It is worth mentioning that NrfA has also been co-crystallised with DMN, data collected, and the structure solved by molecular replacement. However, no extra electron density was observed near NrfH haem 1, and therefore no DMN was present in the menaquinol cavity.

Structure Determination and Refinement

The structure of the NrfHA–HQNO complex was solved by molecular replacement, using the coordinates of the biological unit of the cytochrome *c* nitrite reductase complex ($NrfH_2A_4$, PDB code: 2J7A) as a search model. The refinement was carried out with Refmac5 at 2.8 Å resolution and the manual model building was performed with COOT (Emsley and Cowtan, 2004) (refinement statistics are displayed in Table 9.4).

The initial electron density difference map clearly showed blobs of density in the vicinity of haem 1, which were identified as the inhibitor HQNO. A search in the HIC-Up database provided the HQNO coordinates from the 1KF6 PDB entry. Final R and R_{free} values were 22.0 and 26.1%, respectively. The atomic coordinates and structure factors of the NrfHA–HQNO complex were deposited in the PDB with the accession number **2VR0**. We will not describe in detail the NrfHA:HQNO structure (Rodrigues et al., 2006), but will only focus on the binding site of HQNO, an inhibitor of several menaquinol-oxidising enzymes, including NrfH.

Menaquinol-Binding Site

HQNO binds to NrfH close to haem 1 in a large cavity formed between the haem and the membrane helix that is parallel to the periplasmic interface with an entrance that is notably directed towards the membrane region (Figure 9.20a). This result confirms the quinol-binding site proposed previously, based on the NrfHA X-ray structure (Rodrigues et al., 2006). The polar head of HQNO forms H-bonds to Asp89, the residue which occupies the haem distal ligand position, and to Lys82 and Asn67 (Figure 9.20b). Both Lys82 and Asp89 play crucial roles in menaquinol binding and oxidation; Lys82 is strictly conserved among the family of cytochrome *c* quinol dehydrogenases, whereas Asp89 may be replaced by a histidine or a glutamate. Asp89 is located on the border of the binding cavity and within the polar region of the membrane and may function as a proton shuttle to the periplasm. Lys82 is likely to act as direct proton acceptor for quinol oxidation, whereby the proton may be transported to the solvent region directly or through Asp38 and Lys78 (residues that are partially conserved among NrfH cytochromes) (Figure 9.20b). In particular, the positive charge of Lys78 is maintained in the majority of the proteins. Lys82 is also one of the few residues that are in contact with HQNO hydrophobic tail. The present structure clearly illustrates the need for Gly86 to avoid steric clashes with the menaquinol (the minimum distance between its C_α atom and the naphthoquinone ring is only 3.5 Å).

TABLE 9.4 Diffraction Data Collection and Refinement Statistics for NrfHA–HQNO Complex

Data collection	
Wavelength (Å)	0.934
Space group	$P2_12_12_1$
Unit cell dimensions (Å)	$a=80.1$, $b=189.1$, $c=263.5$
Resolution limits (Å)	29.8–2.8
No. of unique reflections	92,571
Completeness* (%)	86.5 (75.4)
Redundancy*	2.8 (2.5)
I/σ*	9.4 (2.5)
R_{merge}* (%)	11.0 (37.7)
Refinement	
No. of reflections in working set	83,976
No. of reflections in test set	2592
R_{factor}* (%)	22.0 (31.1)
R_{free}* (%)	26.1 (38.3)
No. of non-hydrogen atoms	19,599
No. of non-H protein residues	2272
No. of water molecules	114
Average B values (Å²)	
Protein	32.3
Haems	21.8
HQNO	62.3
Waters	16.0
R.m.s. deviations	
Bond lengths (Å)	0.013
Bond angles (°)	1.42
Chiral volumes (Å³)	0.094

**Values in parenthesis refer to the highest resolution shell (2.9–2.8 Å).*

As concluding remarks, the three-dimensional structure of NrfHA bound to HQNO allowed for the first atomic characterisation of the quinone-binding site of a cytochrome *c* quinol dehydrogenase. In NrfH, the menaquinol-binding cavity is largely polar and has a wider opening to the protein surface than observed in other quinol binding proteins. This may be required to facilitate diffusion of the quinone in and out of this site, since six menaquinol

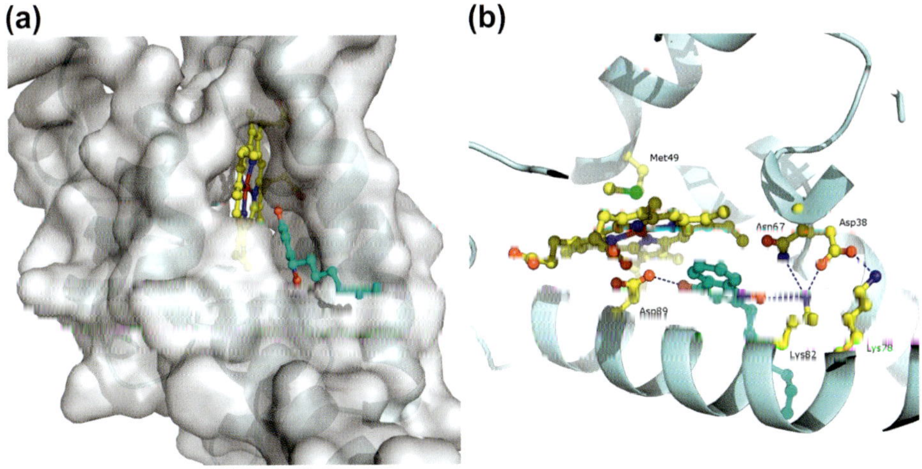

FIGURE 9.20 NrfH molecular surface displaying the entrance for the menaquinol-binding pocket which lies in the vicinity of haem 1 with HQNO in ball-and-stick (a) Close-up view of NrfH haem 1 with HQNO, Asp38, Met49, Asn67, Lys78, Lys82, and Asp89 displayed in ball-and-stick (b) Colour coding as in Figure 9.16a, except for HQNO molecule carbon atoms, which are shown in bright cyan.

molecules have to be oxidised by NrfH in order to provide electrons for the reduction of two nitrite molecules by the NrfA dimer with which NrfH associates. Several characteristics previously described for other quinone-interacting proteins, such as the presence of polar residues H-bonded to the quinone/quinol oxygen atoms, are also observed in NrfH. In addition, the menaquinone/menaquinol-binding site of NrfH and several other respiratory proteins lies in the polar region of the membrane, which would be expected to facilitate proton-transfer reactions (Rodrigues et al., 2006).

GENERAL CONCLUSIONS

Since the determination in the 1960s of the first protein structure (that of myoglobin at 6 Å resolution), macromolecular crystallography walked a long path, evolved, transformed and became one of the most powerful techniques for structure determination. Improvements in solving protein structures with numerous innovations over the past two decades together with the capability of providing structural information at the atomic level, without limitations that other techniques have (e.g. size of the assembly under study), made macromolecular crystallography a fashionable technique. The structures of membrane proteins (e.g. ATP synthase and ion channels), ribosomes and other cellular machines (e.g. RNA polymerase), which resulted in the award of Nobel Prizes to their authors, put crystallography under the spotlight and ranked it as one of the most used techniques in any research centre around the world.

The fact that metalloproteins possess inherent anomalous scattering centres, which circumvents most of the problems in the step of solving the 'phase problem', makes X-ray crystallography a very attractive technique to apply to these biological systems. The three-dimensional structure of a metalloprotein is not *per se* a 'finish line' but rather a 'starting line' for new questions gathered from the insights that the structure provides.

ACKNOWLEDGEMENTS

The authors would like to thank Carlos Frazao, Pedro Matias, Edward Franklin, Miguel Pessanha and Tânia Oliveira for the critical review of the manuscript. We would also like to thank all the members of the Macromolecular Crystallography Unit at ITQB-UNL which provided the crystal photos in Figure 9.3: Ricardo Coelho, Tiago Bandeiras, Isabel Bento, Tânia Oliveira, Célia Romão, Colin McVey and Mário Correia.

REFERENCES

Bamford, V. A., Angove, H. C., Seward, H. E., Thomson, A. J., Cole, J. A., Butt, J. N., Hemmings, A. M., & Richardson, D. J. (2002). Structure and spectroscopy of the periplasmic cytochrome *c* nitrite reductase from *Escherichia coli*. *Biochemistry, 41*, 2921–2931.

Bento, I. M. L., Lopes, G., Carrondo, M. A., & Lindley, P. F. (2005). Dioxygen reduction by multi-copper oxidases; a structural perspective. *Dalton Trans., 7*, 3507–3513.

Blundell, T. L., & Johnson, L. N. (1976). *Protein Crystallography*. London: Academic Press.

Boggon, T. J., & Shapiro, L. (2000). Screening for phasing atoms in protein crystallography. *Structure, 8*, R143–149.

Brito, J. A., Bandeiras, T. M., Teixeira, M., Vonrhein, C., & Archer, M. (2006). Crystallisation and preliminary structure determination of a NADH: quinone oxidoreductase from the extremophile Acidianus ambivalens. *Biochim Biophys Acta, 1764*, 842–845.

Caffrey, M. (2009). Crystallizing membrane proteins for structure determination: use of lipidic mesophases. *Annu Rev Biophys, 38*, 29–51.

Chen, V. B., Arendall, W. B., 3rd, Headd, J. J., Keedy, D. A., Immormino, R. M., Kapral, G. J., Murray, L. W., Richardson, J. S., & Richardson, D. C. (2010). MolProbity: all-atom structure validation for macromolecular crystallography. *Acta Crystallogr. D Biol. Crystallogr., 66*, 12–21.

Collaborative Computational Project, N. (1994). The CCP4 suite: programs for protein crystallography. *Acta. Crystallogr., 50*, 760–763.

Cunha, C. A., Macieira, S., Dias, J. M., Almeida, G., Goncalves, L. L., Costa, C., Lampreia, J., Huber, R., Moura, J. J., Moura, I., & Romao, M. J. (2003). Cytochrome *c* nitrite reductase from Desulfovibrio desulfuricans ATCC 27774. The relevance of the two calcium sites in the structure of the catalytic subunit (NrfA). *J. Biol. Chem., 278*, 17455–17465.

de la Fortelle, E., & Bricogne, G. (1997). Maximum-likelihood heavy-atom parameter refinement for multiple isomorphous replacement and multiwavelength anomalous diffraction methods. *Macromolecular Crystallography, 276*(Pt A), 472–494.

DeLano, W. L. T. (2002). *The PyMOL Molecular Graphics System* (0.98 ed.). San Carlos, CA, USA: DeLano Scientific.

Diederichs, K., & Karplus, P. A. (1997). Improved R-factors for diffraction data analysis in macromolecular crystallography. *Nat. Struct. Biol., 4*, 269–275.

Drenth, J. (1999). *Principles of Protein X-Ray Crystallography* (second ed.). Berlin: Wiley-VCH.

Einsle, O., Messerschmidt, A., Huber, R., Kroneck, P. M., & Neese, F. (2002). Mechanism of the six-electron reduction of nitrite to ammonia by cytochrome *c* nitrite reductase. *J. Am. Chem. Soc., 124*, 11737–11745.

Emsley, P., & Cowtan, K. (2004). Coot: model-building tools for molecular graphics. *Acta Crystallogr. D Biol. Crystallogr., 60*, 2126–2132.

Engh, R. A., & Huber, R. (1991). Accurate bond and angle parameters for X-ray protein structure refinement. *Acta. Crystallogr., A47*, 392–400.

Garman, E., & Murray, J. W. (2003). Heavy-atom derivatisation. *Acta Crystallogr. D Biol. Crystallogr., 59*, 1903–1913.

Hendrickson, W. A., & Ogata, C. M. (1997). *Meth. Enzymol., 276*, 494–523.

Hickman, A. B., & Davies, D. R. (2001). Principles of macromolecular X-ray crystallography. *Curr. Protoc. Protein Sci.*, Chapter 17, Unit 17 13.

Holm, L., & Sander, C. (1999). Protein folds and families: sequence and structure alignments. *Nucleic Acids Res., 27*, 244–247.

Hooft, R. W., Vriend, G., Sander, C., & Abola, E. E. (1996). Errors in protein structures. *Nature, 381*, 272.

Islam, S. A., Carvin, D., Sternberg, M. J., & Blundell, T. L. (1998). HAD, a data bank of heavy-atom binding sites in protein crystals: a resource for use in multiple isomorphous replacement and anomalous scattering. *Acta Crystallogr. D Biol. Crystallogr., 54*, 1199–1206.

Jancarik, J., & Kim, S.-H. (1991). Sparse matrix sampling: a screening method for crystallisation of proteins. *J. Appl. Cryst., 24*, 409–411.

Knablein, J., Neuefeind, T., Schneider, F., Bergner, A., Messerschmidt, A., Lowe, J., Steipe, B., & Huber, R. (1997). Ta6Br(2+)12, a tool for phase determination of large biological assemblies by X-ray crystallography. *J. Mol. Biol., 270*, 1–7.

Laskowski, R. A., MacArthur, M. W., Moss, D. S., & Thornton, J. M. (1993). PROCHECK: a program to check the stereochemical quality of protein structures. *J. Appl. Cryst., 26*, 283–291.

Lovell, S. C., Davis, I. W., Arendall, W. B., 3rd., de Bakker, P. I., Word, J. M., Prisant, M. G., Richardson, J. S., & Richardson, D. C. (2003). Structure validation by Calpha geometry: phi, psi and Cbeta deviation. *Proteins, 50*, 437–450.

McCoy, A. J., Grosse-Kunstleve, R. W., Storoni, L. C., & Read, R. J. (2005). Likelihood-enhanced fast translation functions. *Acta Crystallogr. D Biol. Crystallogr., 61*, 458–464.

McVey, C. E., Amblar, M., Barbas, A., Cairrao, F., Coelho, R., Romao, C., Arraiano, C. M., Carrondo, M. A., & Frazao, C. (2006). Expression, purification, crystallization and preliminary diffraction data characterization of Escherichia coli ribonuclease II (RNase II). *Acta Crystallogr Sect F Struct Biol Cryst Commun, 62*, 684–687.

Murshudov, G. N., Vagin, A. A., Lebedev, A., Wilson, K. S., & Dodson, E. J. (1999). Efficient anisotropic refinement of macromolecular structures using FFT. *Acta Crystallogr. D Biol. Crystallogr., 55*, 247–255.

Oliveira, T. F., Vonrhein, C., Matias, P. M., Venceslau, S. S., Pereira, I. A., & Archer, M. (2008). Purification, crystallisation and preliminary crystallographic analysis of a dissimilatory DsrAB sulfite reductase in complex with DsrC. *J. Struct. Biol., 164*, 236–239.

Otwinowski, Z., & Minor, W. (1997). Processing of X-ray diffraction data collected in oscillation mode. In C. W. CarterJr., & R. M. Sweet (Eds.), *Methods in enzymology: macromolecular crystallography* (pp. 307–326).

Page, C. C., Moser, C. C., & Dutton, P. L. (2003). Mechanism for electron transfer within and between proteins. *Curr. Opin. Chem. Biol., 7*, 551–556.

Pereira, I. A., LeGall, J., Xavier, A. V., & Teixeira, M. (2000). Characterisation of a heme *c* nitrite reductase from a non-ammonifying microorganism, desulfovibrio vulgaris hildenborough. *Biochim. Biophys. Acta, 1481*, 119–130.

Pinho, F. G., Romao, C. V., Pinto, A. F., Saraiva, L. M., Huber, H., Matias, P. M., Teixeira, M., & Bandeiras, T. M. (2010). Cloning, purification, crystallisation and X-ray crystallographic analysis of Ignicoccus hospitalis neelaredoxin. *Acta Crystallogr. Sect. F Struct. Biol. Cryst. Commun., 66*, 605–607.

Rodrigues, M. L., Oliveira, T. F., Pereira, I. A., & Archer, M. (2006). X-ray structure of the membrane-bound cytochrome *c* quinol dehydrogenase NrfH reveals novel haem coordination. *Embo. J., 25*, 5951–5960.

Rodrigues, M. L., Scott, K. A., Sansom, M. S., Pereira, I. A., & Archer, M. (2008). Quinol oxidation by *c*-type cytochromes: structural characterization of the menaquinol binding site of NrfHA. *J Mol Biol, 381*, 341–350.

Rodrigues, M. L., Oliveira, T., Matias, P. M., Martins, I. C., Valente, F. M., Pereira, I. A., & Archer, M. (2006). Crystallisation and preliminary structure determination of the membrane-bound complex cytochrome c nitrite reductase from Desulfovibrio vulgaris Hildenborough. *Acta Crystallogr. Sect. F Struct. Biol. Cryst. Commun., 62*, 565–568.

Rodrigues, M. L., Pereira, I. C., & Archer, M. (2011). The NrfH cytochrome c quinol dehydrogenase. In A. Messerschmidt (Ed.), *Handbook of metalloproteins* (pp. 89–102). Chichester, UK: John Wiley & Sons.

Romao, M. J., Archer, M., Moura, I., Moura, J. J., LeGall, J., Engh, R., Schneider, M., Hof, P., & Huber, R. (1995). Crystal structure of the xanthine oxidase-related aldehyde oxido-reductase from D. gigas. *Science, 270*, 1170–1176.

Romao, C. V., Ladakis, D., Lobo, S. A., Carrondo, M. A., Brindley, A. A., Deery, E., Matias, P. M., Pickersgill, R. W., Saraiva, L. M., & Warren, M. J. (2011). Evolution in a family of chelatases facilitated by the introduction of active site asymmetry and protein oligomerization. *Proc Natl Acad Sci U S A, 108*, 97–102.

Rossmann, M. G., & Arnold, E. (2006). *International Tables for Crystallography. Volume F: Crystallography of Biological Macromolecules.*

Rupp, B. (2010). *Biomolecular Crystallography. Principles, Practice, and Application to Structural Biology.* Garland Science.

Chapter 10

Genetic and Molecular Biological Approaches for the Study of Metals in Biology

Robert L. Robson
School of Biological Sciences, AMS Building, University of Reading Whiteknights, Reading, Berkshire, RG6 6AJ, United Kingdom

Chapter Outline

INTRODUCTION AND AIMS

The closely interrelated fields of genetics and molecular biology provide a wealth of elegant and powerful approaches and techniques which can provide definitive answers to biological questions and which are highly complementary to other techniques covered in this book. The range of tools and resources is expanding rapidly driven by the need to respond to the grand scientific challenges posed by human health, the biosphere, the environment and diminishing natural resources and man's insatiable curiosity to understand more about the world.

This chapter does not set out to provide a cookbook of recipes: there are quite extensive manuals referenced herein that fulfil that need. Rather this chapter lays out some key considerations and approaches or strategies to particular types of questions, introduces some basic concepts and methodologies in genetics and molecular biology especially for those whose primary areas of expertise and interest is not in biology. The chapter closes by illustrating the way that these techniques have been applied over the years to one system, nitrogen fixation, a complex and important metalloenzyme system.

Practical Approaches to Biological Inorganic Chemistry, 1st Edition. http://dx.doi.org/10.1016/B978-0-444-56351-4.00010-5.

BASIC GENETICS AND MOLECULAR GENETICS: ORIGINS AND DEFINITIONS

The Origins, Evolution and Speciation

Most biologists believe that life on Earth today evolved from a single life form which came into existence as the Earth's crust formed about 4.3 billion years ago. This view has gained acceptance because of the unity observed in all life forms: e.g. the commonality of basic macromolecular components (DNA, RNA, polypeptides etc.) and the underlying universal genetic code. Since Haeckel drew the first "tree" of life in 1866, biologists have visualised the relationships between organisms in the form of a tree based on comparisons of form and function. Now, molecular tools (comparison of the sequences of stable RNA molecules [16S and 18S rRNAs]) are commonly used to deduce phylogenetic relationships and construct pathways of evolution and a very approximate clock of when major developments may have occurred. This has led to a number of interesting hypotheses, e.g. that the earliest organisms may have been thermophiles and hyperthermophiles consistent with Darwin's idea that life may have "emerged in warm little ponds".

The basic unit of life is the species. However, the term is interpreted in different ways by different groups of biologists. For some groups of organisms, it is relatively easy to define a species as "an organism which is capable of sexual reproduction and producing fertile offspring". However, in organisms that do not reproduce sexually such a definition is not useful and other criteria need to be applied. For example, in bacteria members of the same species have ~95% DNA sequence identity.

The total number of recorded species is ~1.5 million. At least half of these are insects, with ants being the dominant group. Surprisingly, we only know ~5000 species of bacteria but there might have been expected to have been far more given they are the oldest extant forms of life. However, while we can be reasonably sure that there are no more large animals to be discovered, it has been estimated from studies of environmental samples that we may know only ~1% of all existing bacteria and that the total sum of species on Earth may be in the order of ~10 million.

Grouping the Species. Classification, Taxonomy, Phylogeny

Early classifications grouped organisms into two kingdoms: Plants and Animals. When Van Leeuwenhoek first reported his "little animalcules" the new Kingdom of the Protista was born. Fungi belonged to the Plant Kingdom for many years but they were proposed as a separate kingdom in 1944. The Protozoans or protists, and the Bacteria were reclassified later as separate Kingdoms. In 1937, Edouard Chatton proposed a major distinction between organisms which contained a nuclear membrane (the so-called Eukaryotes consisting of Animals, Plants, Fungi and Protozoans) and those organisms which do not (the Prokaryotes: the bacteria) (see in Sapp, 2005). This is a fundamental division in biology with many implications (Figure 10.1 and see below). However, a new classification scheme which overarches the Kingdoms and which resolves life into three domains was proposed by Woese et al. (1990) on the basis of the comparison of sequences of the 16S ribosomal RNAs (rRNA) (Figure 10.2). One domain contains all the eukaryotes (or Eukarya), the other two domains are prokaryotes. For a long time, we had known about a group of prokaryotes, e.g. the CH_4-forming methanogens, with unusual properties including ether-linked and often branched membrane lipids and insensitivity to many antibiotics. This group is so distinctive that it formed the basis of a separate domain called the Archaea. The other prokaryotes which include most of the well-known bacteria, pathogens etc, form the Eubacteria domain. Several features of the Archaea more closely resemble the properties of the Eukarya than the Eubacteria. It seems likely that Eukarya are more likely to have evolved from the Archaea rather than the Eubacteria.

The Fundamental Molecular Biological Information Molecules: DNA and RNA

DNA is a polymer in which the bases adenine, cytosine, guanine, and thymine are attached to a deoxyribose-phosphate backbone (Figure 10.3). Its commonly adopted structure is that of an anti-parallel double helix held

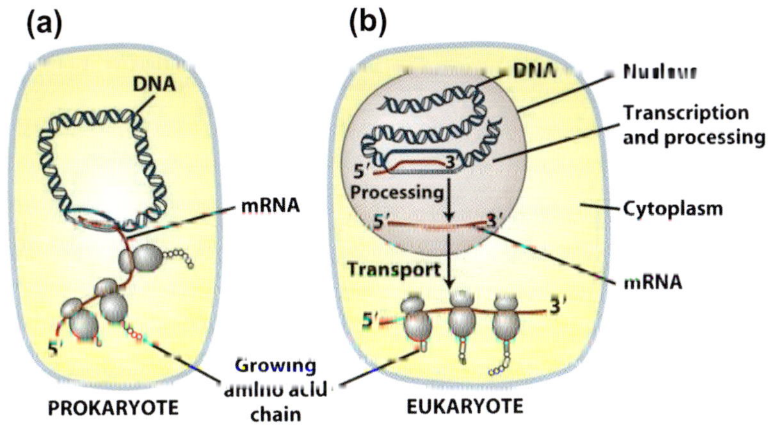

FIGURE 10.1 Prokaryotes and eukaryotes *(Source: apbio-werle wikispaces).*

together by hydrogen bonds between the inward facing bases (A to T and G to C) creating major and minor grooves in the helix. However, DNA can adopt other structures under special circumstance such as cruciforms, triple stranded forms, and bends and kinks produced by runs of particular bases (e.g. A-T-rich regions). In situ, DNA is also supercoiled (negatively or positively) either through the action of specific enzymes (topoisomerases) or by being "wound" onto a protein scaffold formed by histones or histone-like proteins. In cells, DNA exists in a complex with proteins (histones in eukaryotes) and there are a number of DNA-binding proteins which can create sharp bends and loops which in some cases allow interactions between proteins bound to other loci in the vicinity. Also, bases can be modified, often by methylation, and such modifications are biologically highly significant, e.g. in producing resistance to cleavage by restriction endonucleases in bacteria, or by controlling the

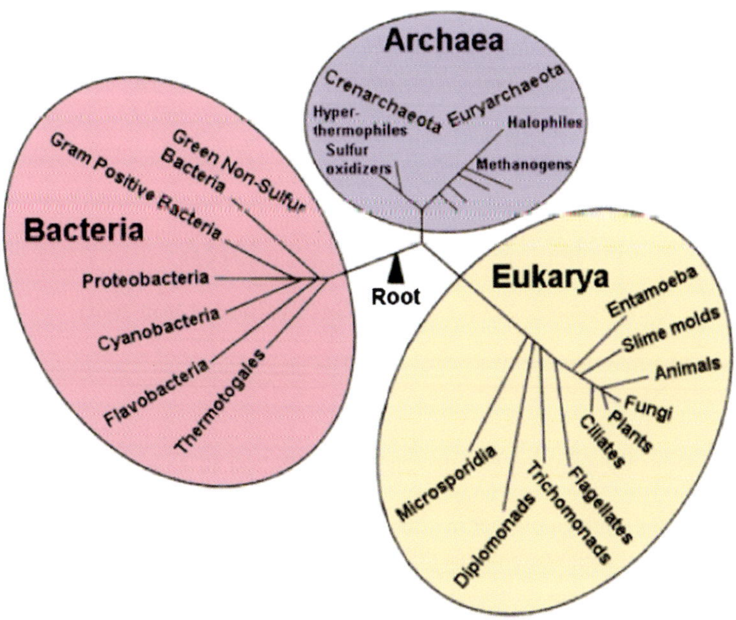

FIGURE 10.2 The three domains of life *(from blogs-scientificamerican.com).*

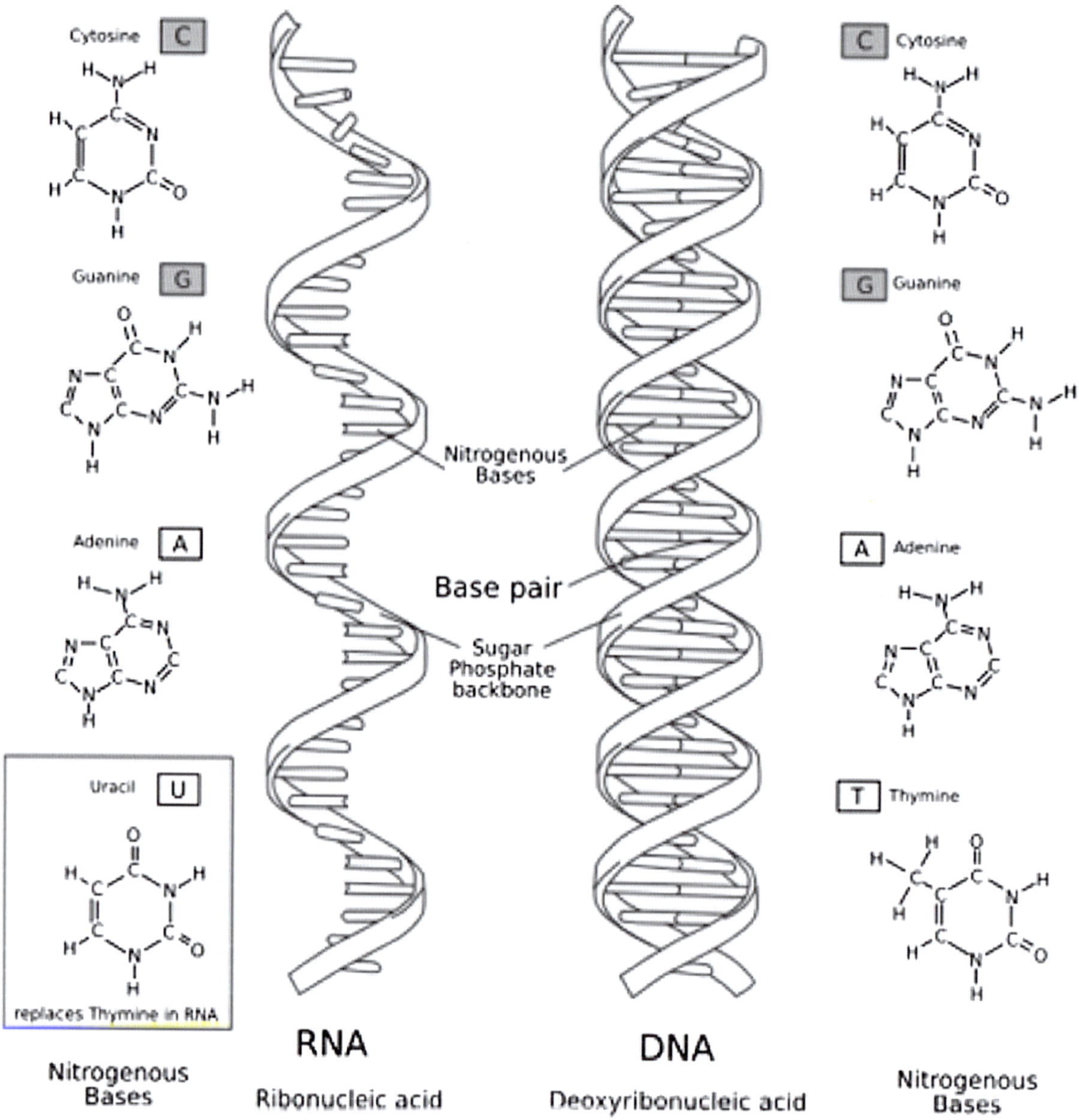

FIGURE 10.3 DNA and RNA: structures and building blocks (*Source: pro-profs.com*).

expression of genes or large groups of genes in cell differentiation in Eukarya. All of these variations have biological significance.

RNA is generally a single-stranded polymer in which the bases adenine, cytosine, guanine, and uracyl (in place of thymine) are attached to a ribose-phosphate backbone (Figure 10.3). However, RNA has the propensity to form complex intramolecular structures, e.g. in transfer RNA (tRNAs) or in ribosomes and spliceosomes. Messenger RNAs can also form secondary structures which can be of considerable functional significance. Again RNA contains modified bases, e.g. the several modifications found in tRNAs. RNA is generally divided into two broad groups: the relatively unstable messenger RNA and the stable RNAs involved in the protein synthesis machinery, e.g. tRNAs and rRNA and RNA splicing by the RNA components of spliceosome assembly in Eukarya. It is generally accepted that the larger rRNA genes are good molecules to use for identification, for constructing phylogenetic trees and for reconstructing evolutionary pathways.

All organisms have genomes composed of DNA except some viruses where the genomes are solely RNA, e.g. the animal pathogens such as Polio, Influenza, and of course the retroviruses such as HIV.

The Central Dogma

The path of information flow in Biology was initially thought to be unidirectional from DNA to DNA to RNA to Protein. But the discovery of retroviruses and the enzyme they contain called reverse transcriptase which synthesises DNA from an RNA template showed that information flows back from RNA to DNA. The processes by which the information flows and the basic enzyme systems which are involved are illustrated in Figure 10.4.

All mechanisms of DNA synthesis (both replication and reverse transcription) require appropriate primers. For example, in the case of the retroviruses, such as HIV in which the enzyme, reverse transcriptase, acts as both an RNA-dependent and DNA-dependent DNA polymerase, the primer for DNA synthesis is a tRNA molecule hijacked from the previous host cell and encapsulated into the virus upon assembly. By contrast, transcription does not require a primer.

The Genetic Code

The genetic code (Figure 10.5) with its triplet basis is organised into a number of family groups which are broadly related to amino acid side chain functionality and/or space filling capacity such that on a statistical basis, mutations tend to cause "functional" substitutions. Recent analysis suggests that the code is exquisitely evolved towards conserving functionality in proteins. ATG encodes methionine but with appropriate translation initiation signals upstream it is used as the translation initiation codon, although CTG, and GTG are also used quite frequently in prokaryotes. Three codons, TAA, TAG, TGA signal the termination of translation. Crick's wobble

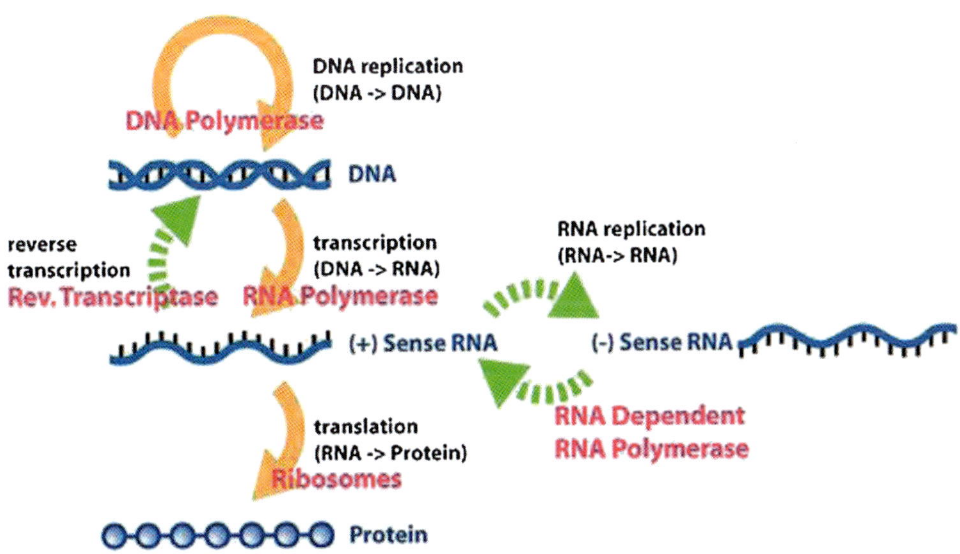

FIGURE 10.4 The central dogma of biology *(Source: en-wikipedia)*.

First letter	Second letter: U	C	A	G	Third letter
U	UUU, UUC Phe; UUA, UUG Leu	UCU, UCC, UCA, UCG Ser	UAU, UAC Tyr; UAA Stop; UAG Stop	UGU, UGC Cys; UGA Stop; UGG Trp	U C A G
C	CUU, CUC, CUA, CUG Leu	CCU, CCC, CCA, CCG Pro	CAU, CAC His; CAA, CAG Gln	CGU, CGC, CGA, CGG Arg	U C A G
A	AUU, AUC, AUA Ile; AUG Met	ACU, ACC, ACA, ACG Thr	AAU, AAC Asn; AAA, AAG Lys	AGU, AGC Ser; AGA, AGG Arg	U C A G
G	GUU, GUC, GUA, GUG Val	GCU, GCC, GCA, GCG Ala	GAU, GAC Asp; GAA, GAG Glu	GGU, GGC, GGA, GGG Gly	U C A G

FIGURE 10.5 The universal genetic code *(Source: bio.miami.edu).*

hypothesis (Crick, 1966) proposed that the triplet code is organised in order of importance: position 2 > position 1 > position 3.

The integrity of the genetic code is as much a function of the cognate tRNAs (often there are several tRNA genes for a single codon) and the modified bases in the vicinity of the anticodon which influence the codon–anticodon pairing specificity. The fidelity of translation is not only controlled by the codon/anticodon pair but also the amino-acyl tRNA synthetases which couple the amino acid residue to the cognate tRNA. This is well illustrated by the incorporation of novel amino acid into proteins during protein synthesis in vitro by preparing synthetically aminoacylated tRNA species (Noren et al., 1989).

Different organisms have different 'codon biases,' i.e. some codons are used in preference to others. This is particularly true in the prokaryotes. Sometimes the bias is extreme especially where the G + C% contents of the genome are either very high or very low. Codon bias can be a reason why a particular gene does not express well in a foreign host. Codon bias is an important consideration in designing oligonucleotide primers to identify/amplify genes from genomic DNA especially from organisms with AT- or GC-rich genomes.

The genetic code is not completely universal. Slight variant genetic codes are found in the small genomes present in a few animal mitochondria and in some bacteria which also have small genomes, e.g. Mycoplasmas. These variants may have evolved first through certain codons becoming redundant and therefore available for reassignment. The non-standard codes mean that expression of genes from such backgrounds in other organisms will lead to aberrant proteins.

What is a gene?

A gene is region of DNA, or RNA in RNA viruses, which codes for a gene product. In the case of DNA, this will be an RNA molecule (a transcript) which may or may not be translated into a polypeptide. However, a gene is not simply the region which is ultimately transcribed and/or translated but it also includes all the necessary signals which make it functional within its host, i.e. transcription/translational initiation and termination signals, and post-transcriptional processing signals (e.g. splicing sites) and regulatory loci.

How Big Are Genes and Genomes?

The average polypeptide has a molecular mass (M^r) of ~33 kDal. Since the average molecular mass of amino acid residues is 108, then the average polypeptide is about 300 residues long. Given the triplet code, the average gene for the average gene product is therefore about a 900 base pairs (bp) or ~1 kilobase (Kb) long. So if you know that the genome of a bacterium is 1000 kb or 1 megabase (Mb) then it will encode approximately 1000 genes.

A genome is an organism's complete set of DNA and contains all of the information needed to build and maintain that organism. In animal cells, the genome includes the nuclear-borne chromosomes and the mitochondrial DNA and in plant cells it also includes DNA present in the plastids (chloroplasts) Mitochondria and plastids have evolved from prokaryotes which have become internal symbionts of eukaryotic cells.

Genome sizes vary greatly. Some viruses may have genomes of just a few kilo bases (Kb). The smallest bacterial genome characterised is that of the highly specialised intracellular parasite *Mycoplasma genitalium* (580 kb). *Escherichia coli* has a genome of 4.3 Mb, yeast's is 13.5 Mb. Amongst the multicellular organisms, the nematode, *Caenorhabditis elegans* has a genome of 100 Mb, the higher plant, *Arabidopsis thaliana* a genome of 100 Mb, and the human genome is ~3000 Mb.

Gene density is that fraction of a genome which encodes gene products. This can be very high (>95%) in viruses, a little less so in bacteria (>80%) but is often far lower especially in higher Eukaryotes. Until very recently this "extra" or "junk" DNA was considered to have little functional relevance possibly acting as "spacer" DNA. However recent research with the human genome reveals that this so called "junk" DNA almost certainly plays an important role in regulating gene expression.

Replicons

A replicon is an autonomously replicating DNA molecule. A chromosome can be defined as a replicon which contributes genes essential for the organism to grow and survive under all conditions, e.g. genes required for the major metabolic, biosynthetic, and macromolecular processes, e.g. ribosome formation, replication, transcription etc.

Usually, prokaryotes have a single circular chromosome which may be present in several copies but exceptions include the metabolically versatile autotroph *Rhodobacter capsulatus* which has two circular chromosomes and the causative agent of Lymes Disease, *Borrelia burgdorferi* which is remarkable in comprising a single linear chromosome of 910 kb and 21 other linear and circular replicons that add an additional 533 kb base pairs of DNA.

Most prokaryotes also contain additional and usually much smaller replicons called plasmids. Sometimes (e.g. in the case of several species of *Rhizobium*), these additional replicons are very large (>1 Mb) and are known as megaplasmids. Plasmids are generally thought to be not essential for survival of the organism in the laboratory and can be eliminated (cured). However, they are required for survival and competitiveness of the organism in its natural environment. Examples would include the megaplasmids in Rhizobia which contain genes for the formation of nitrogen-fixing nodules of leguminous plants.

However, plasmids are usually smaller (3 to 200 kb). There are a great many different plasmids. Important and very well-characterised examples, include drug-resistance plasmids (R(esistance)-factors) such as those found in *Pseudomonas aeruginosa*, which carry resistance genes for a number of antibiotics (tetracycline, ampicillin, streptomycin, kanamycin); plasmids which confer the ability to metabolise particular substrates (e.g. toluene, naphthalene and camphor degrading pathways in pseudomonads) and also plasmids which confer the ability to kill other cells, e.g. the small colicin-encoding plasmids (e.g. the ColE1-plasmid) in *E. coli* from which some of the first recombinant cloning vectors were constructed.

Incompatibility. Two different plasmids with similar or identical replication mechanisms usually cannot co-exist in the same cell. They are said to be incompatible and to belong to the same incompatibility group. However, plasmids of different incompatibility groups can co-exist.

Host range. Plasmids cannot replicate in all hosts. "Narrow host range" plasmids are specific for one host or closely related organisms because they depend heavily on the host replication machinery. Often this means that the origin of replication (*ori*) is only recognised by that host cell's replication machinery. An example is the ColE1 group of plasmids which are only capable of replicating in *E. coli* and closely related members of the *Enterobacteriaceae* such as *Salmonella pneumoniae* or *Klebsiella aerogenes*. However, broad host-range plasmids present greater potential risks for genetic containment and these will need to be taken into account when designing research strategies.

Conjugation/mobilisation. Conjugative plasmids are those capable of transferring themselves to other members of the same or other species. These plasmids carry several *tra* (transfer) genes, and usually a small locus (*mob*) on which those genes act. This is a useful property where one is cloning, e.g. constructing a gene library for the organism of interest first in *E. coli* but want to transfer the recombinant plasmids in the library to the organism under study where those genes can potentially function. Some cloning systems have been developed to separate *tra* and *mob* loci onto compatible plasmids to produce what is known as a binary plasmid system. A classical example is the plasmid pRK2013. This plasmid carries an antibiotic resistance and also R1 *tra* genes but not a *mob* site therefore is non self-transmissible. However, if a cloning vector carrying that *mob* site is introduced into a strain carrying pRK2013, then the *tra* genes will operate to transfer that vector to a new host. The *tra* genes are then said to function in *trans*. This means that the "mobilisable" cloning vector does not need to carry the often large number of *tra* genes and so minimising its size. Also, since the recombinant DNA carried by the cloning vector is not self-transmissible it minimises the risk of unwanted genetic transfer events.

Copy number. Plasmids vary considerably in their replication capacity measured as the number of copies in the cell (copy number). Some high copy number plasmids, e.g. the common ColE1-derived plasmids can exist in >200 copies per cell. They are said to show relaxed replication. This means that they are easy to isolate in bulk. Also the presence of many copies of genes of interest in a host can be useful for overexpression work, and for finding genes in libraries etc. High copy number recombinant plasmids can prove unstable where genes or segments of DNA they carry are toxic. For example, DNA that is A + T rich can often be difficult to clone into high copy number vectors in *E. coli*. Other plasmids have much more stringent replication and may be present in just a few copies per cell. Plasmids with low copy numbers can be useful for constructing vectors where genes or their products are toxic or cause unwanted effects, e.g. titration of activators etc.

Chromosomes in Eukaryotes. Nuclear-encoded genes in eukaryotes are organised into pairs (diploidy) of linear chromosomes which have a very complex structures (see Figure 10.6). Ploidy is the term given to the number of copies of chromosomes in an organism. Human cells except for germ cells (sperm and egg) are diploid and contain two sets of nuclear autosomes (not the chromosomes sex-determining X and Y chromosomes). Germ cells have half the chromosome number and though they are strictly monoploid, they are actually called "haploid" plus either the X or the Y chromosome. Some higher organisms are polyploids and contain multiple copies of

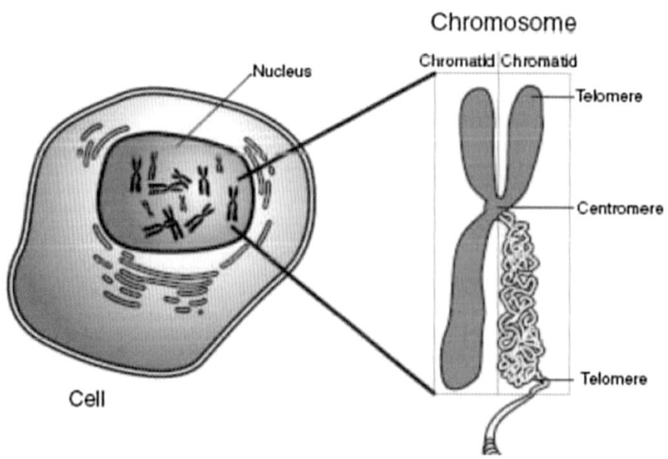

FIGURE 10.6 Eukaryotic chromosomes *(Source: blog-chron.com).*

chromosomes. Polyploid plants are often larger and more vigorous. A fascinating example is the wheat used for making bread, *Triticum aestivum*, which has 42 chromosomes and was derived some 8000 years ago.

Gene Organisation

Gene organisation is markedly different in viruses and prokaryotes as compared to eukaryotes. In viruses, prokaryotes, mitochondria and plastids, the genes are organised into operons which contain often functionally related genes which are co transcribed from a single promoter (Figure 10.7). The gaps between genes are often short and some genes even overlap so that the start codon of one gene overlaps with the stop codon of the preceding gene (e.g. ----ATGA). Not only does this maximise gene density but it ensures translational coupling where ribosomes terminating translation of the first gene do not disengage as normal but move on to translate the second cistron so preserving equivalence in the levels of the polypeptides produced. Not all operons contain more than one gene, some contain a single gene (monocistronic) but multigene operons (polycistronic) predominate.

An unavoidable consequence of polycistronic operons is that mutations in the proximal genes may affect distal genes. This is especially the case where "foreign" DNA, e.g. a transposon is inserted in the proximal genes so that transcription is often terminated at that site causing truncation of the mRNA and loss of the translation products of the distal genes. In such cases, the phenotype of such mutants is not simply a consequence of the loss of function of the gene in which the insertion occurs. It is for this reason that "in frame deletion" mutants, where the reading frame of the mutated gene is not disturbed, are helpful in the precise analysis of individual gene function. But even such mutants may lead to destabilisation of the message or the loss of secondary and hitherto unsuspected promoters.

In Eukaryotes, each gene has its own promoter. This is true even in simple unicellular organisms such as yeast in which the genome is not substantially larger than those found in some bacteria.

A number of reasons have been proposed as to why gene organisations in viruses, prokaryotes and eukaryotes is so different. These include the need for an operon structure and a much higher degree of coding efficiency in microscopic life forms coupled with the selective advantage that a greater mobility or promiscuity of whole functional units can offer. Against that, placing each gene under its own promoter as in the Eukarya provides greater potential for subtlety in expression of each individual gene and the capacity for the reassortment and recombination within or between different alleles via the sexual process provides huge potential for generating variation in the populations of organisms.

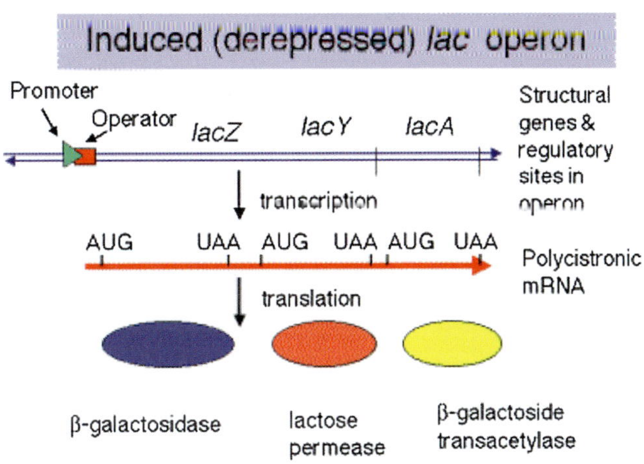

FIGURE 10.7 The *lac* operon in *Escherichia coli (Source: bx.psu.com).*

Insertion Elements, Transposons, and Repetitive DNA

Transposons. Transposons were first described in maize by McClintock (1950) and subsequently they have been found in almost every organism. They are so-called "mobile" or "promiscuous" genetic elements which can effectively "cut and paste" or "copy and paste" themselves independently of the replication of the whole genome of an organism either within the same genome (*cis*) or into another genome in the cell (*trans*). There are two basic classes of transposons: Class 1 or retro-transposons replicate via an RNA intermediate while Class 2 or DNA transposons replicate without an RNA intermediate. Most encode the transposition apparatus which in the simplest case is a transposase which usually recognises specific sequences in the genome as insertion targets. Some elements are however dependent on a transpose gene at a separate locus. Transposons are also usually bordered by short repeat sequences at either end which are a consequence of the transposition mechanism. Some transposons carry one or more additional genes. Notable examples in bacteria being those which carry one or more antibiotic resistance genes, e.g. Tn3 which is nearly 5 kb and carries a *bla gene* for β-lactamase (ampicillin resistance) (Figure 10.8), Tn5 which carries genes for resistance to streptomycin and kanamycin and Tn7 which is a large complex transposon of 14 kb encoding genes for trimethoprim and streptomycin/spectinomycin resistance. These and similar transposons have been responsible in part for the rapid spread of antibiotic resistance genes between bacteria but they have also been very useful for genetic analysis.

Insertion Sequences (IS elements). These are the simplest forms of transposons. Many do not carry a transposase gene but can be acted on *in trans* by transposases located elsewhere in the genome of the organism. Examples include *copia*, one of 30 different elements in *Drosophila* which make up an astonishing 30% of its genome, *Alu* in humans, and the Ty elements in Yeast.

Repetitive DNA. DNA which is present in many copies in excess of the replicon number. Repetitive DNA includes insertion sequences, satellite and microsatellite DNA. It is these sequences which can be used for genetic fingerprinting analysis in humans and other animals and also plants (so-called "bar coding").

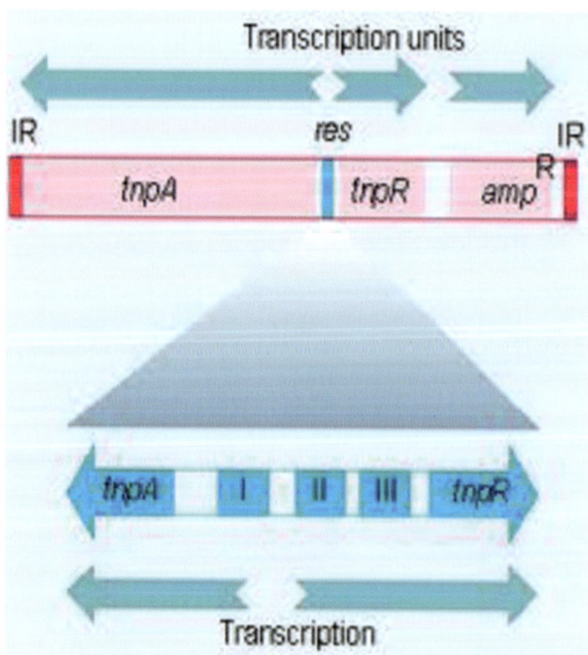

FIGURE 10.8 Transposon Tn*3* *(Source: flylib.com)*.

How DNA Moves and Can be Moved Around between Organisms: Transformation, Transduction, Conjugation

Transformation. Transformation is the natural or forced uptake of naked DNA into cells. A few organisms contain natural DNA uptake mechanisms. Examples include the eubacteria in the genera *Haemophilus*, *Bacillus*, *Azotobacter*, and *Thermus*. These organisms develop the capacity to take up DNA (known as "competence") at a particular stage of growth or in a particular medium. Other cells can be "forced" to take up DNA in a number of ways. For *E. coli*, incubation in high concentrations of divalent cations (e.g. Ca^{2+}) followed by a heat shock in the presence of the DNA (e.g. a plasmid) induces the DNA to enter the cell. A more general and often highly efficient method is electroporation which involves suspending cells in very low ionic strength water and then passing a high-voltage pulse through the suspension in the presence of the DNA (Figure 10.9). The ballistic method involves blasting cells with tungsten bullets (ballotini) coated with the DNA. This method can even succeed in delivering DNA through plant cells and into chloroplasts.

Transduction. Transduction is the process by which DNA is acquired by a virus (bacteriophage in bacteria) and carried to another cell and is then integrated into the host genome. In "generalised transduction", the viral particle picks up "by mistake" any virus-sized piece of DNA. In specialised transduction, the virus genome integrates into the host genome and when it excises for independent replication it does so imprecisely and carries with it a piece of the host genome next to the insertion site. Bacteriophage λ in *E. coli* is a very well-understood and well-exploited example of the latter.

Conjugation. Conjugation is seen in many organisms, bacteria, fungi, and protozoans and it is a sexual process. In terms of this discussion, conjugation in bacteria involves the natural transfer of one plasmid to another organism (Figure 10.10). One of the earliest systems to have been identified was in *E. coli* where Hfr strains carry the F plasmid and are capable of mating with a "female" strain during which the whole chromosome can be transferred. It takes about 60 min for the whole process and that is why the position of genes on the chromosome of *E. coli* are mapped in minutes (relative to the start of conjugation) from a fixed point of the chromosome where the transfer originates.

Homologous Recombination

Homologous recombination is a type of genetic recombination in which nucleotide sequences are exchanged between two similar or identical molecules of DNA. It is most widely used by cells to accurately repair harmful breaks that occur on both strands of DNA, known as double-strand breaks. Homologous recombination also produces new combinations of DNA sequences during meiosis, the process by which eukaryotes make gamete cells. These new combinations of DNA produce genetic variation in offspring, which in turn enables populations to adapt during the course of evolution. Homologous recombination is also used in horizontal gene transfer to exchange genetic material between different strains and species of bacteria and viruses.

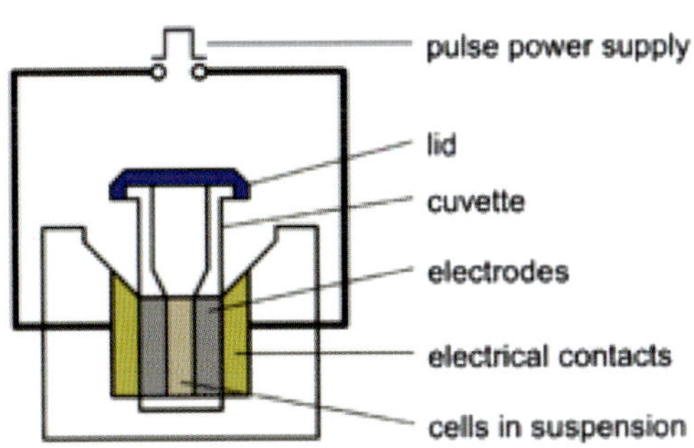

FIGURE 10.9 Electroporation cell *(Source: en-wikipedia.org).*

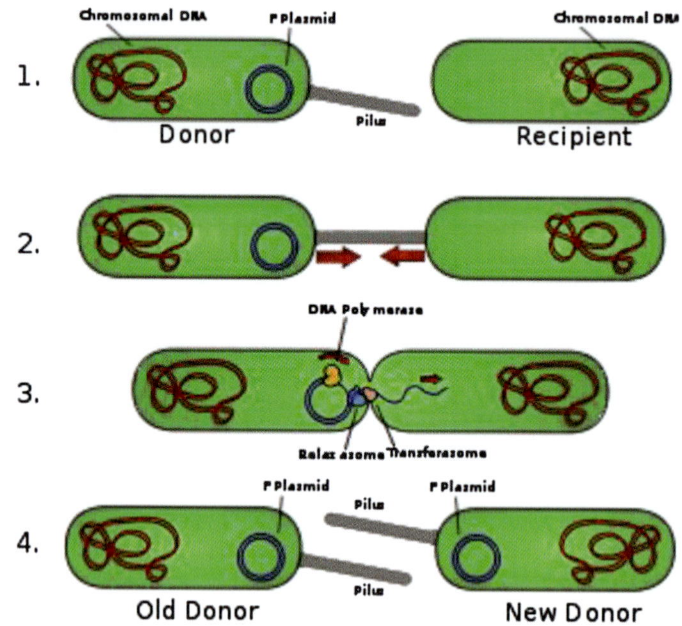

FIGURE 10.10 Bacterial conjugation *(Source: simple-wikipedia.org).*

Although homologous recombination varies widely among different organisms and cell types, most forms involve the same basic steps (Figure 10.11). After a double-strand break occurs, sections of DNA around the 5' ends of the break are cut away in a process called *resection*. In the *strand invasion* step that follows, an overhanging 3' end of the broken DNA molecule then "invades" a similar or identical DNA molecule that is not broken. After strand invasion, one or two cross-shaped structures called Holliday junctions connect the two DNA molecules. Depending on how the two junctions are cut by enzymes, the type of homologous recombination that occurs in meiosis results in either chromosomal crossover or non-crossover. Homologous recombination that occurs during DNA repair tends to result in non-crossover products, in effect restoring the damaged DNA molecule as it existed before the double-strand break.

Homologous recombination is conserved across all three domains of life as well as viruses, suggesting that it is a nearly universal biological mechanism. Homologous recombination is crucially important in molecular biology for introducing genetic changes into the genomes of target organisms. The transfer of a particular mutation from an engineered recombinant construct into the genome of the organism under study involves recombination events on either side of the target location (so-called "double recombination").

Promoters, Transcription Initiation, and Transcriptional Regulation

What is a promoter? A promoter is simply a region of DNA at which RNA polymerase binds and at which transcription by RNA polymerase is initiated. However, that simple definition belies the complexity of the process of transcription initiation and the complexity of many promoter regions and the interplay between transcription initiation and transcriptional regulation.

Transcription initiation. The process of transcription initiation is complex and involves several steps including reversible binding of the RNA polymerase to the promoter to form a complex, isomerisation and DNA strand separation (open complex formation). Binding of RNA polymerase and other molecules to DNA can be demonstrated by DNA protection and footprinting techniques. The "simplest" system is seen in prokaryotes and

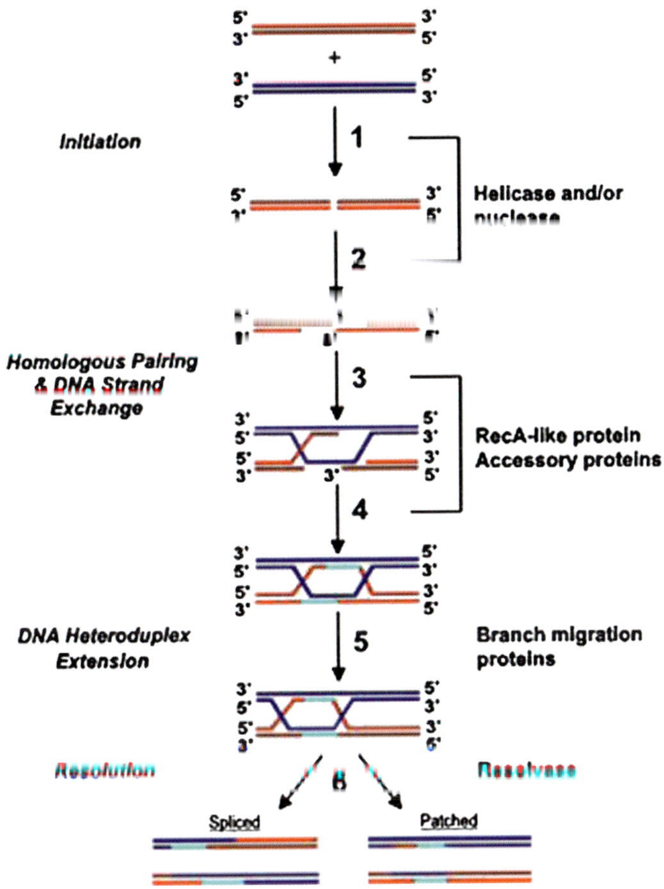

FIGURE 10.11 Homologous recombination *(Source: Biosscience.org)*.

especially bacteria where there is a single RNA polymerase core enzyme and interchangeable subunits (σ subunits) which help to define the specificity of the RNA polymerase binding. For example, in *E. coli*, σ^{70} (so defined because of its molecular mass) is the "housekeeping" σ, the form which transcribes the basic gene set required for normal growth and division. In Archaea, RNA polymerases generally contain more subunits and other initiation factors are required for transcription. In eukaryotes, the situation is more complex still with three types of RNA polymerase for nuclear-encoded genes (I for the larger rRNAs, II for mRNAs and III for tRNAs and 5S rRNA). Many more initiation factors are involved in the formation of the transcription initiation complex. Also mitochondria and plastids contain a fourth RNA polymerase which resembles the bacterial enzyme.

Transcriptional regulation. This is probably the most frequently encountered level at which gene expression is controlled and it represents a huge area to summarise briefly. Transcriptional regulation involves a great diversity of regulatory proteins which often have conserved fold patterns (helix-turn-helix, leucine zippers, zinc fingers) that recognise major or minor grooves of the DNA onto which sequence specific recognition is imposed.

Activators are those proteins required for the expression of certain genes. Often they bind to the DNA in the vicinity of the promoter (classically called **operator** regions) and interact with the RNA polymerase to alter the kinetics of the transcription initiation events, e.g. stimulate open complex formation. It is not necessary that they bind immediately adjacent to the promoter, or even that the activator binds upstream of the promoter because DNA easily forms loops which bring the two DNA-bound proteins together. Also it is possible to have complex arrangements where several activators can bind to the "operator" regions. In Eukaryotes, these regulatory regions can be quite extensive regions of DNA. Activators either have sensor domains or interact with one or more partner protein components which have sensor domains. Two component regulator systems are common in prokaryotes and other organisms. Mutations to activators or to the region of DNA at which they bind usually lead to loss of expression of the genes under their control.

Repressors generally bind to DNA in the vicinity of the promoter and block RNA polymerase binding. Repressor-binding sites often overlap or are close to the promoter. Loss of a repressor gene function leads to constitutive expression.

Translation Initiation

In order for an mRNA to be read, it must be recognised by the ribosome. The signals are markedly different in prokaryotes and eukaryotes.

In prokaryotes, a sequence in the mRNA ~10 bases upstream of the translation initiation codon (usually ATG) is important for translation initiation. The so-called Shine and Dalgarno sequence is a purine-rich motif with the consensus of 5'-GAGGAGA'-3' which is complementary to the 3' end of 15S rRNA and postulated to be important for recognition between the ribosome and the mRNA (Shine and Dalgarno, 1975). A few instances of mRNAs which start directly at the translation initiation codon and yet are still translated are known which suggests that sequences 3' to the translation initiation codon may be important too.

In eukaryotes, mRNAs have to undergo processing of three types before they are translated: capping, tailing and splicing (Figure 10.12). Capping involves the addition of 7-methylguanylate at the 5' end of the message in an unusual 5'-5' phosphotriester bond. It appears to be essential for translation and also for message stability. Tailing involves cleavage by a sequence-specific ribonuclease at the 3' end of the mRNA and then the addition of a polyadenylate tract at the end catalysed by the enzyme polyadenylate polymerase. Tailing is also important for message stability. In addition, mRNA undergoes splicing out of untranslated regions (introns: so called because they **int**errupt the gene) and the remaining sequences (exons: so called because they are the part of the message which is **ex**pressed) are rejoined to create a coherent mature message. This highly complex organisation provides some flexibility for the production of variant proteins ("splice" variants) through different splicing pathways in different tissues. The spliceosome is the remarkably complex protein/RNA assembly that carries out the

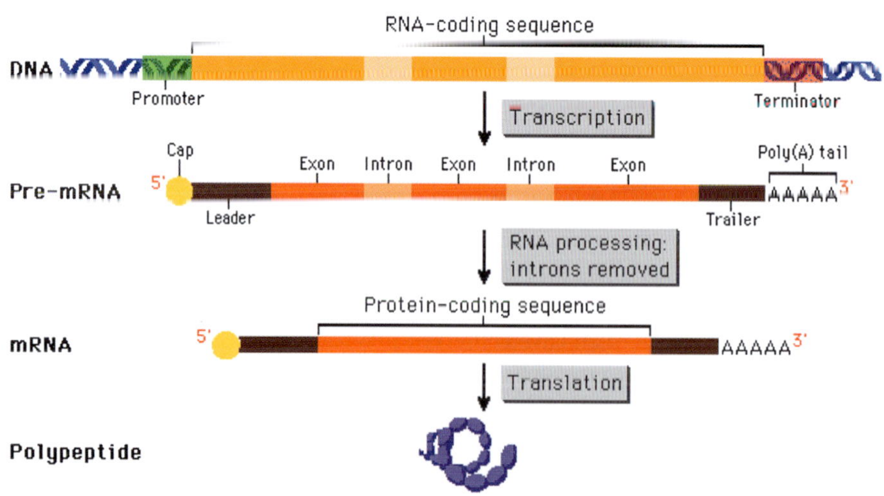

FIGURE 10.12 Processing of mRNA in eukaryotes *(Source: phschool.com).*

recognition of the specific sequences surrounding the splicing sites, and the cleavage and the re-joining of the exons. Other splicing events occur in the processing of rRNAs which are not spliceosome-dependent. These are the so-called type I and type II self-splicing reactions.

The processing of transcripts in eukarya adds additional complexity to the process of cloning specific genes. Some genes especially in the higher eukarya contain many introns and can be arranged over very long distances in the genome. Usually, cloning the expressed component of the gene involves starting with the isolation of the mature mRNA and its reverse transcription into DNA (so-called copy DNA or complementary DNA (cDNA)). However, the polyadenylate tails of the mature mRNAs proved very useful because mature mRNAs can be isolated using poly-T affinity columns and a poly-T primer can be used as a universal primer for synthesis of the copy DNA by reverse transcriptase.

SETTING UP: REGULATIONS, EQUIPMENT, METHODS AND RESOURCES

Regulation and Approvals

If you are considering doing recombinant DNA work for the first time in your laboratory or even handling genetically manipulated (GM) organisms obtained from another lab you will need to meet local genetic manipulation regulations and requirements. These will vary from country to country but almost certainly you will need to consider whether your lab is suitably designed and equipped or whether it might require refurbishment potentially on a significant scale. Generally, the requirements are determined on the basis of the risk of release and spread of organisms/genes which are potentially harmful to man, the biosphere and the environment more generally. Usually, there are three categories of risk: Category I requiring basic facilities; Category II requiring a higher level of containment and Category III being required for pathogens and toxin genes. For each of these categories, laboratory fit out requirements are specified in all countries. The basic presumption is one of over – rather than under-provision. Even in the lowest risk category, it must be possible to contain and effectively decontaminate any lab spills. Therefore, lab benches and surfaces must be sealed so that they can be decontaminated and the flooring must be chemically resistant and will contain no unsealed joins and should overlap up the walls and benches to contain potential liquid leaks. The lab should also have negative air pressure so that the air flows into, but not out of, the lab in order to avoid airborne contamination of corridors or the external environment. Waste disposals sinks will have to release aqueous waste into approved ducts or sewers. Detailed records will need to be kept of the disposal or decontamination of live recombinant organisms or DNA. If working with known pathogens then there will be a need to work in a special containment laboratory (e.g. a Category III lab) which will have very stringent design specifications for air flow, access, security, storage, and record keeping. If you need to use recombinant technologies rarely in your research, these are very powerful reasons for going to work in a lab which has the required standard and approvals from the appropriate regulatory bodies.

Approvals/authorisations from your institution or from a host institution for each project will almost certainly be needed before you even start to do recombinant DNA work. Most institutions have GM safety officers who will be able to advise you and take you through the process of obtaining approval. Obtaining approval is not necessarily onerous but does require appropriate forethought and planning and could mean that you will not be given permission to start the project. Also, unauthorised working with GM organisms or on recombinant DNA projects could potentially put at risk the institution-wide licence of your institution or that of a host institution.

Once the lab/project is approved, just some of the basic equipment required or required access to include: fume hood, autoclave, orbital shakers, −20 freezer, −70 freezer with storage trays and boxes, fridges, laminar flow cabinet, ice machine, micro-centrifuge, refrigerated bench centrifuge (1 l scale), thermocycler, agarose gel electrophoresis boxes, electrophoresis power packs, electroblotter, UV illuminator and camera, bacteriological incubator, electroporator, pipettors (a range), phase contrast microscope, water baths and heat blocks, timers, solvent cupboard, microwave oven, still or equivalent for the production of high-quality water, glass bottles for autoclaving reagents and media, tube racks, pipettors, glass pipettes and pipette cans, autoclave and reagent bottles, disposable pipettes, vacuum pump for aspiration, computers with broadband internet access. Consumable needs will depend on the project but include molecular biology enzymes (e.g. restriction enzymes, DNA polymerases, ribonucleases), antibiotics, nucleotides, solvents, agarose, DNA and plasmid purification kits, a range of plastic ware, e.g. flip cap and screw cap microfuge tubes, a range of pipettor tips, petri dishes, 50 ml tubes, and bacteriological/genetic manipulation disposal bags.

APPROACHES AND SYSTEMS

Defining the goals and devising appropriate and effective strategies to achieve them is crucial for any scientific project. This is particularly true in molecular biology because everything looks straightforward in principle but rushing headlong into a project without careful forethought and planning is likely to be expensive and frustrating. This is true whether the aim of a project is simple or ambitious. For example, the cloning of a single gene in order to express, purify and study its gene product may look easy on paper. The cloning part can be achieved in just a few days when all resources and expertise are available. However, whether the protein will express to the desired level and be correctly folded will be a matter of fortune. This is especially true for metalloproteins where there may be a requirement to manufacture and insert specific metal centres. Even if one does succeed in overexpressing the protein in an apparently native and active state, whether it is an artefact of the system used to produce it can only be judged by reference to the native protein. Laboratories "practised in this art" will have alternative cloning strategies available involving different host strains, expression vectors, and/or cultivation conditions etc. At the other end of the scale, a more complex project might involve the identification and analysis of genes for a novel process in an organism not previously studied. Work on such projects will probably take years and require significant human resources and laboratory facilities to yield results. A project of this type will involve some or all of the following: the isolation and characterisation of mutants: isolation and characterisation of proteins involved, determining protein sequences for DNA primer design; raising antibodies to the protein(s): construction of gene libraries to isolate the genes and flanking DNA, and DNA sequencing and bioinformatics analysis, development of a genetic system for the organism and site-directed mutagenesis to explore gene functions and possibly regulatory analysis. In projects of this complexity, it is probably becoming cost-effective to start by having the complete genome of the organism determined especially in the cases of prokaryotes.

Model Systems

Genetic research has tended to focus on particular organisms with important properties (e.g. pathogens; producers of commercially important substances, e.g. antibiotics, solvents; genetic tractability, and/or they are relatively easy to cultivate). This results in a critical mass of researchers working on the organism with the attendant accumulation of resources and know how. The sharing of expertise and resources and the exchange of researchers between laboratories is very much accepted practice and funding agents around the world have been very forward thinking and supported massive national/international collaborative projects, resource centres (e.g. stock collections; databases and facilitating web-based software). Some research funders will often look more favourably on research proposals involving model organisms rather than relatively little studied systems especially where they offer little novelty.

Wherever possible, it is advisable to work with a model organism, or closely related species. Of course, there is often no alternative but to work on a system which has unique properties but one must recognise the challenges. These can often be lessened by carefully defining the objectives and good project planning. It is also advisable not to start a genetic and molecular genetic project without prior experience. It will save a lot of time and expense to acquire the "art" from experienced practitioners or indeed if the project is of a limited scope (e.g. cloning and expressing a single protein/enzyme) then probably that part of the project is best done in a lab where the experience, equipment and materials reside. It could save a lot of time and expense in buying basic equipment, consumables and molecular biology kits which you may use rarely.

The classic model organism is the Gram-ve bacterium *E. coli* for which there are thousands of strains (both commensal and pathogenic), strain collections, a knock-out mutation collection (The Keio Collection: Baba et al., 2006), the complete genome sequences of different strains, very well-studied and exploited viruses (bacteriophage), a host for gene libraries, cloning and expression vectors, a number of genetic systems and expertise that can be found almost all over the world. There are many other model systems including the bacterium *Bacillus subtilis* (a Gram+ve bacterium); the unicellular eukaryotes, *Chlamydomonas reinhardtii* (a unicellular green alga), *Saccharomyces cerevisiae* (the budding yeast used in brewing, fermentation and the production of vitamins and biochemicals), *Schizomyces pombei* (the fission yeast), and amongst multicellular organisms, *Drosophila melanogaster* (the fruit fly), *C. elegans* (a nematode); *Danio rerio* (zebra fish) and among the mammals (*Musca domestica*, the mouse). These model organisms have been effectively "domesticated" in a manner akin to man's early domestication and breeding of animals and crops.

MOLECULAR BIOLOGY TOOLS AND METHODS

This Chapter is not intended to be a "cookbook". Instead it provides a summary of various methods and techniques, something of their history and their application. There are however a number of excellent manuals available, two of the most comprehensive being:

a. **Molecular Cloning: A Laboratory Manual**, (Sambrook and Russell, 2000)
b. **Short Protocols in Molecular Biology**, (Ausubel et al., 2002).

Preparation of DNA

There are basically two different methods for preparing DNA depending on whether the DNA is chromosomal (large linear or circular molecules with little or moderate supercoiling) or plasmid (usually small circular molecules often with a high degree of negative supercoiling). Furthermore, if it is necessary to have pure DNA it is important to recognise that it exists in cells complexed with specific proteins (histones or histone-like proteins) and these must be removed usually by protease digestion. Also, methods that purify DNA will also purify RNA too and therefore the RNA must also be removed usually by ribonuclease digestion. Clearly, any proteases or ribonucleases used must be DNAase free. For many polymerase chain reaction (PCR) amplifications, the DNA need not be pure and it may be only necessary to boil up a small volume of a cell suspension to release sufficient DNA to serve as a template.

Chromosomal DNA. DNA is a relatively fragile molecule. The aim of methods is to avoid as little breakage (or shearing) of the DNA as possible. Key factors which assist the recovery of long DNA molecules are the ease with which the cell can be disrupted and the absence of endogenous endonucleases released during the cell breakage. All solutions and glassware or plasticware must be sterile to ensure that there are no contaminating nucleases present and no contaminating DNA either. Some cells break easily, e.g. animal cells and some bacterial cells, but others are much more robust. For example, most plant cells are very difficult to break and need extreme techniques such as first freezing the plant tissue in liquid nitrogen and/or grinding up in a pestle and mortar and sometimes with pure sand or glass particles (ballotini). The basic method derives from that published by Marmur (1961). The cells are ruptured (lysed) as gently as possible sometimes with a visco-protectant such as sucrose present. A mixture of high salt (sodium acetate) and isopropanol is layered over the top of the lysate. The nucleic acid/protein complex and RNA will precipitate at the interface and gradually rise up in strands through the upper layer. These strands can then be hooked or spooled out on a very fine glass hook made from a disposable glass pipette. The precipitate containing the DNA is then redissolved by gentle rolling in a buffer containing DNAase-free protease and ribonuclease. Once the DNA has come into solution, it is then rolled very gently with a phenol/chloroform mixture into which the peptide fragments partition. The tube is then left to stand to allow the phases to separate, the upper aqueous phase is removed and the DNA reprecipitated from it with the salt and isopropanol mixture and collected by spooling again. The, by now, pure DNA should then redissolve easily in dilute buffer and any traces of phenol/chloroform removed by ether extraction. Purity of the DNA can be checked by measuring the ratio of absorption at both 260 and 280 nm (after Warburg and Christian, 1942). For long-term storage, it can be freeze dried or more usually kept in aliquots as alcohol precipitates. Both storage methods preserve the DNA as long molecules. Such molecules can then can be easily further fragmented by endonuclease activity or physical breakage.

Plasmid DNA. Methods for the purification of plasmids from prokaryotes generally exploit the relative resistance to denaturation (DNA strand separation) in the presence of mild alkali of highly supercoiled DNA molecules as compared to relaxed chromosomal molecules. In the standard method (Birnboim and Doly, 1979), which is available commercially in kit form, organisms are lysed in detergent and mild alkali to which DNAase-free RNAase can be added. RNA is hydrolysed by ribonuclease and the alkali denatures the released chromosomal DNA and protein which can then be precipitated by the addition of high concentrations of sodium acetate. Following centrifugation, the plasmid DNA is precipitated from the supernatant by the addition of an equal volume of isopropanol. Very large plasmids (megaplasmids) can also be isolated in this way but require controlled lysis in very mild alkaline conditions.

Agarose Gel Electrophoresis

Horizontal submerged agarose gel electrophoresis is the standard and simple technique used for resolving DNA and RNA molecules of different lengths (Southern, 1975). Agarose gels are prepared from high-quality molten agarose solution usually in a TRIS-based buffer poured into a former in an electrophoresis box with a sample comb (Figure 10.13). After the agarose is set, the box is flooded with buffer and the comb is removed carefully leaving behind wells in the gel into which the DNA samples mixed with a buffer containing bromophenol blue dye and sucrose to increase the density are loaded. The gel box is coupled up to a power pack and a voltage of about 10 V/cm applied across the electrodes for between 30 minutes and 2 h depending on the size of the gel and the degree of resolution required. The nucleic acids can be visualised after staining with ethidium bromide (which is usually placed in the agarose gel mix at the start) and illuminated with short wave UV transilluminator. Ethidium bromide is a carcinogen and appropriate precautions must be taken. Alternative proprietary stains are available

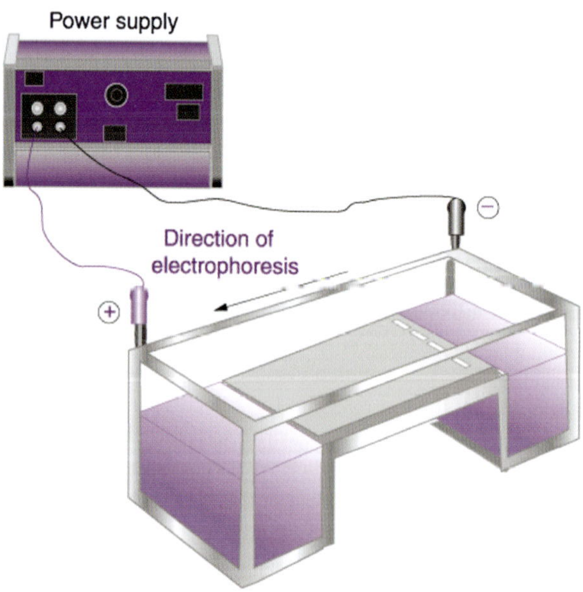

FIGURE 10.13 Agarose gel electrophoresis equipment *(Source: biochem.arozona.edu).*

from commercial sources. DNA fragments can be sized using DNA marker sets (ladders) containing linear DNA fragments of known different lengths most usually between ~150 bp and ~25 Kb depending on the agarose concentration and buffer used (Figure 10.14). Supercoiled and nicked plasmid DNA have different electrophoretic properties compared to linear DNA but linear DNA ladders can be used to size circular DNA molecules just as long as they have been calibrated against intact plasmid DNAs. Often plasmids and especially high copy number examples can give rise to multiple bands which can be confusing. This is because they form concatamers of interlocked circular molecules due to the failure of the molecules to separate (resolve) after replication. Digestion with restriction enzymes should simplify the picture.

RNA molecules such as mRNAs can also be separated in agarose gels but require denaturing conditions produced by the inclusion of formaldehyde in the gel and running buffer so that the molecules do not form secondary structures.

Pulse-Field/Orthogonal Electrophoresis

Generally, very large DNA molecules do not resolve well in a linear field in agarose gels even if the gel concentration is lowered. This is because the long molecules tend to get trapped in the carbohydrate network. This problem is resolved by applying either a pulsed field, where the current is reversed briefly or better still where the current is applied in several different directions but with a net forward direction (Figure 10.15). This allows the trapped molecules to wriggle through the agarose. In this way, it is possible to resolve very long molecules including large megaplasmids and even bacterial chromosomes several megabases long (Schwartz and Cantor, 1984). This requires of course that the DNA is not broken on extraction from the cell and this can be achieved in some cases by embedding the cells in small agarose blocks, lysing them in situ so that the DNA is released and unfolds gently and treating them with protease and if necessary restriction enzymes in situ. The blocks are then placed into the wells and a pulse field is applied. Separation can take as long as 10 to 24 h and it is usually necessary to run such gels in a cold room.

Blotting Techniques

Blotting techniques are common throughout molecular biology. In principle, they involve the capillary or electrophoretic transfer (blotting) of the DNA, RNA or proteins that have been separated in a gel to a porous cellulose nitrate or nylon membrane filter.

Southern blotting. Southern blotting is used to transfer DNA from an agarose gel onto a filter (Southern, 1975). The membrane captures the pattern of DNA molecules produced during electrophoresis and after drying it can then be probed with DNA or RNA probes to detect the presence and location of specific sequences. The

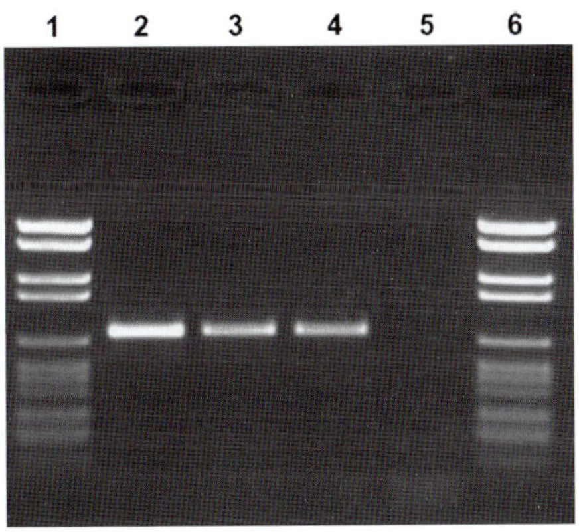

FIGURE 10.14 Agarose gel: visualising DNA *(Source: aegis.org).*

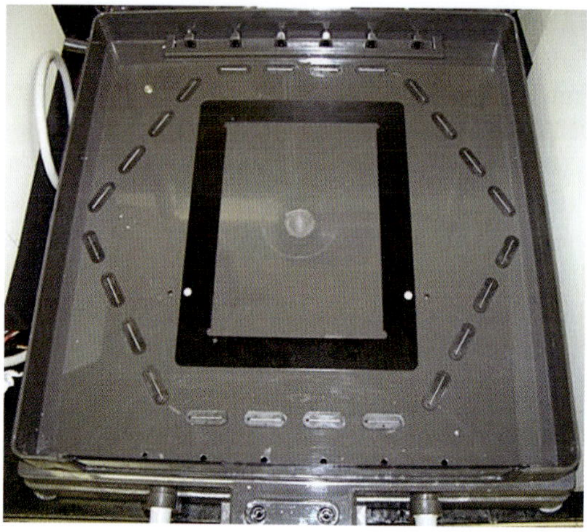

FIGURE 10.15 Pulse field gel electrophoresis apparatus *(Source: forcewearinc.com).*

technique involves placing the gel on to the filter cut to exactly the same size as the gel which is then overlayed with a stack of 3 MM chromatography paper rectangles and a wad of paper towels the same size as the gel. The whole assembly is then placed gel side down on a much longer piece of 3 MM chromatography paper the ends of which are submerged in buffer in an electrophoresis tank. A weight such as a flat glass bottle is placed on top of the stack (Figure 10.16). Over a period of hours, buffer is drawn up by a wicking action from the gel box tank through the gel and through the filter to the stack of filter paper and towels above. The denatured single-stranded DNA is carried along with the flow of buffer but binds to the filter. The technique requires very even wicking of buffer

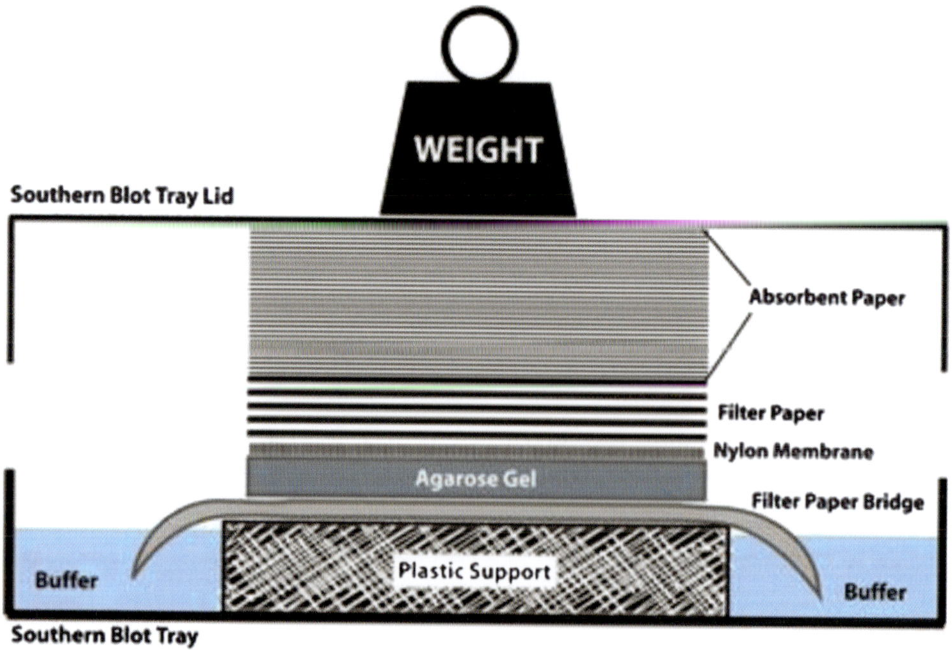

FIGURE 10.16 Southern blotting *(Source: gbiosceinces.com).*

through the gel so that the pattern of DNA fragments in the gel is perfectly reproduced on the membrane. Also, care must also be taken to avoid any bubbles between any layers of the sandwich. Nowadays, electroblotting is generally used for this process because it is quicker and generally more reproducible. In this technique, the filter paper – agarose gel–membrane filter–filter paper sandwich – is clamped between two plastic mesh plates and the whole assembly placed in an electro-blotter tank full of electrophoresis buffer (Figure 10.17). When a voltage is applied across the terminals, DNA is electrophoresed out of the gel and binds onto the membrane.

Northern blotting. Northern blotting is used to transfer RNA from agarose gels onto a nylon or cellulose nitrate membrane. This is essentially the same process as for Southern blotting but with the crucial difference that the agarose gel is prepared with formaldehyde to limit secondary structures forming in the RNA (Alwine et al., 1977). Sizing of RNA molecules will require an RNA ladder. It is particularly important in this technique to ensure that all reagents are not contaminated with RNase and it is advisable to reserve gel boxes solely for running RNA gels.

Detection of specific DNA or RNA sequences by hybridisation. Hybridisation in molecular biology involves use of a specific labelled single-stranded DNA or RNA molecule/fragment of between 50 and 2 kb (probe) to detect homologous polynucleotide sequences usually on Southern or Northern blots or in situ, e.g. on arrayed chromosomes. Originally, ^{32}P-labelled probes were made from the template DNAs by end labelling or nick translation using DNA polymerase I with ^{32}P-labelled deoxynucleotide triphosphates (dNTPs). More recently, non-radioactive methods are recommended because the technologies are safe, no hazardous radioactive waste storage is required, the probes are long lasting and hybridisation solutions can be reused and the exposure times are generally shorter. The two main non-radioactive methods are the biotin–streptavidin and the digoxigenin–antidigoxigenin system. In the biotin system, biotin is incorporated into the probe during its synthesis using biotinylated dNTPs (e.g. biotin-7-dATP). The incorporated biotin is detected directly by avidin or streptavidin or an anti-biotin antibody conjugated to a fluorochrome or an enzyme such as alkaline phosphatase or horseradish peroxidase (Figure 10.18). The digoxigenin-anti-digoxigenin system uses digoxigenin (DIG). In both systems, the probe is detected with either chromogenic (colorimetric) substrates, by fluorescence or chemiluminescence. DIG-labelled probes can be made using digoxigenin-11-dUTP by a variety of methods including nick-translation using DNA polymerase I, random priming using Klenow polymerase and hexa-oligonucleotide primers.

Hybridisation is usually carried out in a sealed bag which will contain the Southern or Northern blot, and hybridisation fluid containing the labelled probe and other components including unlabelled DNA which should be highly unlikely to bind the probe or the target on the filter (salmon sperm DNA is commonly used) and a mix of components and conditions which can be adjusted to provide for hybridisation only where there is a real identity between the probe and the target sequence (stringent conditions) or even between similar but non-identical

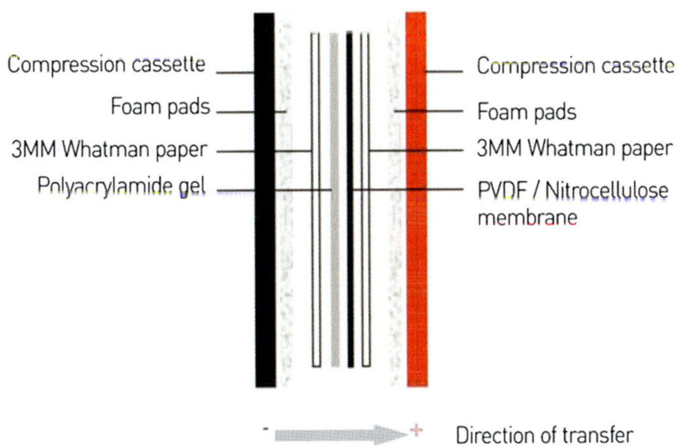

FIGURE 10.17 Electroblotting *(Source: alphametrix.de).*

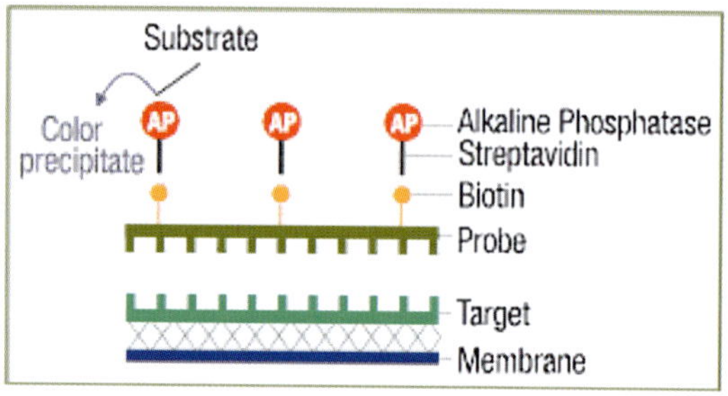

FIGURE 10.18 Detection of hybridisation using biotinylated probes *(Source: fermentas.com).*

sequences (non-stringent). Duplex DNA normally melts at around ~94 °C depending on the base-composition. Incubation conditions just below the melting temperature provide high specificity. The lower the temperature the less stringent is the experiment. A number of factors govern the hybridisation reaction include ionic strength, base composition and temperature etc. Formamide is often added to the hybridisation fluid to lower the melting temperature in a controlled way and to avoid the use of hot ovens or baths while the reaction is taking place. After hybridisation, which usually takes several hours, the filter is washed, blocked to avoid non-specific binding of the detection system, and then developed using the appropriate detection system.

Western blots. This is a standard technique analogous to Southern blotting but for transferring polypeptides which have been separated by SDS-polyacrylamide gel electrophoresis (Laemmli, 1970) to a nylon membrane while preserving their spatial relationships. Usually, this is done using unstained gels with the transfer using an electroblotter as for Southern blotting but with appropriate buffers and transfer conditions (Towbin et al., 1979; Burnette, 1981). The presence of specific proteins can be detected usually with specific antibodies (Figure 10.19).

Molecular Cloning/Recombinant DNA Technology

The origins of recombinant DNA technology or molecular cloning trace back to the early 1970s (Jackson et al., 1972; Lobban and Kaiser, 1973; Cohen et al., 1973) and the first patent was awarded in 1980 (in Hughes, 2001).

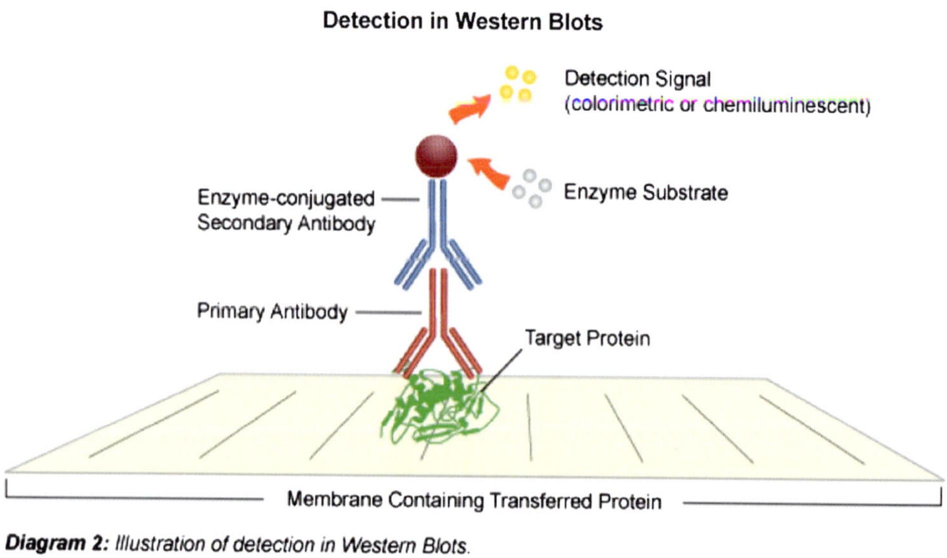

FIGURE 10.19 Antibody detection of specific proteins on western blots *(Source: leinco.com).*

In essence, it was a simple but elegant and revolutionary idea: the use of purified DNA ligase to join DNA molecules end-to end in vitro. Just prior to that, DNA ligase had been purified for the first time from bacteriophage T4 (T4 DNA Ligase) (Weiss and Richardson, 1967), and restriction endonucleases (restriction enzymes) were becoming available (Arber and Linn, 1969). These allowed DNA to be cleaved at specific short sequences to create 5' or 3' overhangs (so called "sticky ends") and then joined (ligated) using DNA ligase to any other DNA molecule with complementary sticky ends. A whole range of restriction enzymes with different target sequences have been purified from different prokaryotes. The use of plasmids or viruses as vectors into which the DNA could be ligated, and their introduction by transformation into bacteria meant that DNA from any source could be "farmed" in a bacterial host. The use of antibiotic resistance genes in the recombinant plasmid allowed those bacteria that had acquired the plasmid by transformation to be selected on medium containing the antibiotic. Early cloning vectors such as pBR322 (Figure 10.20) (Bolivar et al., 1977) and variants thereof used the inactivation of antibiotic resistance to detect cloning events and were the workhorses of molecular cloning from 1977 until more sophisticated vectors appeared which significantly cut down the work involved in identifying recombinant molecules.

A widely used method for the detection of inserts in plasmids is "blue-white" screening which allows for identification of successful cloning events through the colour of the resultant bacterial colony produced by the hydrolysis of the substrate X-gal. The method is based on the principle of α-complementation of a mutant *lacZ* gene (*lacZΔM15*) for β-galactosidase deleted for residues 11−41 by a peptide comprising residues 3−90 of β-galactosidase (the α-complementing fragment; Ullman et al., 1967; Langley et al., 1975). The widely used pUC series of cloning vectors (Vieira and Messing, 1982) were the first plasmid vectors to use this screening method (Figure 10.21). They contain a series of adjacent restriction endonuclease sites (a multiple cloning site or MCS) in the gene for the α-complementing fragment. The system requires the correct host *E. coli* strain (JM109, DH5α, XL1-Blue) which carries the *lacZΔM15* mutation. Transformants are plated onto nutrient agar containing X-Gal and IPTG a non-metabolisable inducer of *lacZ*: most (but not all!) recombinant plasmids will give rise to white colonies (Figure 10.22).

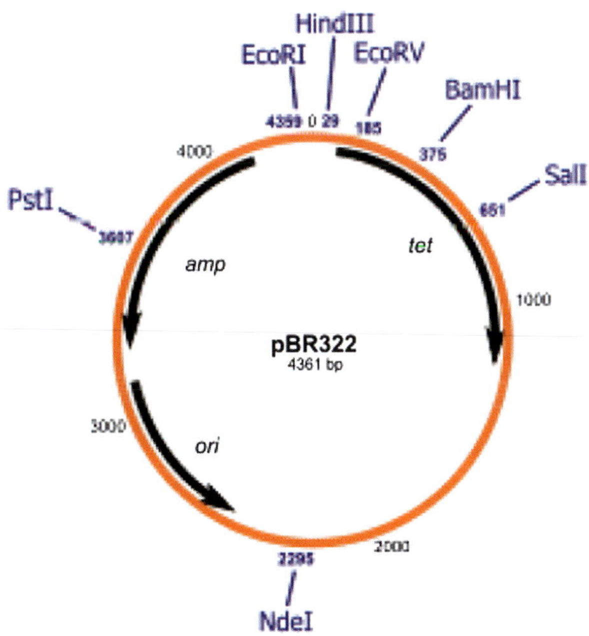

FIGURE 10.20 The general cloning vector pBR322 *(Source: en-wikipedia.org).*

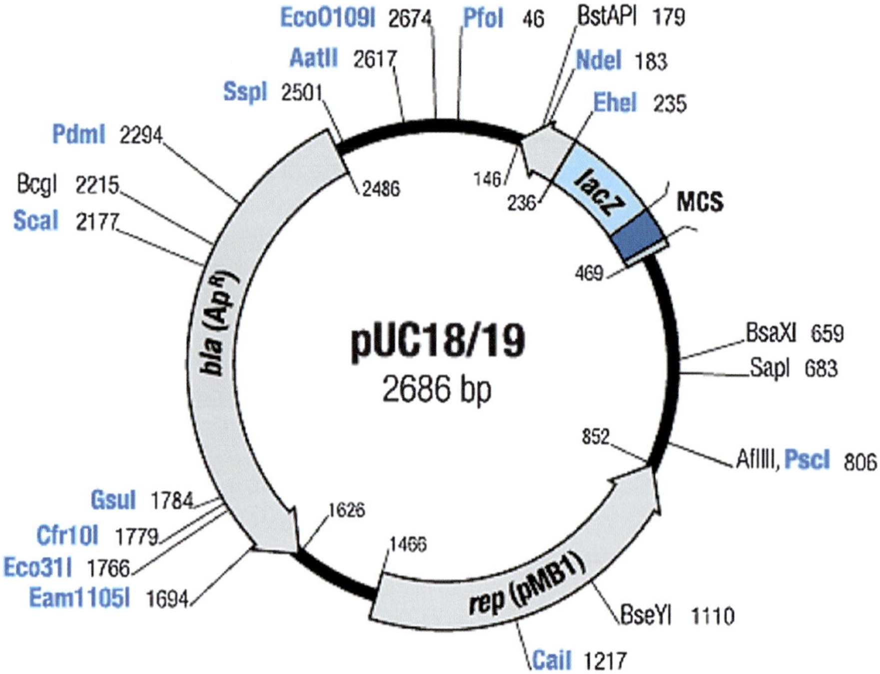

FIGURE 10.21 The general cloning vectors pUC18/19 *(Source: lofstrand.com).*

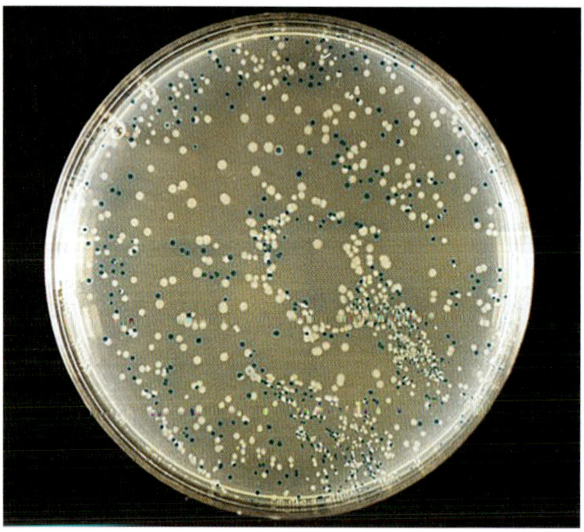

FIGURE 10.22 Blue white selection *(Source: sigmaaldrich.com).*

The Polymerase Chain Reaction

PCR has been one of the many methodologies to have revolutionised molecular biology. Invented and first patented by Mullis (1985) at Cetus, it enabled specific DNA sequences to be synthesised and amplified exponentially in vitro using the following components: the template DNA, a pair of specific oligonucleotide primers designed from the sequence of the DNA to be amplified, purified DNA polymerase and the four deoxynucleotide substrates. The mixture is subjected to a three-step sequence (cycle) of heating at >90 °C to denature the DNA strands, a short *annealing* step at a much lower temperature designed to preferentially allow the oligonucleotide primers (added in considerable excess) to bind to their target sequences on the template but to minimise reannealing of the template, and then an extension/synthesis step at a higher temperature in which the DNA polymerase extends the primers (Figure 10.23). Initially, the technique used a DNA polymerase preparation (the

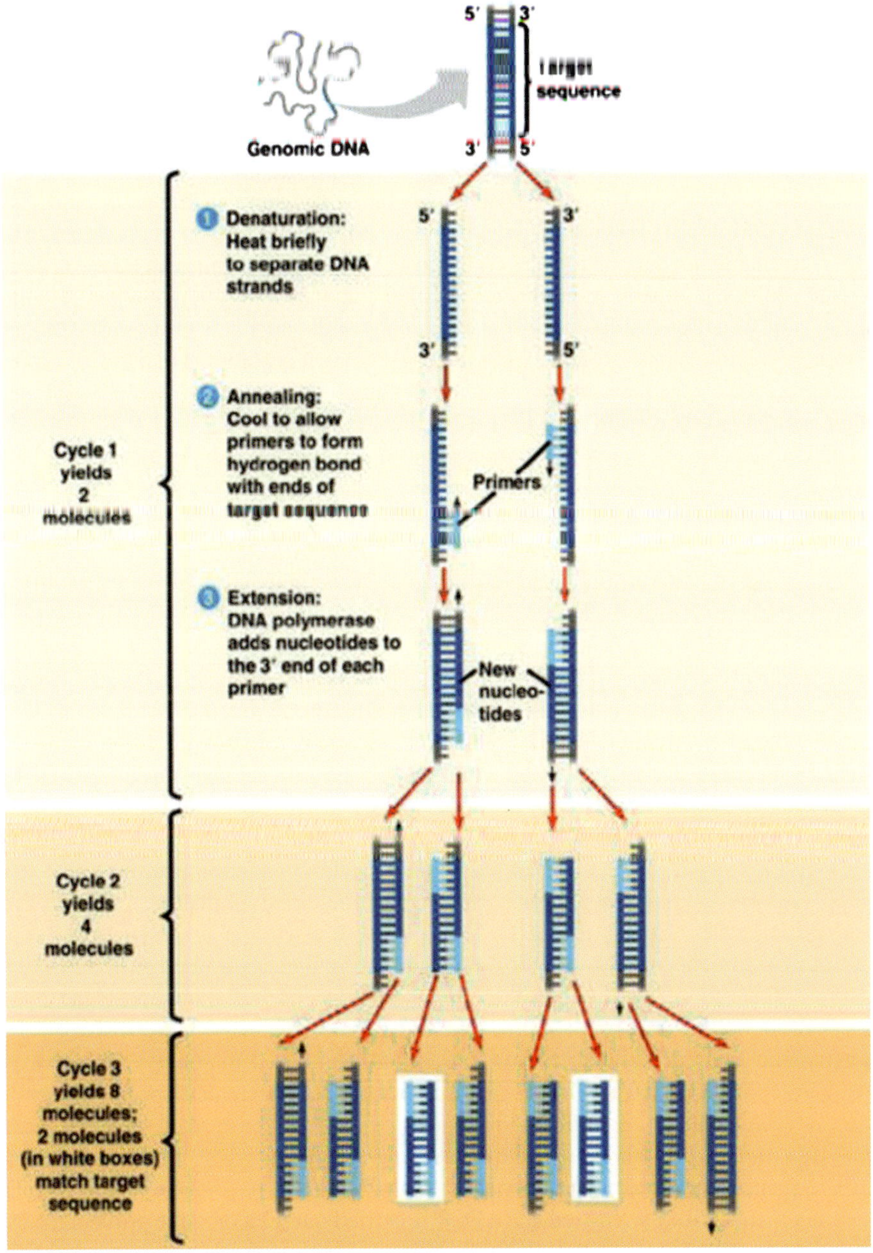

FIGURE 10.23 The polymerase chain reaction *(Source: schoolworkhelper.net)*.

Klenow fragment) from *E. coli* (Klenow and Henningsen, 1970) but this could not survive the denaturation step so had to be added each cycle. But in 1985, use of the thermostable DNA polymerase from the thermophilic bacterium *Thermus aquaticus* (Brock and Freeze, 1969) overcame the enzyme stability problem (Saiki et al., 1985; 1988) and made way for the explosion in the use of this technique. More recently, DNA polymerases from hyperthermophilic prokaryotes are used which are even more stable and have other desirable attributes such as greater processivity allowing longer sequences to be amplified and in some cases proof-reading capacity to reduce errors introduced during DNA synthesis. The PCR has had a somewhat controversial history in terms of patents and litigation (e.g. Bartlett and Stirling, 2003).

The PCR has many applications from the precision cloning of a single gene or part of a gene, to the construction of gene fusions and use in diagnosis and forensics. The maximum length of DNA molecules that can be synthesised is ~10 Kb. For some purposes, e.g. cloning, it is important to check the sequence of the DNA that has been amplified and cloned. Errors can be introduced into the sequences depending on the DNA polymerase used.

The design of the primers is an important consideration. The primer pair is designed according to the target sequence to be amplified. These can be highly specific where the sequence is known or could have a degree of redundancy introduced during primer synthesis which might allow the cloning of genes from an organism where there is no sequence available but gene sequences from closely related organisms can be used. The primers can be synthesised to create additional sequences engineered at either end of the amplified product. A common addition is a restriction enzyme site. The use of different sites at either ends of the molecule ensures directional cloning of the target sequence into the vector. Where protein expression is the goal, it is common that the 5' end primer will carry an *Nde*I site which allows the amplified product to be cloned into the commonly used pT7-5/7 and pET series of protein expression vectors (see Protein Expression section).

There are a large number of variants of PCR and even a scientific journal solely devoted to PCR technologies and methodologies. Some of the more common variants are:

a. **Nested PCR** is intended to mitigate against contamination in products due to the amplification of unexpected primer-binding sites. It increases the specificity of DNA amplification, by reducing background due to non-specific amplification of DNA. Two different sets of primers are used in two successive PCRs. In the first reaction, one pair of primers is used to generate DNA products, which besides the intended target, may still consist of non-specifically amplified DNA fragments. The product(s) are then used in a second PCR with a set of primers whose binding sites are completely or partially different from and located 3' to each of the primers used in the first reaction. Nested PCR is often more successful in specifically amplifying long DNA fragments than conventional PCR, but it requires more detailed knowledge of the target sequences (Sing et al., 1999).

b. **Touchdown PCR (Step-down PCR)** like nested PCR, aims to reduce nonspecific background but by gradually lowering the annealing temperature as PCR cycling progresses (Don et al., 1991). The annealing temperature during the initial cycles is usually a few degrees (3−5 °C) above the T_m of the primers used, while in the later cycles, it is a few degrees (3−5 °C) below the primer T_m. The higher temperatures give greater specificity for primer binding, and the lower temperatures permit more efficient amplification from the specific products formed during the initial cycles (Cox et al., 1991).

c. **Inverse PCR** is commonly used to identify the flanking sequences around genomic inserts. It involves a series of DNA digestions and self ligation, resulting in known sequences at either end of the unknown sequence (Ochman et al., 1988).

d. **Overlap-extension PCR or splicing by overlap extension** is used to splice together two or more DNA fragments that contain complementary sequences. To splice two DNA molecules, special primers are used at the ends to be joined. For each molecule, the primer at the end to be joined is constructed such that it has a 5' overhang complementary to the end of the other molecule. Following annealing when replication occurs, the DNA is extended by a new sequence that is complementary to the molecule it is to be joined to. Once both DNA molecules are extended in such a manner, they are mixed and a PCR is carried out with only the primers for the far ends. The overlapping complementary sequences introduced will serve as primers and the two sequences will be fused. This method has an advantage over other gene splicing techniques in not requiring restriction sites (Higuchi et al., 1988).

e. **Quantitative PCR (Q-PCR)** quantitatively measures starting amounts of DNA, cDNA, or RNA. Q-PCR is commonly used to determine whether a DNA sequence is present in a sample and the number of copies. Q-PCR has a very high degree of precision. The methods use fluorescent dyes, such as Sybr Green, EvaGreen or fluorophore-containing DNA probes to measure the amount of amplified product in real time. It is also sometimes abbreviated to RT-PCR (real-time PCR) or RQ-PCR.

f. **Reverse transcription PCR (RT-PCR)** is used to make DNA copies (cDNA) of RNA. Reverse transcriptase reverse transcribes RNA into a DNA copy, which is then amplified by PCR. RT-PCR is widely used in expression profiling, to determine the expression of a gene or to identify the sequence of an RNA transcript, including transcription start and termination sites. If the genomic DNA sequence of a gene is known, RT-PCR can be used to map the location of exons and introns in the gene. The 5' end of a gene (corresponding to the transcription start site) is typically identified by RACE-PCR (Rapid Amplification of cDNA Ends) (Tan et al., 1994, VanGuilder et al., 2008).

DNA Sequencing

Two quite different but universal methods for DNA sequencing were both published in 1977. Both relied on radioactive labelling and high-resolution polyacrylamide gel electrophoresis to separate and detect single-stranded DNA molecules differing by a single base. The Maxam and Gilbert method (or "chemical sequencing") (Maxam and Gilbert, 1977) depends on radioactively end-labelling one of the two DNA strands and subjecting the molecule to several different chemical treatments to cleave the DNA molecules between individual bases. The Sanger or chain termination method (Sanger et al., 1977) involves carrying out DNA synthesis reactions from the single-stranded template DNA to be sequenced and the use of a radiolabelled deoxynucleotide to enable the newly synthesised DNA strands to be visualised by autoradiography. The method involves the use of four chain-terminating dideoxynucleotides which give rise to a ladder of DNA products which vary in length according to the position of the relevant base.

Maxam–Gilbert sequencing rapidly became popular since purified DNA could be used directly, while the initial Sanger method required that each short fragment to be sequenced needed to be cloned for production of single-stranded DNA. However, with the improvement of the chain-termination method (see below), Maxam-Gilbert sequencing has fallen out of favour due to its technical complexity prohibiting its use in standard molecular biology kits, extensive use of hazardous chemicals, and difficulties with scale-up.

A major step forward in DNA sequencing methods was the introduction of dye terminator sequencing (Smith et al., 1985, 1986). The introduction of even more efficient DNA sequencing methods (so-called "second-generation" DNA sequencing) continues apace (see below). Nowadays, most labs use commercial DNA sequencing services which provide rapid turn round times and high quality assurance and which are both cost and time effective.

The **Maxam and Gilbert** method is no longer used for general DNA sequencing but still has specialist applications, e.g. the methylation interference assay used to map DNA-binding sites for DNA-binding proteins.

a. The method requires radioactive labelling at one 5' end of the DNA (typically by a kinase reaction using gamma-^{32}P ATP) and purification of the DNA fragment to be sequenced. Chemical treatment generates breaks at a small proportion of one or two of the four nucleotide bases in each of four reactions (G, A + G, C, C + T). For example, the purines (A + G) are depurinated using formic acid, the guanines (and to some extent the adenines) are methylated by dimethyl sulfate, and the pyrimidines (C + T) are methylated using hydrazine. The addition of salt (sodium chloride) to the hydrazine reaction inhibits the methylation of thymine for the C-only reaction. The modified DNAs are then cleaved by hot piperidine at the position of the modified base. The concentration of the modifying chemicals is controlled to introduce on average one modification per DNA molecule. Thus, a series of labelled fragments is generated, from the radiolabeled end to the first cleavage in each molecule. The fragments in the four reactions are electrophoresed side by side in denaturing acrylamide gels for size separation. To visualise the fragments, the gel is exposed to X-ray film for autoradiography, yielding a series of dark bands each corresponding to a radiolabeled DNA fragment, from which the sequence may be read.

Chain-termination sequencing was the most commonly used method for DNA sequencing until relatively recently. Chain-termination methods greatly simplified DNA sequencing and kits are commercially available. Limitations include non-specific binding of the primer to the DNA affecting accurate read-out of the DNA sequence, and DNA secondary structures affecting the fidelity of the sequence.

a. The classical chain-termination method requires a single-stranded DNA template, a DNA primer, a DNA polymerase, normal dNTPs, and chain-terminating nucleotides (dideoxy NTPs: ddNTPs) that lack a 3'-OH group required for the formation of a phosphodiester bond between two nucleotides. The DNA sample is divided into four separate sequencing reactions, containing all four of the standard deoxynucleotides (dATP, dGTP, dCTP and dTTP) and the DNA polymerase. To each reaction is added only one of the four dideoxynucleotides (ddATP, ddGTP, ddCTP, or ddTTP). Extension off the primer bound to the template results in DNA fragments of varying length.

b. The newly synthesised and labelled DNA fragments are heat denatured, and separated by size by gel electrophoresis with each of the four reactions run in one of four individual lanes (lanes A, T, G, C) on a denaturing polyacrylamide-urea gel capable of a resolution of just one nucleotide. The DNA bands are then visualised by autoradiography or UV light, and the DNA sequence can be directly read off the X-ray film (Figure 10.24, left).

Dye-terminator sequencing. In dye-terminator sequencing, the reaction is essentially as in Sanger sequencing except that each of the four dideoxynucleotide chain terminators is labelled with fluorescent dyes, each of which emit light at different wavelengths. Owing to its greater expediency and speed, dye-terminator sequencing is now the mainstay in automated sequencing. Its limitations include dye effects due to differences in the incorporation of the dye-labelled chain terminators into the DNA fragment, resulting in unequal peak heights and shapes in the electronic DNA sequence trace chromatogram after capillary electrophoresis (Figure 10.24, right: Figure 10.25a). This problem has been addressed with the use of modified DNA polymerase enzyme systems and dyes that minimise incorporation variability, as well as methods for eliminating "dye blobs". The dye-terminator

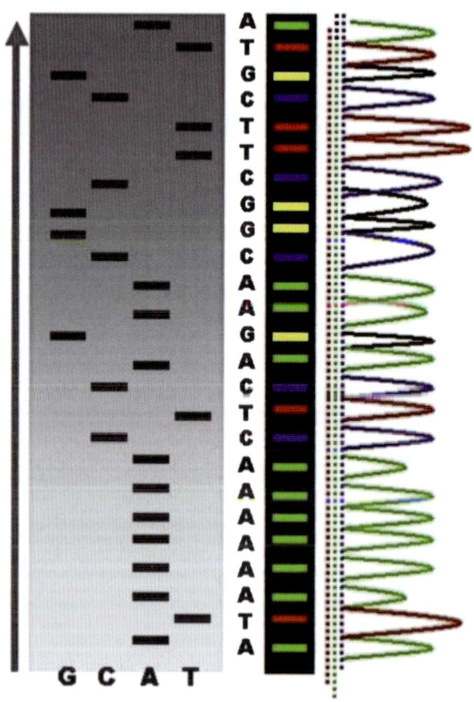

FIGURE 10.24 DNA sequencing: chain termination and dye termination outputs *(Source: dna-rna.net).*

sequencing method, along with automated high-throughput DNA sequence analysers, is now being used for the vast majority of sequencing projects.

Next Generation and High-Throughput DNA Sequencing

Currently, DNA sequencing is hugely efficient enabling the sequencing, assembly and interpretation of small genomes to be achieved in a relatively short time and at relatively low cost and sequencing of different members of

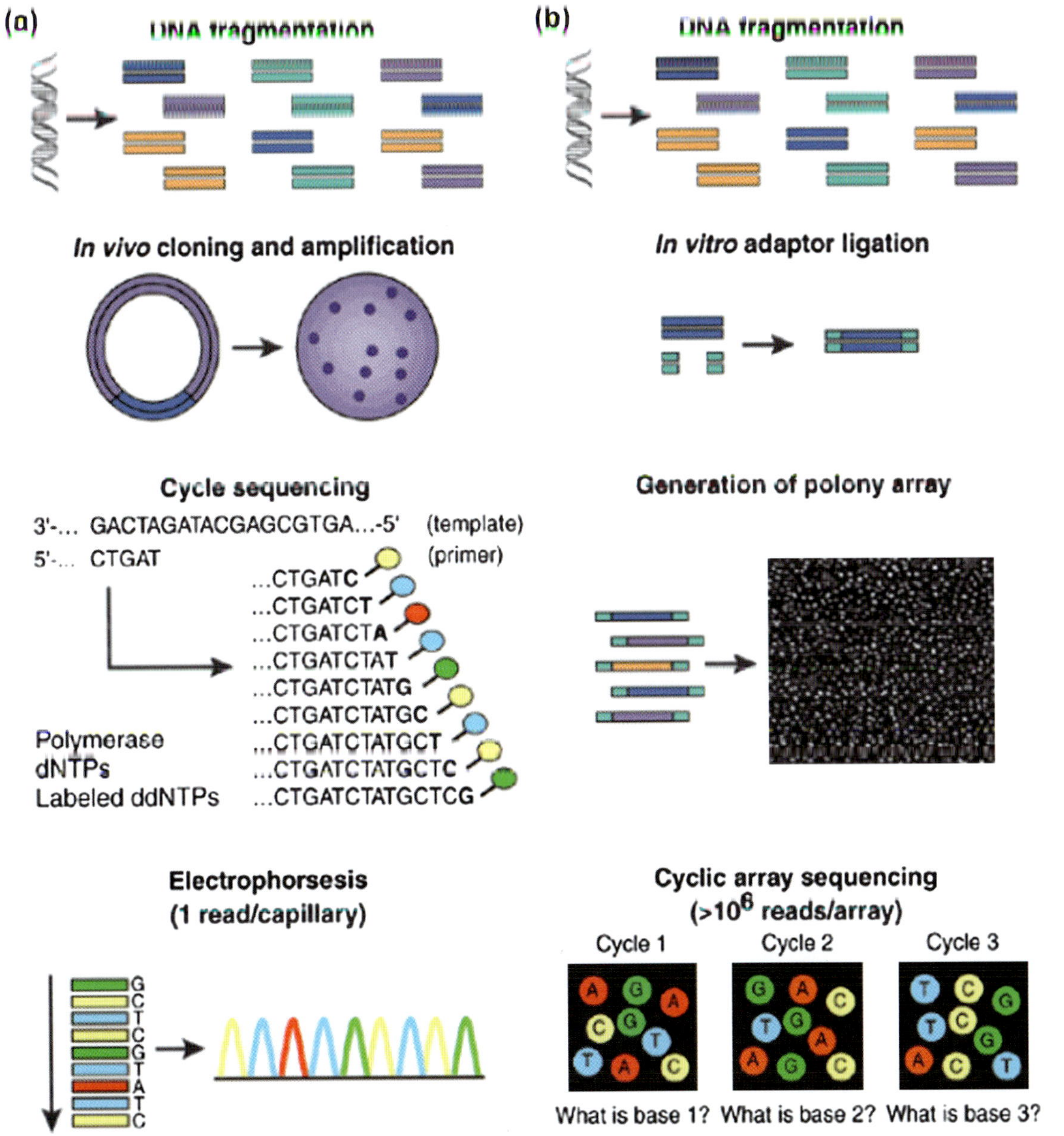

FIGURE 10.25 High-throughput DNA Sequencing: dye termination versus second-generation sequencing *(Source: Shendure and Ji, 2008)*.

the same species to be carried out extremely quickly. However, the search for ever more efficient systems continues with large prizes being available for achieving particular goals. A number of competing technologies are available (Mardis, 2008). Current examples include pyrosequencing which is particularly useful for the repetitive sequencing of alleles (Nyrén, 2007).

More recently, a number of new technologies (second-generation sequencing) are emerging where the process is even more efficient and productive. Most involve binding short DNA fragments to microbeads which are trapped onto arrays which can be read by fluorescence readers (Figure 10.25b). The DNA fragments bound to the beads are denatured to produce single-stranded DNA template which is then replicated enzymatically but in a process which adds one fluorescent base at a time. The addition of that base then prevents a further base from being added. As each of the four different bases carries a different fluor the base which has been added can be determined. The fluorescence reader detects the fluorescence of the base added at each bead in the array, then the fluorescent moiety is removed and the cycle of adding the next fluorescent base can occur. This methodology is particularly powerful for identifying differences between genomes of individuals. It is a sufficiently powerful that it can be used to detect a single base mutation in a complete bacterial genome within 24 h.

GENETIC AND MOLECULAR GENETIC METHODS

Cloning Vectors and Hosts

General purpose cloning vectors. General purpose cloning vectors are usually small circular DNA molecules (~3 to 5 kb in length) which replicate independently to high copy number when introduced into a suitable host, are easy to purify and into which fragments of DNA can be inserted in vitro, e.g. pBR322 and pUC18/19 (Figures 10.20 and 10.21). These plasmids have many uses including shotgun cloning of small (~600 bp) random fragments in large scale and genome sequencing projects. They usually comprise the following: an origin of replication derived from a naturally occurring colicin-producing or antibiotic-resistance plasmid isolated from a member of the enterobacteriaceae such as *E. coli* and which usually will allow the molecule to be replicated to a "high copy number" in *E. coli*. It will also contain an antibiotic resistance gene (or marker) which will enable the detection of the presence of the vector in the host organism; a "multiple cloning site" where several restriction enzyme sites are bunched together and usually at the 5'-end of a gene which is inactivated when a fragment is cloned into the site.

Broad host-range vectors. These are composed of similar elements to the general cloning vectors but can replicate in a relatively wide range of bacteria. However, they are usually larger DNA molecules because they carry all genes necessary for their own for replication. Examples include derivatives of RP4 which can replicate in both *E. coli* and a range of other Gram-ve bacteria, e.g. pseudomonads, rhizobia, rhodobacters, alcaligenes. For cloning in other bacteria, e.g. Gram+ve bacteria different vectors are required based on compatible replication systems.

Expression vectors. These vectors are commonly used in the over expression of proteins. In principle, they are based on general cloning vectors described above but in addition they contain an insert which contains a strong inducible promoter and translation initiation site behind which the gene of interest is inserted in-frame. The most common expression vectors are those based on a system originally described by Tabor and Richardson (1985) which can yield strikingly high levels of protein production.

Suicide vectors. These specialised vectors are used to deliver specific mutations into the genomes of organisms, e.g. in the production of site-directed knockout mutants. Usually, these are binary systems useful in Gram-ve eubacteria. The DNA fragment carrying the desired mutation is constructed in vitro in a vector which carries the *mob* site. They are first propagated in an *E. coli* host and then introduced into the target organism by conjugation in tripartite mating with a second *E. coli* strain carrying a tra^{+} plasmid which will mobilise any plasmid containing a *mob* site.

Shuttle vectors. Shuttle vectors are specialised cloning vectors which have more than one origin or replication that enable them to replicate in two quite different organisms: e.g. a Gram -ve organism such as *E. coli* and a Gram

+ve organism such as *B. subtilis*, or a prokaryote such as *E. coli* and a very commonly studied eukaryote such as Yeast. In this way, advantage can be made of performing cloning and other manipulations in *E. coli* but transferring the final construct to the organism of interest.

Gene Libraries

The aim of gene libraries is to capture fragments representing the whole genome of an organism in a collection of recombinant plasmids carried individually in members of a host organism, usually either *E. coli* or yeast. Different types of gene libraries can be constructed depending on need. They include libraries of short fragments in *E. coli* as a host which are mainly used for genomic sequencing projects and libraries containing very large DNA fragments hundreds of kilo base long in either *E. coli* or in yeast used for long-range physical mapping and sources of segments of the genome where there may be gaps in the sequencing data and they provide a very efficient way of storing a whole genome. Gene libraries can also be constructed in highly restricted host range vectors which may be important for genetic containment or in mobilisable broad host-range plasmid vectors which are very useful for genetic analysis in some Gram-ve bacteria.

Whatever the starting genome or vector to be used, constructing a gene library involves: 1) preparation of high-molecular weight DNA from the target organism; 2) its fragmentation to a greater or lesser extent depending on need; 3) purification of fragments of the desired size rage; 4) cloning of the fragments into a vector in such a way as to minimise multiple genome fragments being cloned into a single vector molecule; 5) transformation into the host organism, and the isolation and storage of individual clones (the library); 6) validation of the library as being representative of the genome of the target organism with a high order of redundancy and 7) long-term storage of the library.

High-molecular weight genomic DNA used to be fragmented using restriction enzymes which recognise 4 bp sequences and which, statistically, will occur very frequently in the genome. This method produces fragments which are easily cloned into a compatible cloning site in a vector but are non-random by definition. Nowadays, it is more usual to prepare DNA fragments by physical shearing where the breakages occur at random: greater shearing produces shorter fragments.

Physical breakage produces fragments with "ragged" or short single-stranded overhangs. These are end-repaired or filled-in enzymically with DNA polymerase to produce blunt ends. The inclusion of a fragment-sizing step at this stage ensures that the fragments to be cloned will be relatively uniform in length. Usually, this involves separating the fragments by regular agarose gel electrophoresis for small fragments or by pulse field electrophoresis for much larger fragments. Special agaroses (e.g. low-melting point agarose) are used in this separation so that the DNA can be recovered from the agarose. The fragments are then blunt-end ligated into the desired vector. Where feasible a common procedure is to use the double adaptor method (Andersson et al., 1996) in which the end-repaired fragments are ligated to oligonucleotide adaptors creating long 12-base overhangs. The use of non-phosphorylated oligonucleotides at this step prevents formation of adaptor dimers and ensures efficient ligation of the insert to the adaptor. The vector is digested with appropriate restriction enzymes so that they produce ends which are complementary to the overhangs created in the fragment digest. Following the annealing of insert to vector, the DNA is directly used for transformation without a ligation step. This protocol produces no chimeric clones and a high proportion (~99%) of clones contain an insert.

Libraries Intended for Genome DNA Sequencing

For the greatest efficiency in genome sequencing and analysis projects, it is usual to prepare two or three types of gene libraries. One library will contain very small fragments of up to 1 to 2 kb to be used primarily for initial DNA shotgun sequencing from one end of the fragment insert or from both ends as in the so-called "double shotgun" method which offers significant advantages in terms of efficiency and sequence assembly to produce "contigs". A second library will contain longer fragments of say 5 to 10 kb and is used to confirm sequences assembled from the shotgun results (contigs) and to hunt for and fill in gaps. A further library containing even larger fragments may be required where there is significant repetition in the genome under study or large gaps to be completed. Sequences are all produced by commercial automated high-throughput sequencing techniques on different platforms. Sequences are assembled and annotated using software packages.

Cosmid Libraries

Cosmids (Collins and Hohn, 1978) are cloning vectors used to construct libraries of intermediate fragment length (e.g. up to ~50 kb) and which are characterised by containing the ~200 bp *cosN* sequence of phage λ. This is the target site for the linearisation of the circular phage λ genome by a specific λ-encoded terminase. The formation of infective phage capsids requires that the DNA be linear and of a relatively specific length of ~49 kb to be encapsulated. The terminase cuts the genome within the *cosN* site to produces 12 bp sticky or "cohesive" ends. In addition to the *cosN* site, these vectors contain an origin of replication (*ori*) either for bacterial or mammalian cells and some selectable marker (e.g. antibiotic/drug resistance). Given that there are no other specific DNA requirements for capsid assembly apart from the DNA fragment length and the *cos* site, any doubled-stranded DNA molecule can be packaged into infective phage particles. The phage particle simply serves as a highly efficient DNA delivery vehicle. The cloning capacity of a cosmid vector is inversely related to the size of the vector itself. In the cloning protocol, two vector arms are generated and these are ligated to genomic DNA fragments of the required length. The assembly of the capsids is carried out in vitro and started by adding the ligation products to a mix of λ packaging extracts prepared from two *E. coli* strains carrying different mutant λs: one defective in head assembly and the other defective in tail assembly. After the packaging reaction, the mix is used to infect *E. coli* and the recombinant DNA bearing clones are selected on the appropriate drug or antibiotic for which the cosmid carries resistance.

Mobilisable and Broad Host-Range Vectors and Cosmids

Broad host-range and mobilisable vectors including cosmid vectors are extremely useful in identifying genes through genetic complementation including interspecific complementation and also for cloning regions around the sites of insertion of transposons. Early examples include the cosmid pLAFR1 for use in Gram-ve bacteria (Friedman et al., 1982) which is a relatively large vector of 21.6 kb constructed by inserting the *cosN* site and a *mob* site into the broad host-range P1 incompatibility group vector pRK290. Cosmid libraries constructed with this vector will accommodate DNA inserts of 20 to 30 kb with *E. coli* as a host. When mixed with a culture of a second *E. coli* strain containing the Tra^{+} helper plasmid pRK2013, the library can be mass mated into a recipient organism. The helper plasmid first transfers to the library clones and will work *in trans* on any plasmid containing the *mob* locus such as pLAFR1 to mobilise it into the intended recipient. In this way, it is possible to identify specific clones from the library which will restore functions to mutants of the recipient strain. In the initial use, recombinant cosmids carrying *Rhizobium meliloti* genome fragments were identified which would restore function to (complement) auxotrophic mutants of *R. meliloti* and therefore carry the relevant genes. The individual cosmids could then be isolated and their inserts studied further. A functionally similar cosmid but of only 13 kb which can carry larger fragments was described by Selveraj and Iyer (1985).

Fosmid libraries. Fosmids are similar to cosmids in containing the *cosN* site but contain the bacterial F-plasmid origin of replication which provides low copy number control and therefore offers greater stability compared to high copy gene libraries constructed in copy number vectors. They are particularly useful for constructing stable libraries from complex genomes (Kim et al., 1992).

Bacterial Artificial Chromosomes

Bacterial Artificial Chromosome (BAC) libraries are based on *E. coli* and its single-copy plasmid F factor. A BAC vector is capable of maintaining very large genomic fragments of >300 kb and even up to 1 Mb (Shizuya et al., 1992).

BAC libraries have proved to be very useful for preparing stable libraries of very complex genomes and this has facilitated further physical and genetic analysis. As discussed above, high-molecular weight genomic DNA is cut with a restriction enzyme and then fractionated using pulse field gel electrophoresis and extracted from the agarose. The BAC vector is digested with the same restriction enzyme which cuts at the single cloning site and the vector is then treated with phosphatase to prevent self-ligation. DNA fragments of the target organism are then ligated to the prepared vector and the ligation mix electroporated into the *E. coli* host with a surprisingly high frequency.

Yeast Artificial Chromosomes

The construction of yeast artificial chromosomes was first described by Murray and Szostak in 1983. These vectors comprise the [illegible] replication origins of chromosomes that will replicate and be stably maintained in yeast. They are used for cloning and physical mapping of large DNA fragments of between 100 and 3000 kb and are particularly valuable for cloning large genes from eukaryotes which can extend over large regions. Moreover, they are particularly effective for cloning and expression of genes which require complex post-transcriptional processing and/or which encode proteins which require post-translation modification.

cDNA Libraries

Typically, many genes in eukaryotes have complex structures consisting of segments of the DNA which are not expressed in the mature message (the introns) and those segments that are expressed (the exons). Some genes can contain a number of introns which can be long and extend the gene over considerable distance in the chromosome. The introns are spliced out by the spliceosomes resulting in a mature message which is also capped and tailed with poly AAAA tail. "Copy" or cDNA libraries are essential in order to isolate and study genes and their products from Eukarya. A cDNA library is a collection of recombinant plasmids which contain copies of the mature mRNAs or coding regions of all the expressed genes. They are prepared by isolation of the mRNA from the organism/tissue and the copying of the RNA into a double-stranded DNA molecule by reverse transcriptase. The double-stranded DNA products are then cloned into a suitable vector. Clearly, a cDNA library only represents those genes that are expressed in that organism or tissue at the time the mRNA is isolated. The composition of the library is also biased in favour of abundant mRNAs. The preparation of a cDNA libraries is an essential starting point for the overexpression of proteins from eukaryotes.

Protein Overexpression and Purification

The overexpression of proteins is frequently used in research projects where the goal is to study protein structure and function. This section will compare and contrast two of the most frequently used systems.

The T7 RNA Polymerase-T7 Promoter System in *E. coli*

T7 is a bacteriophage which infects *E. coli*. On infection, it injects its DNA genome into the host and proceeds to hijack its macromolecular synthesis systems to produce new phage particles which are eventually released when the host bursts. The virus encodes its own RNA-Polymerase (T7-polymerase) which is a single polypeptide which recognises promoters present only in the phage genome and which exhibits a remarkably high degree of processivity. Exploitation of this system for overexpression of foreign proteins was first described by Tabor and Richardson (1985). Their elegant system comprised two compatible plasmid constructs: pGP1-2 carries the T7 polymerase gene under the control of the phage λP_L promoter and the gene for temperature sensitive phage λ repressor (cI857) placed under the control of the *lacZ* promoter; pT7-1 contains the strong Ø10 T7 RNA polymerase promoter, just upstream of a multiple cloning site into which a target gene can be cloned. In the presence of the lactose inducer IPTG, the temperature-sensitive λ repressor is produced which at 30 °C is active and prevents expression of the T7-polymerase. However, when the culture temperature is raised to 42 °C the λ repressor becomes inactive. This switches on expression of the T7 polymerase which in turn drives transcription from the Ø10 promoter and expression of the target gene. Transcriptional selectivity can be further enhanced by adding rifampicin to the culture to shut down the host's own RNA polymerase.

Studier and Moffatt (1986) described a similar system that was later developed and patented and now widely used and available from several biotech companies (Studier et al., 1990). This system comprises the pET family of expression vectors which contain a T7 promoter and a means of providing directional cloning for the target gene (Figure 10.26). This cloning site comprises an *Nde*I restriction site just a few bases downstream of a ribosome-binding site and the Ø10 promoter and contains a 3'-ATG-5' translation initiation codon so that the open reading frame (ORF) of the target gene can be inserted into a closely coupled transcription/translation arrangement. This optimises high levels of protein production. The distal cloning site is a *BamH*1 site which can accommodate the sticky ends produced by other restriction enzymes such as *Bgl*II. The ORF of the target gene is first amplified by PCR from the source DNA using a pair of primers one of which engineers an *Nde*I site at the 3' end of ORF and a BamHI site at the 5' end. A notable feature of this system is that it provides several options for ensuring no expression of a target gene until required as some proteins can be toxic even if only a few molecules of the protein were to be produced inadvertently. Therefore, the first cloning stage is usually conducted in an *E. coli* strain lacking the T7 polymerase gene. For overexpression, the recombinant plasmid is transformed into a special *E. coli* host strain (BL21 (DE3)). This strain carries a λ DE3 lysogen (i.e. an insert into the chromosome) that has the λ phage 21 immunity region, the *lacI* gene and the *lacUV5*-driven T7-RNA polymerase cassette. In this system, when the expression plasmid is present in the host and the *lac* inducer IPTG is added to cultures, the *lacUV*5 promoter is derepressed allowing overexpression of the T7-polymerase. This in-turn induces expression of the target gene cloned into the pET expression plasmid. The *E. coli* BL21 host also lacks the *lon*-encoded protease that can degrade proteins during subsequent purification. The still more sophisticated host BL21-Gold (DE3) carries the plasmid pLysS Cam^R which expresses T7-lysozyme which binds to and inhibits transcription by T7 polymerase. This effectively silences expression from any T7polymerase-dependent promoter in the host until addition of IPTG drives up polymerase expression levels. The intracellular T7-lysozyme aids gentle lysis which is useful when the overexpressed protein is potentially susceptible to more robust cell rupture methods. Where extremely toxic genes are being expressed, the system also allows the T7 polymerase gene to be introduced into the producer strain on a λ phage.

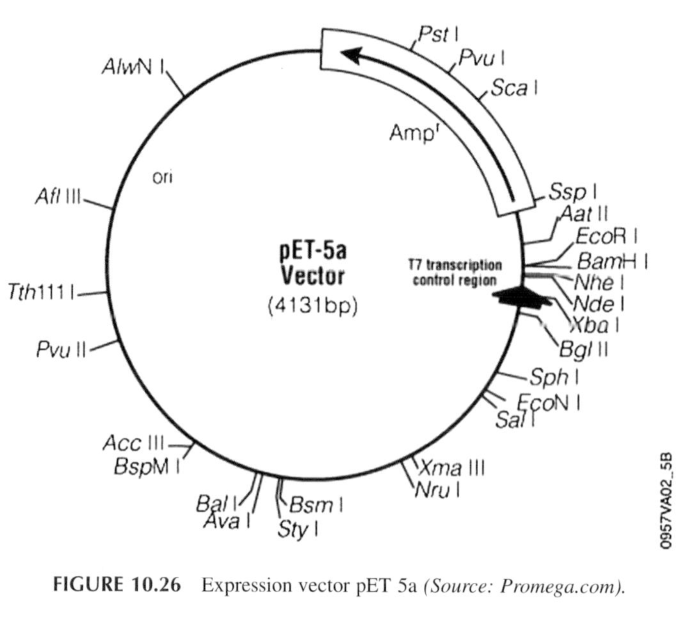

FIGURE 10.26 Expression vector pET 5a *(Source: Promega.com).*

The *Pichia Pastoris* System

While *E. coli* is by far and away the most frequently used host for protein expression it has some limitations. These include the inability to produce disulfide bonds and the inability to glycosylate proteins (Cregg et al., 2009). The methylotrophic yeast *Pichia pastoris* has these capabilities and can be grown easily to a very high cell density on simple defined medium with methanol as sole C source. This is especially useful for isotopic labelling for NMR experiments. *Pichia* expression systems exploit one of the two alcohol oxidase gene (AOX1) promoters to drive expression of the target gene. As in the *E. coli* T7 polymerase system, commercial kits are available, one of which allows a translational fusion to be created between the secretion signal of the α mating factor of *S. cerevisiae* and the ORF of the target gene. This elegant system causes the expressed protein product to be secreted into the medium which is a potential aid to purification. Protein production is not necessarily so consistent in *Pichia* and several clones may need to be tested for effective expression. One drawback of *Pichia* is that induction of expression may take several days of growth of the host compared to a matter of hours in *E. coli*.

Tags for Protein Purification, Correct Folding, Improved Stability

It is now straightforward to engineer gene fusions that produce "tagged" target proteins where the "tag" aids correct folding, greater stability and purification. One of the most common methodologies is to create "his-tagged" proteins, where a poly-histidine linker peptide is engineered at the N-terminus of the target protein. Given the high affinity of clusters of histidinyl residues for divalent cations and especially Ni^{2+}, purification of his-tagged proteins can sometimes be achieved in a one-step process using Ni^{2+} affinity chromatography. Whereas this can be very useful for many proteins, it is perhaps less useful when working with metallo-proteins unless the tag is removed by specific proteolytic digestion following the purification step. Even so, the exposure to relatively high levels of Ni^{2+} means that the proteins are often contaminated with Ni^{2+}. Other affinity tags may prove more useful in metallo protein work such as the chitin binding protein (CBP), maltose binding protein (MBP), and glutathione-S transferase (GST). The latter two tags can overcome the common problem of insolubility of overexpressed proteins. Other tags such as thioredoxin (TRX) and poly(NANP) are also used to overcome insolubility problems which may have many causes including non-specific aggregation and misfolding which may be due to a lack of a correct chaperonin complex in the host. Use of polyanionic amino acid tags, e.g. FLAG-tag have been developed to alter the chromatographic properties of proteins so that they can be more easily resolved during purification steps. Tags which have proved useful for immunoprecipitation experiments and Western blotting use protein sequences which encode highly immunogenic epitopes often derived from virus sequences. These include the HA-, the V5- and the cmyc-tags. Fluorescent tags especially the green fluorescent protein has proven to be remarkably useful in protein expression and localisation studies.

While different tags are extremely useful, the obvious caution is that the tag might alter the properties of the target protein so that it no longer behaves like the native protein. Therefore, experimental findings need to be interpreted with caution.

Mutagenesis

Mutants: General Considerations

It is crucial to understand the potential biological impacts and experimental value of different types of mutations especially when planning a mutagenesis strategy. The many types of mutations include point mutants, deletions, insertions, polar/non-polar, lethal, conditional lethals etc. Point mutations are those which affect a single nucleotide pair. Naturally occurring point mutations found in different alleles (e.g. in human populations) are known as single nucleotide polymorphisms (SNPs). Such mutations occur spontaneously but can also be induced by a range of chemical and physical treatments (see below). Point mutations which have no discernible impact on the cell function are known as silent mutations. However, point mutations can often cause subtle phenotypic change, e.g. affecting the catalytic and/or regulatory properties of the protein. For this reason, point mutations are very useful for fine mapping of structure/function relationships in regulatory sequences (e.g. promoters) or in proteins. Deletion or insertion mutations tend to have major impacts on gene function and in the case of prokaryotes can affect not only the gene in which the mutation has occurred through polarity effects. Many genes are essential and their disruption is lethal. However, it is often possible to isolate mutations in such genes that affect the function only under particular growth conditions such as temperature: where the mutation is silent under the permissive growth conditions but expressed under the restrictive growth conditions. This approach has been essential in the study of very complex processes such as cell division and the cell cycle more generally. Of course, diploid organisms carry two copies of most genes except those borne on the sex chromosomes. Therefore, it is possible to maintain such lethal mutation in a heterozygous state in the cell and to expose them in a homozygous state only after sexual reproduction. Even in haploid organisms such as prokaryotes, it is possible to provide a wild-type copy of the gene on a plasmid in what is known as a partial diploid (merodiploid).

Chemical and Physical Mutagenesis

Mutations occur naturally sometimes due to slight errors in DNA replication and/or repair processes or as a result of environmental factors or even intracellular activities that produce reactive species that damage DNA. The induction of mutations in organisms in the laboratory by physical and chemical means stretches back to the work of Hermann Muller with his work on the effects of X-rays on *Drosophila* in 1927 (Muller, 1927). Lewis Stadler and colleagues demonstrated the mutagenic effects of X-rays and UV light on cereal plants (Stadler, 1928; Stadler and Sprague, 1936) and Auerbach, Robson and Carr showed that mustard gas induced mutations in *Drosophila* (Auerbach et al., 1947). Now we know that chemicals such polycyclic aromatic hydrocarbons, alkylating agents such as N-nitrosamines, intercalators and many other types of agents damage DNA in a wide variety of different ways producing characteristic kinds of mutations.

From a research point of view, the use of chemical and physical mutagenesis has largely been replaced by more targeted (site-directed) mutagenesis protocols that aim to generate specific mutations. Nevertheless, much has been learned from mutants produced by chemical and physical mutagenesis. Many of the thousands of mutants of common host organisms such as *E. coli*, yeast and *Drosophila* were produced by these means. Indeed, genome sequencing on lab-trained strains of *E. coli* such as K12 reveal the battering that they have received from successive rounds of mutagenesis over the years. However, the randomness of mutations created by such techniques often throw up the unexpected and opens up new areas for study. Moreover, chemical and physical mutagenesis are probably the only recourse where no system exists to engineer the desired mutant strains, e.g. through site-directed mutagenesis approaches.

The usual protocol for producing mutants using chemical or physical agents is to expose the organism to the mutagen at a level that statistically induces only one mutation in the genome of only some individuals in the population. This mitigates against the induction of multiple mutations in an individual organism which can produce complex and potentially misleading phenotypes. In any event, it is essential to check for multiple mutations if feasible by performing complementation analysis involving the introduction of the wild-type gene into the mutant to look for restoration of the fully wild-type phenotype or to cross the mutation back into a wild-type organism and to check the phenotype.

To produce controlled mutagenesis, the usual approach is first to determine a dose response curve for the mutagen measuring the levels of kill produced with increasing exposure to the agent. Once this is determined then mutagenesis should be carried out at levels of exposure to the agent which cause death of a minority of the population (e.g. 10 to 25%). Once the mutagenesis step has been carried out, it is essential to "outgrow" the population for several generations in non-selective conditions to allow the mutation to segregate from the wild-type copies through cell division. This enables the mutation to express biochemically or physiologically which is essential before searching for, or attempting to select, the desired mutants.

Transposable Elements and Their Use Mutagenesis

Transposons were described in an earlier section. For the purpose of this section, it is important to know only that transposons create mutations wherever they insert or where they delete (Figure 10.27) and if the transposon carries a selectable marker, e.g. an antibiotic/drug resistance gene, then individuals carrying such mutations can be easily selected and the location of the insertion can be easily identified because it has become physically "tagged". The DNA flanking site can be easily isolated and characterised (Kleckner et al., 1977).

In bacterial systems, the usual protocol for transposon mutagenesis is to introduce the transposon into the target organism via a vector (a suicide vector) which has a limited capacity to replicate and sometimes under particular culture conditions. The marker gene carried by the transposon will be lost unless the transposon copies itself into the genome of the organism. Colonies of the organism which survive on antibiotic-containing agar are those which must have acquired the transposon in their genomes (e.g. see Morales and Sequira, 1985).

More recently, a number of sophisticated systems have been developed which allow the transposition into the target DNA to be carried out in vitro. One system which is of potentially wide application in bacteria is the GAMBIT method (genomic analysis and mapping by in vitro transposition) (Akerly et al., 1998). In this system, originally applied to *Haemophilus* and *Streptococcus*, the transposition event is performed in vitro with the transposase from the *mariner*-family transposon *himar1* from the horn fly. This enzyme mediates transposition in vitro without other cellular factors and has very little insertion site specificity. The target DNA can be fragments of the whole genome or the insertion can be targeted at a specific region of the genome by using large fragments carrying that region (~10 kb) synthesised by extended PCR. The transposons used in the original work were artificial mini-transposons carrying antibiotic resistance genes. In this method, the mutated DNA was transformed into the host using its natural competence systems and the mutated organisms selected on agar plates with the

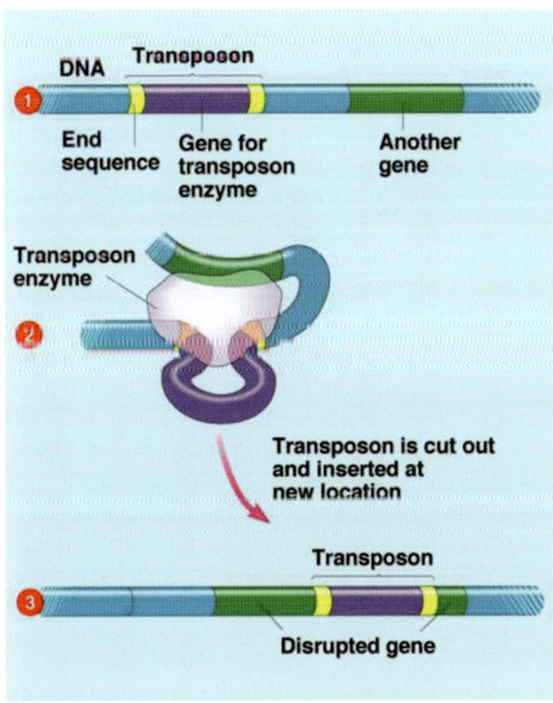

FIGURE 10.27 Transposition: cut and paste mode *(Source: chrisdellovedova.com)*.

appropriate antibiotics. An essential requirement of this protocol and many other types of similar mutational strategies is the need for the transposon to recombine into the genome through double homologous recombination.

Site-Directed Mutagenesis

Site-directed mutagenesis is the targeting of mutations to specific loci. It involves to the knockout of a single gene or clusters of genes (gene knockouts or deletions) or the mutation of a single base (known as point mutations). These techniques are extremely powerful for analysis of gene function and protein structure/function relationships. Many different approaches are available. Site-directed mutations can be created in a number of ways. Nowadays, starting with the DNA sequence of the target gene and its flanking regions, the mutation is created using the PCR to amplify two fragments, one being the flanking sequence upstream of the desired target and the other being the downstream flanking region. For the purposes of knocking out the gene in the chromosome of the organism, the fragments need to be sufficiently long (~200 bp at least) to allow efficient gene exchange via homologous recombination on either side of the desired mutation. The two arms are also synthesised with linker extensions containing restriction enzyme sites so that they can be cloned sequentially into a suitable vector and at the same time allowing the creation of a restriction site at the point of deletion into which an antibiotic resistance gene (or some other marker) can be inserted. The construct needs to be introduced into the target organisms by transformation or electroporation or using a suicide vector. Presumptive mutants are selected on a medium containing the appropriate antibiotic. Ideally, the construct needs to be linearised (except in the case of a suicide vector) to avoid a single recombination into the genome which would result in integration of the whole construct into the chromosome and producing a potentially misleading genotype and phenotype. Several counter-selection systems (e.g. the *sacB* system) have been developed to "force" the recombination and ensure that the vector is eliminated (Reyrat et al., 1998). Failure to confirm the mutation could lead to spurious results and conclusions about the possible function of the target gene. This can be done either by hybridisation with a suitable probe or more usually using the PCR with primers designed to distinguish unequivocally between the mutant and wild-type genotypes. Deletions can also be constructed without the need to construct two separate flanking fragments by using overlap extension PCR (see the PCR section).

Delitto perfetto mutagenesis. An elegant method with wide applicability for producing markerless (sometimes called "scar-less") mutations in the genome of a target organism was first reported for yeast by Storici and Resnick (2003). The first step in this two-step technique involves using a gene cassette consisting of both a selectable marker (e.g. a drug or antibiotic resistance gene), and a counter-selectable marker, e.g. in yeast the *KlURA3* or *GAL1/10*-p53 genes which when present in yeast prevent growth in media containing 5-flouroorotic acid or galactose, respectively. In a yet more sophisticated variant, the cassette carries a recombinant *GAL1*-I-*Sce*I construct in which the *Sce*I gene encodes the so-called homing endonuclease under the control of the *GAL1* promoter. For this technique, it is essential to check that a *Sce*I target sequence is not present in the genome of the organism under study. Using overlap extension PCR, the cassette is extended at either end with sequences amplified from the genomic DNA of the organism which flank the intended site of the mutation. The construct is then electroporated into yeast and antibiotic/drug-resistant transformants are selected. These will have the entire cassette integrated into the yeast chromosome at the required locus. In the second step, starting again with the yeast genomic DNA, the same flanking regions are amplified and fused by overlap PCR to create a single linear molecule which is then transformed into the strain constructed in the first step. Now the transformants are grown in the counter-selective conditions and the survivors will be those in which the cassette has been lost through double homologous recombination. This technique, suitably modified, has been applied to several prokaryotic systems (e.g. Kristich et al., 2008).

Site-Directed Point Mutants

There are a number of methods of producing mutations which affect only one base or a few bases in a gene. In 1985, Kunkel introduced a very elegant and effective technique which reduces the need to select for the mutants (Kunkel, 1985). The vector DNA into which the target gene to be mutated is first cloned in to the phage-based replicon

M13mp2 from which single-stranded DNA was produced. The recombinant vector is then propagated in a *dut* and *ung* strain of *E. coli* resulting in DNA which contains some uracil residues. The single-stranded DNA is isolated and used as the template for mutagenesis. An oligonucleotide containing the desired mutation is used for primer extension in vitro and the heteroduplex DNA formed contained the template strand unmutated and containing uracil instead of thymines, and the newly synthesised strand mutated but containing no uracil. The DNA is then treated with uracil deglycosidase which removes the uracil from the template and then with alkali which specifically degrades the strand that contained the uracil. The surviving mutated strand is then transformed back into *E. coli*. Various elaborations of this technique have been developed including the use of plasmid-based systems some of which are available commercially as kits containing DNA polymerases with greater processivity (Figure 10.28) and the use of two oligonucleotide primers one designed by the experimenter to create the desired mutation and the other one provided in the kit which corrects a mutation in an antibiotic resistance gene in the starting vector. This allows the synthesised strand carrying both mutations to be selected for when transformed into the host.

One of the most efficient systems for studying protein structure and function is where the gene can be expressed in an active form in an expression vector and the site-directed mutagenesis can be carried out in the same vector. This allows mutants to be created very rapidly and one can move straight to over-expression and characterisation of the mutant protein. This is an ideal situation but where complex metallo-proteins are concerned it is rare that such an approach will be feasible.

BIOINFORMATICS

Bioinformatics is the application of computing and informatics to biology. The explosive growth in this area of science has been driven by a number of important factors. These include amazing developments in the technologies and the introduction of highly robotic and industrial-scale operations for genomics, transcriptomics, proteomics, and metabolomics (see below). These developments have been supported by governments and commercial concerns in North America, Europe and Asia. The convention requiring academic and other workers to deposit and release sequence and other data upon or before publication to open source databases has also been crucial as has the ease of access to databanks, web-based search engines and other bioinformatics softwares. The databases and other resources that we see today stem from the foresight and philosophies of the pioneers in this field who first applied the emerging developments in information technology and the worldwide web to biology. The following section provides some of the more commonly used and immensely useful web-based resources.

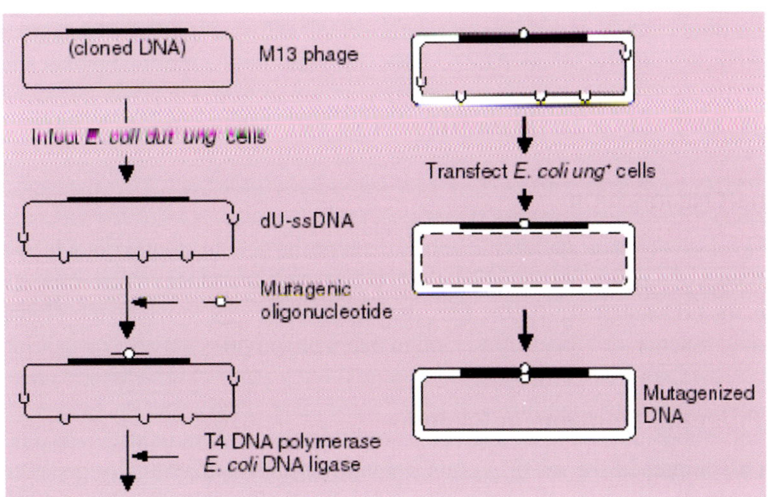

FIGURE 10.28 Kunkel method for site-directed point mutagenesis *(Source: catalog.takara-bio.co.jp)*.

General Bioinformatics Web Sites

Web sites which provide a huge range of information and links include those of: the National Center for Biotechnology Information (NCBI) site (http://www.ncbi.nlm.nih.gov/) and the EMBL European Bioinformatics Institute (EBI) which are gateways to vast resources of biomedical and genomic information, software, and publications. The EXPASY Life Science Directory (http://www.expasy.ch/alinks.html) contains a huge list of sites listed by category. Pedro's tools (http://www.biophys.uni-duesseldorf.de/BioNet/Pedro/research_tools.html) is also a huge resource for links to other databases, search engines and methods.

Sequence Searching Sites

The NCBI site listed above gives access to the Basic Local Alignment Tool (BLAST) search sites for proteins, DNA and RNA sequences originally developed by Altschul et al. (1990). The EBI also offers another powerful suite of sequence searches tools known as FASTA (standing for FAST-ALL) (http://www2.ebi.ac.uk/fasta3/). In both the BLAST and FASTA sites, sequences are easily copied and pasted into dialogue boxes on the web page. The query is submitted and the software searches all sequences deposited in current databases and matches are returned rapidly in descending order listing the closest matches first. The default parameters for the search are entirely adequate but it is possible to adjust these should the need arise.

Multiple Sequence Alignment

The ability to create multiple alignments of polypeptide or nucleic acid sequences is immensely useful. For example, in proteins it allows the identification of conserved residues or domains. One site for creating multiple alignments that has proved immensely powerful is INRA's MULTALIN software (http://bioinfo.genotoul.fr/multalin/) (Corpet, 1988). To use this site, the first operation is to make a "file of files" containing the sequences of interest. Each sequence should be in a predetermined text file format. This should be in the PIR/FASTA format in which the first line starts with the > symbol which is immediately followed by a short unique identifier of not more than eight characters. The following lines contain the amino acid sequence in the single character nomenclature. The next sequence should then start on the next line and in this way it is possible to stack up many sequences. These can be pasted into the dialogue box on the web page and submitted to the server. The alignment is normally returned in a few minutes. As with sequence searches, it is possible to adjust the parameters of the search but the default option is sufficient for most needs.

ClustalW (Chenna et al., 2003) is another powerful set of software tools for making multiple alignments (http://www.ebi.ac.uk/Tools/clustalw2/index.html). This programme is more sophisticated. For example, it enables you to adjust alignments by eye. Also, the alignment output from ClustalW can be used directly to input into phylogeny programmes such as the Felsenstein package called PHYLIP (see below).

Comparative Gene Organisation

The following site is exclusively dedicated to the comparison of the genetic context in which any specified gene is located within the genomes of prokaryotes for which the genome sequence has been determined and annotated (http://www.microbesonline.org/) (Dehal et al., 2010). This site shows the extent to which the organisation of genes that encode similar or related functions has been conserved in many prokaryotes but also reveals interesting differences.

Identification of Potential Domains in Proteins

ProDom is based on a comprehensive set of protein domain families automatically generated from the UniProt Knowledge Database (http://prodom.prabi.fr) (Servant et al., 2002). The software allows the search for a sequence of interest you want to explore (for example by entering a gene name) and this will display the domains within that protein and other proteins which contain similar domains but often with quite different overall activities and functions. This software also allows you to submit a new protein sequence, e.g. a new gene product you have discovered but have no idea what it does. ProDom analysis can provide clues as to the possible activities of unknown proteins and therefore ideas for further experiments.

Genome Sites

There are numerous sites that relate to total genome sequencing projects. Some are specific to major genome projects, e.g. human, mouse, drosophila etc. Some have been developed by major labs which carry out many projects. All sites are interlinked to varying degrees. One of the most productive organisations has been the J Craig Venter Institute (JCVI) (http://www.jcvi.org/). Other sites well worth exploring are those of the Department of Energy Joint Genome Institute (http://www.jgi.doe.gov/) and the Wellcome Trust Sanger Institute (http://www.sanger.ac.uk/).

Cross-relational Databases for Genomes and Metabolic and Other Pathways

For a truly amazing cross-relational database which links genomes and metabolic, regulatory and many other pathways, see the Kyoto Encyclopedia of Genes and Genomes (http://www.genome.ad.jp/kegg/) and in particular the search engine for biochemical and other pathways (http://www.genome.jp/kegg/pathway.html#metabolism) for all organisms for which the genome has been sequenced. As an example of how to use this software, on this page, scroll down to 'energy metabolism' and click on 'photosynthesis'. The page opens to reveal a reference photosynthesis system. At the top of the page on the left find a pull down menu box containing 'REFERENCE PATHWAY'. Pull down the menu which lists all the organisms for which genome sequences have been determined (totally or partially). From the list, select ANABENA and then click on the grey EXEC button to the right. When the page loads up, you will see what has been inferred to be present in this cyanobacterium. The genes now lit up in green on the bars below have all been found in *Anabena* and passing the cursor over each reveals its function and clicking on it will allow you to link to pages which will give you much more detail even down to the level of protein structures if determined.

Molecular Phylogenies and Tree Drawing Programmes

The construction of phylogenetic and evolutionary relationships between organisms based on the alignment and comparison of macromolecular sequences (DNA, RNA, protein) is firmly established as the basis for constructing evolutionary trees stemming back to the work of Lane et al. (1985). Nowadays, rRNAs are the molecules most commonly used for this purpose. In an earlier section, the ClustalW web software (http://www2.ebi.ac.uk/clustalw/) was highlighted as a tool for generating multiple sequence alignments in a format which could be input into a powerful commonly used software package for generating such trees known as PHYLIP (for Phylogeny Inference Package). For example, it allows you to choose many different parameters and different methods of making alignments. Pull down the menu under 'Tree type' and find there several alternative methods of alignments, e.g. nj, neighbour joining; dist, distance matrix. The multiple sequence alignment can then be uploaded into web-based PHYLIP software. This is a much used and sophisticated molecular phylogenetics package and you should click on 'documentation' to learn more about it. PHYLIP software is accessible at the following URL: http://bioweb.pasteur.fr/seqanal/phylogeny/phylip-uk.html. The PHYLIP software returns outputs which contains data with file extensions of .dnd (the tree) and .aln (the alignment). These can be uploaded into a number of different tree drawing softwares. To construct trees using the data with .dnd from ClustalW and many other file extensions, an effective programme called TREEVIEW can be downloaded from the following web site: http://taxonomy.zoology.gla.ac.uk/rod/treeview.html.

Many similar and related web software sites for phylogenetic and evolutionary analysis are listed here: (http://evolution.genetics.washington.edu/phylip/software.html).

Visualisation of Molecular Structures

There are a number of web-based softwares which support the visualisation on macromolecular structures. A commonly used example is the user-sponsored molecular visualisation system on an open source foundation at http://www.pymol.org/.

THE OMICS REVOLUTION

In recent years, sophisticated technologies have been developed that allow the, almost industrial scale, sequencing and annotation of complete genomes and the analysis of large numbers of genes, RNAs and proteins. Sequencing is becoming so cheap now that it is probably the stepping off point of choice for many molecular biological studies of organisms. The handling and analysis of such data are underpinned by **Bioinformatics** or Biocomputing.

Genomics

Genomics has largely been covered by the paragraphs on DNA sequencing above.

Transcriptomics

Transcriptomics is one of the elements of 'postgenomics'. It involves the analysis of gene expression by measuring and comparing the abundance of mRNAs for individual genes. One way in which this is done is to produce gene microarrays on glass slides which consist of fixed microspots of DNAs for each gene in the genome produced by the PCR. Total RNA can be isolated from cells under study and then fluorescently labelled and hybridised to the microarrays. Those genes which are highly expressed will bind higher amounts of mRNA and this can be measured by fluorescent detectors which read across the complete microarray. In this way, it is possible to compare mRNA abundances for the complete set of genes for cells exposed to different conditions or healthy or diseased cells. Usually, statistical analysis is vital because there are many steps in the process where errors could occur. Again this is a methodology which is greatly enhanced by robotics.

Proteomics

Proteomics is the second postgenomics technique but one which attempts to look at gene expression at the protein or polypeptide level. In this technique, total polypeptides are isolated from cells and then subjected to 2D electrophoresis in a slab gel. This technique displays the polypeptides when stained as individual spots which can then be subjected to image analysers to record and measure relative abundances. Spots of polypeptides of interest, e.g. those which are up-regulated under a particular condition, are cut out of the gel, digested with specific proteases and then subjected to MALDI-TOF Mass spectrometry. From this, total molecular masses can be determined for component peptides and these can be compared with a database deduced from all the genes identified in the genome sequencing project. In this way, the polypeptide can be ascribed to a particular gene. Again through the use of array techniques, high-speed readers, and computing it is possible to analyse a highly complex mixture of proteins, to identify all the genes which encode them and to move to a functional analysis of those genes based on the rapidly growing database of known gene functions.

Structural Genomics

This is a highly ambitious concept which attempts to provide high-throughput structural determinations of proteins. For example, one objective might be to determine the structures for all human proteins. In practice, there are many reasons why this may be impossible. Also, we may not need to know the structures for all proteins. The idea is to over express whole sets of genes from a specific genome. Proteins produced in great abundance can be purified easily and then high-throughput crystallisation trials established. Once crystals are obtained, the diffraction qualities can be assessed at high speed and then further structural analysis and solving can proceed. Once the structure is obtained, it may be possible to deduce function from structure. Even if the function is not know, if the protein is a potential drug target, chemists can then try to design possible ligands to potential active sites.

Omniomics

The more biological sequences, structures, functions that are determined it becomes increasingly possible to carry out automated comparisons of the genetic potential of [illegible] organisms or groups of organisms. Omniomics attempts to link all the databases into automated intelligent networks that can carry out high-speed analysis and comparison. That is, as genome sequences emerge from the sequencers the whole process of analysis down to the level of gene function and expression can be produced automatically.

Metabolomics

Metabolomics attempts to analyse and identify metabolites within cells and interrogate changes in them. The use of NMR of complex cell metabolite mixtures together with neural networks and other artificial intelligence systems is providing ways to probe complex mixtures and the changes occurring within them.

Economics

None of the above technologies is cheap. The equipment and the consumables costs are high. However, there are research institutes around the world and biotech and pharmaceutical companies which are equipped with large numbers of capillary sequencers, robotics, arrayers, etc. It is difficult to predict what the next quantum leaps will be but obviously much of the work is driven one way or another by the need to find cures for diseases.

ILLUSTRATIVE EXAMPLES IN THE GENETICS AND MOLECULAR BIOLOGY OF N_2 FIXATION

The table in this section is intended to illustrate the application of many of the techniques described in this chapter to a complex metalloenzyme systems as exemplified by selected studies which over 40 years have exploited different technological developments as they have emerged and leading to our current state of understanding of the process of biological N_2 fixation catalysed by the nitrogenase enzyme family. It is not the intention here to provide a detailed account of the several nitrogenase enzyme systems now known, their biochemistry, nor the *nif*, *vnf*, and *anf genes* which encode them. These are described in considerable detail in the references provided which represent only a small fraction of the studies published on this important and highly complex biological process.

Development	Illustrative publications
Ground breaking paper on nitrogenase assay, purification, and characterisation.	Bulen and LeComte (1966)
Early mutants affecting nitrogen fixation.	Streicher et al. (1971), Shah et al. (1973)
Mobilisation of functional *nif* genes to *Klebsiella* and *Salmonella*.	Dixon and Postgate (1972), Dixon et al. (1976), Cannon et al. (1974)
Earliest application of recombinant DNA technology to *nif* genes.	Riedel et al. (1977)
Physical and genetic maps of *K. pneumoniae nif* genes.	Riedel et al. (1979)
Early mapping of nif mutants of *A. vinelandii.*	Bishop and Brill (1977)
Detection of *nif* structural genes in different N_2-fixers using interspecific DNA hybridisation.	Ruvkun and Ausubel (1980)

(Continued)

Development	Illustrative publications
Protein sequencing of nitrogenase proteins.	Tanaka et al. (1977), Hausinger and Howard (1982)
DNA sequencing as applied to *nif* genes.	Mevarech et al. (1980), Sunderesan and Ausubel (1981), Scott et al. (1981), Mazur and Chui (1982)
Crystal structures of MoFe nitrogenase components.	Georgiadis et al. (1992), Kim et al. (1993)
Knockout mutants dissect gene function and reveal new nitrogenase systems.	Eady et al. (1987), Robson et al. (1987), Chisnell et al. (1988)
Protein overexpression reveals functional analysis.	Evans et al. (1991), Zheng et al. (1998), Curatti et al. (2007)
Mechanistic overview of nitrogenase.	Howard and Rees (2006)

REFERENCES

Alwine, J. C., Kemp, D. J., & Stark, G. R. (1977). Method for detection of specific RNAs in agarose gels by transfer to diazobenzyloxymethyl-paper and hybridization with DNA probes. *Proc. Natl. Acad. Sci. USA, 74*(12), 5350–5354.

Akerly, B. J., Rubin, E. J., Camilli, A., Lampe, D. J., Robertson, H. M., & Mekalanos, J. J. (1998). Systematic identification of essential genes by in vitro mariner mutagenesis. *Proc. Natl. Acad. Sci. USA, 95*, 8927–8932.

Arber, W., & Linn, S. (1969). DNA modification and restriction. *Annu. Rev. Biochem., 38*, 467–500.

Altschul, S. F., Gish, W., Miller, W., Myers, E. W., & Lipman, D. J. (1990). Basic local alignment search tool. *J. Mol. Biol., 215*, 403–410.

Andersson, B., Wentland, M. A., Ricafrente, J. Y., Liu, W., & Gibbs, R. A. (1996). A "double adaptor" method for improved shotgun library construction. *Anal. Biochem., 236*, 107–113.

Auerbach, C., Robson, J. M., & Carr, J. G. (1947). Chemical production of mutations. *Science, 105*, 243–247.

Ausubel, F. M., Brent, R., Kingston, R. E., Moore, D. D., Seidman, J. G., Smith, J. A., & Struhl, K. (2002). *Short Protocols in Molecular Biology* (fifth ed.). New York, NY: John Wiley & Sons, Inc.

Baba, T., Tara, T., Hasegawa, M., Takai, Y., Okumura, Y., Baba, M., Kirill, A., Datsenko, K. A., & Tomita, M. (2006). Construction of *Escherichia coli* K-12 in-frame, single-gene knockout mutants: the Keio collection. *Mol. Syst. Biol., 2*, 1–11.

Bartlett, J. M. S., & Stirling, D. (2003). A short history of the polymerase chain reaction. *PCR Protoc., 226*, 3–6.

Birnboim, H. C., & Doly, J. (1979). A rapid alkaline extraction procedure for screening recombinant plasmid DNA. *Nucleic Acids Res., 7*(6), 1513–1523.

Bishop, P. E., & Brill, W. (1977). Genetic analysis of *Azotobacter vinelandii* mutant strains unable to fix nitrogen. *J. Bacteriol., 130*, 954–956.

Bolivar, F., Rodriguez, R. L., Greene, P. J., Betlach, M. C., Heynecker, H. L., Boyer, H. W., Crosa, J. H., & Falkow, S. (1977). Construction and characterization of new cloning vehicle. II. A multipurpose cloning system. *Gene, 2*, 95–113.

Brock, T. D., & Freeze, H. (1969). *Thermus aquaticus*, a nonsporulating extreme thermophile. *J. Bact., 98*(1), 289–297.

Bulen, W. A., & LeComte, J. R. (1966). The nitrogenase system from Azotobacter: two enzyme requirement for N_2 reduction, ATP-dependent H_2 evolution and ATP hydrolysis. *Proc. Natl. Acad. Sci. USA, 56*, 979–986.

Burnette, W. N. (1981). 'Western blotting': electrophoretic transfer of proteins from sodium dodecyl sulfate—polyacrylamide gels to unmodified nitrocellulose and radiographic detection with antibody and radioiodinated protein A. *Anal. Biochem., 112*(2), 195–203.

Cannon, F. C., Dixon, R. A., Postgate, J. R., & Primrose, S. B. (1974). Plasmids formed in Nitrogen-fixing *E. coli-Klebsiella pneumoniae* hybrids. *J. Gen. Microbiol., 80*, 241–251.

Chisnell, J. R., Premakumar, R., & Bishop, P. E. (1988). Purification of a second alternative nitrogenase from a *nifHDK* deletion strain of *Azotobacter vinelandii. J. Bacteriol., 170*, 27–33.

Cohen, S., Chang, A., Boyer, H., & Helling, R. (1973). Construction of biologically functional bacterial plasmids in vitro. *Proc. Natl Acad. Sci. USA, 70*, 3240–3244.

Collins, J., & Hohn, B. (1978). Cosmids: a type of plasmid gene-cloning vector that is packageable in vitro in bacteriophage lamda heads. *Biotechnology, 24*, 193–197.

Chenna, R., Sugawara, H., Koike, T., Lopez, R., Gibson, T. J., Higgins, D. G., & Thompson, J. D. (2003). *Nucleic Acids Res., 31*, 3497–3500.

Corpet, F. (1988). Multiple sequence alignment with hierarchical clustering. *Nucleic Acids Res., 16*, 10881–10890.

Cregg, J. M., Tolstorukov, I., Kusari, A., Sunga, J., Madden, K., & Chappell, T. (2009). Expression in the yeast *Pichia pastoris. Methods Enzymol., 463*, 169–189.

Crick, F. H. C. (1966). Codon-anticodon pairing: the wobble hypothesis. *J. Mol. Biol., 19*, 548–555.

Curatti, L., Hernandez, J. A., Igarashi, R. Y., Soboh, B., Zhao, D., & Rubio, L. M. (2007). *In vitro* synthesis of the iron-molybdenum cofactor of nitrogenase from iron, molybdenum, and homocitrate using purified proteins. *Proc. Natl. Acad. Sci. USA, 104*, 17626–17631.

Dehal, P. S., Joachimiak, M. P., Price, M. N., Bates, J. T., Baumohl, J. K., Chivian, D., Friedland, G. D., Huang, K. H., Keller, K., Novichkov, P. N., Dubchak, I. L., Alm, E. J., & Arkin, A. P. (2010). MicrobesOnline: an integrated portal for comparative and functional genomics. *Nucleic Acids Res., 38*, D396–D400.

Dixon, R. A., Cannon, F. C., & Kondorosi, A. (1976). Construction of a P plasmid carrying nitrogen-fixation genes from *Klebsiella pneumoniae*. *Nature, 260*, 268–271.

Dixon, R. A., & Postgate, J. R. (1972). Genetic transfer of nitrogen fixation from *Klebsiella pneumoniae* to *Escherichia coli*. *Nature, 237*, 102–103.

Don, R., Cox, P., Wainwright, D., Baker, K., & Mattick, J. (1991). 'Touchdown' PCR to circumvent spurious priming during gene amplification. *Nucleic Acids Res., 19*(14), 4008.

Eady, R. R., Robson, R. L., Richardson, T. H., Miller, R. W., & Hawkins, M. (1987). The vanadium nitrogenase of *Azotobacter chroococcum*. Purification and properties of the VFe-protein. *Biochem. J., 244*, 197–207.

Evans, D., Jones, R., Woodley, P. R., Wilborn, J. R., & Robson, R. L. (1991). Nucleotide sequence and genetic analysis of the *Azotobacter chroococcum nifUSVWZM* gene cluster including a new gene *nifP* which encodes a serine acetyl transferase. *J. Bacteriol., 173*, 5457–5469.

Friedman, A. M., Long, S. R., Brown, S. E., Buikema, W. J., & Ausubel, F. M. (1982). Construction of a broad host range cosmid cloning vector and its sue in the genetic analysis of *Rhizobium* mutants. *Gene, 18*, 289–296.

Hughes, S. (2001). Making dollars out of DNA. The first major patent in biotechnology and the commercialization of molecular biology, 1974–1980. *Isis, 92*(3), 541–575.

Jackson, D., Symons, R., & Berg, P. (1972). Biochemical method for inserting new genetic information into DNA of Simian Virus 40: circular SV40 DNA molecules containing lambda phage genes and the galactose operon of *Escherichia coli*. *Proc. Natl. Acad. Sci. USA, 69*, 2904–2909.

Langley, K. E., Villarejo, M. R., Fowler, A. V., Zamenhof, P. J., & Zabin, I. (1975). Molecular basis of beta-galactosidase alpha-complementation. *Proc. Natl Acad. Sci. USA, 72*, 1254–1257.

Lobban, P., & Kaiser, A. (1973). Enzymatic end-to end joining of DNA molecules. *J. Mol. Biol., 78*(3), 453–471.

Georgiadis, M. M., Komiya, H., Chakrabarti, P., Woo, D., Kornuc, J. J., & Rees, D. C. (1992). Crystallographic structure of the nitrogenase iron protein from *Azotobacter vinelandii*. *Science, 257*, 1653–1659.

Hausinger, R. P., & Howard, J. B. (1982). The amino acid sequence of the nitrogenase iron protein form *Azotobacter vinelandii*. *J. Biol. Chem.*, 2483–2490.

Higuchi, R., Krummel, B., & Saiki, R. (1988). A general method of *in vitro* preparation and specific mutagenesis of DNA fragments: study of protein and DNA interactions. *Nucleic Acid. Res., 16*, 7351–7367.

Howard, J. B., & Rees, D. C. (2006). How many metals does it take to fix N_2? A mechanistic overview of biological nitrogen fixation. *Proc. Natl Acad. Sci. USA, 103*, 17088–17093.

Kim, J., Woo, D., & Rees, D. C. (1993). X-ray crystal structure of the nitrogenase molybdenum-iron protein from *Clostridium pasteurianum* at 3.0 Å resolution. *Biochemistry, 32*, 7104–7115.

Kim, U.-J., Shizuya, H., Pieter, J., de Jong, P. J., Bruce Birren, B., & Simon, M. I. (1992). Stable propagation of cosmid sized human DNA inserts in an F factor based vector. *Nucleic Acids Res., 20*, 1083–1085.

Kleckner, N., Roth, J., & Bolstein, D. (1977). Genetic engineering in vivo using translocable drug resistant elements. *J. Mol. Biol., 116*, 125–159.

Klenow, H., & Henningsen, I. (1970). Selective elimination of the exonuclease activity of the deoxyribonucleic acid polymerase from *Escherichia coli* B by limited proteolysis. *Proc. Natl Acad. Sci. USA, 65*, 168–175.

Kristich, C. J., Nguyen, V. T., Le, T., Barnes, A. M. T., Grindle, S., & Dunny, G. M. (2008). Development and use of an efficient system for random mariner transposon mutagenesis to identify novel genetic determinants of biofilm formation in the core *Enterococcus faecalis* genome. *Appl. Environ. Micobiol., 74*, 3377–3386.

Kunkel, T. A. (1985). Rapid and efficient site-specific mutagenesis without phenotypic selection. *Proc. Natl Acad. Sci. USA, 82*, 488–492.

Laemmli, U. K. (1970). Cleavage of structural proteins during the assembly of the head of bacteriophage T4. *Nature, 227*, 680–685.

Lane, D. J., Pace B., Olsen G. J., Stahl, D. A., Sogin, M. L., & Pace, N. R. (1985). Rapid determination of 16S ribosomal RNA sequences for phylogenetic analyses. *Proc. Natl. Acad. Sci. USA. 82*, 6955–6959.

Mardis, E. R. (2008). Next-generation DNA sequencing methods. *Annu. Rev. Genomics Hum. Genet., 9*, 387–402.

Marmur, J. (1961). A procedure for the isolation of deoxyribonucleic acid from microorganisms. *J. Mol. Biol., 3*, 208–218.

Maxam, A. M., & Gilbert, W. (1977). A new method for sequencing DNA. *Proc. Natl Acad. Sci. USA, 74*(2), 560–564.

Mazur, B. J., & Chui, C.-F. (1982). Sequence of the gene coding for the β-subunit of dinitrogenase from the blue-green alga *Anabaena. Proc. Natl Acad. Sci. USA., 79*, 6782–6786.

McClintock, B. (1950). The origin and behavior of mutable loci in maize. *Proc. Natl Acad. Sci. USA, 36*, 344–355.

Mevarech, M., Rice, D., & Haselkorn, E. R. (1980). Nucleotide sequence of a cyanobacterial *nifH* gene coding for nitrogenase reductase. *Proc. Natl Acad. Sci. USA, 77*, 6476–6480.

Morales, V. M., & Sequira, L. (1985). Suicide vector for transposon mutagenesis in *Pseudomonas solanacearum. J. Bacteriol., 163*, 1263–1264.

Muller, H. J. (1927). Artificial transmutation of the gene. *Science, 66*, 84–87.

Mullis, K. B. (1985). Process for amplifying nucleic acid sequences. U.S. Patent 4,683, 202.

Noren, C. J., Anthony-Cahill, S. J., Griffith, M. C., & Schultz, P. G. (1989). A general method for site-specific incorporation of unnatural amino acids into proteins. *Science, 244*, 182–188.

Nyrén, P. (2007). The history of pyrosequencing. *Methods Mol. Biol., 373*, 1–14.

Ochman, H., Gerber, A. S., & Hartl, D. L. (1988). Genetic applications of an inverse polymerase chain reaction. *Genetics, 120*, 621–623.

Riedel, G. E., Ausubel, F. M., & Cannon, F. C. (1977). Recombinant plasmid that carries part of the nitrogen fixation (*nif*) gene cluster of *Klebsiella pneumoniae. Proc. Natl. Acad. Sci. USA, 74*, 2963–2967.

Riedel, G. E., Ausubel, F. M., & Cannon, F. C. (1979). Physical map of chromosomal nitrogen fixation (nif) genes of *Klebsiella pneumoniae. Proc. Natl. Acad. Sci. U.S.A., 76*, 2866–2870.

Reyrat, J. M., Pelicic, V., Gicqual, B., & Rappuoli, R. (1998). Counterselectable markers: untapped tools for bacterial genetics and pathogenesis. *Infect. Immun., 66*, 4011–4017.

Robson, R. L., Woodley, P. R., Pau, R. N., & Eady, R. R. (1987). Structural genes for the vanadium nitrogenase from *Azotobacter chroococcum. EMBO J., 8*, 1217–1224.

Ruvkun, G. B., & Ausubel, F. M. (1980). Interspecies homology of nitrogenase genes. *Proc. Natl Acad. Sci. USA, 77*, 191–195.

Sanger, F., Nicklen, S., & Coulson, A. R. (1977). DNA sequencing with chain-terminating inhibitors. *Proc. Natl Acad. Sci. USA, 74*(12), 5463–5467.

Saiki, R. K., et al. (1985). Enzymatic amplification of β-globin genomic sequences and restriction site analysis for diagnosis of sickle cell anemia. *Science, 230*, 1350–1354.

Saiki, R., Gelfand, D., Stoffel, S., Scharf, S., Higuchi, R., Horn, G., Mullis, K., & Erlich, H. (1988). Primer-directed enzymatic amplification of DNA with a thermostable DNA polymerase. *Science, 239*(4839), 487–491.

Sambrook, J., & Russell, D. (2000). *Molecular Cloning: A Laboratory Manual.* Coldspring Harbor Laboratory Press.

Sapp, J. (2005). The prokaryote-eukaryote dichotomy: meanings and mythology. *Microbiol. Mol. Biol. Rev., 69*, 292–305.

Schwartz, D. C., & Cantor, C. R. (1984). Separation of yeast chromosome-sized DNAs by pulsed field gradient gel electrophoresis. *Cell, 37*, 67–75.

Scott, K. F., Rolfe, B. G., & Shine, J. (1981). Biological nitrogen fixation: primary structure of the gene coding for the *Klebsiella pneumoniaen nifH* and *nifD* genes. *J. Mol. Appl. Genet., 1*, 71–81.

Selveraj, G., & Iyer, V. N. (1985). A small mobilisable Inc P group plasmid vector packageable into bacteriophage lambda *in vitro. Plasmid, 13*, 70–74.

Servant, F., Bru, C., Carrère, S., Courcelle, E., Gouzy, J., Peyruc, D., & Kahn, D. (2002). ProDom: automated clustering of homologous domains. *Brief. Bioinform., 3*, 246–251.

Shah, V. K., Davis, L. C., Gordon, J. K., Orme-Johnson, W. H., & Brill, W. J. (1973). Nitrogenaseless mutants of *Azotobacter vinelandii* activities, cross reactions and EPR spectra. *Biochim. Biophys. Acta, 292*, 246–255.

Shendure, J., & Ji, H. (2008). Next-generation DNA sequencing. *Nat. Biotech., 26*, 1135–1145.

Shine, J., & Dalgarno, L. (1975). Determinant of cistron specificity in bacterial ribosomes. *Nature, 254*, 34–38.

Shizuya, H., Birren, B., Kim, U. J., Mancino, V., Slepak, T., Tachiiri, Y., & Simon, M. (1992). Cloning and stable maintenance of 300 kilobase-pair fragments of human DNA in *Escherichia coli* using an F-factor-based vector. *Proc. Natl. Acad. Sci. U.S.A., 89*, 8794–8797.

Sing, A., Roggenkamp, A., & Autenrieth, I. B. (1999). *Pneumocystis carinii* carriage in immunocompetent patients with primary pulmonary disorders as detected by single or nested PCR. *J. Clin. Microbiol., 38*, 1461–1467.

Smith, L. M., Fung, S., Hunkapiller, M. W., Hunkapiller, T. J., & Hood, L. E. (1985). The synthesis of oligonucleotides containing an aliphatic amino group at the 5' terminus: synthesis of fluorescent DNA primers for use in DNA sequence analysis. *Nucleic Acids Res., 13*(7), 2399–2412.

Smith, L. M., Sanders, J. Z., Kaiser, R. J., et al. (1986). Fluorescence detection in automated DNA sequence analysis. *Nature, 321*(6071), 674–679.

Southern, E. M. (1975). Detection of specific sequences among DNA fragments separated by gel electrophoresis. *J. Mol. Biol., 98*, 503–508.

Stadler, L. J. (1928). Mutations in barley induced by X-rays and radium. *Science, 68*(1756), 186–187.

Stadler, L. J., & Sprague, G. F. (1936). Genetic effects of ultra-violet radiation in maize. I. Unfiltered radiation. *Proc. Natl Acad. Sci. USA, 22*(10), 572–578.

Storici, F., & Resnick, M. A. (2003). Delitto perfetto targeted mutagenesis in yeast with oligonucleotides. *Genet. Eng., 25*, 189–207.

Streicher, S., Gurney, E., & Valentine, R. C. (1971). Transduction of the nitrogen fixation genes in *Klebsiella pneumoniae*. *Proc. Natl Acad. Sci. USA, 68*, 1174–1177.

Studier, F. W., & Moffatt, B. A. (1986). Use of bacteriophage T7 RNA polymerase to direct selective high-level expression of cloned genes. *J. Mol. Biol., 189*, 113–130.

Studier, F. W., Rosenberg, A. H., Dunn, J. J., & Dubendorff, J. W. (1990). Use of T7 RNA polymerase to direct expression of cloned genes. *Methods Enzymol., 185*, 60–89.

Sunderesan, V., & Ausubel, F. M. (1981). Nucleotide sequence of the gene coding for the nitrogenase iron protein from *Klebsiella pneumoniae*. *J. Biol. Chem., 256*, 2808–2812.

Tabor, R., & Richardson, C. C. (1985). A bacteriophage T7 RNA polymerase/promoter system for controlled exclusive expression of specific genes. *Proc. Natl Acad. Sci. USA, 82*, 1074–1078.

Tan, X., Sun, X., Gonzalez, C. F., & Hsueh, W. (1994). PAF and TNF increase the precursor of NF-kappa B p50 mRNA in mouse intestine: quantitative analysis by competitive PCR. *Biochim. Biophys. Acta., 1215*, 157–162.

Tanaka, M., Haniu, M., Yasunobu, K. T., & Mortenson, L. E. (1977). The amino acid sequence of *Clostridium pasteurianum* Fe-protein: a component of nitrogenase. *J. Biol. Chem., 252*, 7093–7100.

Towbin, H., Staehelin, T., & Gordon, J. (1979). Electrophoretic transfer of proteins from polyacrylamide gels to nitrocellulose sheets: procedure and some applications. *Proc. Natl. Acad Sci. USA, 76*(9), 4350–4354.

Ullmann, A., Jacob, F., & Monod, J. (1967). Characterization by in vitro complementation of a peptide corresponding to an operator-proximal segment of the beta-galactosidase structural gene of *Escherichia coli*. *J. Mol. Biol., 24*(2), 339–343.

VanGuilder, H. D., Vrana, K. E., & Freeman, W. M. (2008). Twenty-five years of quantitative PCR for gene expression analysis. *Biotechniques, 44*(5), 619–626.

Vieira, J., & Messing, J. (1982). The pUC plasmids, an M13mp7-derived system for insertion mutagenesis and sequencing with synthetic universal primers. *Gene, 19*(3), 259–268.

Warburg, O., & Christian, W. (1942). Isolation and crystallization of enolase. *Biochem. Z, 310*, 384–421.

Weiss, B., & Richardson, C. C. (1967). Enzymatic breakage and joining of deoxyribonucleic acid, I. Repair of single strand breaks in DNA by an enzyme system from *Escherichia coli* infected with T4 bacteriophage. *Proc. Natl Acad. Sci. USA, 57*(4), 1021–1028.

Woese, C., Kandler, O., & Wheelis, M. (1990). Towards a natural system of organisms: proposal for the domains Archaea, Bacteria, and Eucarya. *Proc. Natl. Acad. Sci. USA, 87*, 4576–4579.

Zheng, L., Cash, V. L., Flint, D. H., & Dean, D. R. (1998). Assembly of iron-sulfur clusters Identification of an iscSUA-hscBA-fdx gene cluster from *Azotobacter vinelandii*. *J. Biol. Chem., 273*, 13264–13272.

Index

Page references followed by "f" indicate figure, "t" indicate table, and "b" indicate box.

F

G

H

N

O

P

S

T

U

V

W

X

Y

Z